London Mathematical Society Lecture Note Series: 339

Groups St Andrews 2005

Volume 1

Edited by

C.M. Campbell
University of St Andrews

M.R. Quick
University of St Andrews

E.F. Robertson
University of St Andrews

G.C. Smith
University of Bath

CAMBRIDGE
UNIVERSITY PRESS

CAMBRIDGE UNIVERSITY PRESS
Cambridge, New York, Melbourne, Madrid, Cape Town, Singapore, São Paulo

Cambridge University Press
The Edinburgh Building, Cambridge CB2 2RU, UK

Published in the United States of America by Cambridge University Press, New York

www.cambridge.org
Information on this title: www.cambridge.org/9780521694698

Printed in the United Kingdom at the University Press, Cambridge

A catalogue record for this publication is available from the British Library

ISBN-13   978-0-521-69469-8 paperback
ISBN-10   0-521-69469-8 paperback

LONDON MATHEMATICAL SOCIETY

Managing Editor: Professor N.J. Hitchin, ... Giles, Oxford
OXI 3LB, United Kingdom

The titles below are available from booksellers, or from Cambridge University Press at
www.cambridge.org/mathematics

Contents of Volume 1

Contents of Volume 2

Introduction

Groups St Andrews 2005 was held in the University of St Andrews from 30 July to 6 August 2005. This was the seventh in the series of Groups St Andrews group theory conferences organised by Colin Campbell and Edmund Robertson of the University of St Andrews. The first three were held in St Andrews and subsequent conferences held in Galway, Bath and Oxford before returning to St Andrews in 2005. We are pleased to say that the conference was, we believe, a success having been attended by 230 participants from 37 countries. The lectures and talks were given in the Mathematical Institute and the School of Physics and Astronomy of the University of St Andrews. Accommodation was provided in New Hall and in Fife Park.

The Scientific Organising Committee of Groups St Andrews 2005 was: Colin Campbell (St Andrews), Nick Gilbert (Heriot-Watt), Steve Linton (St Andrews), John O'Connor (St Andrews), Edmund Robertson (St Andrews), Nik Ruskuc (St Andrews), Geoff Smith (Bath). The Committee received very valuable support from our Algebra colleagues at St Andrews, both staff and postgraduate students. Once again, we believe that the support of the two main British mathematics societies, the Edinburgh Mathematical Society and the London Mathematical Society has been an important factor in the success of these conferences.

The main speakers at the meeting were Peter J Cameron (Queen Mary, London), Rostislav I Grigorchuk (Texas A&M), John C Meakin (Nebraska-Lincoln) and Akos Seress (Ohio State). Additionally there were seven one-hour invited speakers together with an extensive programme of over a hundred seminars; a lot to fit into a week! As has become the tradition, all the main speakers have written substantial articles for these Proceedings. Each volume begins with two such articles. All papers have been subjected to a formal refereeing process comparable to that of a major international journal. Publishing constraints have forced the editors to exclude some very worthwhile papers, and this is of course a matter of regret.

It is hoped that the next conference in this series will be held in 2009 and will be Groups St Andrews in Bath. As Colin and Edmund will have retired before Groups St Andrews 2009, they look forward to this conference with special interest and anticipate a slightly different role. These Proceedings do, however, mark the first 25 years of Groups St Andrews conferences. In addition to the mathematics, illustrated by these Proceedings, the seven conferences have contained a wide selection of social events. "Groups St Andrews tourism" has taken us to a variety of

interesting and scenic venues in Scotland, Ireland, England and Wales. Bus trips
have included Kellie Castle, Loch Earn and Loch Tay, Falkland Palace and Hill
of Tarvit, Crathes Castle and Deeside, Loch Katrine and the Trossachs, House of
Dun, Connemara and Kylemore Abbey, the Burren and the Cliffs of Moher, Tin-
tern Abbey and Welsh Valleys, the Roman Baths in Bath, Salisbury Cathedral,
Rufus Stone and the New Forest, Stonehenge, Wells Cathedral and the Cheddar
Gorge, Blenheim Palace, Glamis Castle. We have been on boats on Loch Katrine,
the Thames, and Galway Bay to the Aran Islands. There have also been: musi-
cal events with participants as the musicians, Scottish Country Dance evenings,
barn dances, piano recitals, organ recitals, theatre trips, whisky tasting, putting,
chess, walks along the Fife Coast, walks round Bath, and walks round Oxford. All
these have provided opportunities for relaxation, but also opportunities to continue
mathematical discussions. We hope that the cricket match (and balloon trip) post-
poned from 1997 will take place only 12 years late at Groups St Andrews 2009 in
Bath!

Thanks to those authors who have contributed articles to these Proceedings. We
have edited their articles to produce some uniformity without, we hope, destroying
individual styles. For any inconsistency in, and errors introduced by, our edit-
ing we take full responsibility. Our final thanks go to Roger Astley and the rest
of the Cambridge University Press team for their assistance and friendly advice
throughout the production of these Proceedings.

Colin Campbell
Martyn Quick
Edmund Robertson
Geoff Smith

ASPECTS OF INFINITE PERMUTATION GROUPS

PETER J. CAMERON

School of Mathematical Sciences, Queen Mary, University of London, Mile End Road, London E1 4NS, UK
Email: `P.J.Cameron@qmul.ac.uk`

Abstract

Until 1980, there was no such subject as 'infinite permutation groups', according to the Mathematics Subject Classification: permutation groups were assumed to be finite. There were a few papers, for example [10, 62], and a set of lecture notes by Wielandt [72], from the 1950s.

Now, however, there are far more papers on the topic than can possibly be summarised in an article like this one.

I shall concentrate on a few topics, following the pattern of my conference lectures: the random graph (a case study); homogeneous relational structures (a powerful construction technique for interesting permutation groups); oligomorphic permutation groups (where the relations with other areas such as logic and combinatorics are clearest, and where a number of interesting enumerative questions arise); and the Urysohn space (another case study). I have preceded this with a short section introducing the language of permutation group theory, and I conclude with briefer accounts of a couple of topics that didn't make the cut for the lectures (maximal subgroups of the symmetric group, and Jordan groups).

I have highlighted a few specific open problems in the text. It will be clear that there are many wide areas needing investigation! I have also included some additional references not referred to in the text.

1 Notation and terminology

This section contains a few standard definitions concerning permutation groups. I write permutations on the right: that is, if g is a permutation of a set Ω, then the image of α under g is written αg.

The *symmetric group* $\mathrm{Sym}(\Omega)$ on a set Ω is the group consisting of all permutations of Ω. If Ω is infinite and c is an infinite cardinal number not exceeding Ω, the *bounded symmetric group* $\mathrm{BSym}_c(\Omega)$ consists of all permutations moving fewer than c points; if $c = \aleph_0$, this is the *finitary symmetric group* $\mathrm{FSym}(\Omega)$ consisting of all *finitary* permutations (moving only finitely many points). The *alternating group* $\mathrm{Alt}(\Omega)$ is the group of all even permutations, where a permutation is even if it moves only finitely many points and acts as an even permutation on its support.

Assuming the Axiom of Choice, the only non-trivial normal subgroups of $\mathrm{Sym}(\Omega)$ for an infinite set Ω are the bounded symmetric groups and the alternating group.

A *permutation group* on a set Ω is a subgroup of the symmetric group on Ω. As noted above, we denote the image of α under the permutation g by αg. For the most part, I will be concerned with the case where Ω is countably infinite.

The permutation group G on Ω is said to be *transitive* if for any $\alpha, \beta \in \Omega$, there exists $g \in G$ with $\alpha g = \beta$. For $n \leq |\Omega|$, we say that G is *n-transitive* if, in its induced action on the set of all n-tuples of distinct elements of Ω, it is transitive: that is, given two n-tuples $(\alpha_1, \ldots, \alpha_n)$ and $(\beta_1, \ldots, \beta_n)$ of distinct elements, there exists $g \in G$ with $\alpha_i g = \beta_i$ for $i = 1, \ldots, n$. In the case when Ω is infinite, we say that G is *highly transitive* if it is n-transitive for all positive integers n. If a permutation group is not highly transitive, the maximum n for which it is n-transitive is its *degree of transitivity*. (Of course, the condition of n-transitivity becomes stronger as n increases.)

The bounded symmetric groups and the alternating group are all highly transitive. We will see that there are many other highly transitive groups!

The subgroup $G_\alpha = \{g \in G : \alpha g = \alpha\}$ of G is the *stabiliser* of α. Any transitive action of a group G is isomorphic to the action on the set of right cosets of a point stabiliser, acting by right multiplication.

The permutation group G on Ω is called *semiregular* (or *free*) if the stabiliser of any point of Ω is the trivial subgroup; and G is *regular* if it is semiregular and transitive. Thus, a regular action of G is isomorphic to the action on itself by right multiplication.

A transitive permutation group G on Ω is *imprimitive* if there is a G-invariant equivalence relation on Ω which is not trivial (that is, not the relation of equality, and not the relation with a single equivalence class Ω). If no such relation exists, then G is *primitive*. A G-invariant equivalence relation is called a *congruence*; its equivalence classes are *blocks of imprimitivity*, and the set of blocks is a *system of imprimitivity*. A block or system of imprimitivity is non-trivial if the corresponding equivalence relation is. A non-empty subset B of Ω is a block if and only if $B \cap B^g = B$ or $\emptyset$ for all $g \in G$.

A couple of simple results about primitivity:

Proposition 1.1   (a) *The transitive group G on Ω is primitive if and only if G_α is a maximal proper subgroup of G for some (or every) $\alpha \in \Omega$.*

(b) *A 2-transitive group is primitive.*

(c) *The orbits of a normal subgroup of a transitive group G form a system of imprimitivity. Hence, a non-trivial normal subgroup of a primitive group is transitive.*

In connection with the last part of this result, we say that a transitive permutation group is *quasiprimitive* if every non-trivial normal subgroup is transitive.

Let G be a group, and S a subset of G. The *Cayley graph* $\mathrm{Cay}(G, S)$ is the (directed) graph with vertex set G, having directed edges (g, sg) for all $g \in G$ and $s \in S$. If $1 \notin S$, then this graph has no loops; if $s \in S \Rightarrow s^{-1} \in S$, then it is an undirected graph (that is, whenever (g, h) is an edge, then also (h, g) is an edge, so we can regard edges as unordered pairs). It is easy to see that $\mathrm{Cay}(G, S)$ is connected if and only if S generates G. Most importantly, G, acting on itself by right multiplication, is a group of automorphisms of $\mathrm{Cay}(G, S)$. In this situation,

G acts regularly on the vertex set of $\mathrm{Cay}(G, S)$. Conversely, if a graph Γ admits a group G as a group of automorphisms acting regularly on the vertices, then Γ is isomorphic to a Cayley graph for G. (Choose a point $\alpha \in \Omega$, and take S to be the set of elements s for which $(\alpha, \alpha s)$ is an edge.)

2 The random graph

2.1 B-groups

According to Wielandt [71], a group A is a *B-group* if any primitive permutation group G which contains the group A acting regularly is doubly transitive; that is, if any overgroup of A which kills all the A-invariant equivalence relations necessarily kills all the non-trivial A-invariant binary relations. The letter B stands for Burnside, who showed that a cyclic group of prime-power but not prime order is a B-group. The proof contained a gap which was subsequently fixed by Schur, who invented and developed Schur rings for this purpose. The theory of Schur rings (or S-rings) is connected with many topics in representation theory, quasigroups, association schemes, and other areas of mathematics; historically, it was an important source of ideas in these subjects. The theory of S-rings and its connection with representation theory is described in Wielandt's book.

Following the classification of finite simple groups, much more is known about B-groups, since indeed it is known that primitive groups are comparatively rare. For example, the set of numbers n for which there exists a primitive group of degree n other than S_n and A_n has density zero [21], and hence the set of orders of non-B-groups has density zero. (However, there are non-B-groups of every prime power order, and a complete description is not known.)

One could then ask:

Are there any infinite B-groups?

Remarkably, no example of an infinite B-group is known. One of the most powerful nonexistence theorems is the following result. A *square-root set* in a group X is a set of the form

$$\sqrt{a} = \{x \in X : x^2 = a\};$$

it is *non-principal* if $a \neq 1$. A slightly weaker form of this theorem (using a stronger form of the condition, and concluding only that A is not a B-group) was proved by Graham Higman; the form given here is due to Ken Johnson and me [20].

Theorem 2.1 *Let A be a countable group with the following property:*

A cannot be written as the union of finitely many translates of non-principal square-root sets together with a finite set.

Then A is not a B-group. More precisely, there exists a primitive but not 2-transitive group G which contains a regular subgroup isomorphic to each countable group satisfying this condition.

Note that the condition of Theorem 2.1 is not very restrictive: any countable abelian group of infinite exponent satisfies the condition; and for any finite or countable group A, the direct product of A with an infinite cyclic group satisfies it, so the group G of the theorem embeds every countable group as a semiregular subgroup.

2.2   The Erdős–Rényi Theorem

The group G of the last subsection is the automorphism group of the remarkable *random graph* R (sometimes known as the *Rado graph*). In the rest of this section we consider this graph and some of its properties.

The reason for the name is the following theorem of Erdős and Rényi [28]:

Theorem 2.2 *There exists a countable (undirected simple) graph R with the property that, if a graph X on a fixed countable vertex set is chosen by selecting edges independently at random with probability $1/2$ from the unordered pairs of vertices, then* $\mathrm{Prob}(X \cong R) = 1$.

Proof   The proof depends on the following graph property denoted by $(*)$:

> Given two finite disjoint sets U, V of vertices, there exists a vertex z joined to every vertex in U and to no vertex in V.

Now the theorem is immediate from the following facts:
1. $\mathrm{Prob}(X$ satisfies $(*)) = 1$;
2. Any two countable graphs satisfying $(*)$ are isomorphic.

To prove (1), we have to show that the event that $(*)$ fails is null. This event is the union of countably many events, one for each choice of the pair (U, V) of sets; so it is enough to show that the probability that no z exists for given U, V is zero. But the probability that n given vertices $z_1, \dots, z_n$ fail to satisfy $(*)$, where $|U \cup V| = k$, is $(1 - 1/2^k)^n$, which tends to 0 as $n \to \infty$.

Claim (2) is proved by a simple back-and-forth argument. Any partial isomorphism between two countable graphs satisfying $(*)$ can be extended so that its domain or range contains one additional point. Proceeding back and forth for countably many steps, starting with the empty partial isomorphism, we obtain the desired isomorphism. □

This is a fine example of a non-constructive existence proof: almost all countable graphs have property $(*)$, but no example of such a graph is exhibited! There are many constructions: here are a few. In each case, to show that the graph is isomorphic to R, we have to verify that $(*)$ holds.
1. Let M be a countable model of the Zermelo–Fraenkel axioms of set theory. (The existence of such a model is *Skolem's paradox*.) Thus M consists of a countable set carrying a binary relation $\in$. Form a graph by symmetrising this relation: that is, $\{x, y\}$ is an edge if $x \in y$ or $y \in x$. This graph is isomorphic to R. [Indeed, not all the ZF axioms are required here; it is enough to have the empty set, pair set, union, and foundation axioms.]

2. As a specialisation of the above, we have Rado's model of *hereditarily finite set theory* (satisfying all the ZF axioms except the axiom of infinity): the set of elements is $\mathbb{N}$, and $x \in y$ if and only if the xth digit in the base 2 expansion of y is equal to 1. This is the construction of R in [54].

3. Let $\mathbb{P}_1$ be the set of primes congruent to 1 mod 4; for $p, q \in \mathbb{P}_1$, join p to q if and only if p is a quadratic residue mod q. (By quadratic reciprocity, this relation is symmetric.) Verification of $(*)$ uses Dirichlet's Theorem on primes in arithmetic progressions.

2.3 Properties of R

A graph Γ is said to be *homogeneous* if every isomorphism between finite induced subgraphs extends to an automorphism of Γ.

Theorem 2.3 *R is homogeneous.*

Proof Given an isomorphism between finite induced subgraphs, the back-and-forth method of the preceding proof extends it to an automorphism. □

We now consider various properties of the group $G = \mathrm{Aut}(R)$. By homogeneity, G acts transitively on the vertices, (oriented) edges, and (oriented) non-edges of R; so, acting on the vertex set, it is a transitive group with (permutation) rank 3 (that is, three orbits on ordered pairs of vertices). Moreover, G is primitive on the vertices. For suppose that $\equiv$ is a congruence on the vertex set which is not the relation of equality, so that there are distinct vertices v, v' with $v \equiv v'$. Suppose that $\{v, v'\}$ is an edge (the argument in the other case is similar). Since G is transitive on edges, it follows that for every edge $\{w, w'\}$, we have $w \equiv w'$. Now let u, u' be non-adjacent vertices. Choosing $U = \{u, u'\}$ and $V = \emptyset$ in property $(*)$, we find a vertex z joined to u and u'. Thus $u \equiv z \equiv u'$, so $u \equiv u'$. Thus $\equiv$ is the universal congruence.

Thus, a group which can be embedded as a regular subgroup of G (that is, a group A for which some Cayley graph is isomorphic to R) is not a B-group. Now Theorem 2.1 is proved by showing that, if A is a group satisfying the hypotheses of the theorem, then a random Cayley graph for A is isomorphic to R with probability 1, and hence that almost all Cayley graphs for A are isomorphic to R.

As an example of the proof technique, I show:

Proposition 2.4 *R has $2^{\aleph_0}$ non-conjugate cyclic automorphisms.*

Proof If a graph Γ has a cyclic automorphism σ, then we can index the vertices by integers so that σ acts as the cyclic shift $x \mapsto x+1$. Then the graph Γ is determined by the set $S = \{x \in \mathbb{Z} : x > 0, x \sim 0\}$, where $\sim$ is the adjacency relation in the graph Γ. Indeed, $\Gamma = \mathrm{Cay}(\mathbb{Z}, S \cup (-S))$. Furthermore, it is an easy calculation to show that, if the same graph Γ has two cyclic automorphisms σ_1 and σ_2, giving rise to sets S_1 and S_2, then $S_1 = S_2$ if and only if σ_1 and σ_2 are conjugate in $\mathrm{Aut}(\Gamma)$. So the theorem will be proved if we can show that there are $2^{\aleph_0}$ different sets S for which the resulting graph Γ is isomorphic to R.

We do this by choosing the elements of S independently at random from the positive integers, and showing that the probability that $(*)$ fails is zero. Suppose that we are testing, for a given pair (U, V), whether there is a vertex z "correctly joined". Of course, we must exclude the elements of $U \cup V$ from consideration. We must discard all z for which we have already decided about the membership of any element of the form $|z - u|$ or $|z - v|$ in S, for $u \in U$ and $v \in V$: there are finitely many such. We also discard any element z for which an equation of the form $z - w_1 = w_2 - z$ holds, for $w_1, w_2 \in U \cup V$; again there are only finitely many such z. So for all but finitely many z, the events $z \sim w$ for $w \in U \cup V$ are independent, and the probability that such a z is correctly joined is non-zero.

Now the argument proceeds as in the proof of the Erdős–Rényi theorem. Since the event $\Gamma \cong R$ has probability 1, it certainly has cardinality $2^{\aleph_0}$. $\square$

An example of a countable group for which no Cayley graph is isomorphic to R is

$$A = \langle a, b : b^4 = 1, b^{-1}ab = a^{-1} \rangle.$$

Every element of the form $a^i b^j$ in this group with j odd is a square root of b^2, and the remaining elements form a translate of this square root set. Hence, in a Cayley graph for A, we cannot find a point joined to 1 and b but not to b^2 or b^3.

Problem 2.5 Is the above group A a B-group?

2.4 Properties of Aut(R)

A number of properties of the group $G = \mathrm{Aut}(R)$ are known:
  (a) It has cardinality $2^{\aleph_0}$. (This follows from Proposition 2.4, or from (c) below.)
  (b) It is simple [64].
  (c) It has the strong small index property (see below) [34, 18].

Truss proved that G is simple by showing that, given any two non-identity elements $g, h \in G$, it is possible to write h as the product of five conjugates of $g^{\pm 1}$. Subsequently [66] he improved this: only three conjugates of g are required. This is best possible.

A permutation group on a countable set X is said to have the *small index property* if every subgroup of index less than $2^{\aleph_0}$ contains the pointwise stabiliser of a finite set; it has the *strong small index property* if every such subgroup lies between the pointwise and setwise stabilisers of a finite set. Hodges *et al.* [34] showed that G has the small index property, and Cameron [18] improved this to the strong small index property.

Truss [65] found all the cycle structures of automorphisms of R. In particular, R has cyclic automorphisms. (This is the assertion that the infinite cyclic group is a regular subgroup of G.)

A number of subgroups of G with remarkable properties have been shown to exist. Here are two examples, due to Bhattacharjee and Macpherson [6, 7]. The first settles a question of Peter Neumann.

Theorem 2.6 *There exist automorphisms f, g of R such that*

(a) *f has a single cycle on R, which is infinite,*

(b) *g fixes a vertex v and has two cycles on the remaining vertices (namely, the neighbours and non-neighbours of v),*

(c) *the group $\langle f, g \rangle$ is free and is transitive on vertices, edges, and non-edges of R, and each of its non-identity elements has only finitely many cycles on R.*

This theorem is proved by building the permutations f and g as limits of partial maps constructed in stages. Of course the existence of automorphisms satisfying (a) and (b) follows from Truss' classification of cycle types, but much more work is required to achieve (c).

Theorem 2.7 *There is a locally finite group G of automorphisms of R which acts homogeneously (that is, any isomorphism between finite subgraphs can be extended to an element of G).*

This theorem uses a result of Hrushovski [35] on extending partial automorphisms of graphs.

Various other subgroups acting homogeneously on R can be constructed. For example, using either of the explicit descriptions (numbers 2 and 3) of R given in Section 2.2, the group of automorphisms given by recursive (or primitive recursive) functions (on $\mathbb{N}$ or $\mathbb{P}_1$ respectively) acts homogeneously.

We noted that $\mathrm{Aut}(R)$ embeds all finite or countable groups. On a similar note, Bonato *et al.* [8] showed that the endomorphism monoid of R embeds all finite or countable monoids.

2.5 Topology

I now turn to properties of G as a topological group. There is a natural topology on the symmetric group of countable degree, the topology of pointwise convergence: a basis of neighbourhoods of the identity is given by the stabilizers of finite tuples. (The intuition is that two permutations are close together if they agree on a large finite subset of the domain.) This topology is derived from either of the following two metrics. (We identify the permutation domain with $\mathbb{N}$.)

(a) $d(g, g) = 0$ and $d(g, h) = 1/2^n$ where n is such that $ig = ih$ for $i < n$ but $ng \neq nh$, for $g \neq h$.

(b) $d'(g, g) = 0$ and $d'(g, h) = 1/2^n$ where n is such that $ig = ih$ and $ig^{-1} = ih^{-1}$ for $i < n$ but either $ng \neq nh$ or $ng^{-1} \neq nh^{-1}$, for $g \neq h$.

(Here $1/2^n$ could be replaced by any decreasing sequence tending to zero.) The advantage of the metric d' is that the symmetric group is a complete metric space for this metric. (The cycles $(1, 2, \ldots, n)$ form a Cauchy sequence for d which does not converge to a permutation.)

The closed and open subgroups of the symmetric group can be characterised as follows. A *first-order structure* consists of a set carrying a collection of relations and functions (of various positive arities) and constants; it is *relational* if there are no functions or constants. Thus, a graph is an example of a relational structure (with a single binary relation). Just as for graphs, a relational structure M is *homogeneous* if any isomorphism between finite induced substructures can be extended to an automorphism of M.

Theorem 2.8   (a) *A subgroup H of the symmetric group is open if and only if it contains the stabiliser of a finite tuple.*

(b) *A subgroup H of the symmetric group is closed if and only if it is the full automorphism group of some first-order structure M; this structure can be taken to be a homogeneous relational structure.*

The first part of this theorem gives an interpretation of the small index property:

Proposition 2.9 *A permutation group G of countable degree has the small index property if and only if any subgroup of G with index less than $2^{\aleph_0}$ is open in G.*

Thus, if a closed group has the small index property, then its topology can be recovered from its group structure: the subgroups of index less than $2^{\aleph_0}$ form a basis of open neighbourhoods of the identity.

For the second part of the theorem, we define the *canonical relational structure* associated with a permutation group G as follows: For each orbit O of G on n-tuples of points of the domain Ω, we take an n-ary relation R_O, and specify that R_O holds precisely for those n-tuples which belong to the orbit O. The resulting structure is easily seen to be homogeneous. Now the closure of G is the automorphism group of its canonical relational structure.

A subgroup H of G is dense in G if and only if G and H have the same orbits on Ω^n for all $n \in \mathbb{N}$. In particular, a subgroup H of $\mathrm{Aut}(R)$ is dense in $\mathrm{Aut}(R)$ if and only if it acts homogeneously on R (as described earlier). Thus, for example, the second theorem of Bhattacharjee and Macpherson asserts that $\mathrm{Aut}(R)$ has a locally finite dense subgroup.

A remarkable classification of closed subgroups of the symmetric group of countable degree has been given by Bergman and Shelah [5]. Call two subgroups G_1 and G_2 of $\mathrm{Sym}(\Omega)$ *equivalent* if there exists a finite subset U of $\mathrm{Sym}(\Omega)$ such that $\langle G_1 \cup U \rangle = \langle G_2 \cup U \rangle$.

Theorem 2.10 *If Ω is countable, there are just four equivalence classes of closed subgroups of $\mathrm{Sym}(\Omega)$. They are characterised by the following conditions, where $G_{(\Gamma)}$ denotes the pointwise stabiliser of the subset Γ:*

(a) *For every finite subset Γ of Ω, the subgroup $G_{(\Gamma)}$ has at least one infinite orbit on Ω.*

(b) *There exists a finite subset Γ such that all the orbits of $G_{(\Gamma)}$ are finite, but none such that the cardinalities of these orbits have a common upper bound.*

(c) *There exists a finite subset Γ such that the orbits of $G_{(\Gamma)}$ have a common upper bound, but none such that $G_{(\Gamma)} = 1$.*

(d) *There exists a finite subset Γ such that $G_{(\Gamma)} = 1$.*

Note that $\mathrm{Aut}(R)$ falls into class (a) of this theorem; in fact, for any finite set Γ, all the orbits of $G_{(\Gamma)}$ outside Γ are infinite, and have the property that the induced subgraph is isomorphic to R and $G_{(\Gamma)}$ acts homogeneously on it. (If $|W| = n$, then the pointwise stabiliser of W has 2^n orbits outside W: for each subset U of W, there is an orbit consisting of the points witnessing condition $(*)$ for the pair (U, V), where $V = W \setminus U$.) So there is a finite set B such that $\langle \mathrm{Aut}(R), B \rangle = \mathrm{Sym}(R)$. (This fact follows from Theorem 1.1 of [43]. Galvin [30] showed that we can take B to consist of a single element. As we will see in the final section of the paper, these results are not specific to $\mathrm{Aut}(R)$.)

Recall that a subset of a metric space is *residual* if it contains a countable intersection of dense open sets. The *Baire category theorem* states that a residual set in a complete metric space is non-empty. Indeed, residual sets are 'large'; their properties are analogous to those of sets of full measure in a measure space. (For example, the intersection of countably many residual sets is dense.)

Now let G denote any closed subgroup of the symmetric group of countable degree. The element $g \in G$ is said to be *generic* if its conjugacy class is residual in G. A given group has at most one conjugacy class of generic elements. There may be no such class: for example, if G is discrete, then the only residual set is G itself.

As an example, we show that a permutation having infinitely many cycles of each finite length and no infinite cycles is generic in the symmetric group. Obviously the set P of such permutations is a conjugacy class. It suffices to show that, for each n, the set P_n of permutations of $\mathbb{N}$ having at least n cycles of length i for $i = 1, \ldots, n$ and in which the point n lies in a finite cycle is open and dense: for the intersection of the sets P_n is obviously P.

If $g \in P_n$, then there is a finite set X_g such that any permutation agreeing with g on X_g is in P_n. This set is an open ball; so P_n is open.

Any open ball is defined by a finite partial permutation h of $\mathbb{N}$, and consists of all permutations which extend h. Let h be any finite partial permutation. Then there is a finite extension of h which is a permutation of a finite set. By adjoining some more cycles, we may assume that this extension has at least n cycles of length i for $1 \leq i \leq n$ and that n lies in a finite cycle. Thus, the open ball defined by h meets P_n. So P_n is dense.

The first part of the following theorem is due to Truss [65]; the second is due to Hodges *et al.* [34], and is crucial for proving the small index property for $\mathrm{Aut}(R)$.

Theorem 2.11 (a) *The group $\mathrm{Aut}(R)$ contains generic elements. Such an element has infinitely many cycles of any given finite length and no infinite cycles.*

(b) *For any positive integer n, the group $\mathrm{Aut}(R)^n$ contains generic elements. Such an n-tuple generates a free subgroup of $\mathrm{Aut}(R)$, all of whose orbits are finite.*

2.6 Reducts of R

We conclude this section with Thomas' theorem [60] about the reducts of R. For my purposes here, a *reduct* of a structure A is a structure B on the same set with $\mathrm{Aut}(B) \geq \mathrm{Aut}(A)$. If our structures are first-order, then their automorphism groups are closed subgroups of the symmetric group on A; so we are looking for closed overgroups of $\mathrm{Aut}(A)$.

An *anti-automorphism* of a graph Γ is a permutation of the vertices mapping Γ to its complement. A *switching automorphism* is a permutation mapping Γ to a graph equivalent to it under switching. (Here *switching* with respect to a set X of vertices consists in replacing edges between X and its complement by non-edges, and non-edges by edges, leaving edges within or outside X unchanged.) A *switching anti-automorphism* is a permutation mapping Γ to a graph equivalent to its complement under switching. Now Thomas' Theorem asserts:

Theorem 2.12 *The closed subgroups of $\mathrm{Sym}(R)$ containing $\mathrm{Aut}(R)$ are:*

- $\mathrm{Aut}(R)$;
- *the group of automorphisms and anti-automorphisms of R;*
- *the group of switching automorphisms of R;*
- *the group of switching automorphisms and anti-automorphisms of R;*
- $\mathrm{Sym}(R)$.

3 Homogeneous relational structures

3.1 Fraïssé's Theorem

For finite permutation groups, as is well known, a consequence of the Classification of Finite Simple Groups is the fact that the only finite 6-transitive permutation groups are the symmetric and alternating groups. What about the infinite case? Analogues of the "geometric" multiply-transitive finite permutation groups such as projective general linear groups give groups which are at most 3-transitive. Can we achieve higher degrees of transitivity?

I do not know any trivial construction which produces an infinite permutation group with any prescribed degree of transitivity. If we take Ω to be countable, and let g_i be any permutation fixing i points and permuting the others in a single cycle, then it is clear that $\langle g_0, \ldots, g_{n-1} \rangle$ is n-transitive; but it may be $(n+1)$-transitive, or even highly transitive. A sufficiently clever choice of the permutations might give a group which is not $(n+1)$-transitive. However, it is more straightforward to ensure that the group is not $(n+1)$-transitive by letting it preserve a suitable $(n+1)$-ary relation, and to choose this relation to have an n-transitive automorphism group.

A theorem of Fraïssé guarantees the existence of suitable structures. I now describe this theorem.

We work in the context of relational structures over a fixed relational language (that is, the relations are named, and relations with the same name in different structures have the same arity; moreover, the induced substructure on a subset of a relational structure is obtained by restricting all relations to the subset). Fraïssé defined the *age* Age(M) of a relational structure M to be the class of all finite relational structures embeddable (as induced substructure) in M. Thus, the age of the random graph R consists of all finite graphs.

As noted earlier, a relational structure is homogeneous if every isomorphism between finite substructures extends to an automorphism. Thus, the random graph R is homogeneous. Another homogeneous relational structure, which was Fraïssé's motivating example, is the set $\mathbb{Q}$ of rational numbers as ordered set. (It is easy to see that any order-preserving map between two n-tuples of rationals can be extended to a piecewise-linear order-preserving map on the whole of $\mathbb{Q}$.)

The following straightforward result allows us to recognise ages. The *joint embedding property* of a class $\mathcal{C}$ asserts that, for all $B_1, B_2 \in \mathcal{C}$, there exists $C \in \mathcal{C}$ such that both B_1 and B_2 can be embedded in C.

Proposition 3.1 *Let $\mathcal{C}$ be a class of finite relational structures over a fixed relational language. Then $\mathcal{C}$ is the age of a countable relational structure M if and only if the following conditions hold:*

  (a) *$\mathcal{C}$ is closed under isomorphism;*

  (b) *$\mathcal{C}$ is closed under taking induced substructures;*

  (c) *$\mathcal{C}$ contains only finitely many structures up to isomorphism;*

  (d) *$\mathcal{C}$ has the joint embedding property.*

An age $\mathcal{C} = \text{Age}(M)$ can be represented by a rooted tree: the nodes on level n are the structures in $\mathcal{C}$ on the set $\{0, 1, \ldots, n-1\}$, and the children of a structure are all its one-point extensions. Now the elements of the boundary of $\mathcal{C}$ (the infinite paths starting at the root) represent, in a natural way, structures N on the set $\mathbb{N}$ satisfying Age(N) $\subseteq$ Age(M). (Fraïssé calls such structures N *younger* than M.) I note in passing that the set of such paths has a natural hypermetric (the distance between two paths being a decreasing function of the number of steps for which they coincide), and is complete in this metric; so we can talk about residual properties of structures younger than M. A simple example: of the structures younger than M, a residual set have the same age as M.

Fraïssé [29] gave a necessary and sufficient condition for a class $\mathcal{C}$ of finite structures to be the age of a homogeneous relational structure. This involves the *amalgamation property*, the assertion that if $A, B_1, B_2 \in \mathcal{C}$ and $f_i : A \to B_i$ are embeddings for $i = 1, 2$, then there is a structure $C \in \mathcal{C}$ and embeddings $g_i : B_i \to C$ for $i = 1, 2$, forming a commutative diagram (that is, $f_1 g_1 = f_2 g_2$). Note that I allow the empty structure as a substructure; hence the amalgamation property as stated here includes the joint embedding property. (This depends on the fact that there is only one empty set, which holds because we do not allow relations of arity 0.) Note also that $A f_1 g_1 \subseteq B_1 g_1 \cap B_2 g_2$; but it is not required that the images of B_1 and B_2 in C intersect just in the image of A. (Later we say that the *strong amalgamation property* holds if this condition can be required.)

Theorem 3.2 *Let $\mathcal{C}$ be a class of finite relational structures over a fixed relational* *language. Then $\mathcal{C}$ is the age of a countable homogeneous relational structure M if and only if the following conditions hold:*

(a) *$\mathcal{C}$ is closed under isomorphism;*

(b) *$\mathcal{C}$ is closed under taking induced substructures;*

(c) *$\mathcal{C}$ contains only finitely many structures up to isomorphism;*

(d) *$\mathcal{C}$ has the amalgamation property.*

If the conditions (a)–(d) hold, then the countable homogeneous structure M is unique up to isomorphism.

A class $\mathcal{C}$ satisfying (a)–(d) of the theorem is called a *Fraïssé class*, and the unique countable homogeneous structure M with age $\mathcal{C}$ is its *Fraïssé limit*. For example, the classes of finite graphs and finite linearly ordered sets are Fraïssé classes; their Fraïssé limits are respectively the random graph and the ordered set of rationals.

A more general form of Fraïssé's Theorem has been found by Hrushovski, who has used it to produce some remarkable examples in model theory (such as a counterexample to Lachlan's pseudoplane conjecture). These structures have interesting automorphism groups which have not been much investigated in their own right.

Hrushovski's method relies on the fact that we do not have to require all instances of the amalgamation property in the class $\mathcal{C}$, but only those where A is "closed" in a suitable sense. I will not attempt further explanation here; there is a good exposition of this theory by Wagner [70].

3.2   Applications

As a first application of Fraïssé's Theorem, I show that groups with any desired degree of transitivity exist. Let $\mathcal{C}_n$ be the class of $(n+1)$-*uniform hypergraphs*: these are structures whose edges are sets of $n+1$ points, rather than two points as in the case of graphs. (As relational structure, a hypergraph has one $(n+1)$-ary relation, which holds only if its arguments are all distinct, and is completely symmetric.) It is easy to see that $\mathcal{C}_n$ is a Fraïssé class; let R_n be its Fraïssé limit, and $G_n = \mathrm{Aut}(R_n)$. Then

(a) G_n is n-transitive. This is because, by definition, the induced substructure on any n-set is trivial (with no edges), and so any bijection between n-sets is an isomorphism of induced substructures; by homogeneity, such an isomorphism extends to an automorphism of R_n.

(b) G_n is not $(n+1)$-transitive. This is because no automorphism can map $n+1$ points forming an edge to $n+1$ points not forming an edge.

Of course, R_1 is the random graph R. In a similar way, R_n is the random $(n+1)$-uniform hypergraph. It has a number of properties which resemble or extend those of R. Here are a few:

(a) Every countable group can be embedded as a regular subgroup of R_n if $n > 1$.

(b) All reducts of R_n are known (Thomas [61]).

(c) There are explicit constructions of R_n.

In fact, the hypergraphs R_n were constructed by Rado in the same paper in which he considered the graph R. I outline the construction. There is a nice bijection F_n from the set of n-element subsets of $\mathbb{N}$ to the set $\mathbb{N}$, defined as follows: if $x_0 < x_1 < \cdots < x_{n-1}$, then

$$F_n(\{x_0, x_1, \ldots, x_{n-1}\}) = \binom{x_0}{1} + \binom{x_1}{2} + \cdots + \binom{x_{n-1}}{n}.$$

Now, for $x_0 < \cdots < x_{n-1} < y$, we take $\{x_0, x_1, \ldots, x_{n-1}, y\}$ to be an edge of R_n if and only if the mth digit in the base 2 expansion of y is equal to 1, where $m = F_n(\{x_0, x_1, \ldots, x_{n-1}\})$. This reduces to Rado's construction of R when $n = 1$.

Further interesting groups can be obtained by exploiting also a method due to Tits [63]. (He first used this to show that the free group of countable rank has BN-pairs of all spherical types.)

Theorem 3.3 *Suppose that G is a permutation group on a countable set Ω, and N a normal subgroup of G having the same orbits as G. Suppose that G/N contains a non-abelian free subgroup. Then G has a subgroup F, which is a free group of countable rank, having the same orbits on Ω as G and satisfying $F \cap N = 1$.*

Proof Let $f_1 N, f_2 N, \ldots$ be free generators of a free subgroup of countable rank in G/N. Enumerate all pairs (α_i, β_i) of points for which α_i and β_i lie in the same G-orbit on Ω, and choose elements $n_i \in N$ such that $\alpha_i f_i n_i = \beta_i$. Then it is easy to show that $F = \langle f_1 n_1, f_2 n_2, \ldots \rangle$ is the required subgroup.   □

The first application, due to Kantor, shows that a free group F of countable rank has a faithful highly transitive action. We take Ω to be the union, over all n, of the sets of n-tuples of distinct elements of a fixed countable set X; let G be the symmetric group on X, and N the finitary symmetric group consisting of permutations of finite support. Then a subgroup of G has the same orbits on Ω as G if and only if it is highly transitive on X. Embed F into G by means of its regular representation; then $F \cap N = 1$, so F embeds into G/N. The conditions of the theorem are easily checked.

This action of F has the further property that it is *cofinitary*, that is, any non-identity element fixes only finitely many points. For any non-identity element of the subgroup F is the product of a non-identity element of the regular copy of F and a finitary permutation.

The application of the theorem is limited by the requirement that G should have a normal subgroup which is 'not too big and not too small'. For example, since $\mathrm{Aut}(R)$ is simple, we cannot apply this directly to R. Often it is possible to refine the structure with extra relations. Here is an example.

Theorem 3.4 *For every positive integer k, there is a permutation group which is k-transitive, in which the stabiliser of any $k + 1$ points is the identity.*

This theorem is best possible, since another theorem due to Tits [62] and Hall [31] independently shows that, for $k \geq 4$, there is no *sharply k-transitive group*, that is no k-transitive group in which the stabiliser of any k points is the identity. The groups of the theorem are "higher analogues" of Frobenius and Zassenhaus groups.

Proof Take the class $\mathcal{C}$ of finite structures each of which consists of a colouring of the $(k+1)$-subsets of its domain with a fixed set of countably many colours. Then $\mathcal{C}$ is a Fraïssé class; let M be its Fraïssé limit. Let G be the group of automorphisms of M (where automorphisms are allowed to permute the colours), and N the subgroup fixing all the colours. Then G/N is the symmetric group on the set of colours, and certainly contains a free group of countable rank acting regularly. Moreover, both G and N are k-transitive. Applying the theorem, we find a k-transitive copy of F inside G. The construction shows that F acts regularly on the set of colours; so the stabiliser of any $(k+1)$-set, and *a fortiori* any $(k+1)$-tuple, is trivial. $\qquad\square$

Other strange permutation groups can be constructed with 'bare hands', where permutations generating the required group are built up in stages. Such methods require a great deal of ingenuity. Many beautiful examples have been given by Macpherson. We have already seen two of his theorems with Bhattacharjee [6, 7]; here is another result [42].

Theorem 3.5 *For any $k \geq 1$, the group of order-automorphisms of the rational numbers contains a free subgroup F of countable rank with the properties that for any two k-tuples $x_1 < x_2 < \cdots < x_k$ and $y_1 < y_2 < \cdots < y_k$, there is a unique element of F mapping the first to the second. (In other words, F is "sharply k-set transitive".)*

This result answered an open question on scale type in the theory of measurement, an area with close links with various aspects of permutation group theory.

Macpherson also showed that $\mathrm{Aut}(R_n)$ contains an automorphism fixing i points and acting as a single cycle on the rest, for $0 \leq i \leq n-1$. As we saw at the start of this section, the group generated by such permutations is obviously n-transitive; in this case it cannot be $(n+1)$-transitive, since it leaves the hypergraph R_n invariant. So, at least in principle, the construction of groups with any degree of transitivity hinted at there can be made to work. Macpherson further showed that we can take the group generated by these automorphisms to be free.

3.3 Homogeneous structures

A lot of effort has gone into classifying the homogeneous structures of various types. One important result is due to Lachlan and Woodrow [39]:

Theorem 3.6 *A countable homogeneous (undirected) graph is one of the following:*

   (a) *the disjoint union of m complete graphs of size n, where $mn = \aleph_0$;*

   (b) *complements of (a);*

   (c) *the Fraïssé limit of the class of graphs containing no complete subgraph of size n, for $n \geq 3$;*

(d) *complements of* (c);

(e) *the random graph* R.

Other classes in which the homogeneous structures have been determined include directed graphs (Cherlin [25], the case of tournaments having been done earlier by Lachlan [38]) and posets (Schmerl [57]).

The Fraïssé limit of the class of K_n-free graphs is known as the *Henson graph* H_n, since it was first constructed by Henson [32]. One of Henson's results was that H_3 admits cyclic automorphisms but H_n does not if $n > 3$. This can be extended as follows.

The triangle-free graph H_3 has the property that it admits many regular subgroups, and so is a Cayley graph for many different groups. The proof is similar to that for R. However, for $n > 3$, it can be shown that H_n is not a normal Cayley graph for any countable group (where a Cayley graph $\text{Cay}(G, S)$ is *normal* if it is invariant under both left and right translation by G, or in other words, if S is a normal subset of G). In particular, it is not a Cayley graph of any countable abelian group.

Problem 3.7 Is H_n a Cayley graph for $n > 3$?

The graph H_3 also gives us another highly transitive group, namely the stabiliser of a vertex v in $\text{Aut}(H_3)$, acting on the set of neighbours of v.

3.4   Some recent connections

I conclude this section with a brief account of two recent occurrences of homogeneous structures and Fraïssé classes in very different parts of mathematics: Ramsey theory and topological dynamics.

I begin with a simple form of the finite version of Ramsey's Theorem.

Theorem 3.8 *For any two natural numbers a, b with $a < b$, there exists a natural number c with $b < c$ having the property that, if the a-element subsets of a c-element set are coloured red and blue in any manner, then there exists a b-element subset, all of whose a-element subsets have the same colour.*

In order to convert this theorem into a theory, we might take a fixed type of relational structure, replace "set" and "subset" by "structure" and "substructure", and ask whether the analogue of Ramsey's Theorem holds. To formalise this, let $\mathcal{C}$ be a class of finite relational structures. For $A, B \in \mathcal{C}$, we denote by $\binom{B}{A}$ the class of all substructures of B isomorphic to A. We say that $\mathcal{C}$ is a *Ramsey class* if the following holds:

> For any $A, B \in \mathcal{C}$ with A embeddable in B, there exists $C \in \mathcal{C}$ with B embeddable in C such that, if the elements of $\binom{C}{A}$ are coloured red and blue in any manner, there exists $B' \in \binom{C}{B}$ such that all elements of $\binom{B'}{A}$ have the same colour.

It is natural to assume that $\mathcal{C}$ is isomorphism-closed and substructure-closed. Usually we assume that it is an age.

Various deep theorems assert that the classes of graphs and of triangle-free graphs are Ramsey classes. Taking this hint, Nešetřil [47] showed the following result:

Theorem 3.9 *Let $\mathcal{C}$ be the age of a countable structure (that is, let it satisfy conditions (a)–(d) of Proposition 3.1). If $\mathcal{C}$ is a Ramsey class, then it is a Fraïssé class (that is, it satisfies conditions (a)–(d) of Theorem 3.2).*

Proof Suppose that $\mathcal{C}$ is a Ramsey class but not a Fraïssé class. Then the amalgamation property fails in $\mathcal{C}$; that is, there exist embeddings $f_1 : A \to B_1$, $f_2 : A \to B_2$ which cannot be amalgamated. However, $\mathcal{C}$ is an age, so we can find $C \in \mathcal{C}$ into which B_1 and B_2 can be embedded. (Without loss, we assume that B_1 and B_2 are substructures of C.) Now by the Ramsey property, we can find D such that, if the elements of $\binom{D}{A}$ are coloured red and blue in any manner, there is a monochromatic copy C' of C.

Colour the elements of $\binom{D}{A}$ as follows: the structure $A' \cong A$ is red if it is contained in a structure $B' \cong B_1$ inside D; otherwise it is blue. Now find $C' \cong C$ inside D such that all copies of A in C' have the same colour. This colour must be red, since we know that C contains at least one copy of A which is inside a copy of B_1 (within C), namely Af_1. So all copies of A in C' are red. But we know there is a copy of B_2 in C; so the image Af_2 of A under its embedding in B_2 is contained in $B' \cong B_1$ (within D). Then $B' \cup B_2$ is an amalgam, contrary to assumption. □

This suggests a programme for determining the Ramsey classes of some given type. (It is reasonable to look only for classes which are ages.)
- Find the Fraïssé classes of that type.
- Decide which of them are Ramsey classes.

Before the last topic, we make a small digression. Some relational structures naturally carry a total order (for example, those for which one of the relations is interpreted as a total order in the structure). Often, if there is not a natural order on the structure, we can add one.

We say that a Fraïssé class $\mathcal{C}$ has the *strong amalgamation property* if, in the statement of the a.p., we may assume that $B_1 g_1 \cap B_2 g_2 = A f_1 g_1$, that is, the intersection of the images of B_1 and B_2 in the amalgam is no larger than it has to be. It can be shown that, if $\mathcal{C}$ is a Fraïssé class with Fraïssé limit M, then $\mathcal{C}$ has the strong amalgamation property if and only if $G = \mathrm{Aut}(M)$ has the property that the stabiliser of finitely many points in G has no finite orbits on the remaining points.

If $\mathcal{C}_1$ and $\mathcal{C}_2$ are Fraïssé classes with the strong a.p., then we can form a class $\mathcal{C}_1 \wedge \mathcal{C}_2$ whose members consist of a $\mathcal{C}_1$-structure and a $\mathcal{C}_2$-structure on the same set. Then $\mathcal{C}_\infty \wedge \mathcal{C}_\in$ is a Fraïssé class, and indeed has the strong a.p., since we can amalgamate the two types of structure independently. In particular, since the class of total orders has the strong a.p., we see that if $\mathcal{C}$ has the strong a.p., then there is a class $\mathcal{C}^<$ whose elements are totally ordered versions of $\mathcal{C}$-structures. So, in this case, if we do not have an order already, we can impose one.

Now we can discuss the result of Kechris, Pestov, and Todorcevic [37]. In this section, all topological spaces and groups will be assumed to be Hausdorff.

A topological group G is *extremely amenable* if, whenever G acts continuously on a compact space, it has a fixed point.

This condition is stronger than the usual forms of amenability which have been studied, which merely require the existence of an invariant measure in this situation. Indeed, a theorem of Veech [68] shows that no locally compact group can be extremely amenable, since such a group always has a free action on a compact space. However, extremely amenable groups do exist. The remarkable theorem of Kechris *et al.* asserts the following. Here a closed subgroup of the symmetric group $\mathrm{Sym}(\Omega)$ has the topology of pointwise convergence (the induced topology from $\mathrm{Sym}(\Omega)$). Recall that any closed subgroup of $\mathrm{Sym}(\Omega)$, for countable Ω, is the automorphism group of some homogeneous relational structure M on Ω.

Theorem 3.10 *Let G be the automorphism group of a countable homogeneous relational structure M. Then the following are equivalent:*
  (a) *G is extremely amenable;*
  (b) *the structure on M includes a total order, and $\mathrm{Age}(M)$ is a Ramsey class.*

In particular, $\mathrm{Aut}(\mathbb{Q}, <)$ is extremely amenable, as are $\mathrm{Aut}(R^<)$ and $\mathrm{Aut}(H_n^<)$ for $n \geq 3$, where H_n is Henson's graph.

Even in cases where the structure on M does not include a total order, the methods of the paper allow the possible minimal G-flows (minimal continuous actions on compact spaces) to be determined. However, I cannot tell more of this fascinating story here!

4   Oligomorphic permutation groups

4.1   Definition and basic properties

Let G be a permutation group on a set Ω. (Usually we assume that Ω is infinite.) We say that G is *oligomorphic* if it has only finitely many orbits on Ω^n for all natural numbers n. (By convention, Ω^0 is a single point.)

There are three convenient numbers with the property that, if one is finite (for some value of n) then so are the others:

  F_n^* : the number of G-orbits on Ω^n;

  F_n : the number of G-orbits on n-tuples of distinct elements of Ω;

  f_n : the number of G-orbits on n-element subsets of Ω.

If we need to specify the group under consideration, we write $F_n^*(G)$, etc.

The first two sequences determine each other, since

$$F_n^* = \sum_{k=1}^{n} S(n,k) F_k$$

for $n \geq 1$, where $S(n,k)$ is the *Stirling number* of the second kind, the number of partitions of an n-set into k parts. Because of this, we concentrate on the sequence F_n. The permutation group G is highly transitive if and only if $F_n(G) = 1$ for all $n \in \mathbb{N}$. If this holds, then $F_n^*(G) = B(n)$, the nth Bell number (the number of partitions of an n-set).

The second and third sequences clearly satisfy

$$f_n \leq F_n \leq n! f_n,$$

but neither determines the other in general.

Let G be oligomorphic. We denote by $F^*(x)$ and $F(x)$ the exponential generating functions

$$F^*(x) = \sum_{n \geq 0} \frac{F_n^* x^n}{n!}, \qquad F(x) = \sum_{n \geq 0} \frac{F_n x^n}{n!}.$$

It follows from our observation about the relation between these sequences that $F^*(x) = F(e^x - 1)$.

Let $f(x)$ be the ordinary generating function

$$f(x) = \sum_{n \geq 0} f_n x^n.$$

All of these are specialisations of a multi-variable power series which we call the *modified cycle index* of G, defined by

$$\tilde{Z}(G) = \sum_{A \in \mathcal{P}_f(\Omega)/G} Z(G_A^A),$$

where $\mathcal{P}_f(\Omega)/G$ denotes a set of orbit representatives for G acting on the finite subsets of Ω; G_A^A denotes the (finite) permutation group induced on A by its setwise stabiliser in G; and $Z(H)$ is the ordinary cycle index of the finite permutation group H, defined by

$$Z(H) = \frac{1}{|H|} \sum_{h \in H} s_1^{c_1(h)} s_2^{c_2(h)} \cdots,$$

where $c_i(h)$ is the number of i-cycles of h and $s_1, s_2, \ldots$ are indeterminates.

As a matter of notation, we denote the result of substituting an expression t_i for s_i (for all i) in a polynomial or formal power series F in variables $s_1, s_2, \ldots$ by $F(s_i \leftarrow t_i)$.

If G is finite, the modified cycle index gives nothing new:

$$\tilde{Z}(G) = Z(G; s_i \leftarrow s_i + 1).$$

But the modified cycle index is defined for any oligomorphic permutation group, although there is no hope of defining the ordinary cycle index of an infinite permutation group.

Theorem 4.1

$$f_G(x) = \tilde{Z}(G; s_i \leftarrow x^i)$$
$$F_G(x) = \tilde{Z}(G; s_1 \leftarrow x, s_i \leftarrow 0 \text{ for } i > 0)$$

4.2 General properties

It is easy to see that passing from an oligomorphic group G to its closure does not change the modified cycle index of G, and hence does not change the sequences F_n and f_n associated with G.

Furthermore, if G is an oligomorphic permutation group on Ω, then there exists an oligomorphic permutation group G' on a countable set Ω' which has the same modified cycle index (and hence the same numbers F_n and f_n) as G. This follows from the downward Löwenheim–Skolem theorem of model theory. For in a suitable countable first-order language we can state that G is a group, that G acts on Ω, and that G has f_n orbits on n-subsets of Ω with the appropriate group induced on a set in the ith orbit, for all n. This theory has a model, by assumption, and so has a countable model. (Indeed, we may assume that G' is countable as well; but if we replace G' by its closure, this is no longer true.)

A first-order theory is $\aleph_0$-*categorical* if it has a unique countable model (up to isomorphism). It turns out that the closed oligomorphic permutation groups of countable degree are precisely the automorphism groups of countable models of $\aleph_0$-categorical theories; this is the celebrated theorem of Engeler, Ryll-Nardzewski, and Svenonius.

Theorem 4.2 *Let M be a countable first-order structure. Then $\mathrm{Th}(M)$ is $\aleph_0$-categorical if and only if $\mathrm{Aut}(M)$ is oligomorphic. If this happens, then every n-type over $\mathrm{Th}(M)$ is realised in M, and the set of realising n-tuples is an orbit of $\mathrm{Aut}(M)$ on M^n.*

Here, an n-*type* is a set S of n-variable formulae in the language of $\mathrm{Th}(M)$ which is maximal with respect to the property that

$$\mathrm{Th}(M) \cup \{(\exists x_1, \ldots, x_n)\phi(x_1, \ldots, x_n) : \phi \in S\}$$

is consistent. It is *realised* in M if there exists a realising n-tuple $(a_1, \ldots, a_n)$ (with $a_i \in M$ for all i) such that $\phi(a_1, \ldots, a_n)$ is true in M.

So the sequences $F_n^*(G)$ for oligomorphic permutation groups G are precisely the sequences counting types in $\aleph_0$-categorical theories.

Example 4.3   1. According to Cantor's theorem, the ordered set $\mathbb{Q}$ of rational numbers is characterised as the unique countable dense total order without endpoints. All the conditions except countability are first-order. Thus $\mathbb{Q}$ as ordered set is $\aleph_0$-categorical, and $A = \mathrm{Aut}(\mathbb{Q}, <)$ is oligomorphic. Indeed, it is easy to see that A is transitive on n-element subsets of $\mathbb{Q}$ for all n, and has $n!$ orbits on ordered n-tuples of distinct elements of $\mathbb{Q}$.

2. Condition (∗), which we met in the discussion of the random graph R, is a conjunction of first-order sentences (one for each pair of values of $|U|$ and $|V|$). Thus, R is $\aleph_0$-categorical, and $\mathrm{Aut}(R)$ is oligomorphic. Indeed, $F_n(\mathrm{Aut}(R))$ is the number of *labelled* n-vertex graphs (namely $2^{n(n-1)/2}$), and $f_n(\mathrm{Aut}(R))$ is the number of *unlabelled* n-vertex graphs (that is, graphs up to isomorphism).

In general, let $\mathcal{C}$ be an isomorphism-closed class of finite structures. When we refer to *unlabelled* structures in $\mathcal{C}$, we mean isomorphism classes of $\mathcal{C}$-structures; when we refer to *labelled* structures in $\mathcal{C}$, we mean structures with underlying set $\{1, \ldots, n\}$. (For example, there are four unlabelled graphs and eight labelled graphs on 3 vertices). Now, if $\mathcal{C}$ is the age of a countable homogeneous relational structure M, then the numbers of unlabelled and labelled n-element structures in $\mathcal{C}$ are $f_n(\mathrm{Aut}(M))$ and $F_n(\mathrm{Aut}(M))$ respectively. So our orbit-counting problems generalise standard combinatorial enumeration problems for Fraïssé classes.

In this connection, I remark that an age is a *species* in the sense of Joyal [36]. If $\mathcal{C} = \mathrm{Age}(M)$ where M is homogeneous, then the modified cycle index of $\mathrm{Aut}(M)$ is the cycle index of the species $\mathcal{C}$, in Joyal's terminology.

4.3 Highly set-transitive groups

Fraïssé classes give rise to a wide class of combinatorial enumeration problems, and raise the possibility of general results about the behaviour of the counting functions for such classes, which we now discuss. First, note that, unlike other counting problems in group theory (such as word growth or subgroup growth of finitely generated groups), there is no upper bound to the growth rate of $(f_n(G))$ or $(F_n(G))$. For given any sequence (a_n), however rapidly growing, consider a relational language with a_n relations of arity n, for all n, and let $\mathcal{C}$ be the class of finite structures in which each n-ary relation is an n-uniform hypergraph. (This condition ensures that there are only finitely many n-element structures in $\mathcal{C}$, up to isomorphism.) Then $\mathcal{C}$ is a Fraïssé class, and its Fraïssé limit M satisfies $f_n(\mathrm{Aut}(M)) \geq 2^{a_n}$ for all n.

Indeed, it is the slowest growth that is the most interesting!

Of course, the slowest possible growth is no growth at all. Obviously $F_n(G) = 1$ for all n if and only if G is highly transitive; so the only closed group with this property is the symmetric group (though there are many other highly transitive groups, some of which we have seen). For $f_n(G)$, things are more interesting. We say that a permutation group G on Ω is *highly set-transitive* if $f_n(G) = 1$ for all n. Now the following holds [11]:

Theorem 4.4 *Let G be a closed highly sest-transitive permutation group of countable degree. Then G is one of the following five groups:*

A : *the group of order-preserving permutations of $\mathbb{Q}$;*

B : *the group of order-preserving or order-reversing permutations of $\mathbb{Q}$;*

C : *the group of permutations preserving the circular order on the set of complex roots of unity;*

D : the group of permutations preserving or reversing the circular order on the set of complex roots of unity;

S : the symmetric group.

This theorem has wider implications. It follows that any highly set-transitive group of countable degree is contained as a dense subgroup in one of these five groups. Moreover, using the downward Löwenheim–Skolem theorem, it shows that a highly set-transitive but not highly transitive group of arbitrary infinite degree preserves or reverses a linear or circular order. Also, the theorem shows that the only reducts of A are the groups A, B, C, D, S.

Throughout the remainder of this article, the symbols A, B, C, D, S will be used in the sense of this theorem.

The modified cycle index for each of these groups can be calculated easily, since there is just one orbit on n-sets for each n, and the group induced on an n-set by its setwise stabiliser is respectively the trivial group, the cyclic group of order 2 with at most one fixed point (if $n > 1$, the regular cyclic group of order n, and the dihedral group of order $2n$ (if $n > 2$). In particular, we have the following:

G	$F_G(x)$	$f_G(x)$
A	$1/(1-x)$	$1/(1-x)$
C	$1 - \log(1-x)$	$1/(1-x)$
S	e^x	$1/(1-x)$

Note the similarity between the conclusions of Theorem 4.4 and Thomas' Theorem 2.12 on the reducts of $\mathrm{Aut}(R)$. The analogy goes deeper, since the circular order can be produced by a switching-like construction, as John McDermott pointed out to me. (It was this insight which sparked my interest in infinite permutation groups in the 1970s.)

4.4 Direct and wreath products

Next I turn to some constructions of new oligomorphic groups from old.

Let G_1 and G_2 be permutation groups on Ω_1 and Ω_2 respectively. The direct product $G_1 \times G_2$ has two natural actions:

- the *intransitive action* on the disjoint union of Ω_1 and Ω_2, where

$$\alpha(g_1, g_2) = \alpha g_i \quad \text{if } \alpha \in \Omega_i, \quad i = 1, 2.$$

- the *product action* on the Cartesian product of Ω_1 and Ω_2, where

$$(\alpha_1, \alpha_2)(g_1, g_2) = (\alpha_1 g_1, \alpha_2 g_2).$$

If G_1 and G_2 are oligomorphic, then $G_1 \times G_2$ is oligomorphic in each of these actions. For the intransitive action, things are simple; the modified cycle index is multiplicative:

$$\tilde{Z}(G_1 \times G_2) = \tilde{Z}(G_1)\tilde{Z}(G_2).$$

From this, of course, it follows that the generating functions $F(x)$ and $f(x)$ for orbits on ordered tuples or unordered sets are also multiplicative.

The product action is more complicated. Here, the number of orbits on arbitrary n-tuples is multiplicative:

$$F_n^*(G_1 \times G_2) = F_n^*(G_1)F_n^*(G_2).$$

This relation does not easily translate into a relation for the generating functions. The composition of modified cycle index works as follows [19]: define a 'multiplication' $\circ$ of indeterminates by

$$s_i \circ s_j = (s_{\mathrm{lcm}(i,j)})^{\gcd(i,j)},$$

and extend multiplicatively to arbitrary monomials and then additively to arbitrary polynomials. Then

$$Z(G \times H) = Z(G) \circ Z(H).$$

It is possible to deduce the multiplicativity of F_n^* from this.

Example 4.5   1. For the group $G = S \times S$ in the product action, the asymptotic behaviour of the sequence $(f_n(G))$ is not known.

2. Consider the group $G = A \times A$ in the product action. Then $f_n(G)$ is equal to the number of zero-one matrices with n ones, having no row or column consisting entirely of zeros. The asymptotics of this sequence have been worked out [22]: we have

$$f_n(G) \sim \frac{n!}{4}e^{-\frac{1}{2}(\log 2)^2}\frac{1}{(\log 2)^{2n+2}}.$$

Now we turn to wreath products. The definition of the (unrestricted) wreath product $G_1 \operatorname{Wr} G_2$ depends on the action of G_2 on Ω_2: the base group consists of the Cartesian product of Ω_2 copies of G_1, and the top group G_2 acts on the base group by permuting the indices of the factors. Thus, elements of the base group are functions $f : \Omega_2 \to G_1$, and

$$f^{g_2}(\alpha_2) = f(\alpha_2 g_2^{-1}).$$

There are two natural actions of the wreath product:
- the *imprimitive action* on $\Omega_1 \times \Omega_2$: the base group acts by

$$(\alpha_1, \alpha_2)f = (\alpha_1 f(\alpha_2), \alpha_2),$$

and the top group acts on the second coordinate.
- the *power action* (also called the product action) on the set $\Omega_1^{\Omega_2}$ of functions $\phi : \Omega_2 \to \Omega_2$, where the base group acts by

$$(\phi f)(\alpha_2) = \phi(\alpha_2)f(\alpha_2),$$

and the top group acts by

$$(\phi g_2)(\alpha_2) = \phi(\alpha_2 g_2^{-1}).$$

If G_1 and G_2 are oligomorphic, then $G_1 \operatorname{Wr} G_2$ in the imprimitive action is oligomorphic. Its modified cycle index is obtained from that of G_2 by substituting $\tilde{Z}(s_i, s_{2i}, \ldots) - 1$ for s_i, for $i = 1, 2, \ldots$. It follows that

$$F_{G_1 \operatorname{Wr} G_2}(x) = F_{G_2}(F_{G_1}(x) - 1);$$

however, $f_{G_1 \operatorname{Wr} G_2}$ is not expressible in terms of f_{G_1} and f_{G_2} alone. (It can be obtained from the modified cycle index of G_2 by substituting $f_{G_1}(x^i) - 1$ for s_i for $i = 1, 2, \ldots$.)

Exercise: Show that the groups $C \operatorname{Wr} S$ and A have the same numbers of orbits on ordered n-tuples of distinct elements for all n. (Combinatorially, the orbits of these two groups correspond to permutations and linear orders of $\{1, \ldots, n\}$ respectively; any permutation can be written as a disjoint union of cycles.)

The power action of $G_1 \operatorname{Wr} G_2$ is oligomorphic if and only if G_1 is oligomorphic and G_2 is a finite permutation group acting on a finite set. (If Ω_2 is finite, then $G_1 \operatorname{Wr} G_2$ contains the direct product of $|\Omega_2|$ copies of G_1 in the product action, and hence is oligomorphic. But if Ω_2 is infinite, then pairs of functions (ϕ, ϕ') agreeing in i positions lie in different orbits for $i = 1, 2, \ldots$.)

The problem of expressing the modified cycle index of $G_1 \operatorname{Wr} G_2$ (in the power action) in terms of those of the factors has not been solved.

4.5 Growth rates

One of the most basic facts about the sequences (f_n) and (F_n) is that they are non-decreasing. In the case of (F_n), this is clear: there is a map from $(n+1)$-tuples to n-tuples (taking the initial segment) which commutes with the action of the group, and is onto. We see that $F_n = F_{n+1}$ if and only if G is $(n+1)$-transitive.

The inequality $f_n \leq f_{n+1}$ is more subtle: two proofs are known, one using linear algebra, the other Ramsey's theorem. The characterisation of the case of equality here is not known. It is known that a primitive group with $f_n = f_{n+1}$ must be 2-set-transitive (that is, $f_2 = 1$), and that f_n is bounded by a function of n; it is conjectured that there is an absolute bound on n for which this equality can hold in a primitive group [23]. (We will soon see an example of a primitive group with $f_6 = f_7 = 2$.)

For the symmetric group S, we have $f_n(S) = F_n(S) = 1$ for all n. Thus, $f_S(x) = (1-x)^{-1}$. So if S^k denotes the direct product of k copies of S with the intransitive action, then $f_{S^k}(x) = (1-x)^{-k}$, from which we see by the negative binomial theorem that

$$f_n(S^k) = \binom{n+k-1}{k-1},$$

a polynomial in n with degree $k-1$. Examples of transitive but imprimitive groups with polynomial growth of (f_n) can also easily be constructed. However, for the group $G = S \operatorname{Wr} S$ (with the imprimitive action), we see that $f_n(G) = p(n)$, the number of partitions of n; for this imprimitive group, we have fractional exponential growth: precisely,

$$p(n) \sim \frac{1}{4n\sqrt{3}} e^{\pi\sqrt{2n/3}}.$$

What happens for primitive groups? The first result asserts the existence of a wide gap, between constant and exponential growth.

Theorem 4.6 *There is an absolute constant $c > 1$ such that, if G is primitive but not highly set-transitive, then*

(a) *the numbers of orbits on n-sets satisfy $f_n \geq c^n / p(n)$ for some polynomial p;*

(b) *the numbers of orbits on n-tuples of distinct elements satisfy $F_n \geq c^n n! / p(n)$ for some polynomial p.*

The first part of the theorem is due to Macpherson [40], who gave the value $c = \sqrt[5]{2} = 1.148\ldots$. The second part is due to Merola [46], who also improved Macpherson's constant to $c = 1.324\ldots$. No counterexamples are known to the conjecture that both parts hold with $c = 2$.

The growth rates of the sequences (f_n) and (F_n) appear to have a great deal of structure; little is known, but examples suggest some conjectures. One of the most challenging conjectures is to prove that various 'obvious' limits, such as $\lim(\log f_n)/(\log n)$ (for polynomial growth), $\lim(\log \log f_n)/(\log n)$ (for fractional exponential growth), or $\lim(\log f_n)/n$ (for exponential growth), actually exist. For primitive groups with exponential growth, as noted above, we have no examples where $\lim(f_n)^{1/n}$ is strictly less than 2 (assuming that it is not 1); and indeed, only countably many values for this limit are known.

Example 4.7 Here are two examples of primitive groups where f_n grows exponentially.

1. A *tournament* is a directed graph in which each pair of distinct vertices is joined by a directed edge in just one direction. A tournament is said to be a *local order* if it does not contain a 4-point subtournament consisting of a vertex dominating or dominated by a 3-cycle. (In a tournament, we say that x dominates y if there is an edge from x to y; this is extended to sets of vertices by saying that A dominates B if every vertex in A dominates every vertex in B.) The finite local orders form a Fraïssé class, whose Fraïssé limit T is more easily described as follows. Take a countable dense set of points on the unit circle containing no antipodal pair of points. (If we consider the set of all complex roots of unity, and randomly choose one out of each pair $\{\omega, -\omega\}$, then the resulting set is dense with probability 1.) Now put a directed edge from x to y if the shorter arc from x to y is in the anticlockwise direction. The result is the universal homogeneous local order T. We see that $\mathrm{Aut}(T)$ is 2-set transitive, hence primitive; and $f_n(\mathrm{Aut}(T))$ is equal to the number of isomorphism types of n-vertex local order, which is asymptotically $2^{n-1}/n$. (In fact, by taking the larger group G of permutations which preserve or reverse the edge directions, we obtain $f_n(G) \sim 2^{n-2}/n$, the slowest known growth rate for a primitive but not highly set-transitive group.

2. A *boron tree* is a finite tree in which every vertex has valency 1 or 3. (Boron trees describe the analogue of hydrocarbons in a boron-based chemistry.) On the set of leaves of a boron tree, we can define a 4-place relation as follows. Given four points a, b, c, d, there is a unique partition into two sets of size 2 such that the paths joining vertices in the same set do not intersect. Write $R(a, b; c, d)$ if this partition is $ab \mid cd$. (See Figure 1). The relational structures obtained in this way form a Fraïssé class; the boron tree is uniquely recoverable from the quaternary relation on the set of leaves. Thus, if G is the automorphism group of the Fraïssé limit, then $f_n(G)$ is equal to the number of boron trees with $2n - 2$ vertices, which is asymptotically $An^{-5/2}c^n$, where $c = 2.483\ldots$. Note in passing that this group is 3-transitive but not 4-transitive, is 5-set transitive, and has $f_6 = f_7 = 2$. (All these assertions can be proved by drawing diagrams like Figure 1.)

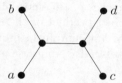

Figure 1. A boron tree

It has been observed that most of the primitive oligomorphic groups in which (f_n) exhibits exponential growth (including the above examples) are associated either with linear or circular orders or with trees (or some combination of these). Recently, a conceptual framework has been found which includes all these examples [33]: they are automorphism groups of $\aleph_0$-categorical, weakly o-minimal structures (those which are linearly ordered so that every definable set is a finite union of convex subsets). However, we are far from having a result asserting that any group with exponential growth of (f_n) is of this form!

Known primitive groups at first appear to show that there is a gap between exponential and factorial growth, and that there is a spectrum of possible values of $\lim(f_n/n!)^{1/n}$ (that is, values of c for which the growth is roughly $c^n n!$). Some results have been established under model-theoretic hypotheses related to stability, such as the strict order property, by Macpherson [41]. However, the group $G = S \operatorname{Wr} S_2$, with the power action, has the property that $f_n(G)$ grows faster than exponentially but slower than factorially. The precise asymptotic behaviour is not known. A similar remark implies to the symmetric group in its action on 2-sets. (For this group G, $f_n(G)$ is the number of graphs with n edges and no isolated vertices, up to isomorphism.)

Problem 4.8 What is the asymptotic behaviour of $f_n(G)$ for the two groups G in the preceding paragraph?

I end with one further problem.

Problem 4.9 Is the set of F-sequences of oligomorphic groups closed under point-wise multiplication?

The answer is "no" in general. For if G is the stabiliser of a point in the infinite symmetric group, then $F_n(G) = n+1$ for all n. There is no group G' with $F_n(G) = (n+1)^2$. For such a group would have four orbits on Ω, and hence at least 12 orbits on ordered pairs of distinct elements of Ω, contradicting $F_2(G') = 9$.

For transitive but imprimitive groups, I suspect that the answer is also "no". Let G be the group fixing a partition of Ω into infinitely many sets of size 2. It is easy to see that $F_n(G)$ is equal to the number of solutions of $g^2 = 1$ in the symmetric group of degree n. The square of this sequence cannot be realised by a primitive group, by Merola's Theorem, since it grows faster than $n!$ but slower than $c^n n!$ for any $c > 1$. I conjecture that no imprimitive group realises this sequence either.

For F-sequences of primitive groups, the problem is open.

Note that the set of F^*-sequences of oligomorphic groups is closed under point-wise product (by taking the direct product of the groups, in the product action), while the set of F-sequences of automorphism groups of Fraïssé limits of classes with the strong a.p. is also closed under pointwise product (by taking the wedge of the Fraïssé classes).

4.6 Graded algebras

The numbers f_n have another interpretation, as the coefficients in the Hilbert series of a graded algebra.

Given a set Ω, the *reduced incidence algebra* of subsets of Ω is defined as follows. The nth homogeneous component V_n is the space of functions from $\binom{\Omega}{n}$ to $\mathbb{C}$, with pointwise addition. Multiplication is the map $V_n \times V_m \to V_{n+m}$ defined as follows: for $f_1 \in V_n$, $f_2 \in V_m$, and $X \in \binom{\Omega}{n+m}$, we have

$$(f_1 f_2)(X) = \sum_{Y \in \binom{X}{n}} f_1(Y) f_2(X \setminus Y).$$

It is easily verified that

$$A = \bigoplus_{n \geq 0} V_n$$

is a commutative associative $\mathbb{N}$-graded algebra.

Now if G is a permutation group on Ω, then there is a natural action of G on V_n, and hence on A; we let

$$A^G = \bigoplus_{n \geq 0} V_n^G$$

be the algebra of A-fixed points. Clearly $\dim(V_n^G) = f_n$ if this number is finite.

The structure of A is known in some cases:

1. If $G = S$ (or indeed if G is highly set-transitive), then A^G is the polynomial algebra in a single generator e, the constant function with value 1 in V_1.

2. If $G = G_1 \times G_2$ in its intransitive action, then $A^G = A^{G_1} \otimes_{\mathbb{C}} A^{G_2}$. Hence, for $G = S^n$ in the intransitive action, A^G is a polynomial ring in n generators of degree 1.

3. If $G = S\mathrm{Wr}H$ in its imprimitive action, where H is a finite permutation group, then A^G is isomorphic to the algebra of invariants of the linear group H. In particular, if $G = S \mathrm{Wr} S_n$, then A^G is a polynomial ring in generators of degree $1, 2, \ldots, n$.

4. Suppose that $G = \mathrm{Aut}(M)$, where M is the Fraïssé limit of a class $\mathcal{C}$ of finite structures. If there are notions of 'connectedness' and 'involvement' in $\mathcal{C}$ satisfying a few simple axioms, then A^G is a polynomial algebra generated by the connected structures in $\mathcal{C}$. (For details, see [15].) Here are three special cases.

   - If $\mathcal{C}$ consists of all finite graphs, so that M is the random graph, then the polynomial generators of A^G correspond to the connected graphs.

   - Suppose that elements of $\mathcal{C}$ are totally ordered sets coloured with m colours. Any such structure is coded by a word in an m-letter alphabet. Then A^G is the *shuffle algebra* [55, p. 24], and the polynomial generators are the *Lyndon words*. (The fact that the shuffle algebra is polynomial was proved by Radford [53].)

   - Let $G = H \mathrm{Wr} S$, in its imprimitive action. Then A^G is a polynomial algebra with $f_n(H)$ generators of degree n for each n; the connected structures are those contained within a block of imprimitivity.

In general, little is known about the structure of A^G. An old conjecture of mine asserts that, if G has no finite orbits on Ω, then A^G is an integral domain. (This conclusion would have some implications for the growth of the sequence $(f_n(G))$.) It is known that the property that A^G is an integral domain is preserved by passing to overgroups or transitive extensions [16]. Maurice Pouzet (personal communication) has a proof of the general conjecture, but it has not yet been published.

If A^G is a polynomial algebra with a_n homogeneous generators of degree n for each n, then the sequences (a_n) and (f_n) determine each other:

$$\sum_{n \geq 0} f_n x^n = \prod_{k \geq 1} (1 - x^k)^{-a_k}.$$

There are some permutation groups G for which the structure of A^G is not known, but the numbers a_n computed from this formula on the assumption that it is a polynomial algebra have a combinatorial interpretation. Any such group provides an interesting puzzle, whether or not A^G turns out to be a polynomial algebra: if it is, we have to match the generators with the combinatorial objects counted by (a_n); if not, we have to explain the apparent coincidence.

For example, let G be the group of switching automorphisms of the random graph R (see Theorem 2.12). It is known that $f_n(G)$ is equal to the number of *even* graphs (graphs with all vertex degrees even) on n vertices [45, 12]; so a_n is equal to the number of *Eulerian* graphs (connected even graphs) on n vertices. Now $A^G \subseteq A^{\mathrm{Aut}(R)}$, and $A^{\mathrm{Aut}(R)}$ is a polynomial ring, so A^G is an integral domain; but it is not known whether it is itself a polynomial ring.

Problem 4.10 For the above group G, is A^G a polynomial ring?

5   The Urysohn space

We have seen that the techniques used by Erdős and Rényi to prove their remarkable theorem about the random graph had been developed by Fraïssé more than ten years earlier for arbitrary relational structures. In fact, Fraïssé was not the first to do this.

In a posthumously-published paper, P. S. Urysohn [67] constructed a remarkable metric space:

Theorem 5.1 *There exists a metric space* $\mathbb{U}$*, unique up to isometry, with the following properties:*

- $\mathbb{U}$ *is a Polish space (a complete separable metric space);*
- $\mathbb{U}$ *is homogeneous (every isometry between finite subspaces extends to an isometry of* $\mathbb{U}$*);*
- $\mathbb{U}$ *is universal (every Polish space can be embedded isometrically into* $\mathbb{U}$*).*

It is interesting to compare this theorem with Fraïssé's. The theorem of Fraïssé deals only with countable objects. A universal Polish space clearly cannot be countable, since it must embed the real numbers. The replacement for countability is the condition of separability. (A metric space is *separable* if it has a countable dense subset).

With hindsight, can we construct $\mathbb{U}$ by Fraïssé's method? Clearly we can't do so directly, because it is uncountable. (Indeed, there are uncountably many two-element metric spaces.) However, we can proceed as follows. Let $\mathcal{C}$ be the class of *rational metric spaces*, those for which all distances are rational. It is relatively easy to show that $\mathcal{C}$ is a Fraïssé class. We denote its Fraïssé limit by $Q\mathbb{U}$, and call it the *rational Urysohn space*. Then $\mathbb{U}$ is the completion of $Q\mathbb{U}$.

Vershik [69] showed the analogue of the Erdős–Rényi theorem, namely, that $\mathbb{U}$ is the random Polish space. Of course, the interpretation of this statement requires some care. Suppose that we build a countable metric space by adding one point at a time. At the nth stage, we have points $a_0, \ldots, a_{n-1}$, and wish to add a point a_n with specified distances from the existing points. The vector $(x_0, \ldots, x_{n-1})$, where $x_i = d(a_n, a_i)$, must satisfy the conditions

$$|x_i - x_j| \leq d(a_i, a_j) \leq x_i + x_j$$

as well as $x_i \geq 0$ for all i. (We allow non-strict inequality, but this will be very unlikely). The set of such admissible vectors then lies in an unbounded region of the positive orthant of $\mathbb{R}^n$ defined by these inequalities. Vershik showed that, if the vector of distances is chosen randomly from this set at each stage according to any of a wide range of natural probability distributions, then the completion of the resulting countable space is isometric to $\mathbb{U}$ with probability 1.

In the paper [24], the authors attempted to apply some of the methods used to find subgroups of $\mathrm{Aut}(R)$ to the isometry group of $\mathbb{U}$. I now describe some of the results of this paper and some open problems. Let $\mathrm{Isom}(M)$ denote the isometry group of the metric space M. We note that $\mathrm{Isom}(Q\mathbb{U})$ is a dense subgroup of $\mathrm{Isom}(\mathbb{U})$ (in the topology induced by the product topology on $\mathbb{U}^{\mathbb{U}}$).

Bounded and unbounded isometries An isometry g of a metric space M is *bounded* if there is a number c such that $d(x, xg) \leq c$ for all $x \in M$. The bounded isometries form a normal subgroup $\mathrm{BIsom}(M)$ of the group of all isometries. In the case of the Urysohn space, we have

$$1 \neq \mathrm{BIsom}(\mathbb{U}) \neq \mathrm{Isom}(\mathbb{U}),$$

but $\mathrm{BIsom}(\mathbb{U})$ is dense in $\mathrm{Isom}(\mathbb{U})$.

Problem 5.2 Is it true that
   (a) $\mathrm{BIsom}(\mathbb{U})$ and $\mathrm{Isom}(\mathbb{U})/\mathrm{BIsom}(\mathbb{U})$ are simple groups, and
   (b) $\mathrm{Isom}(\mathbb{U})$ is simple as a topological group?

The Tits method The factor group $\mathrm{Isom}(Q\mathbb{U})/\mathrm{BIsom}(Q\mathbb{U})$ contains a non-abelian free subgroup. Hence the method of Tits can be used to show that

Theorem 5.3 $\mathrm{Isom}(\mathbb{U})$ *contains a dense free subgroup.*

Cyclic automorphisms A method like that for R (but significantly more complicated) shows:

Theorem 5.4 $Q\mathbb{U}$ *has an isometry σ which permutes all its points in a single cycle: indeed, there are uncountably many conjugacy classes of such isometries.*

Such a σ can be regarded as an isometry of $\mathbb{U}$, all of whose orbits are dense. (This suggests the possibility of finding interesting dynamical systems here.) Moreover, the closure of the group generated by such a σ is an abelian group acting transitively on $\mathbb{U}$; so the space $\mathbb{U}$ has an abelian group structure (indeed, many such structures).

The closure of an infinite cyclic group is abelian, but is not necessarily torsion-free. It is not known what sort of torsion these groups may have.

Problem 5.5 Describe the possible structure of the closure of a cyclic group of isometries of $\mathbb{U}$ with dense orbits.

Other regular groups The results obtained here are much more restricted. The sum total of our knowledge is the following: The countable elementary abelian 2-group acts regularly on $Q\mathbb{U}$, but the countable elementary abelian 3-group cannot act on $\mathbb{U}$ with a dense orbit.

The latter fact is rather easy to prove, as follows. Suppose that we have such an action of this group A. Since the stabiliser of a point in the dense orbit is trivial, we can identify the points of the orbit with elements of A (which we write additively).

Choose $x \neq 0$ and let $d(0, x) = \alpha$. Then $\{0, x, -x\}$ is an equilateral triangle with side α. Since $\mathbb{U}$ is universal and A is dense, there is an element y such that $d(x, y), d(-x, y) \approx \frac{1}{2}\alpha$ and $d(0, y) \approx \frac{3}{2}\alpha$. (The approximation is to within a given ϵ chosen smaller than $\frac{1}{6}\alpha$). Then the three points $0, y, x - y$ form a triangle with sides approximately $\frac{3}{2}\alpha$, $\frac{1}{2}\alpha$, $\frac{1}{2}\alpha$, contradicting the triangle inequality.

An open problem is to find other examples of groups which do (or don't) have a regular action on $Q\mathbb{U}$ (or another countable dense subset of $\mathbb{U}$).

Very recently, Nešetřil [48] has shown that finite metric spaces form a Ramsey class. It follows from the results of Kechris *et al.* [37] that the isometry group of the "ordered" version of $\mathbb{U}$ is extremely amenable.

6 Further topics

6.1 Finitary permutation groups

A permutation group is *finitary* if its elements move only finitely many points. The finitary permutations of Ω form a group, the *finitary symmetric group*, which is locally finite and highly transitive, and is normal in the symmetric group on Ω. Indeed, if Ω is countable, then the only normal subgroups of the symmetric group, apart from itself and the trivial group, are the finitary symmetric group and the *alternating group* (consisting of all the finitary even permutations).

Wielandt [72], extending a theorem of Jordan to the infinite, showed:

Theorem 6.1 *A primitive infinite permutation group which contains a non-identity finitary permutation must contain the alternating group (and hence is highly transitive).*

This result is usually referred to as the Jordan–Wielandt Theorem.

Extending further, Neumann [49, 50] developed a structure theory for finitary permutation groups. This has been further extended to finitary linear groups and even to finitary automorphism groups of modules over arbitrary rings.

6.2 Jordan groups

This topic for finite permutation groups dates from the nineteenth century, although definitive results had to await the Classification of Finite Simple Groups. The picture is much richer in the infinite case.

Let G be a transitive permutation group on Ω. A subset Δ of Ω, containing more than one point, is a *Jordan set* if G contains a subgroup H which fixes the complement of Δ pointwise and acts transitively on Ω. Of course, if G is $(n + 1)$-transitive, then any set whose complement contains just n points is a Jordan set; such a Jordan set is called *improper*. A *Jordan group* is a transitive permutation group having at least one proper Jordan set. Usually we consider only primitive Jordan groups.

For example, the Jordan–Wielandt Theorem, discussed in the last subsection, shows that a primitive group with a finite Jordan set must contain the alternating group. Dually, Neumann [51] showed that a primitive Jordan group with a cofinite proper Jordan set is a projective or affine group over a finite field or is highly transitive. (This paper is recommended for its treatment of finite Jordan groups.)

These examples, however, give no hint of the richness of general Jordan groups. For example, the group of homeomorphisms of a manifold (of dimension at least 2) is a Jordan group: any open ball is a Jordan set, since we can permute its points transitively while fixing its boundary. This group is highly transitive. To avoid such examples, we make the further restriction that our Jordan group is not highly transitive.

Example 6.2   1. The group A of order-preserving permutations of the rational numbers is a Jordan group: the stabiliser of the point α is isomorphic to $A \times A$ (where the factors act transitively on the Jordan sets $\{x : x < \alpha\}$ and $\{x : x > \alpha\}$.

   2. The group associated with boron trees in the last section is a Jordan group. We can think of the points as being the ends of a dense treelike object; given an internal vertex, the ends fall into three classes each admitting a transitive group fixing the others pointwise.

   3. As in the finite case, a projective or affine group (of finite or infinite dimension over a finite or infinite field) is a Jordan group, the Jordan sets being complements of subspaces.

A remarkable classification of the infinite Jordan groups which are primitive but not highly transitive was achieved by Adeleke and Macpherson [1]. They show that such a group preserves either a relational structure of a special types or a "limit" of such structures. The types are linear, circular and semilinear orders, general betweenness and separation relations, C- and D-relations, and Steiner systems. (See the memoir of Adeleke and Neumann [2] for a detailed description and axiomatisation of the types of relations which can appear.)

Note that there is a big overlap between these groups and two classes we have mentioned before: the primitive oligomorphic groups G for which $(f_n(G))$ grows exponentially; and the automorphism groups of the $\aleph_0$-categorical, weakly o-minimal structures. It is not clear why these overlaps occur.

6.3   Maximal subgroups of $\mathrm{Sym}(\Omega)$

In this section, unlike most of the rest of the paper, we do not assume that Ω is countable.

The O'Nan–Scott Theorem is usually stated as a result about finite primitive (or quasiprimitive) permutation groups. Yet in its original form [58], it was a classification of maximal subgroups of finite symmetric (or alternating) groups into several types: intransitive (subset stabilisers); imprimitive (partition stabilisers); affine; wreath products with product action; diagonal groups; and almost simple groups. Now we know much more about which groups of each type are actually maximal.

For infinite symmetric groups, things are more complicated: a classification based on the socle is not possible since there may be no minimal normal subgroup; and indeed, there may be subgroups of the symmetric group which lie in no maximal subgroup.

The first difficulty arises from the fact that $\mathrm{Sym}(\Omega)$ has non-trivial normal subgroups. If H is a maximal subgroup of $\mathrm{Sym}(\Omega)$ and $\aleph_0 \leq c \leq |\Omega|$, then either $\mathrm{BSym}_c(\Omega) \leq H$ or $\mathrm{BSym}_c(\Omega)H = \mathrm{Sym}(\Omega)$. In the first case, H is highly transitive; in the second, Macpherson and Neumann [43] showed that there is a subset Δ with $|\Delta| < c$ whose setwise stabiliser in Ω induces the symmetric group on its complement.

It is clear that the setwise stabiliser of a finite set is a maximal subgroup of $\mathrm{Sym}(\Omega)$. In 1967, Richman [56] demonstrated two further classes of maximal subgroups: the setwise stabiliser of a finite set of equivalent ultrafilters; and the almost stabiliser of a partition into a finite number of sets of equal cardinality. (A *filter* is a family of sets closed under intersection and closed upwards but not containing the empty set; an *ultrafilter* is a filter which is maximal with respect to inclusion. An ultrafilter is *principal* if it consists of all sets containing a particular point of Ω. Two ultrafilters are *equivalent* if one can be mapped to the other by a permutation of Ω. Thus Richman's first class of groups includes the setwise stabilisers of finite sets. A partition π is *almost stabilised* by g if, for all $A \in \pi$, there exists $B \in \pi$ such that the symmetric difference of Ag and B is finite.)

These two classes of maximal subgroups correspond roughly to the maximal intransitive and imprimitive subgroups of finite symmetric groups. Subsequently, several other classes of maximal subgroups of symmetric groups, corresponding in a similar rough sense to other O'Nan–Scott types, have been established [9, 26].

Peter Neumann raised the question: is every proper subgroup H of $\mathrm{Sym}(\Omega)$ contained in a maximal subgroup? Macpherson and Praeger showed that this is the case if H is not highly transitive. However, Baumgartner *et al.* [4] showed that it is consistent with the ZFC axioms for set theory that there exists a subgroup T of $\mathrm{Sym}(\Omega)$, where $|\Omega| = \kappa$, for which the subgroups between T and $\mathrm{Sym}(\Omega)$ form a well-ordered chain of order type κ^+. (This result is proved assuming GCH. Of course, all the intermediate groups are highly transitive. For independence results, see [59]. Note also that Macpherson and Neumann proved that a chain of proper subgroups with union $\mathrm{Sym}(\Omega)$ must have length greater than $|\Omega|$, so in a sense the result of Baumgartner *et al.* is best possible.)

If H is a maximal subgroup of $\mathrm{Sym}(\Omega)$, then $\langle H, g \rangle = \mathrm{Sym}(\Omega)$ for all $g \notin H$. Galvin [30] proved the remarkable result that if H is any subgroup such that $\langle H, B \rangle = \mathrm{Sym}(\Omega)$ for some set B with $|B| \leq |\Omega|$, then there is *some* element g such that $\langle H, g \rangle = \mathrm{Sym}(\Omega)$. Moreover, the order of g can be chosen to be any preassigned even number greater than 2. (We saw earlier, while discussing the the Bergman–Shelah Theorem 2.10, that the hypothesis of Galvin's theorem hold if Ω is countable, H is closed in $\mathrm{Sym}(\Omega)$, and the stabiliser of a finite tuple of points in H has at least one infinite orbit on the remaining points.)

References

[1] S. A. Adeleke and H. D. Macpherson, Classification of infinite primitive Jordan groups, *Proc. London Math. Soc.* (3) **72** (1996), 63–123.

[2] S. A. Adeleke and P. M. Neumann, Relations related to betweenness: their structure and automorphisms, *Memoirs Amer. Math. Soc.* **131**, no. 623, American Mathematical Society, Providence, RI, 1988.

[3] S. A. Adeleke and P. M. Neumann, Primitive permutation groups with primitive Jordan sets, *J. London Math. Soc.* (2) **53** (1996), 209-229.

[4] J. E. Baumgartner, S. Shelah and S. Thomas, Maximal subgroups of infinite symmetric groups, *Notre Dame J. Formal Logic* **34** (1993), 1–11.

[5] G. M. Bergman and S. Shelah, Closed subgroups of the infinite symmetric group, preprint.

[6] M. Bhattacharjee and H. D. Macpherson, Strange permutation representations of free groups, *J. Austral. Math. Soc.* **74** (2003), 267–285.

[7] M. Bhattacharjee and H. D. Macpherson, A locally finite dense group acting on the random graph, *Forum Math.* **17** (2005), 513–517.

[8] A. Bonato, D. Delić and I. Dolinka, All countable monoids embed into the monoid of the infinite random graph, to appear.

[9] M. Brazil, J. Covington, T. Penttila, C. E. Praeger and A. R. Woods, Maximal subgroups of infinite symmetric groups, *Proc. London Math. Soc.* (3) **68** (1994), 77–111.

[10] M. Brown, Weak n-homogeneity implies weak $(n-1)$-homogeneity, *Proc. Amer. Math. Soc.* **10** (1959), 644–647.

[11] P. J. Cameron, Transitivity of permutation groups on unordered sets, *Math. Z.* **48** (1976), 127–139.

[12] P. J. Cameron, Cohomological aspects of two-graphs, *Math. Z.* **157** (1977), 101–119.

[13] P. J. Cameron, *Oligomorphic Permutation Groups*, London Math. Soc Lecture Notes **152**, Cambridge University Press, Cambridge, 1990.

[14] P. J. Cameron, The random graph, pp. 331–351 in *The Mathematics of Paul Erdős*, (J. Nešetřil and R. L. Graham, eds.), Springer, Berlin, 1996.

[15] P. J. Cameron, The algebra of an age, pp. 126–133 in *Model Theory of Groups and Automorphism Groups* (ed. David M. Evans), London Mathematical Society Lecture Notes **244**, Cambridge University Press, Cambridge, 1997.

[16] P. J. Cameron, On an algebra related to orbit-counting, *J. Group Theory* **1** (1998), 173–179.

[17] P. J. Cameron, Homogeneous Cayley objects, *European J. Combinatorics* **21** (2000), 745–760.

[18] P. J. Cameron, The random graph has the strong small index property, *Discrete Math.* **291** (2005), 41–43.

[19] P. J. Cameron, D. A. Gewurz and F. Merola, Product action, preprint.

[20] P. J. Cameron and K. W. Johnson, An essay on countable B-groups, *Math. Proc. Cambridge Philos. Soc.* **102** (1987), 223–232.

[21] P. J. Cameron, P. M. Neumann and D. N. Teague, On the degrees of primitive permutation groups, *Math. Z.* **180** (1982), 141–149.

[22] P. J. Cameron, T. J. Prellberg, and D. S. Stark, Asymptotic enumeration of incidence matrices, preprint.

[23] P. J. Cameron and S. Thomas, Groups acting on unordered sets, *Proc. London Math. Soc.* (3) **59** (1989), 541–557.

[24] P. J. Cameron and A. M. Vershik, Some isometry groups of Urysohn space, preprint.

[25] G. Cherlin, The classification of countable homogeneous directed graphs and countable homogeneous n-tournaments, *Memoirs Amer. Math. Soc.* **621**, American Mathematical Society, Providence, RI, 1998.

[26] J. Covington, H. D. Macpherson and A. D. Mekler, Some maximal subgroups of the symmetric groups, *Quart. J. Math. Oxford* (2) **47** (1996), 297–311.

[27] M. Droste, Structure of partially ordered sets with transitive automorphism groups, *Memoirs Amer. Math. Soc.* **57** (1985).

[28] P. Erdős and A. Rényi, Asymmetric graphs, *Acta Math. Acad. Sci. Hungar.* **14** (1963), 295–315.

[29] R. Fraïssé, Sur certains relations qui généralisent l'ordre des nombres rationnels, *C. R. Acad. Sci. Paris* **237** (1953), 540–542.

[30] F. Galvin, Generating countable sets of permutations, *J. London Math. Soc.* (2) **51** (1995), 232–242.

[31] M. Hall Jr., On a theorem of Jordan, *Pacific J. Math.* **4** (1954), 219–226.

[32] C. W. Henson, A family of countable homogeneous graphs, *Pacific J. Math.* **38** (1971), 69–83.

[33] B. Herwig, H. D. Macpherson, G. Martin, A. Nurtazin and J. Truss, On $\aleph_0$-categorical weakly o-minimal structures, *Ann. Pure Appl. Logic* **101** (2000), 65–93.

[34] W. A. Hodges, I. M. Hodkinson, D. Lascar and S. Shelah, The small index property for ω-stable, ω-categorical structures and for the random graph, *J. London Math. Soc.* (2) **48** (1993), 204–218.

[35] E. Hrushovski, Extending partial isomorphisms of graphs, *Combinatorica* **12** (1992), 411–416.

[36] A. Joyal, Une theorie combinatoire des séries formelles, *Advances Math.* **42** (1981), 1–82.

[37] A. S. Kechris, V. G. Pestov, and S. Todorcevic, Fraïssé limits, Ramsey theory, and topological dynamics of automorphism groups, *Geometric and Functional Analysis* **15** (2005), 106–189.

[38] A. H. Lachlan, Countable homogeneous tournaments, *Trans. Amer. Math. Soc.* **284** (1984), 431–461.

[39] A. H. Lachlan and R. E. Woodrow, Countable ultrahomogeneous undirected graphs, *Trans. Amer. Math. Soc.* **262** (1980), 51–94.

[40] H. D. Macpherson, Growth rates in infinite graphs and permutation groups, *Proc. London Math. Soc.* (3) **51** (1985), 285–294.

[41] H. D. Macpherson, Infinite permutation groups of rapid growth, *J. London Math. Soc.* (2) **35** (1987), 276–286.

[42] H. D. Macpherson, Sharply multiply homogeneous permutation groups, and rational scale types, *Forum Math.* **8** (1996), 501–507.

[43] H. D. Macpherson and P. M. Neumann, Subgroups of infinite symmetric groups, *J. London Math. Soc.* (2) **42** (1990), 64–84.

[44] H. D. Macpherson and C. E. Praeger, Subgroups of infinite symmetric groups, *J. London Math. Soc.* (2) **42** (1990), 85–92.

[45] C. L. Mallows and N. J. A. Sloane, Two-graphs, switching classes, and Euler graphs are equal in number, *SIAM J. Appl. Math.* **28** (1975), 876–880.

[46] F. Merola, Orbits on n-tuples for infinite permutation groups, *Europ. J. Combinatorics* **22** (2001), 225–241.

[47] J. Nešetřil, Ramsey classes and homogeneous structures, *Combinatorics, Probability and Computing* **14** (2005), 171–189.

[48] J. Nešetřil, Metric spaces are Ramsey, to appear.

[49] P. M. Neumann, The lawlessness of finitary permutation groups, *Arch. Math.* **26** (1975), 561–566.

[50] P. M. Neumann, The structure of finitary permutation groups, *Arch. Math.* **27** (1976), 3–17.

[51] P. M. Neumann, Some primitive permutation groups, *Proc. London Math. Soc.* (3) **50**

(1985), 265–281.

[52] V. Pestov, Ramsey–Milman phenomenon, Urysohn metric spaces, and extremely amenable groups, *Israel J. Math.* **127** (2002), 317–357.

[53] D. E. Radford, A natural ring basis for the shuffle algebra and an application to group schemes, *J. Algebra* **58** (1979), 432–454.

[54] R. Rado, Universal graphs and universal functions, *Acta Arith.* **9** (1964), 331–340.

[55] C. Reutenauer, *Free Lie Algebras*, London Math. Soc. Monographs (New Series) **7**, Oxford University Press, Oxford, 1993.

[56] F. Richman, Maximal subgroups of infinite symmetric groups, *Canad. Math. Bull.* **10** (1967), 375–381.

[57] J. H. Schmerl, Countable homogeneous partially ordered sets, *Algebra Universalis* **9** (1979), 317–321.

[58] L. L. Scott, Representations in characteristic p, *Proc. Symp. Pure Math.* **37** (1980), 319–331.

[59] S. Shelah and S. Thomas, The cofinality spectrum of the infinite symmetric group, *J. Symbolic Logic* **62** (1997), 902–916.

[60] S. R. Thomas, Reducts of the random graph, *J. Symbolic Logic* **56** (1991), 176–181.

[61] S. R. Thomas, Reducts of random hypergraphs, *Ann. Pure Appl. Logic* **80** (1996), 165–193.

[62] J. Tits, Généralisation des groupes projectifs basée sur leurs propriétés de transitivité, *Acad. Roy. Belgique Cl. Sci. Mem.* **27** (1952).

[63] J. Tits, *Buildings of Spherical Type and Finite BN-Pairs*, Lecture Notes in Math. **382**, Springer–Verlag, Berlin, 1974.

[64] J. K. Truss, The group of the countable universal graph, *Math. Proc. Cambridge Philos. Soc.* **98** (1985), 213–245.

[65] J. K. Truss, Generic automorphisms of homogeneous structures, *Proc. London Math. Soc.* (3) **65** (1992), 121–141.

[66] J. K. Truss, The automorphism group of the random graph: four conjugates good, three conjugates better, *Discrete Math.* **268** (2003), 257–271.

[67] P. S. Urysohn, Sur un espace metrique universel, *Bull. Sci. Math.* **51** (1927), 1–38.

[68] W. Veech, Topological dynamics, *Bull. Amer. Math. Soc.* **83** (1977), 775–830.

[69] A. M. Vershik, The universal and random metric spaces, *Russian Math. Surv.* **356** (2004), No. 2, 65–104.

[70] F. O. Wagner, Relational structures and dimensions, pp. 153–180 in *Automorphisms of First-Order Structures* (ed. R. W. Kaye and H. D. Macpherson), Oxford University Press, Oxford, 1994.

[71] H. Wielandt, *Finite Permutation Groups*, Academic Press, New York, 1964.

[72] H. Wielandt, *Unendliche Permutationsgruppen*, Lecture Notes, Universität Tübingen, 1959. [English translation by P. Bruyns included in Wielandt's collected works: H. Wielandt, *Mathematische Werke: Mathematical Works*, Volume 1 (Bertram Huppert and Hans Schneider, eds.), Walter de Gruyter, Berlin, 1994.]

SELF-SIMILARITY AND BRANCHING IN GROUP THEORY

ROSTISLAV GRIGORCHUK and ZORAN ŠUNIĆ

Department of Mathematics, Texas A&M University, MS-3368, College Station, TX 77843-3368, USA
Email: grigorch@math.tamu.edu, sunic@math.tamu.edu

Contents

Introduction

The idea of self-similarity is one of the most basic and fruitful ideas in mathematics of all times and populations. In the last few decades it established itself as the central notion in areas such as fractal geometry, dynamical systems, and statistical physics. Recently, self-similarity started playing a role in algebra as well, first of all in group theory.

Regular rooted trees are well known self-similar objects (the subtree of the regular rooted tree hanging below any vertex looks exactly like the whole tree). The self-similarity of the tree induces the self-similarity of its group of automorphisms and this is the context in which we talk about self-similar groups. Of particular interest are the finitely generated examples, which can be constructed by using finite automata. Groups of this type are extremely interesting and usually difficult to study as there are no general means to handle all situations. The difficulty of study is more than fairly compensated by the beauty of these examples and the wealth of areas and problems where they can be applied.

Branching is another idea that plays a major role in many areas, first of all in Probability Theory, where the study of branching processes is one of the main directions.

Partially supported by NSF grants DMS-0308985 and DMS-0456185

The idea of branching entered Algebra via the so called branch groups that were introduced by the first author at the Groups St Andrews Conference in Bath 1997.

Branch groups are groups that have actions "of branch type" on spherically homogeneous rooted trees. The phrase "of branch type" means that the dynamics of the action (related to the subnormal subgroup structure) mimics the structure of the tree. Spherically homogenous trees appear naturally in this context, both because they are the universal models for homogeneous ultra-metric spaces and because a group is residually finite if and only if it has a faithful action on a spherically homogeneous tree.

The importance of the choice of the "branch type" action is reflected in the fact that it is the naturally opposite to the so called diagonal type. While any residually finite group can act faithfully on a rooted homogeneous tree in diagonal way, the actions of branch type are more restrictive and come with some structural implications.

Actions of branch type give rise to many examples of just-infinite groups (thus answering a question implicitly raised in [21] on existence of exotic examples of just infinite groups) and to a number of examples of "small groups" (or atomic groups) in the sense of S. Pride [78].

The ideas of self-similarity and branching interact extremely well in group theory. There is a large intersection between these two classes of groups and this article demonstrates some important features and examples of this interaction.

This survey article is based on the course of four talks that were given by the first author at the Groups St Andrews 2005 Conference (although we do indicate here some new examples and links). We hope that it will serve as an accessible and quick introduction into the subject.

The article is organized as follows.

After a quick overview of several self-similar objects and basics notions related to actions on rooted trees in Section 1 and Section 2, we define the notion of a self-similar group in Section 3 and explain how such groups are related to finite automata. Among the examples we consider are the Basilica group, the 3-generated 2-group of intermediate growth known as "the first group of intermediate growth" and the Hanoi Towers groups H^k, which model the popular (in life and in mathematics) Hanoi Towers Problem on k pegs, $k \geq 3$.

Section 4 contains a quick introduction to the theory of iterated monodromy groups developed by Nekrashevych [70, 71]. This theory is a wonderful example of application of group theory in dynamical systems and, in particular, in holomorphic dynamics. We mention here that the well known Hubbard Twisted Rabbit Problem in holomorphic dynamics was recently solved by Bartholdi and Nekrashevych [12] by using self-similar groups arising as iterated monodromy groups.

Section 5 deals with branch groups. We give two versions of the definition (algebraic and geometric) and mention some of basic properties of these groups. We show that the Hanoi Group $\mathcal{H}^{(3)}$ is branch and hence the other Hanoi groups are at least weakly branch.

Section 6 and Section 7 deal with important asymptotic characteristics of groups such as growth and amenability. Basically, all currently known results on groups of intermediate growth and on amenable but not elementary amenable groups are based on self-similar and/or branch groups. Among various topics related to amenability we discus (following the article [19]) the question on the range of Tarski numbers and amenability of groups generated by bounded automata and their generalizations, introduced by Said Sidki.

In the last sections we give an account of the use of Schreier graphs in the circle of questions described above, related to self-similarity, amenability and geometry of Julia sets and other fractal type sets, substitutional systems and the spectral problem. We finish with an example of a computation of the spectrum in a problem related to Sierpiński gasket.

Some of the sections end with a short list of open problems.

The subject of self-similarity and branching in group theory is quite young and the number of different directions, open questions, and applications is growing rather quickly. We hope that this article will serve as an invitation to this beautiful, exciting, and extremely promising subject.

1   Self-similar objects

It is not our intention in this section to be very precise and define the notion of self-similarity in any generality. Rather, we provide some examples of objects that, for one reason or another, may be considered self-similar and that will be relevant in the later sections.

The unit interval $I = [0, 1]$ is one of the simplest mathematical objects that may be considered self-similar. Indeed, I is equal to the union of two intervals, namely $[0, 1/2]$ and $[1/2, 1]$, both of which are similar to I. The similarities between I and the intervals $I_0 = [0, 1/2]$ and $I_1 = [1/2, 1]$ are the affine maps ϕ_0 and ϕ_1 given by

$$\phi_0(x) = \frac{x}{2}, \qquad \phi_1(x) = \frac{x}{2} + \frac{1}{2}.$$

The free monoid $X^* = X_k^*$ over the alphabet $X = X_k = \{0, \dots, k-1\}$ is another example of an object that may be considered self-similar (we usually omit the index in X_k). Namely

$$X^* = \{\emptyset\} \cup 0X^* \cup \dots \cup (k-1)X^*, \tag{1.1}$$

where $\emptyset$ denotes the empty word and, for each letter x in X, the map ϕ_x defined by

$$\phi_x(w) = xw,$$

for all words w over X, is a "similarity" from the set X^* of all words over X to the set xX^* of all words over X that start in x.

Both examples so far may be considered a little bit imperfect. Namely, the intersection of I_0 and I_1 is a singleton (so the decomposition $I = I_0 \cup I_1$ is not disjoint), while the union decomposing X^* involves an extra singleton (that is not similar to X^* in any reasonable sense). However, in both cases the apparent imperfection can be easily removed.

For example, denote by J the set of points on the unit interval that is the complement of the set of diadic rational points in I. Thus $J = I \setminus D$ where

$$D = \left\{ \frac{p}{q} \;\middle|\; 0 \leq p \leq q, \;\; p \in \mathbb{Z}, \;\; q = 2^m, \text{ for some non-negative integer } m \right\}$$

Then $\phi_0(J) = J_0$ and $\phi_1(J) = J_1$ are sets similar to J, whose disjoint union is J.

Before we discuss how to remove the apparent imperfection from the decomposition (1.1) we define more precisely the structure that is preserved under the similarities ϕ_x, $x \in X$.

The free monoid X^* has the structure of a rooted k-ary tree $\mathcal{T} = \mathcal{T}^{(k)}$. The empty word $\emptyset$ is the root, the set X^n of words of length n over X is the level n, denoted $\mathcal{L}_n$, and every vertex u in $\mathcal{T}$ has k children, namely ux, $x \in X$. Figure 1 depicts the ternary rooted tree. Let $u = x_1 \ldots x_n$ be a word over X. The set uX^*

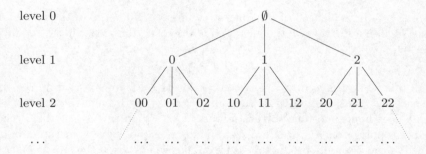

Figure 1. The ternary rooted tree

of words over X that start in u is a subtree of $\mathcal{T}$, denoted $\mathcal{T}_u$, which is canonically isomorphic to the whole tree through the isomorphism ϕ_u defined as the composition $\phi_u = \phi_{x_1} \ldots \phi_{x_n}$ (see Figure 2). In particular, each ϕ_x, $x \in X$, is a canonical tree isomorphism between the tree $\mathcal{T}$ and the tree $\mathcal{T}_x$ hanging below the vertex x on the first level of $\mathcal{T}$.

The boundary of the tree $\mathcal{T}$, denoted $\partial \mathcal{T}$, is an ultrametric space whose points are the infinite geodesic rays in $\mathcal{T}$ starting at the root. In more detail, each infinite geodesic ray in $\mathcal{T}$ starting at the root is represented by an infinite (to the right) word ξ over X, which is the limit $\xi = \lim_{n \to \infty} \xi_n$ of the sequence of words $\{\xi_n\}_{n=0}^{\infty}$ such that ξ_n is the word of length n representing the unique vertex on level n on the ray. Denote the set of all infinite words over X by X^{ω} and, for u in X^*, the set of infinite words starting in u by uX^{ω}. The set X^{ω} of infinite words over X decomposes as disjoint union as

$$X^{\omega} = 0X^{\omega} \cup \cdots \cup (k-1)X^{\omega}.$$

Defining the distance between rays ξ and ζ by

$$d(\xi, \zeta) = \frac{1}{k^{|\xi \wedge \zeta|}}$$

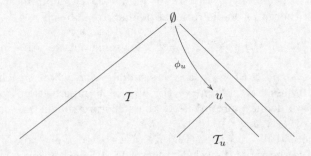

Figure 2. Canonical isomorphism between $\mathcal{T}$ and $\mathcal{T}_u$

where $|\xi \wedge \zeta|$ is the length of the longest common prefix $\xi \wedge \zeta$ of the rays ξ and ζ turns X^ω into a metric space, denoted $\partial \mathcal{T}$. Moreover, for each x in X, the map ϕ_x given by

$$\phi_x(w) = xw,$$

for all infinite words w in X^ω, is a contraction (by a factor of k) from $\partial \mathcal{T}$ to the subspace $\partial \mathcal{T}_x$ consisting of those rays in $\partial \mathcal{T}$ that pass through the vertex x. The space $\partial \mathcal{T}$ is homeomorphic to the Cantor set. Its topology is just the Tychonov product topology on $X^\mathbb{N}$, where X has the discrete topology. A measure on $\partial \mathcal{T}$ is defined as the Bernoulli product measure on $X^\mathbb{N}$, where X has the uniform measure.

The Cantor middle thirds set C is a well known self-similar set obtained from I by removing all points whose base 3 representation necessarily includes the digit 1. In other words, the open middle third interval is removed from I, then the open middle thirds are removed from the two obtained intervals, etc. (The first three steps in this procedure are illustrated in Figure 3.) In this case C is the disjoint

Figure 3. The first three steps in the construction of the middle thirds Cantor set

union of the two similar subsets $C_0 = \phi_0(C)$ and $C_1 = \phi_1(C)$ where the similarities ϕ_0 and ϕ_1 are given by

$$\phi_0(x) = \frac{x}{3}, \qquad \phi_1(x) = \frac{x}{3} + \frac{2}{3}.$$

A homeomorphism between the boundary of the binary tree $\partial \mathcal{T}^{(2)}$ and the Cantor middle thirds set is given by

$$x_1 x_2 x_3 \cdots \leftrightarrow 0.(2x_1)(2x_2)(2x_3)\ldots\,.$$

The number of the right is the ternary representation of a point in C (consisting solely of digits 0 and 2).

Another well known self-similar set is the Sierpiński gasket. It is the set of points in the plane obtained from the set of points bounded by an equilateral triangle by successive removal of the middle triangles. A set of points homeomorphic to the Sierpiński gasket is given in Figure 15.

The set in Figure 15 is the Julia set of a rational post-critically finite map on the Riemann Sphere. Julia sets of such maps provide an unending supply of self-similar subsets of the complex plane. For example, the Julia set of the quadratic map $z \mapsto z^2 - 1$ is given in Figure 4.

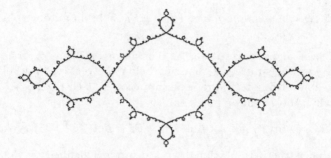

Figure 4. Julia set of the map $z \mapsto z^2 - 1$

In fact, some of the other examples we already mentioned are (up to homeomorphism) also Julia sets of quadratic maps. Namely, the interval $[-2, 2]$ is the Julia set of the quadratic map $z \mapsto z^2 - 2$, while the Julia set of the quadratic map $z \mapsto z^2 + c$, for $|c| > 2$, is a Cantor set.

2   Actions on rooted trees

The self-similarity decomposition

$$X^* = \{\emptyset\} \cup 0X^* \cup \ldots (k-1)X^*$$

of the k-ary rooted tree $\mathcal{T} = \mathcal{T}^{(k)}$, defined over the alphabet $X = \{0, \ldots, k-1\}$, induces the self-similarity of the group of automorphisms $\mathsf{Aut}(\mathcal{T})$ of the tree $\mathcal{T}$. Namely, each automorphism of $\mathcal{T}$ can be decomposed as

$$g = \pi_g \ (g_0, \ldots, g_{k-1}), \tag{2.1}$$

where π_g is a permutation in $\mathsf{S}_k = \mathsf{Sym}(X_k)$, called the *root permutation* of g and, for x in X, g_x is an automorphism of $\mathcal{T}$, called the *section* of g at x. The root permutation and the sections of g are uniquely determined by the relation

$$g(xw) = \pi_g(x)g_x(w),$$

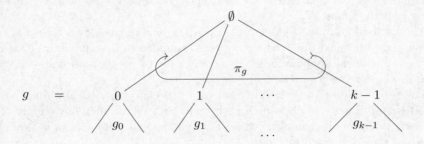

Figure 5. Decomposition $g = \pi_g\,(g_0, \ldots, g_{k-1})$ of an automorphism of $\mathcal{T}$

for x a letter in X and w a word over X. The automorphisms g_x, x in X, represent the action of g on the subtrees $\mathcal{T}_x$ hanging below the vertices on level 1, which are then permuted according to the root permutation π_g (see Figure 5).

Algebraically, $\mathsf{Aut}(\mathcal{T})$ decomposes as

$$\mathsf{Aut}(\mathcal{T}) = \mathsf{S}_k \ltimes (\mathsf{Aut}(\mathcal{T}) \times \cdots \times \mathsf{Aut}(\mathcal{T})) = \mathsf{S}_k \ltimes \mathsf{Aut}(\mathcal{T})^X = \mathsf{S}_k \wr_X \mathsf{Aut}(\mathcal{T}). \quad (2.2)$$

The product $\wr_X$ is the permutational wreath product defined by the permutation action of S_k on X, i.e., S_k acts on $\mathsf{Aut}(\mathcal{T})^X$ by permuting the coordinates (we usually omit the subscript in $\wr_X$). For $f, g \in \mathsf{Aut}(\mathcal{T})$, we have

$$\begin{aligned}
gh &= \pi_g\,(g_0, \ldots, g_{k-1})\,\pi_h\,(h_0, \ldots, h_{k-1}) \\
&= \pi_g \pi_h\,(g_0, \ldots, g_{k-1})^{\pi_h}(h_0, \ldots, h_{k-1}) \\
&= \pi_g \pi_h\,(g_{\pi_h(0)}h_0, \ldots, g_{\pi_h(k-1)}h_{k-1}).
\end{aligned}$$

Thus

$$\pi_{gh} = \pi_g \pi_h \qquad \text{and} \qquad (gh)_x = g_{h(x)}h_x,$$

for x in X. The decomposition (2.2) can be iterated to get

$$\mathsf{Aut}(\mathcal{T}) = \mathsf{S}_k \wr \mathsf{Aut}(\mathcal{T}) = \mathsf{S}_k \wr (\mathsf{S}_k \wr \mathsf{Aut}(\mathcal{T})) = \cdots = \mathsf{S}_k \wr (\mathsf{S}_k \wr (\mathsf{S}_k \wr \ldots)).$$

Thus $\mathsf{Aut}(\mathcal{T})$ has the structure of iterated permutational wreath product of copies of the symmetric group S_k.

The sections of an automorphism g of $\mathcal{T}$ are also automorphisms of $\mathcal{T}$ (describing the action on the first level subtrees). The sections of these sections are also automorphisms of $\mathcal{T}$ (describing the action on the second level subtrees) and so on. Thus, we may recursively define the sections of g at the vertices of $\mathcal{T}$ by

$$g_{ux} = (g_u)_x,$$

for x a letter and u a word over X. By definition, the section of g at the root is g itself. Then we have, for any words u and v over X,

$$g(uv) = g(u)g_u(v).$$

Definition 1 Let G be a group acting by automorphisms on a k-ary rooted tree $\mathcal{T}$. The *vertex stabilizer* of a vertex u in $\mathcal{T}$ is

$$\mathsf{St}_G(u) = \{\, g \in G \mid g(u) = u \,\}.$$

The *level stabilizer* of level n in $\mathcal{T}$ is

$$\mathsf{St}_G(\mathcal{L}_n) = \{\, g \in G \mid g(u) = u, \text{ for all } u \in \mathcal{L}_n \,\} = \bigcap_{u \in \mathcal{L}_n} \mathsf{St}_G(u).$$

Proposition 2.1 *The n-th level stabilizer of a group G acting on a k-ary tree is a normal subgroup of G. The group $G/\mathsf{St}_G(\mathcal{L}_n)$ is isomorphic to a subgroup of the group*

$$\underbrace{\mathsf{S}_k \wr (\mathsf{S}_k \wr (\mathsf{S}_k \wr \cdots \wr (\mathsf{S}_k \wr \mathsf{S}_k) \ldots))}_{n \ copies}$$

and the index $[G : \mathsf{St}_G(\mathcal{L}_n)]$ is finite and bounded above by $(k!)^{1+k+\cdots+k^{n-1}}$.

Since the intersection of all level stabilizers in a group of tree automorphisms is trivial (an automorphism fixing all the levels fixes the whole tree) we have the following proposition.

Proposition 2.2 *Every group acting faithfully on a regular rooted tree is residually finite.*

The group $\mathsf{S}_k \wr (\mathsf{S}_k \wr (\mathsf{S}_k \wr \cdots \wr (\mathsf{S}_k \wr \mathsf{S}_k) \ldots))$ that appears in Proposition 2.1 is the automorphism group, denoted $\mathsf{Aut}(\mathcal{T}_{[n]})$ of the finite rooted k-ary tree $\mathcal{T}_{[n]}$ consisting of levels 0 through n in $\mathcal{T}$.

The group $\mathsf{Aut}(\mathcal{T})$ is a pro-finite group. Indeed, it is the inverse limit of the sequence

$$1 \leftarrow \mathsf{Aut}(\mathcal{T}_{[1]}) \leftarrow \mathsf{Aut}(\mathcal{T}_{[2]}) \leftarrow \ldots$$

of automorphism groups of the finite k-ary rooted trees, where the surjective homomorphism $\mathsf{Aut}(\mathcal{T}_{[n]}) \leftarrow \mathsf{Aut}(\mathcal{T}_{[n+1]})$ is given by restriction of the action of $\mathsf{Aut}(\mathcal{T}_{[n+1]})$ on $\mathcal{T}_{[n]}$.

Since tree automorphisms fix the levels of the tree the highest degree of transitivity that they can achieve is to act transitively on all levels.

Definition 2 A group acts *spherically transitively* on a rooted tree if it acts transitively on every level of the tree.

Let G act on $\mathcal{T}$ and u be a vertex in $\mathcal{T}$. Then the map

$$\varphi_u : \mathsf{St}_G(u) \to \mathsf{Aut}(\mathcal{T})$$

given by

$$\varphi_u(g) = g_u$$

is a homomorphism.

Definition 3 The homomorphism φ_u is called the *projection* of G at u. The image of φ_u is denoted by G_u and called the *upper companion group* of G at u.

The map

$$\psi_n : \mathsf{St}_G(\mathcal{L}_n) \to \prod_{u \in \mathcal{L}_n} \mathsf{Aut}(\mathcal{T})$$

given by

$$\psi_n(g) = (\varphi_u(g))_{u \in \mathcal{L}_n} = (g_u)_{u \in \mathcal{L}_n}$$

is a homomorphism. We usually omit the index in ψ_n when $n = 1$.

In the case of $G = \mathsf{Aut}(\mathcal{T})$ the maps $\varphi_u : \mathsf{St}_{\mathsf{Aut}(\mathcal{T})}(u) \to \mathsf{Aut}(\mathcal{T})$ and $\psi_n : \mathsf{St}_{\mathsf{Aut}(\mathcal{T})}(\mathcal{L}_n) \to \prod_{u \in \mathcal{L}_n} \mathsf{Aut}(\mathcal{T})$ are isomorphisms, for any word u over X and any $n \geq 0$.

The group $\mathsf{Aut}(\mathcal{T})$ is isomorphic to the group $\mathsf{Isom}(\partial \mathcal{T})$ of isometries of the boundary $\partial \mathcal{T}$. The isometry corresponding to an automorphism g of $\mathcal{T}$ is naturally defined by

$$g(\xi) = \lim_{n \to \infty} g(\xi_n),$$

where ξ_n is the prefix of length n of the infinite word ξ over X.

3 Self-similar groups

3.1 General definition

We observed that the self-similarity of the k-ary rooted tree $\mathcal{T}$ induces the self-similarity of its automorphism group. The decomposition of X^* in (1.1) is reflected in the decomposition of tree automorphisms in (2.1) and in the decomposition of $\mathsf{Aut}(\mathcal{T})$ in (2.2). The most obvious manifestation of the self-similarity of $\mathsf{Aut}(\mathcal{T})$ is that all sections of all tree automorphisms are again tree automorphisms. We use this property as the basic feature defining a self-similar group.

Definition 4 A group G of k-ary tree automorphisms is *self-similar* if all sections of all elements in G are elements in G.

Proposition 3.1 *A group G of k-ary tree automorphisms is self-similar if and only if, for every element g in G and every letter x in X, there exists a letter y in X and an element h in G such that, for all words w over X,*

$$g(xw) = yh(w).$$

Definition 5 A self-similar group G of k-ary tree automorphisms is *recurrent* if, for every vertex u in $\mathcal{T}$, the upper companion group of G at u is G, i.e.,

$$\varphi_u(\mathsf{St}_G(u)) = G.$$

Proposition 3.2 *A self-similar group G of k-ary tree automorphisms is recurrent if for every letter x in X the upper companion group of G at x is G, i.e., for every element g in G and every letter x, there exists an element h in the stabilizer of x whose section at u is g.*

Example 1 (Odometer) For $k \geq 2$, define a k-ary tree automorphism a by

$$a = \rho\,(1, \ldots, 1, a),$$

where $\rho = (0\ \ 1\ \ \ldots\ \ k{-}1)$ is the standard cycle that cyclically permutes (rotates) the symbols in X. The automorphism a is called the k-ary *odometer* because of the way in which it acts on the set of finite words over X. Namely if we interpret the word $w = x_1 \ldots x_n$ over X as the number $\sum x_i k^i$ then

$$a(w) = w + 1, \text{ for } w \neq (k-1)\ldots(k-1), \qquad a((k-1)\ldots(k-1)) = 0\ldots0.$$

Thus the automorphism a acts transitively on each level of T, $\mathbb{Z} \cong \langle a \rangle$ and since

$$a^k = (a, a, \ldots, a)$$

the group generated by the odometer is a recurrent group. Therefore $\mathbb{Z}$ has a self-similar, recurrent action on any k-ary tree.

Proposition 3.3 *A k-ary tree automorphism acts spherically transitively on T if and only if it is conjugate in $\mathsf{Aut}(T)$ to the k-ary odometer automorphism.*

3.2   Automaton groups

We present a simple way to construct finitely generated self-similar groups. Let $\pi_1, \ldots, \pi_m$ be permutations in S_k and let $S = \{s^{(1)}, \ldots, s^{(m)}\}$ be a set of m distinct symbols. Consider the system

$$\begin{cases} s^{(1)} = \pi_1(s_0^{(1)}, \ldots, s_{k-1}^{(1)}) \\ \ldots \\ s^{(m)} = \pi_m(s_0^{(m)}, \ldots, s_{k-1}^{(m)}) \end{cases} \tag{3.1}$$

where each $s_x^{(i)}$, $i = 1, \ldots, m$, x in X, is a symbol in S. Such a system defines a unique set of k-ary tree automorphisms, denoted by the symbols in S, whose first level decompositions are given by the equations in (3.1). Moreover, the group $G = \langle S \rangle$ is a self-similar group of tree automorphisms, since all the sections of all of its generators are in G.

The language of finite automata (in fact finite transducers, or sequential machines, or Mealy automata) is well suited to describe the groups of tree automorphisms that arise in this way.

Definition 6 A finite automaton is a quadruple $\mathcal{A} = (S, X, \tau, \pi)$ where S is a finite set, called set of *states*, X is a finite alphabet, and $\tau : S \times X \rightarrow S$ and $\pi : S \times X \rightarrow X$ are maps, called the *transition map* and the *output map* of $\mathcal{A}$.

The automaton $\mathcal{A}$ is *invertible* if, for all s in S, the restriction $\pi_s : X \rightarrow X$ given by $\pi_s(x) = \pi(s, x)$ is a permutation of X.

All the automata in the rest of the text are invertible and we will not emphasize this fact. Each state of an automaton $\mathcal{A}$ acts on words over X as follows. When the automaton is in state s and the current input letter is x, the automaton produces the output letter $y = \pi_s(x) = \pi(s,x)$ and changes its state to $s_x = \tau(s,x)$. The state s_x then handles the rest of the input (a schematic description is given in Figure 6).

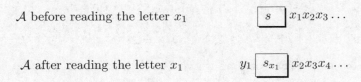

$\mathcal{A}$ before reading the letter x_1

$\mathcal{A}$ after reading the letter x_1

Figure 6. An automaton $\mathcal{A}$ processing an input word $x_1x_2x_3 \ldots$ starting at state s

Thus we have

$$s(xw) = \pi_s(x)s_x(w),$$

for x a letter and w a word over X, and we see that the state s acts on T as the automorphism with root permutation π_s and sections given by the states s_x, x in X.

Definition 7 Let $\mathcal{A}$ be a finite invertible automaton. The *automaton group* of $\mathcal{A}$, denoted $G(\mathcal{A})$, is the finitely generated self-similar group of tree automorphisms $G(\mathcal{A}) = \langle S \rangle$ generated by the states of $\mathcal{A}$.

Example 2 (Basilica group $\mathcal{B}$) Automata are often encoded by directed graphs such as the one in Figure 7. The vertices are the states, each state s is labeled by its corresponding root permutation π_s and, for each pair of a state s and a letter x, there is an edge from s to s_x labeled by x. We use () to label the states with trivial root permutations (sometimes we just leave such states unlabeled). The state 1 represents the identity automorphism of T. Consider the corresponding automaton

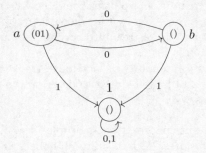

Figure 7. The binary automaton generating the Basilica group

group $\mathcal{B} = \langle a, b, 1 \rangle = \langle a, b \rangle$. The decompositions of the binary tree automorphisms

a and b are given by

$$a = (01)\ (b,\ 1),$$
$$b = \qquad (a,\ 1).$$

Note that we omit writing trivial root permutations in the decomposition. The group $\mathcal{B}$ is recurrent. Indeed we have $\mathsf{St}_\mathcal{B}(\mathcal{L}_1) = \langle b, a^2, a^{-1}ba \rangle$ and

$$b = (a,\ 1),$$
$$a^2 = (b,\ b),$$
$$a^{-1}ba = (1,\ a)$$

which shows that $\varphi_0(\mathsf{St}_\mathcal{B}(\mathcal{L}_1)) = \varphi_1(\mathsf{St}_\mathcal{B}(\mathcal{L}_1)) = \langle a, b \rangle = \mathcal{B}$.

Example 3 (The group $\mathcal{G}$) Consider the automaton group $\mathcal{G} = \langle a, b, c, d \rangle$ generated by the automaton in Figure 8. The decompositions of the binary tree auto-

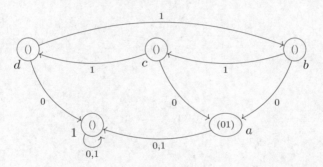

Figure 8. The binary automaton generating the group $\mathcal{G}$

morphisms a, b, c and d are given by

$$a = (01)\ (1,\ 1),$$
$$b = \qquad (a,\ c),$$
$$c = \qquad (a,\ d),$$
$$d = \qquad (1,\ b),$$

The group $\mathcal{G}$ is recurrent. Indeed we have $\mathsf{St}_\mathcal{G}(\mathcal{L}_1) = \langle b, c, d, aba, aca, ada \rangle$ and

$$
\begin{aligned}
b &= (a,\ c), & aba &= (c,\ a), \\
c &= (a,\ d), & aca &= (d,\ a), \\
d &= (1,\ b), & ada &= (b,\ 1),
\end{aligned}
$$

which shows that $\varphi_0(\mathsf{St}_\mathcal{G}(\mathcal{L}_1)) = \varphi_1(\mathsf{St}_\mathcal{G}(\mathcal{L}_1)) = \langle a, b, c, d \rangle = \mathcal{G}$.

Example 4 (Hanoi Towers groups) Let $k \geq 3$. For a permutation α in S_k define a k-ary tree automorphism a_α by

$$a_\alpha = \alpha \, (a_0, \ldots, a_{k-1})$$

where $a_i = 1$ if $i \in \mathsf{Supp}(\alpha)$ and $a_i = a_\alpha$ if $i \notin \mathsf{Supp}(\alpha)$. Define a group

$$\mathcal{H}^{(k)} = \langle \, a_{(ij)} \mid 0 \leq i < j \leq k-1 \, \rangle$$

of k-ary tree automorphisms, generated by a_α corresponding to the transpositions in S_k. For example, the automaton generating $\mathcal{H}^{(4)}$ is given in Figure 9. The state in the middle represents the identity and its loops are not drawn. We note

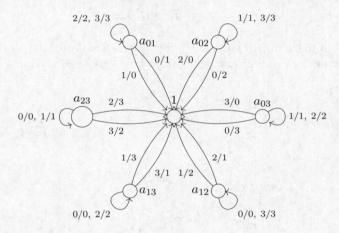

Figure 9. Automaton generating the Hanoi Towers group $\mathcal{H}^{(4)}$ on 4 pegs

that the directed graph in Figure 9 follows another common convention to encode the transition and output function of an automaton. Namely, the vertices are the states and for each pair of a state s and a letter x there is an edge connecting s to s_x labeled by $x|s(x)$.

The effect of the automorphism $a_{(ij)}$ on a k-ary word is that it changes the first occurrence of the symbol i or j to the other symbol (if such an occurrence exists) while leaving all the other symbols intact. A recursive formula for $a_{(ij)}$ is given by

$$a_{(ij)}(iw) = jw, \qquad a_{(ij)}(jw) = iw, \qquad a_{(ij)}(xw) = xa_{(ij)}(w), \text{ for } x \notin \{i,j\}.$$

The group $\mathcal{H}^{(k)}$ is called the *Hanoi Towers group* on k pegs since it models the Hanoi Towers Problem on k pegs (see, for example, [57]). We quickly recall this classical problem. Given n disks of distinct size, labeled $1, \ldots, n$ by their size, and k pegs, $k \geq 3$, labeled $0, \ldots, k-1$, a configuration is any placement of the disks on the pegs such that no disk is placed on top of a smaller disk. Figure 10 depicts a configuration of 5 disks on 3 pegs. In a single step, the top disk from one peg can be moved to the top of another peg, provided the newly obtained placement

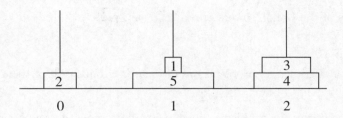

Figure 10. A configuration of 5 disks on 3 pegs

of disks represents a configuration. Initially all n disks are on peg 0 and the goal is to move all the disks to another peg.

Words of length n over X encode the configurations of n disks on k pegs. Namely, the word $x_1 \ldots x_n$ over X encodes the unique configuration of n disks in which disk number i, $i = 1, \ldots, n$, is placed on peg x_i (once the content of each peg is known the order of disks on the pegs is determined by their size). The automorphism $a_{(ij)}$ represents a move between pegs i and j. Indeed, if the symbol i appears before the symbol j (or j does not appear at all) in w then the disk on top of peg i is smaller than the disk on top of peg j (or peg j is empty) and a proper move between these two pegs moves the disk from peg i to peg j, thus changing the first appearance of i in w to j, which is exactly what $a_{(ij)}$ does. If none of the symbols i or j appears, this means that there are no disks on either of the pegs and the automorphism $a_{(ij)}$ acts trivially on such words. In terms of $\mathcal{H}^{(k)}$, the initial configuration of the Hanoi Towers Problem is encoded by the word 0^n and the goal is to find a group element $h \in \mathcal{H}^{(k)}$ written as a word over the generators $a_{(ij)}$ such that $h(0^n) = x^n$, where $x \neq 0$.

For example, the configuration in Figure 10 is encoded by the ternary word 10221 and the three generators $a_{(01)}$, $a_{(02)}$ and $a_{(12)}$ of $\mathcal{H}^{(3)}$, representing moves between the corresponding pegs, produce the configurations

$$a_{(01)}(10221) = 00221, \qquad a_{(02)}(10221) = 12221, \qquad a_{(12)}(10221) = 20221.$$

Let i, j, ℓ be three distinct symbols in X, Since $(ij)(j\ell)(ij)(i\ell)$ is the trivial permutation in S_k the element $h = a_{(ij)} a_{(j\ell)} a_{(ij)} a_{(i\ell)}$ is in the first level stabilizer. Direct calculation shows that

$$\varphi_i(h) = a_{(ij)}, \quad \varphi_j(h) = a_{(j\ell)} a_{(i\ell)}, \quad \varphi_\ell(h) = a_{(ij)}, \quad \varphi_t(h) = h, \text{ for } t \notin \{i, j, \ell\},$$

which implies that every generator of $\mathcal{H}^{(k)}$ belongs to every projection $\varphi_i(\mathsf{St}(\mathcal{L}_1))$, for all $i = 0, \ldots, k-1$.

Proposition 3.4 *The group $\mathcal{H}^{(k)}$ is a spherically transitive, recurrent, self-similar group of k-ary tree automorphisms.*

For latter use, in order to simplify the notation, we set

$$a_{(01)} = a, \qquad a_{(02)} = b, \qquad a_{(12)} = c$$

for the generators of the Hanoi Towers group $\mathcal{H}^{(3)}$ on 3 pegs.

3.3   Problems

We propose several algorithmic problems on automaton groups. All of them (and many more) can be found in [43].

Problem 1 Is the conjugacy problems solvable for all automaton groups?

Problem 2 Does there exist an algorithm that, given a finite automaton $\mathcal{A}$ and a state s in $\mathcal{A}$, decides
  (a) if s acts spherically transitively?
  (b) if s has finite order?

Problem 3 Does there exist an algorithm that, given a finite automaton $\mathcal{A}$, decides
  (a) if $G(\mathcal{A})$ acts spherically transitively?
  (b) if $G(\mathcal{A})$ is a torsion group?
  (c) if $G(\mathcal{A})$ is a torsion free group?

4   Iterated monodromy groups

The notion of iterated monodromy groups was introduced by Nekrashevych. The monograph [71] treats the subject in great detail (for earlier work see [70, 9]). We provide only a glimpse into this area.

Let M be a path connected and locally path connected topological space and let $f : M_1 \to M$ be a k-fold covering of M by an open, path connected subspace M_1 of M. Let t be a base point of M. The set of preimages

$$T = \bigcup_{n=0}^{\infty} f^{-n}(t)$$

can be given the structure of a k-regular rooted tree in which t is the root, the points from $L_n = f^{-n}(t)$ constitute level n and each vertex is connected by an edge to its k preimages. Precisely speaking, some preimages corresponding to different levels may coincide in M_1 and one could be more careful and introduce pairs of the form (x, n), where $x \in f^{-n}(t)$ to represent the vertices of T, but we will not use such notation.

Example 5 Consider the map $f : \mathbb{C}^* \to \mathbb{C}^*$ given by $f(z) = z^2$ (here $\mathbb{C}^* = \mathbb{C} \backslash \{0\}$). This is a 2-fold self-covering of $\mathbb{C}^*$. If we choose $t = 1$ we see that the level n in the binary tree T consist of all 2^n-th roots of unity, i.e.,

$$L_n = \{ e^{(2\pi i/2^n) \cdot m} \mid m = 0, \ldots, 2^n - 1 \}.$$

We can encode the vertex $e^{(2\pi i/2^n) \cdot m}$ at level n by the binary word of length n representing m in the binary system, with the first digit being the least significant one (see Figure 11).

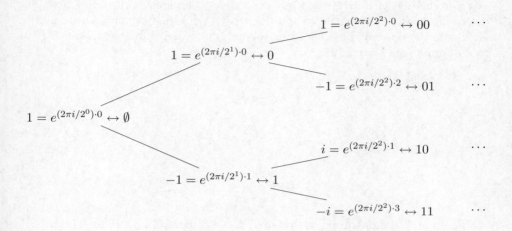

Figure 11. The tree T corresponding to $f(z) = z^2$

Define an action of the fundamental group $\pi(M) = \pi(M, t)$ on the tree T of preimages of t as follows. Define $M_n = f^{-n}(M)$. The n-fold composition map $f^n : M_n \to M$ is a k^n-fold covering of M. For a path α in M that starts at t and a vertex u in $L_n = f^{-n}(t) \subseteq M_n$ denote by $\alpha_{[u]}$ the unique lift (under f^n) of α that starts at u. Let γ be a loop based at t. For u in L_n the endpoint of $\gamma_{[u]}$ must also be a point in L_n. Denote this point by $\gamma(u)$ and define a map $L_n \to L_n$ by $u \mapsto \gamma(u)$. This map permutes the vertices at level n in T and is called the n-th *monodromy action* of γ.

Proposition 4.1 (Nekrashevych) *The map $T \to T$ given by*

$$u \mapsto \gamma(u)$$

is a tree automorphism, which depends only on the homotopy class of γ.

The action of $\pi(M)$ on T by tree automorphisms, called the *iterated monodromy action* of π, is not necessarily faithful.

Definition 8 Let N be the kernel of the monodromy action of $\pi(M)$ on T. The *iterated monodromy group* of the k-fold cover $f : M_1 \to M$, denoted $IMG(f)$, is the group $\pi(M)/N$.

The action of $IMG(f)$ on T is faithful. The classical monodromy actions on the levels of T are modeled within the action of $IMG(f)$ on T.

Example 6 Continuing our simple example involving $f(z) = z^2$, let γ be the loop $\gamma : [0, 1] \to \mathbb{C}^*$ based at $t = 1$ given by

$$\theta \mapsto e^{2\pi i \theta},$$

i.e., γ is the unit circle traversed in the positive (counterclockwise) direction. For a vertex u in L_n (a 2^n-th root of unity) the path $\gamma_{[u]}$ is just the path starting at u, moving in the positive direction along the unit circle and ending at the next 2^n-th root of unity (see Figure 12 for the case $n = 2$). Thus

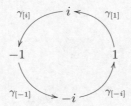

Figure 12. The monodromy action of $\pi(\mathbb{C}^*) = \mathbb{Z}$ on level 2 in T, for $f(z) = z^2$

$$\gamma\left(e^{(2\pi i/2^n)\cdot m}\right) = e^{(2\pi i/2^n)\cdot(m+1)}$$

and we see that on the corresponding binary tree $\mathcal{T}$ the action of γ is the binary odometer action

$$\gamma = (01)\,(1,\gamma).$$

Therefore the action of $\pi(C^*)$ on T is faithful and $IMG(z \mapsto z^2) \cong \mathbb{Z}$.

It is convenient to set a tree isomorphism $\Lambda : \mathcal{T} \rightarrow T$, where $\mathcal{T}$ is the k-ary rooted tree over the alphabet $X = \{0, \ldots, k-1\}$, with an induced action of $IMG(f)$ on $\mathcal{T}$. For this purpose, set $\Lambda(\emptyset) = t$, choose a bijection $\Lambda : X \rightarrow L_1$, let $\ell(\emptyset)$ be the trivial loop 1_t based at t and fix paths $\ell(0), \ldots, \ell(k-1)$ from t to the k points $\Lambda(0), \ldots, \Lambda(k-1)$ (the preimages of t). Assuming that, for every word v of length n, a point $\Lambda(v)$ at level n in T and a path $\ell(v)$ in M connecting t to $\Lambda(v)$ are defined, we define, for $x \in X$, the path $\ell(xv)$ by

$$\ell(xv) = \ell(x)_{[\Lambda(v)]}\ell(v)$$

and set $\Lambda(xv)$ to be the endpoint of this path (the composition of paths in $\pi(M)$ is performed from right to left).

Definition 9 Let γ be a loop in $\pi(M)$. Define an action of γ on $\mathcal{T}$ by

$$\gamma(u) = \Lambda^{-1}\gamma\Lambda(u),$$

for a vertex u in $\mathcal{T}$. This action is called the *standard action* of the iterated monodromy group $IMG(f)$ on $\mathcal{T}$.

We think of Λ as a standard isomorphism between $\mathcal{T}$ and T and, in order to simplify notation, we do not distinguish the vertices in $\mathcal{T}$ from the points in T they represent.

Theorem 4.2 (Nekrashevych) *The standard action of $IMG(f)$ on $\mathcal{T}$ is faithful and self-similar. The root permutation of γ is the 1-st monodromy action of γ on $L_1 = X$. The section of γ at u, for u a vertex in $\mathcal{T}$, is the loop*

$$\gamma_u = \ell(\gamma(u))^{-1} \, \gamma_{[u]} \, \ell(u).$$

Example 7 We calculate now the iterated monodromy group $IMG(f)$ of the double cover map $f : \mathbb{C} \setminus \{0, -1, 1\} \to \mathbb{C} \setminus \{0, -1\}$ given by $z \mapsto z^2 - 1$.

Choose the fixed point $t = (1 - \sqrt{5})/2$ as the base point. The fundamental group $\pi(\mathbb{C} \setminus \{0, -1\})$ is the free group on two generators represented by the loops a and b, where a is the loop based at t moving around -1 in positive direction along the circle centered at -1 (see Figure 13) and b is the loop based at t moving around 0 in the positive direction along the circle center at 0. We have $L_1 = f^{-1}(t) = \{t, -t\}$ and we choose $\Lambda(0) = t$, $\Lambda(1) = -t$. Set $\ell(0)$ to be the trivial loop 1_t at t and $\ell(1)$ to be the path from t to $-t$ that is moving along the top part of the loop b (in direction opposite to b). Let c be the loop based at t^2 traversed in the positive

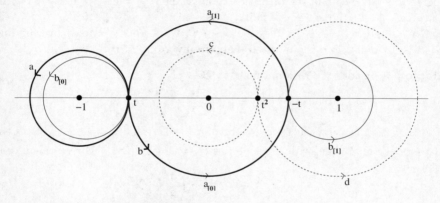

Figure 13. Calculation of the action of $IMG(z \mapsto z^2 - 1)$

direction along the circle centered at 0 and d the loop based at t^2 traversed along the circle centered at 1. Since $t^2 - 1 = t$, c is exactly one unit to the right of a and d is exactly one unit to the right of b. It is easy to see that $a_{[0]}$ is the path from t to $-t$ moving along the bottom part of the loop b and $a_{[1]} = \ell(1)^{-1}$ is the path from $-t$ to t moving along the top part of b. This is because applying $z \mapsto z^2$ to either one of these paths produces the loop c. Further, $b_{[0]}$ is the loop based at t that is entirely in the interior of the loop a, whose radius is chosen in such a way that applying $z \mapsto z^2$ yields the loop d. Similarly, $b_{[1]}$ is the loop based at $-t$ that is entirely in the interior of the loop d, whose radius is chosen in such a way that applying $z \mapsto z^2$ yields the loop d ($b_{[0]}$ and $b_{[1]}$ are symmetric with respect to the origin).

The loop a acts on the first level of $\mathcal{T}$ by permuting the vertices 0 and 1 and

b acts trivially. For the sections we have

$$a_0 = \ell(a(0))^{-1} \, a_{[0]} \, \ell(0) = \ell(1)^{-1} \, a_{[0]} \, \ell(0) = a_{[1]} \, a_{[0]} \, 1_t = b$$
$$a_1 = \ell(a(1))^{-1} \, a_{[1]} \, \ell(1) = \ell(0)^{-1} \, a_{[1]} \, \ell(1) = 1_t \, a_{[1]} \, a_{[1]}^{-1} = 1$$
$$b_0 = \ell(b(0))^{-1} \, b_{[0]} \, \ell(0) = \ell(0)^{-1} \, b_{[0]} \, \ell(0) = 1_t \, a \, 1_t = a$$
$$b_1 = \ell(b(1))^{-1} \, b_{[1]} \, \ell(1) = \ell(1)^{-1} \, b_{[1]} \, \ell(1) = a_{[1]} \, 1_{-t} \, a_{[1]}^{-1} = 1$$

Thus $IMG(z \mapsto z^2 - 1)$ is the self-similar group generated by the automaton

$$a = (01) \, (b, 1)$$
$$b = \qquad (a, 1)$$

and we see that $IMG(z \mapsto z^2 - 1)$ is the Basilica group $\mathcal{B}$.

The Basilica group belongs to the class of iterated monodromy groups of post-critically finite rational functions over the Riemann Sphere $\hat{\mathbb{C}} = \mathbb{C} \cup \{\infty\}$. The groups in this class are described as follows. For a rational function $f : \hat{\mathbb{C}} \to \hat{\mathbb{C}}$ of degree k let C_f be the set of critical points and $P_f = \bigcup_{n=1}^{\infty} (f^n(C_f))$ be the *post-critical set*. If the set P_f is finite, f is said to be *post-critically finite*. Set $M = \hat{\mathbb{C}} \setminus P_f$ and $M_1 = \hat{\mathbb{C}} \setminus f^{-1}(P_f)$. Then $f : M_1 \to M$ is a k-fold covering and $IMG(f)$ is, by definition, the iterated monodromy group of this covering.

Example 8 Consider the rational map $f : \hat{\mathbb{C}} \to \hat{\mathbb{C}}$

$$f(z) = z^2 - \frac{16}{27z}.$$

Denote $\Omega = \{\omega_0, \omega_1, \omega_2\}$, where $\omega_0 = 1$, $\omega_1 = \frac{1}{2} + \frac{\sqrt{3}}{2}i$ and $\omega_2 = \frac{1}{2} - \frac{\sqrt{3}}{2}i$ are the third roots of unity. The critical set of f is $C_f = -\frac{2}{3}\Omega \cup \{\infty\}$. Direct calculation shows that

$$f\left(-\frac{2}{3}\omega_i\right) = \frac{4}{3}\bar{\omega}_i = f\left(\frac{4}{3}\omega_i\right).$$

The post-critical set $P_f = \frac{4}{3}\Omega \cup \{\infty\}$ is finite (f conjugates the points in this set), and we have $M = \hat{\mathbb{C}} \setminus (\frac{4}{3}\Omega \cup \{\infty\})$ and $M_1 = \hat{\mathbb{C}} \setminus (-\frac{2}{3}\Omega \cup \frac{4}{3}\Omega \cup \{0, \infty\})$. The fundamental group $\pi(M)$ is free of rank 3. It is generated by the three loops a, b and c based at $t = 0$ as drawn in the upper left corner in Figure 14. In the figure, the critical points in $-\frac{2}{3}\Omega$ are represented by small empty circles, while the post-critical points in $\frac{4}{3}\Omega$ are represented by small black disks. The 3 inverse images of the base point $t = 0$ are the points in the set $\frac{2\sqrt[3]{2}}{3}\Omega$ and they are denoted by $\Lambda(0)$, $\Lambda(1)$ and $\Lambda(2)$ as in Figure 14. The base point and its three pre-images are represented by small shaded circles in the figure. The paths $\ell(0)$, $\ell(1)$ and $\ell(2)$ are chosen to be along straight lines from the base point $t = 0$. Calculation of the vertex permutations and the sections of the generators a, b and c of $IMG(f)$ then gives

$$a = (01) \, (1, 1, b)$$
$$b = (02) \, (1, a, 1)$$
$$c = (12) \, (c, 1, 1).$$

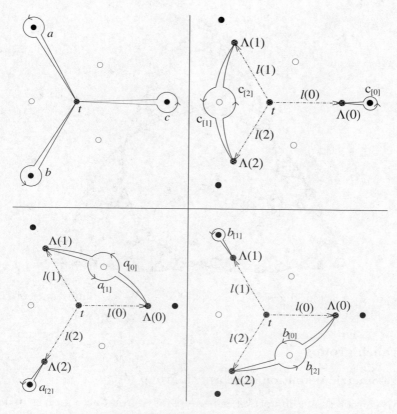

Figure 14. Calculation of the action of $IMG\left(z \mapsto z^2 - \frac{16}{27z}\right)$

The resemblance between $IMG(f)$ and $\mathcal{H}^{(3)}$ is obvious. In fact, for the map $\bar{f} : \hat{\mathbb{C}} \to \hat{\mathbb{C}}$ given by

$$z \to \bar{z}^2 - \frac{16}{27\bar{z}},$$

we have $IMG(\bar{f}) = \mathcal{H}^{(3)}$. Further, if we define

$$g = (12)\,(h, h, h)$$
$$h = \qquad (g, g, g),$$

then g conjugates $IMG(f)$ to $\mathcal{H}^{(3)}$ in $\mathsf{Aut}(\mathcal{T})$. The Julia set of the map f (or $\bar{f}$) is given in Figure 15.

This set is homeomorphic to the Sierpiński Gasket, which is the well known plane analogue of the Cantor middle thirds set. We will return to this set in Section 8, when we discuss Schreier graphs of self-similar groups.

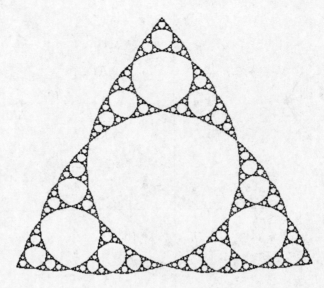

Figure 15. Sierpiński gasket, Julia set of $z \mapsto z^2 - \frac{16}{27z}$, limit space of $\mathcal{H}^{(3)}$

5   Branch groups

5.1   Geometric definition of a branch group

Here we consider some important subgroups of a group acting on a rooted tree leading to the geometric definition of a branch group. We extend our considerations to spherically homogeneous rooted trees (rather than regular).

For a sequence $\bar{k} = \{k_n\}_{n=1}^{\infty}$ of integers with $k_n \geq 2$, define the *spherically homogenous rooted tree* $\mathcal{T} = \mathcal{T}^{(\bar{k})}$ to be the tree in which level n consists of the elements in $\mathcal{L}_n = X_{k_1} \times \cdots \times X_{k_n}$ (recall that X_k is the standard alphabet on k letters $X_k = \{0, \ldots, k-1\}$) and each vertex u in $\mathcal{L}_n$ has k_{n+1} children, namely ux, for $x \in X_{k_{n+1}}$. The sequence $\bar{k}$ is the *degree sequence* of $\mathcal{T}$.

The definition of a vertex and level stabilizer is analogous to the case of regular rooted trees. We define now rigid vertex and level stabilizers.

Definition 10 Let G be a group acting on a spherically homogeneous rooted tree $\mathcal{T}$.

The *rigid stabilizer* of a vertex u in $\mathcal{T}$ is

$$\mathsf{RiSt}_G(u) = \{\, g \in G \mid \text{ the support of } g \text{ is contained in the subtree } \mathcal{T}_u \,\}.$$

The *rigid level stabilizer* of level n in $\mathcal{T}$ is

$$\mathsf{RiSt}_G(\mathcal{L}_n) = \left\langle \bigcup_{u \in \mathcal{L}_n} \mathsf{RiSt}_G(u) \right\rangle = \prod_{u \in \mathcal{L}_n} \mathsf{RiSt}_G(u).$$

If the action of G on $\mathcal{T}$ is spherically transitive (transitive on every level) then all (rigid) vertex stabilizers on the same level are conjugate.

For an arbitrary vertex u of level n we have

$$\mathsf{RiSt}_G(u) \leq \mathsf{RiSt}_G(\mathcal{L}_n) \leq \mathsf{St}_G(\mathcal{L}_n) \leq \mathsf{St}_G(u).$$

The index of $\mathsf{St}_G(\mathcal{L}_n)$ in G finite for every n (bounded by $k_1!(k_2!)^{k_1}\ldots(k_n!)^{k_1\cdots k_{n-1}}$). We make important distinctions depending on the (relative) size of the rigid level stabilizers.

Definition 11 Let G act spherically transitively on a spherically homogeneous rooted tree $\mathcal{T}$.

We say that the action of G is

(a) of *branch type* if, for all n, the index $[G : \mathsf{RiSt}_G(\mathcal{L}_n)]$ of the rigid level stabilizer $\mathsf{RiSt}_G(\mathcal{L}_n)$ in G is finite.

(b) of *weakly branch type* if, for all n, the rigid level stabilizer $\mathsf{RiSt}_G(\mathcal{L}_n)$ is non-trivial (and therefore infinite).

(c) of *diagonal type* if, for some n, the rigid level stabilizer $\mathsf{RiSt}_G(\mathcal{L}_n)$ is finite (and therefore trivial after some level).

Definition 12 A group is called a *branch* (*weakly branch*) group if it admits a faithful branch (weakly branch) type action on a spherically homogeneous tree.

When we say that G is a branch group, we implicitly think of it as being embedded as a spherically transitive subgroup of the automorphism group of $\mathcal{T}$.

Example 9 A rather trivial, but important, example of a branch group is the full group of tree automorphisms $\mathsf{Aut}(\mathcal{T}^{(\overline{k})})$ of the spherically homogeneous tree defined by the degree sequence $\overline{k}$. Indeed, in this case, for any vertex u at level n,

$$\mathsf{RiSt}_G(u) \cong \mathsf{Aut}\big(\mathcal{T}^{(\sigma^n(\overline{k}))}\big),$$

where $\mathsf{Aut}\big(\mathcal{T}^{(\sigma^n(\overline{k}))}\big)$ is the spherically homogeneous rooted tree defined by the n-th shift $\sigma^n(\overline{k})$ of the degree sequence $\overline{k}$ and

$$\mathsf{RiSt}(\mathcal{L}_n) = \mathsf{St}(\mathcal{L}_n).$$

In particular, for a regular k-ary tree $\mathcal{T}$

$$\mathsf{RiSt}_G(u) \cong \mathsf{Aut}(\mathcal{T}) \qquad \text{and} \qquad \mathsf{RiSt}(\mathcal{L}_n) = \mathsf{St}(\mathcal{L}_n).$$

Another example is given by the group of finitary automorphisms of $\mathcal{T}^{(k)}$. This group consists of those automorphisms that have only finitely many non-trivial sections.

When the tree is k-regular another important example is the group of finite state automorphisms. This group consists of those automorphisms that have only finitely many distinct sections.

Definition 13 Let G be a self-similar spherically transitive group of regular k-ary tree and let K and $M_0, \dots, M_{k-1}$ be subgroups of G. We say that K *geometrically contains* $M_0 \times \cdots \times M_{k-1}$ if $\psi^{-1}(M_0 \times \cdots \times M_{k-1})$ is a subgroup of K.

In the above definition ψ is the map $\psi : \mathsf{St}_G(\mathcal{L}_1) \to G \times \cdots \times G$ defined in Section 2.

Definition 14 A self-similar spherically transitive group of automorphisms G of the regular k-ary tree $\mathcal{T}$ is *regular branch group* over its normal subgroup K if K has finite index in G and $K \times \cdots \times K$ (k copies) is geometrically contained in K as a subgroup of finite index.

The group G is *regular weakly branch* group over its non-trivial subgroup K if $K \times \cdots \times K$ (k copies) is geometrically contained in K.

Theorem 5.1 *The Hanoi Towers group $H = \mathcal{H}^{(3)}$ is a regular branch group over its commutator subgroup H'. The index of H' in H is 8, $H/H' = C_2 \times C_2 \times C_2$ and the index of the geometric embedding of $H' \times H' \times H'$ in H' is 12.*

Proof Since the order of every generator in $\mathcal{H}^{(3)}$ is 2, every square in H is a commutator. The equalities

$$(abac)^2 = ([b,c],\ 1,\ 1),$$
$$(babc)^2 = ([a,c],\ 1,\ 1),$$
$$(acacba)^2 = ([a,b],\ 1,\ 1),$$

show that H' geometrically contains $H' \times H' \times H'$. Indeed, since $\mathcal{H}^{(3)}$ is recurrent, for any element h_0 in H there exists an element h in $\mathsf{St}_H(\mathcal{L}_1)$ such that $h = (h_0, *, *)$, where the $*$'s denote elements that are not of interest to us. Since H' is normal we than have that whenever $g = (g_0, 1, 1)$ is in H' so is $g^h = (g_0^{h_0}, 1, 1)$.

Thus H' geometrically contains $H' \times 1 \times 1$. The spherical transitivity of H then implies that the copy of H' at vertex 0 can be conjugated into a copy of H' at any other vertex at level 1. Thus H' geometrically contains $H' \times H' \times H'$.

For the other claims see [48]. We just quickly justify that the index of $H' \times H' \times H'$ in H' must be finite, without calculating the actual index.

Since all the generators of H have order 2, the index of the commutator subgroup H' in H is finite (not larger than 2^3). On the other hand $H' \times H' \times H'$ is a subgroup of the stabilizer $\mathsf{St}(\mathcal{L}_1)$ which embeds via ψ into $H \times H \times H$. The index of $H' \times H' \times H'$ in $H \times H \times H$ is finite, which then shows that the index of $H' \times H' \times H'$ in $\mathsf{St}(\mathcal{L}_1)$ is finite. Since the index of $\mathsf{St}(\mathcal{L}_1)$ in H is 6 we have that the index of $H' \times H' \times H'$ in H (and therefore also in H') is finite. $\square$

Figure 16 provides the full information on the indices between the subgroups of H mentioned in the above proof.

Example 10 The group $\mathcal{G}$ is a regular branch group over the subgroup $K = [a,b]^{\mathcal{G}}$.

The group $IMG(z \mapsto z^2 + i)$ is also a regular branch group. On the other hand, the Basilica group $\mathcal{B} = IMG(z \mapsto z^2 - 1)$ is a regular weakly branch group over its commutator, but it is not a branch group.

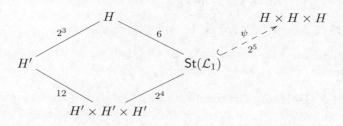

Figure 16. Some subgroups near the top of $\mathcal{H}^{(3)}$

We offer several immediate consequences of the branching property.

Theorem 5.2 (Grigorchuk [47]) *Let G be a branch group of automorphisms of $\mathcal{T}$. Then G is centerless. Moreover the centralizer*

$$C_{\mathsf{Aut}(\mathcal{T})}(G)$$

of G in $\mathsf{Aut}(\mathcal{T})$ is trivial.

Theorem 5.3 (Grigorchuk [47]) *Let N be a non-trivial normal subgroup of a weakly branch group G of automorphisms of $\mathcal{T}$. Then there exists level n such that*

$$(\mathsf{RiSt}_G(\mathcal{L}_n))' \leq N.$$

Corollary 5.4 *Any proper quotient of a branch group is virtually abelian.*

Quotients of branch groups can be infinite abelian groups. For example, this is the case for $\mathsf{Aut}(\mathcal{T})$ whose abelianization is infinite. It was an open question if quotients of finitely generated branch groups can be infinite and this was answered affirmatively in [25].

Theorem 5.5 (Bartholdi, Grigorchuk [10]) *For any branch group G of automorphisms of tree the stabilizer $P = \mathsf{St}_G(\xi)$ of an infinite ray ξ in $\partial\mathcal{T}$ is weakly maximal in G, i.e., P has infinite index and is maximal with respect to this property.*

The following is a consequence of a more general result of Abért stating that if a group G acts on a set X and all stabilizers of the finite subsets of X are different, then G does not satisfy any group identities (group laws).

Theorem 5.6 (Abért [1]) *No weakly branch group satisfies any group identities.*

Proof Consider the action of G on the boundary $\partial\mathcal{T}$. It is enough to show that for any finite set of rays $\Xi = \{\xi_1, \ldots, \xi_m\}$ in $\partial\mathcal{T}$ and any ray ξ in $\partial\mathcal{T}$ that is not in Ξ there exists an element g in G that fixes the rays in Ξ but does not fix ξ. Such an element g can be chosen from the rigid stabilizer $\mathsf{RiSt}_G(u)$ of any vertex u along ξ that is not a vertex on any of the rays in Ξ. $\square$

The above result, with a more involved proof, can also be found in [62].

Theorem 5.7 (Abért [2]) *No weakly branch group is linear.*

A weaker version of the non-linearity result, applying only to branch groups, was proved earlier by Delzant and Grigorchuk.

Theorem 5.8 (Delzant, Grigorchuk [25]) *A finitely generated branch group G has Serre's property (FA) (i.e., any action without inversions of G on a tree has a fixed point) if and only if G is not indicable.*

Under some conditions the branch type action of a branch group is unique up to level deletion/insertion, i.e., we have rigidity results. Let $\mathcal{T}$ be a spherically homogenous rooted tree and $\overline{m}$ an increasing sequence of positive integers (whose complement in $\mathbb{N}$ is infinite). Define a spherically homogeneous rooted tree $\mathcal{T}'$ obtained by deleting the vertices from $\mathcal{T}$ whose level is in $\overline{m}$ and connecting two vertices in $\mathcal{T}'$ if one is descendant of the other in $\mathcal{T}$ and they belong to two consecutive undeleted levels in $\mathcal{T}$. An action of a group G on $\mathcal{T}$ induces an action on $\mathcal{T}'$. We say that the tree $\mathcal{T}'$ is obtained from $\mathcal{T}$ by level deletion and the action of G on $\mathcal{T}'$ is obtained from the action on $\mathcal{T}$ by level deletion.

Theorem 5.9 (Grigorchuk, Wilson [45]) *Let G be a branch group of automorphisms of $\mathcal{T}$ such that*

(1) *the degree sequence $\overline{k}$ consists of primes,*

(2) *each vertex permutation of each element g in G acts as a transitive cycle of prime length on the children below it, and*

(3) *for each pair of incomparable vertices u and v (neither u is in $\mathcal{T}_v$ nor v is in $\mathcal{T}_u$), there exists an automorphism g in $\mathsf{St}_G(u)$ that is active at v, i.e., the root permutation of the section g_v is nontrivial.*

Then

(a) *for any branch type action of G on a tree $\mathcal{T}''$, there exists an action of G on $\mathcal{T}'$ obtained by level deletion such that there exists a G-equivariant tree isomorphism between $\mathcal{T}''$ and $\mathcal{T}'$.*

(b)

$$\mathsf{Aut}(G) = N_{\mathsf{Aut}(\mathcal{T})}(G).$$

The following result provides conditions on topological as well as combinatorial rigidity of weakly branch groups. It uses the notion of a saturated isomorphism. Let G and H be level transitive subgroups of $\mathsf{Aut}(\mathcal{T})$. An isomorphism $\phi : G \to H$ is *saturated* if there exists a sequence of subgroups $\{G_n\}_{n=0}^{\infty}$ such that G_n and $H_n = \phi(G_n)$ are subgroups of $\mathsf{St}_G(\mathcal{L}_n)$ and $\mathsf{St}_H(\mathcal{L}_n)$, respectively, and the action of both G_n and H_n is level transitive on every subtree $\mathcal{T}_u$ hanging below a vertex u on level n in $\mathcal{T}$.

Theorem 5.10 (Lavrenyuk, Nekrashevych [61]; Nekrashevych [71]) *Let G and H be two weakly branch groups acting faithfully on spherically homogeneous rooted trees $\mathcal{T}_1$ and $\mathcal{T}_2$.*

(a) *Any group isomorphism $\phi : G \to H$ is induced by a measure preserving homeomorphism $F : T_1 \to T_2$, i.e., there exists a measure preserving homeomorphism $F : \partial T_1 \to \partial T_2$ such that*

$$\phi(g)(F(w)) = F(g(w)),$$

for all g in G and w in ∂T_1.

(b) *If $T_1 = T_2 = T$ and $\phi : G \to H$ is a saturated isomorphism, then ϕ is induced by a tree automorphism $F : T \to T$.*

Corollary 5.11 (Lavrenyuk, Nekrashevych [61])

(a) *For any weakly branch group G of automorphisms of T the automorphism group of G is the normalizer of G in the group of homeomorphisms of the boundary ∂T of T, i.e.,*

$$\mathsf{Aut}(G) = N_{Homeo(\partial T)}(G).$$

(b) *For any saturated weakly branch group G of automorphisms of T the automorphism group of G is the normalizer of G in $\mathsf{Aut}(T)$, i.e.,*

$$\mathsf{Aut}(G) = N_{\mathsf{Aut}(T)}(G).$$

A spherically transitive group G of tree automorphism is saturated if it has a characteristic sequence of subgroups $\{G_n\}_{n=0}^{\infty}$ such that, for all n, G_n is a subgroup of $\mathsf{St}_G(\mathcal{L}_n)$ and G_n acts transitively on all subtrees T_u hanging below a vertex u on level n.

We note that the first complete description of the automorphism group of a finitely generated branch group was given by Sidki in [80]. More recent examples can be found in [44] and [13].

5.2 Algebraic definition of a branch group

We give here an algebraic version of a definition of a branch group. It is based on the subgroup structure of the group.

Definition 15 A group G is *algebraically branch group* if there exists a sequence of integers $\overline{k} = \{k_n\}_{n=0}^{\infty}$ and two decreasing sequences of subgroups $\{R_n\}_{n=0}^{\infty}$ and $\{V_n\}_{n=0}^{\infty}$ of G such that

(1) $k_n \geq 2$, for all $n > 0$, $k_0 = 1$,

(2) for all n,

$$R_n = V_n^{(1)} \times V_n^{(2)} \times \cdots \times V_n^{(k_1 k_2 \dots k_n)}, \tag{5.1}$$

where each $V_n^{(j)}$ is an isomorphic copy of V_n,

(3) for all n, the product decomposition (5.1) of R_{n+1} is a refinement of the corresponding decomposition of R_n in the sense that the j-th factor $V_n^{(j)}$ of R_n, $j = 1, \dots, k_1 k_2 \dots k_n$ contains the j-th block of k_{n+1} consecutive factors

$$V_{n+1}^{((j-1)k_{n+1}+1)} \times \cdots \times V_{n+1}^{(jk_{n+1})}$$

of R_{n+1},

(4) for all n, the groups R_n are normal in G and

$$\bigcap_{n=0}^{\infty} R_n = 1,$$

(5) for all n, the conjugation action of G on R_n permutes transitively the factors in (5.1),
    and

(6) for all n, the index $[G : R_n]$ is finite.

A group G is *weakly algebraically branch group* if there exists a sequence of integers $\overline{k} = \{k_n\}_{n=0}^{\infty}$ and two decreasing sequences of subgroups $\{R_n\}_{n=0}^{\infty}$ and $\{V_n\}_{n=0}^{\infty}$ of G satisfying the conditions (1)–(5).

Thus the only difference between weakly algebraically branch and algebraically branch groups is that in the former we do not require the indices of R_n in G to be finite. The diagram in Figure 17 may be helpful in understanding the requirements of the definition.

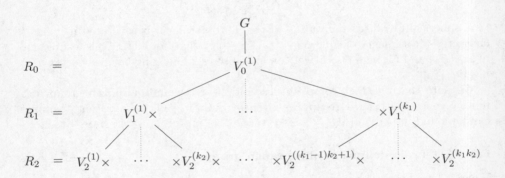

Figure 17. Branch structure of an algebraically branch group

Proposition 5.12 *Every (weakly) branch group is algebraically (weakly) branch group.*

Proof Let G admits a faithful branch type action on a spherically homogeneous rooted tree $\mathcal{T}$. By letting $R_n = \mathsf{RiSt}_G(\mathcal{L}_n)$ and $V_n = \mathsf{RiSt}_G(v_n)$, where v_n is the vertex on level n on some fixed ray $\xi \in \partial \mathcal{T}$ we obtain an algebraically (weakly) branch structure for G. □

Every algebraically branch group G acts transitively on a spherically homogeneous rooted tree, namely the tree determined by the branch structure of G (G acts on its subgroups by conjugation and, by definition, this action is spherically transitive on the tree in Figure 17). However, this action may not be faithful. In particular, it is easy to see that a direct product of an algebraically branch group and a finite abelian group produces an algebraically branch group with non-trivial center. According to Theorem 5.2 such a group cannot have a faithful branch type action.

Remark 1 We note that every infinite residually finite group has a faithful transitive action of diagonal type on some spherically homogeneous tree (even if the group is a branch group and has a branch type action on another tree). Indeed, let $\{G_n\}_{n=0}^{\infty}$ be a strictly decreasing sequence of normal subgroups of G such that (i) $G_0 = G$, (ii) the index $[G_{n-1}, G_n] = k_n$ is finite for all $n > 0$ and (iii) the intersection $\bigcap_{n=1}^{\infty} G_n$ is trivial. The set of all left cosets of all the groups in the sequence G_n can be given the structure of a spherically homogenous tree, called the coset tree of $\{G_n\}$, in which the vertices at level n are the left cosets of G_n and the edges are determined by inclusion (each coset of G_{n-1} splits as a disjoint union of k_n left cosets of G_n, which are the k_n children of G_{n-1}). The group G acts on the coset tree by left multiplication and this action preserves the inclusion relation. Thus we have an action of G on the coset tree by tree automorphisms. The action is transitive on levels (since G acts transitively on the set of left cosets of any of its subgroups) and it is faithful because the only element that can fix the whole sequence $\{G_n\}$ is an element in the intersection $\bigcap_{n=0}^{\infty} G_n$, which is trivial. Let g be an element in the rigid stabilizer of the vertex G_1 in the coset tree. Since G_1 must be fixed by g we have $g \in G_1$. Let h be an element in $G \setminus G_1$. For all $n \geq 1$ we must have $ghG_n = hG_n$. This is because, for all $n \geq 1$, the coset hG_n is contained in hG_1 and is therefore not in the subtree of the coset tree rooted at G_1. However, $ghG_n = hG_n$ if and only if $h^{-1}gh \in G_n$, and the normality of G_n implies that $g \in G_n$. Since the intersection $\bigcap_{n=1}^{\infty} G_n$ is trivial we obtain that $g = 1$ and therefore the rigid stabilizer of G_1 is trivial. The spherical transitivity of the action then shows that $\mathsf{RiSt}_G(\mathcal{L}_1)$ is trivial.

5.3   Just infinite branch groups

Definition 16 A group is *just infinite* if it is infinite and all of its proper quotients are finite.

Remark 2 Equivalently, a group is just infinite if it is infinite and all of its non-trivial normal subgroups have finite index.

Proposition 5.13 *Every finitely generated group has a just infinite quotient.*

Proof The union of normal subgroups of infinite index in a finitely generated group G has infinite index in G (since the subgroups of finite index in G are finitely generated). Therefore, by Zorn's Lemma, there exists a maximal normal subgroup N of infinite index and the quotient G/N is just infinite.   □

Definition 17 A group is *hereditarily just infinite* if all of its subgroups of finite index are just infinite.

Remark 3 Equivalently, a group is hereditarily just infinite if all of its normal subgroups of finite index are just infinite. This is because all subgroups of finite index in a group G contain a normal subgroup of G of finite index.

Another way to characterize hereditarily just infinite groups is by the property that all of their non-trivial subnormal subgroups have finite index.

Theorem 5.14 (Grigorchuk [47]) *Let G be a just infinite group. Then either*

(i) *G is a branch group*

or

G contains a a normal subgroup of finite index of the form $K \times \cdots \times K$, G acts transitively on the factors in $K \times \cdots \times K$ by conjugation, and

(iia) *K is residually finite hereditarily just infinite group*

or

(iib) *K is an infinite simple group.*

The above trichotomy result refines the description of just infinite groups proposed by Wilson in [90].

5.4 Minimality

The notion of largeness was introduced in group theory by Pride in [78]. We say that the group G is *larger* than the group H and we write $G \succeq H$ if there exists a finite index subgroup G_0 of G, finite index subgroup H_0 of H and a finite normal subgroup N of H_0 such that G_0 maps homomorphically onto H_0/N. Two groups are *equally large* if each of them is larger than the other. The class of groups that are equally large with G is denoted by $[G]$. Classes of groups can be ordered by the largeness relation $\succeq$. The class of the trivial group $[1]$ consists of all finite groups. This is the smallest class under the largeness ordering. A class of infinite groups $[G]$ is *minimal* (or atomic) if the only class smaller than $[G]$ is the class $[1]$. An infinite group G is minimal if $[G]$ is a minimal class.

Example 11 Obviously, the infinite cyclic group is minimal. Also, any infinite simple group is minimal.

A fundamental question in the theory of largeness of groups is the following.

Question 1 Which finitely generated groups are minimal?

We note that, since each finitely generated infinite group has a just infinite image, every minimal class of finitely generated groups has a just infinite representative. In the light of the trichotomy result in Theorem 5.14 we see that it is of particular interest to describe the minimal branch groups. The following result provides a sufficient condition for a regular branch group to be minimal.

Theorem 5.15 (Grigorchuk, Wilson [46]) *Let G be a regular branch group over K acting on the k-ary tree $\mathcal{T}$. Assume that K is a subdirect product of finitely many just infinite groups each of which is abstractly commensurable to G. If $\psi_n^{-1}(K \times \cdots \times K)$ is contained in K' for some $n \geq 1$, then G is a minimal group.*

Corollary 5.16 *The following groups are minimal.*
(1) *The group $\mathcal{G}$.*
(2) *The Gupta–Sidki p-groups from [56].*

The following questions were posed in [78] and in [26].

Question 2 (1) Are there finitely generated groups that do not satisfy the ascending chain condition on subnormal subgroups?
(2) Are all minimal finitely generated groups finite-by-D_2-by-finite, where D_2 denotes the class of hereditarily just infinite groups?

The group $\mathcal{G}$ provides positive answer to the first question [39] and negative answer to the second question. The minimality of $\mathcal{G}$ was established only recently in [46]. An example of a minimal branch group answering negatively the second question was constructed by P. Neumann in [72]. In fact, [72] contains examples answering most of the questions from [78] and [26].

5.5 Problems

Most of the following problems appear in [11].

Problem 4 Is the conjugacy problem solvable in all finitely generated branch groups with solvable word problem?

Problem 5 Do there exist finitely presented branch groups?

Problem 6 Do there exist branch groups with Property T?

Problem 7 Let N be the kernel of the action of an algebraically branch group G by conjugation on the spherically homogeneous rooted tree determined by the branch structure of G. What can be said about N?

Problem 8 Are there finitely generated torsion groups that are hereditarily just infinite?

Problem 9 Can a hereditarily just infinite group have the bounded generation property?

Problem 10 Is every maximal subgroup of a finitely generated branch group necessarily of finite index?

Problem 11 Which finitely generated just infinite branch groups are minimal?

The following problem is due to Nekrashevych [71].

Problem 12 For which post-critically finite polynomials f is the iterated monodromy group $IMG(f)$ a branch group?

6 Growth of groups

6.1 Word growth

Let G be a group generated by a finite symmetric set S (symmetric set means that $S = S^{-1}$). The *word length* of an element g in G with respect to S is defined as

$$|g|_S = \min\{\, n \mid g = s_1 \dots s_n, \text{ for some } s_1, \dots, s_n \in S \,\}.$$

The ball of radius n in G is the set

$$B_S(n) = \{\, g \mid |g|_S \leq n \,\}$$

of elements of length at most n in G. The *word growth function* of G with respect to S is the function $\gamma_S(n)$ counting the number of elements in the ball $B_S(n)$, i.e.,

$$\gamma_S(n) = |B_S(n)|,$$

for all $n \geq 0$. The word growth function (often called just growth function) depends on the chosen generating set S, but growth functions with respect to different generating sets can easily be related.

Proposition 6.1 *Let S_1 and S_2 be two finite symmetric generating sets of G. Then, for all $n \geq 0$,*

$$\gamma_{S_1}(n) \leq \gamma_{S_2}(Cn),$$

where $C = \max\{\, |s|_{S_2} \mid s \in S_1 \,\}$.

For two non-decreasing functions $f, g : \mathbb{N} \to \mathbb{N}^+$, where $\mathbb{N}$ is the set of non-negative integers and $\mathbb{N}^+$ is the set of positive integers, we say that f is dominated by g, and denote this by $f \preceq g$, if there exists $C > 0$ such that $f(n) \leq g(Cn)$, for all $n \geq 0$. If f and g mutually dominate each other we denote this by $f \sim g$ and say that f and g have the same *degree of growth*.

Thus any two growth functions of a finitely generated group G have the same degree of growth, i.e., the degree of growth is invariant of the group G. For example, the free abelian group $\mathbb{Z}^m$ of rank m has degree of growth equal to n^m. On the other hand, for $m \geq 2$, the degree of growth of the free group of rank m is exponential, i.e., it is equal to e^n. In general we have the following possibilities. The growth of a finitely generated group G can be

polynomial: $\quad \lim_{n \to \infty} \sqrt[n]{\gamma(n)} = 1, \quad \gamma(n) \preceq n^m, \text{ for some } m \geq 0$

intermediate: $\quad \lim_{n \to \infty} \sqrt[n]{\gamma(n)} = 1, \quad n^m \preceq \gamma(n), \text{ for all } m \geq 0$

exponential: $\quad \lim_{n \to \infty} \sqrt[n]{\gamma(n)} = c > 1, \quad \gamma(n) \sim c^n.$

The class of groups of polynomial growth is completely described. Recall that, by definition, a group G is virtually nilpotent if it has a nilpotent subgroup of finite index.

Theorem 6.2 *A finitely generated group G has polynomial growth if and only if it is virtually nilpotent.*

In that case, $\gamma(n) \sim n^m$, where m is the integer

$$m = \sum_{i=1}^{n} i \cdot \operatorname{rank}_{\mathbb{Q}}(G_{i-1}/G_i),$$

$1 = G_n \leq \cdots \leq G_1 \leq G_0 = G$ *is the lower central series of G and $\operatorname{rank}_{\mathbb{Q}}(H)$ denotes the torsion free rank of the abelian group H.*

The formula for the growth of a nilpotent group appears in [17] as well as in [55]. The other direction, showing that polynomial growth implies virtual nilpotence, is due to Gromov [54].

In [68] Milnor asked if groups of intermediate growth exist. There are many classes of groups that do not contain groups of intermediate growth. Such is the class of linear groups (direct consequence of Tits Alternative [84]), solvable groups [93, 67], elementary amenable groups [22], etc.

The first known example of a group of intermediate growth is the group $\mathcal{G}$.

Theorem 6.3 (Grigorchuk [38, 39]) *The group $\mathcal{G}$ has intermediate growth.*

The group $\mathcal{G}$ was constructed in [36] as an example of a finitely generated infinite 2-group. Other examples followed in [39, 40, 30] and more recently in [14, 11, 28, 18]. All known examples of groups of intermediate growth are either branch self-similar groups or are closely related to such groups.

The best known upper bound on the growth of a group of intermediate growth is due to Bartholdi.

Theorem 6.4 (Bartholdi [6]) *The growth function of $\mathcal{G}$ satisfies*

$$\gamma(n) \preceq e^{n^{\alpha}},$$

where $\alpha = \log 2 / (\log 2 - \log \eta) \approx 0.767$ and $\eta \approx 0.81$ is the positive root of the polynomial $x^3 + x^2 + x - 2$.

We define now the class of groups of intermediate growth introduced in [39]. Each group in this class is defined by an infinite word $\bar{\omega}$ in $\Omega = \{0, 1, 2\}^{\mathbb{N}}$. Set up a correspondence

$$0 \leftrightarrow \begin{bmatrix} 1 \\ 1 \\ 0 \end{bmatrix}, \qquad 1 \leftrightarrow \begin{bmatrix} 1 \\ 0 \\ 1 \end{bmatrix}, \qquad 2 \leftrightarrow \begin{bmatrix} 0 \\ 1 \\ 1 \end{bmatrix} \qquad (6.1)$$

and, for a letter ω in $\{0, 1, 2\}$, define $\omega(b)$, $\omega(c)$ and $\omega(d)$ to be the top entry the middle entry and the bottom entry, respectively, in the matrix corresponding to ω. Let a be the binary tree automorphism defined by

$$a = (01)\,(1, 1).$$

Thus a only changes the first letter in every binary word. For a word $\bar{\omega} = \omega_1\omega_2\dots$ in Ω define binary tree automorphisms $b_{\bar{\omega}}$, $c_{\bar{\omega}}$ and $d_{\bar{\omega}}$ by

$$b_{\bar{\omega}} = (a^{\omega_1(b)}, b_{\sigma(\bar{\omega})}),$$
$$c_{\bar{\omega}} = (a^{\omega_1(c)}, c_{\sigma(\bar{\omega})}),$$
$$d_{\bar{\omega}} = (a^{\omega_1(d)}, d_{\sigma(\bar{\omega})})$$

where $\sigma(\bar{\omega})$ is the shift of $\bar{\omega}$ defined by $\sigma(\bar{\omega}) = \omega_2\omega_3\dots$, i.e, $(\sigma(\bar{\omega}))_n = (\bar{\omega})_{n+1}$, for $n = 1, 2, \dots$. The only possible non-trivial sections of $b_{\bar{\omega}}$, $c_{\bar{\omega}}$ and $d_{\bar{\omega}}$ appear at the vertices along the infinite ray $111\dots$ and the vertices at distance 1 from this ray. Define $G_{\bar{\omega}}$ to be the group

$$G_{\bar{\omega}} = \langle a, b_{\bar{\omega}}, c_{\bar{\omega}}, d_{\bar{\omega}} \rangle$$

of binary tree automorphisms.

Example 12 Let $\bar{\omega}$ be the periodic sequence $\bar{\omega} = (01)^\infty$. The corresponding sequence of matrices is

$$\begin{bmatrix} 1 \\ 1 \\ 0 \end{bmatrix} \begin{bmatrix} 1 \\ 0 \\ 1 \end{bmatrix} \begin{bmatrix} 1 \\ 1 \\ 0 \end{bmatrix} \begin{bmatrix} 1 \\ 0 \\ 1 \end{bmatrix} \dots$$

Since the top row of entries is invariant under the shift σ we have that $b_{\sigma(\bar{\omega})} = b_{\bar{\omega}}$. Therefore

$$b_{\bar{\omega}} = (a, b_{\bar{\omega}}).$$

On the other hand, the shift of the middle row of entries is equal to the bottom row of entries and vice versa. Therefore we have $c_{\sigma(\bar{\omega})} = d_{\bar{\omega}}$ and $d_{\sigma(\bar{\omega})} = c_{\bar{\omega}}$ and

$$c_{\bar{\omega}} = (a, d_{\bar{\omega}}),$$
$$d_{\bar{\omega}} = (1, c_{\bar{\omega}}).$$

Thus $G_{(01)^\infty}$ is a self-similar group defined by the 5 state binary automaton in Figure 18. The similarity with the automaton generating $\mathcal{G}$ is not accidental. In

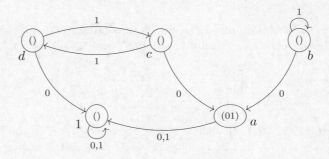

Figure 18. The binary automaton generating the group $G_{(01)^\infty}$

fact, in this context, the group $\mathcal{G}$ is defined by the periodic sequence $012012\dots$.

Theorem 6.5 (Grigorchuk [39]) (a) *For every word $\bar{\omega}$ in Ω, $G_{\bar{\omega}}$ is a spherically transitive group of binary automorphisms such that the upper companion group of $G_{\bar{\omega}}$ at any vertex at level n is $G_{\sigma^n(\bar{\omega})}$.*

(b) *For every word $\bar{\omega}$ in Ω that is not ultimately constant, $G_{\bar{\omega}}$ is not finitely presented.*

(c) *For every word $\bar{\omega}$ in Ω in which all letters appear infinitely often, $G_{\bar{\omega}}$ is a just infinite branch 2-group.*

(d) *For every word $\bar{\omega}$ in Ω that is not ultimately constant, $G_{\bar{\omega}}$ has intermediate growth.*

The groups in this class exhibit very rich range of (intermediate) growth behavior. For example, the set of degrees of growth of these groups contain uncountable chains and anti-chains (under the comparison of degrees of growth given by dominance).

The examples from [39] were generalized to examples of groups of intermediate growth acting on p-ary trees, for p a prime, in [40]. These groups are defined by infinite words over $\{0, 1, \ldots, p\}$. There is a general upper bound on the growth in the case of groups defined by homogeneous sequences. A sequence $\bar{\omega}$ in $\Omega = \{0, 1, \ldots, p\}^{\mathbb{N}}$ defining a group $G_{\bar{\omega}}$ is *r-homogeneous* if every symbol in $\{0, 1, \ldots, p\}$ appears in every subword of $\bar{\omega}$ of length r.

Theorem 6.6 *Let $\bar{\omega}$ be an r-homogeneous sequence.*

(a) (Muchnik, Pak [69]) *In case $p = 2$, the growth of $G_{\bar{\omega}}$ satisfies*

$$\gamma(n) \preceq e^{n^{\alpha}},$$

where $\alpha = \log 2/(\log 2 - \log \eta)$ and η is the positive root of the polynomial $x^r + x^2 + x - 2$.

(b) (Bartholdi, Šunić [14]) *For arbitrary prime p, the growth of $G_{\bar{\omega}}$ satisfies*

$$\gamma(n) \preceq e^{n^{\alpha}},$$

where $\alpha = \log p/(\log p - \log \eta)$ and η is the positive root of the polynomial $x^r + x^{r-1} + x^{r-2} - 2$.

An estimate analogous to the one above holds in the wider context of the so called spinal groups (see [14, 11]).

Groups of intermediate growth appear also as iterated monodromy groups of post-critically finite polynomials.

Theorem 6.7 (Bux, Perez [18]) *The group $IMG(z \mapsto z^2 + i)$ has intermediate growth.*

Most of the proofs that a group has sub-exponential growth are based on a variation of the following contraction themes (see [39, 14, 69, 18]).

Proposition 6.8 (a) *If G is a self-similar group of k-ary tree automorphisms generated by a finite set S and there exist η in $(0,1)$, $\alpha \in (0,1]$ and a constant C such that, for all n, the ratio between the number of elements g in the ball $B_S(n)$ that satisfy*

$$\sum_{x=0}^{k-1} |g_x|_S \leq \eta |g|_S + C$$

and $\gamma_S(n)$ is at least α, then G has sub-exponential growth.

(b) *If G is a self-similar group of k-ary tree automorphisms generated by a finite set S and there exist η in $(0,1)$ and a constant C such that, for all $g \in G$,*

$$\sum_{x=0}^{k-1} |g_x|_S \leq \eta |g|_S + C,$$

then the growth of G satisfies

$$\gamma_S(n) \preceq e^{n^\alpha},$$

where $\alpha = \log k / (\log k - \log \eta)$.

(c) *Let $\Phi = \{G_\lambda\}_{\lambda \in \Lambda}$ be a family of groups acting on spherically homogeneous rooted trees, let Φ be closed for upper companion groups, let each member G_λ of Φ be generated by a finite set S_λ and let there be a uniform bound on the number of generators in S_λ. Let η be a number in $(0,1)$. Assume that for each λ there exists a level m_λ and a constant $C_\lambda \geq 0$ such that, for all elements g in G_λ,*

$$\sum_{u \in \mathcal{L}_{m_\lambda}} |g_u|_{S_{\lambda_u}} \leq \eta |g|_{S_\lambda} + C_\lambda,$$

Then all groups in Φ have sub-exponential growth.

A quite a different approach, both for the lower and upper bounds on the growth, based on the Poisson boundary, is used by Erschler in her work.

Theorem 6.9 (Erschler [27]) *The growth of the group G_ω acting on the binary tree and defined by the sequence $\bar{\omega} = (01)^\infty$ satisfies*

$$e^{n/\ln^{2+\epsilon} n} \leq \gamma(n) \leq e^{n/\ln^{1-\epsilon} n}$$

for any $\epsilon > 0$ and any sufficiently large n.

In many cases, a general lower bound exists for the growth of groups of intermediate growth.

Theorem 6.10 (a) (Grigorchuk [41]) *The degree of growth of any residually-p group that is not virtually nilpotent is at least $e^{\sqrt{n}}$.*

(b) (Lubotzky, Mann [64]) *The degree of growth of any residually nilpotent group that is not virtually nilpotent is at least $e^{\sqrt{n}}$.*

Improvements over these general bounds are given for the particular case of $\mathcal{G}$ in [63] and [7].

Finally, we mention that there are many examples of automaton groups that have exponential growth. Such examples are the lamplighter groups [51, 15, 82], most ascending HNN extensions of free abelian groups [15] (including the Baumslag–Solitar solvable groups $BS(1, m)$ for $m \neq \pm 1$), free groups [33, 88], etc. There are self-similar groups of exponential growth within the class of branch automaton groups. For example, $\mathcal{H}^{(k)}$ has exponential growth, for all $k \geq 3$.

6.2 Uniformly exponential growth

Let G be a group of exponential growth and let

$$\epsilon_G = \inf\{ \sqrt[n]{\gamma_S(n)} \mid S \text{ a finite symmetric generating set of } G \}.$$

We say that G has uniformly exponential growth if $\epsilon_G > 1$. In 1981 Gromov asked if all groups of exponential growth have uniformly exponential growth. Affirmative answer has been obtained for the classes of hyperbolic groups [60], one-relator groups [20, 42], solvable groups (Wilson and Osin [77], independently), linear groups over fields of characteristic 0 [29], etc. In [92] John Wilson showed that the answer is negative in the general case.

Theorem 6.11 (Wilson [92]) *There exist 2-generated branch groups of exponential growth that do not have uniformly exponential growth.*

Further examples of this type appear in [91] and [8].

6.3 Torsion growth

Let G be a finitely generated torsion group and let S be a finite symmetric generating set for G. For $n \in \mathbb{N}$, denote by $\tau(n)$ the largest order of an element in G of length at most n with respect to S. The function τ is called the *torsion growth function* of G with respect to S. The degree of the torsion growth function of G does not depend on the chosen finite generating set, i.e., if τ_1 and τ_2 are two torsion growth functions of G with respect to two finite symmetric generating sets S_1 and S_2, then $\tau_1 \sim \tau_2$.

There exist polynomial estimates for the torsion growth of some groups G_ω from [39] and [40].

Theorem 6.12 (Bartholdi, Šunić [14])   (a) *Let $G_{\bar{\omega}}$ be a p-group defined by a r-homogeneous sequence $\bar{\omega}$. The torsion growth function satisfies*

$$\tau(n) \preceq n^{(r-1)\log_2(p)}.$$

(b) *The torsion growth function of $\mathcal{G}$ satisfies*

$$\tau(n) \preceq n^{3/2}.$$

6.4   Subgroup growth

We describe here a result of Segal [79] that uses branch groups to fill a conjectured gap in the spectrum of possible rates of subgroup growth.

For a finitely generated group G the number of subgroups of index at most n, denoted $s(n)$, is finite. The function s counting the subgroups up to a given index is called the *subgroup growth function* of G. All subgroup growth considerations are usually restricted to residually finite groups, since the lattice of subgroups of finite index in G is canonically isomorphic to the lattice of subgroups of finite index in the residually finite group G/N, where N is the intersection of all groups of finite index in G.

It is shown by Lubotzky, Mann and Segal in [65] that a residually finite group has polynomial subgroup growth (there exists m such that $s(n) \leq n^m$, for all sufficiently large n) if and only if it is a virtually solvable group of finite rank. It has been conjectured in [66] that if a finitely generated residually finite group does not have polynomial subgroup growth then there exists a constant $c > 0$ such $s(n) \geq n^{c \log(n)/(\log\log(n))^2}$, for infinitely many values of n.

Theorem 6.13 (Segal [79]) *There exists a finitely generated just infinite branch group such that, for any $c > 0$, $s(n) \leq n^{c \log(n)/(\log\log(n))^2}$, for all sufficiently large n.*

The construction used by Segal is rather flexible and can be used to answer other problems related to finite images of finitely generated groups.

Theorem 6.14 (Segal [79]) *Let S be any set of finite non-abelian simple groups. There exists a finitely generated just infinite branch group G such that the upper composition factors of G (the composition factors of the finite images of G) are precisely the members of S.*

6.5   Problems

Problem 13   (a) What is the degree of growth of $\mathcal{G}$?

  (b) What is the degree of growth of any group $G_{\bar{\omega}}$, when $\bar{\omega}$ is not ultimately constant?

Problem 14   (a) Is it correct that every group that is not virtually nilpotent has degree of growth at least $e^{\sqrt{n}}$?

  (b) Are there groups whose degree of growth is $e^{\sqrt{n}}$?

Problem 15 (Nekrashevych [71])   For which post-critically finite polynomials f does the iterated monodromy group $IMG(f)$ have intermediate growth?

7   Amenability

In this section we present some basic notions and results concerning amenability of groups and show how branch $(\mathcal{G})$ and weakly branch $(\mathcal{B})$ self-similar groups provide some crucial examples distinguishing various classes of groups related to the notion of amenability.

7.1   Definition

The fundamental notion of amenability is due to von Neuman [87].

Definition 18   A group G is *amenable* if there exists a finitely additive left-invariant probabilistic measure μ defined on all subsets of G, i.e.,
   (i) (μ is defined for all subsets)

$$0 \leq \mu(E) \leq 1,$$

for all subsets E of G,
   (ii) (μ is probabilistic)

$$\mu(G) = 1,$$

   (iii) (μ is left invariant)

$$\mu(E) = \mu(gE),$$

for all subsets E of G and elements g in G,
   (iv) (μ is finitely additive)

$$\mu(E_1 \cup E_2) = \mu(E_1) + \mu(E_2),$$

for disjoint subsets E_1, E_2 of G.

Denote the class of amenable groups by AG. All finite groups are amenable (under the uniform measure, which is the only possible measure in this case). Also, all abelian groups are amenable, but all known proofs rely on the Axiom of Choice (even for the infinite cyclic group).

Remark 4 A group G is amenable if and only if it admits a left invariant mean, i.e., there exists a non-negative left invariant linear functional m on the space of bounded complex valued functions defined on G that maps the constant function 1 to 1. If such a mean exists we may define a measure on G by $\mu(E) = m(f_E)$, where f_E is the characteristic function of the subset E of G.

The next result gives a combinatorial characterization of amenability.

Theorem 7.1 (Følner) *A countable group G is amenable if and only if there exists a sequence of finite subsets $\{A_n\}$ of G such that, for every g in G,*

$$\lim_{n \to \infty} \frac{|gA_n \cap A_n|}{|A_n|} = 1.$$

Definition 19 Let $\Gamma = (V, E)$ be a graph. The *boundary* of a subset A of the vertex set V, denoted $\partial(A)$, is the set of edges in E connecting a vertex in A to a vertex outside of A.

Definition 20 Let $\Gamma = (V, E)$ be a graph of uniformly bounded degree. The *Cheeger constant* of Γ is the quantity

$$ch(\Gamma) = \inf \left\{ \frac{|\partial(A)|}{|A|} \,\middle|\, A \text{ a finite subset of } V \right\}$$

Definition 21 An graph Γ of uniformly bounded degree is *amenable* if its Cheeger constant is 0.

Remark 5 Let G be a finitely generated infinite group with finite generating set S and let $\Gamma = \Gamma(G, S)$ be the Cayley graph of G with respect to S. Then G is amenable if and only if Γ is amenable.

Example 13 The infinite cyclic group $\mathbb{Z}$ is amenable. Indeed, for the sequence of intervals $A_n = [1, n]$, $n = 1, 2, \ldots$, and the symmetric generating set $S = \{\pm 1\}$, we have

$$\frac{|\partial(A_n)|}{|A_n|} = \frac{4}{n}$$

which tends to 0 as n grows.

As was already mentioned all abelian groups are amenable. Indeed Følner criterion can be used to prove the following more general result.

Theorem 7.2 *Every finitely generated group of sub-exponential growth is amenable.*

Since all finitely generated abelian (virtually nilpotent) groups have polynomial growth we obtain the following corollary.

Corollary 7.3 *All virtually nilpotent groups are amenable.*

7.2   Elementary classes

Theorem 7.4 (von Neumann [87]) *The class of amenable groups AG is closed under taking*
  (i) *subgroups,*
 (ii) *homomorphic images,*
(iii) *extensions, and*
(iv) *directed unions.*

Call the constructions (i)–(iv) above elementary constructions.

Definition 22 The class of *elementary amenable groups*, denoted EA, is the smallest class of groups that contains all finite and all abelian groups, and is closed under the elementary constructions.

Theorem 7.5 *The free group F_2 of rank 2 is not amenable.*

The fact that F_2 is not amenable can be easily proved by several different arguments. In particular, one may use the doubling condition of Gromov.

Theorem 7.6 (Gromov doubling condition) *Let* Γ *be a graph of uniformly bounded degree. Then* Γ *is not amenable if and only if there exists a map* $f : V(\Gamma) \to V(\Gamma)$ *such that the pre-image of every vertex has at least 2 elements and the distance between any vertex and its image is uniformly bounded.*

To see now that $F_2 = F(a,b)$ is not amenable just map every vertex $ws^{\pm 1}$ in the Cayley graph of F_2 with respect to $S = \{a,b\}$ to w and leave 1 fixed. Under this map, the distance between every vertex and its image is at most 1 and every vertex has at least 3 pre-images.

Since F_2 is not amenable and AG is closed under subgroups, no group that contains F_2 is amenable. Denote by NF the class of groups that do not contain a copy of the free group F_2 of rank 2 as a subgroup. We have

$$EA \subseteq AG \subseteq NF.$$

Mahlon Day asked in [24] if equality holds in either of the two inclusions above. The question if $AG = NF$ is sometimes referred to as von Neumann Problem. Chou showed in 1980 that $EA \neq NF$ by using the known fact that there exist infinite torsion groups and proving the following result.

Theorem 7.7 (Chou [22]) *No finitely generated torsion group is elementary amenable. Therefore*
$$EA \neq NF.$$

A negative solution to the von Neumann Problem was given by Ol'shanskii [73] in 1980 and later by Adian [3].

Theorem 7.8 (Ol'shanskii [73]) *There exist non-amenable groups that do not contain free subgroups of rank 2. Therefore*

$$AG \neq NF.$$

The examples Ol'shanskii used to show that $AG \subsetneq NF$ are the Tarski monsters he constructed earlier [75].

Theorem 7.9 (Adian [3]) *The free Burnside groups*

$$B(m,n) = \langle\, a_1, \ldots, a_m \mid x^n = 1, \text{ for all } x \,\rangle,$$

for $m \geq 2$ *and odd* $n \geq 665$, *are non-amenable.*

Both Ol'shanskii and Adian used the following co-growth criterion of amenability in their work.

Theorem 7.10 (Grigorchuk [37]) *Let G be a m-generated group presented as F_m/H where $F_m = F(X)$ is the free group of rank m and H is a normal subgroup of F_m. Denote by $h(n)$ the number of words of length n over $X \cup X^{-1}$ that represent elements in H. Then G is amenable if and only if*

$$\limsup_{n \to \infty} \sqrt[n]{h(n)} = 2m - 1.$$

We quote another result of Chou.

Theorem 7.11 (Chou [22]) *No finitely generated group of intermediate growth is elementary amenable.*

This shows that any example of a group of intermediate growth would answer the question if $EG = AG$ negatively (since all such groups are amenable). Such an example was provided in [38, 39].

Theorem 7.12 *The group $\mathcal{G}$ is amenable but not elementary amenable group. Thus*

$$EA \neq AG.$$

In fact, the classes EA, AG and NF are distinct even in the context of finitely presented groups.

Theorem 7.13 (Grigorchuk [49]) *The HNN extension*

$$\tilde{\mathcal{G}} = \langle\, \mathcal{G}, t \mid a^t = aca, \ b^t = d, \ c^t = b, \ d^t = c \,\rangle$$

is a finitely presented amenable group that is not elementary amenable. Thus the class $AG \setminus EA$ contains finitely presented (torsion-by-cyclic) groups.

Theorem 7.14 (Ol'shanskii, Sapir [74]) *The class $NF \setminus AG$ contains finitely presented (torsion-by-cyclic) groups.*

Definition 23 The class of *sub-exponentially amenable groups*, denoted SA, is the smallest class of groups that contains all finitely generated groups of sub-exponential growth and is closed under the elementary constructions.

It is clear that
$$EA \subseteq SA \subseteq AG.$$
The group $\mathcal{G}$ is an example of a group in $SA \setminus EA$.

Theorem 7.15 (a) (Grigorchuk, Żuk [52]) *The Basilica group is not sub-exponentially amenable.*

(b) (Bartholdi, Virág [16]) *The Basilica group is amenable.*

Corollary 7.16
$$EA \subsetneq SA \subsetneq AG \subsetneq NF.$$

We will only prove here that the Basilica group is not sub-exponentially amenable. The proof follows the exposition in [52], but we first introduce the notion of elementary classes.

Definition 24 (Chou [22], Osin [76]) Let $\mathcal{C}$ be a class of groups. For each ordinal α define the *elementary class* $E_\alpha(\mathcal{C})$ as follows. For $\alpha = 0$ define

$$E_0(\mathcal{C}) = \mathcal{C}.$$

For non-limit ordinals of the form $\alpha = \beta + 1$ define $E_\alpha(\mathcal{C}) = E_{\beta+1}(\mathcal{C})$ to be the class of groups that can be obtained either as an extension of a group in $E_\beta(\mathcal{C})$ by a group in $\mathcal{C}$ or as a directed union of a family of groups in $E_\beta(\mathcal{C})$. For limit ordinals α define

$$E_\alpha(\mathcal{C}) = \bigcup_{\beta < \alpha} E_\beta(\mathcal{C}).$$

Finally, define $E(\mathcal{C})$ to be the union of $E_\alpha(\mathcal{C})$ taken over all ordinals.

Theorem 7.17 (Chou [22], Osin [76]) *If $\mathcal{C}$ is closed under taking homomorphic images and subgroups then $E(\mathcal{C})$ is the smallest class of groups containing $\mathcal{C}$ that is closed under all four elementary constructions* (i)–(iv).

Moreover, all elementary classes $E_\alpha(\mathcal{C})$ are closed under taking homomorphic images and subgroups.

This result is proved for the case of elementary amenable groups in [22] and in general in [76].

Proof of Theorem 7.15(a) Let $\mathcal{SG}$ be the class of groups of sub-exponential growth and $\mathcal{C}$ be the closure of $\mathcal{SG}$ under taking subgroups and homomorphic images. We show that $\mathcal{B}$ does not belong to $E_\alpha(\mathcal{C})$ for any cardinal α.

By way of contradiction, assume that α is the smallest ordinal such that $\mathcal{B} \in E_\alpha(\mathcal{C})$. Since $\mathcal{B}$ has exponential growth, we have $\mathcal{B} \notin E_0(\mathcal{C})$ and since $\mathcal{B}$ is finitely generated it cannot be a directed union of its proper subgroups. Thus $\alpha = \beta + 1$ and there exists a short exact sequence

$$1 \to N \to \mathcal{B} \to Q \to 1$$

with N in $E_\beta(\mathcal{C})$. Since $\mathcal{B}$ is weakly branch over its commutator $\mathcal{B}'$, according to Theorem 5.3, there exists a level n such that

$$N \geq (\mathsf{RiSt}_\mathcal{B}(\mathcal{L}_n))' \geq (\mathcal{B}' \times \cdots \times \mathcal{B}')' \geq \mathcal{B}'' \times \cdots \times \mathcal{B}''.$$

Since the elementary class $E_\beta(\mathcal{C})$ is closed for subgroups and homomorphic images [76] we conclude that $\mathcal{B}''$ is in $E_\beta(\mathcal{C})$. The following observations (made in [52]) can then be used to show that $\mathcal{B}$ must also be in $E_\beta(\mathcal{C})$, leading to a contradiction.

Namely, we have $\mathcal{B}'' \leq \gamma_3(\mathcal{B}) \leq \mathsf{St}_\mathcal{B}(L_1)$ (here $\gamma_3(\mathcal{B}) = [[\mathcal{B}, \mathcal{B}], \mathcal{B}]$), $\mathcal{B}''$ geometrically decomposes as

$$\mathcal{B}'' = \gamma_3(\mathcal{B}) \times \gamma_3(\mathcal{B}),$$

and we have the following projections

$$\varphi_1(\gamma_3(\mathcal{B})) = \langle \gamma_3(\mathcal{B}), (a^2)^b \rangle$$
$$\varphi_1(\langle \gamma_3(\mathcal{B}), (a^2)^b \rangle) = \langle \gamma_3(\mathcal{B}), (a^2)^b, b \rangle$$
$$\varphi_0(\langle \gamma_3(\mathcal{B}), (a^2)^b, b \rangle) = \mathcal{B}.$$

□

In fact the four classes EA, SA, AG and NF are separated even within the class of finitely presented groups, since there exists an HNN-extension

$$\tilde{\mathcal{B}} = \langle\, \mathcal{B}, t \mid a^t = b, \; b^t = a^2 \,\rangle$$
$$= \langle\, a, t \mid [[a, a^t], a^t] = 1, \; a^{tt} = a^2 \,\rangle$$

of $\mathcal{B}$ that is finitely presented and amenable.

The group $\tilde{\mathcal{B}}$ has a balanced presentation on 2 generators and 2 relations, just as the Thompson group

$$F = \langle\, a, b \mid b^{aa} = b^{ab}, \; b^{aaa} = b^{abb} \,\rangle.$$

While it is known that F is in NF it is a long standing question if F is amenable.

Question 3 Is the Thompson group F amenable?

We observe that F cannot be realized as a group of k-ary rooted tree automorphisms for any finite k, since F is not residually finite (the commutator of F is not trivial and is contained in all normal subgroups of F). However, F can be realized as a group of homeomorphisms of the boundary $\partial \mathcal{T}$ of the binary rooted tree. Moreover the action of F on $\partial \mathcal{T}$ can be given by a finite asynchronous automaton [43].

7.3 Tarski numbers

Definition 25 A finitely generated group G has a *paradoxical decomposition* if there exist a decomposition

$$G = A_1 \cup \cdots \cup A_m \cup B_1 \cup \cdots \cup B_n,$$

of G into a disjoint union of $m + n$ nonempty sets (with $m, n \geq 1$) and there exist elements a_i, $i = 1, \ldots, m$, and b_j, $j = 1, \ldots, n$, in G such that

$$G = a_1 A_1 \cup \cdots \cup a_m A_m = b_1 B_1 \cup \cdots \cup b_n B_n.$$

The smallest $m + n$ in a paradoxical decomposition of G is called the *Tarski number* of G. Groups that have no paradoxical decomposition have infinite Tarski number.

Denote the Tarski number of a group G by $\tau(G)$. The Tarski number of any group cannot be 3 or less. A proof of the following result of Tarski is provided in [89]. Another proof, based on the Hall–Radon matching theorem, is provided in [19].

Theorem 7.18 (Tarski) *A finitely generated group G is amenable if and only if it has a paradoxical decomposition.*

Theorem 7.19 *A group G contains a copy of the free group F_2 if and only if its Tarski number is 4.*

Theorem 7.20 (Ceccherini-Silberstein, Grigorchuk, de la Harpe [19])

  (a) *There exists a 2-generated non-amenable torsion-free group G whose Tarski number satisfies*

$$5 \le \tau(G) \le 34.$$

  (b) *For $m \ge 2$ and odd $n \ge 665$, the Tarski number of the free Burnside group $B(m,n)$ satisfies the inequalities*

$$5 \le \tau(B(m,n)) \le 14.$$

7.4 Other topics related to amenability

A large class of automaton groups in NF was constructed by Sidki. For a k-ary tree automorphism g define $\alpha_g(n)$ (see [81]), called the *activity number* at level n, to be the number of nontrivial sections at level n, $n = 0, 1, 2, \ldots$. If a k-ary tree automorphism is generated by a state of a finite automaton, then the growth of α_g is either polynomial or exponential. Denote by Pol_k the set of k-ary tree automorphisms g for which α_g grows polynomially. This set forms a subgroup of $\mathsf{Aut}(\mathcal{T}^{(k)})$.

Theorem 7.21 (Sidki [81]) *The group Pol_k does not contain free subgroups of rank 2, for any finite $k \ge 2$.*

It is easy to check if the growth of α_s is exponential or polynomial for a state s of a finite automaton $\mathcal{A}$. Each state t of $\mathcal{A}$ can be classified as active or non-active depending on whether π_t is non-trivial or trivial, respectively. The activity number $\alpha_s(n)$ is equal to the number of paths in the directed graph representing $\mathcal{A}$ starting at s and ending in an active state. A simple criterion, due to Ufnarovskiĭ [85], may be used to see that a k-ary automaton group $G(\mathcal{A})$ is in Pol_k if and only if every state of the automaton $\mathcal{A}$ appears as a vertex in at most one directed cycle from which an active state can be reached.

Definition 26 An automaton $\mathcal{A}$ is *bounded* if all of its states have bounded activity growth.

Example 14 All the states in the automaton generating Basilica group (see Example 2) have bounded activity growth. In fact, it is clear that $\alpha_a(n), \alpha_b(n) \leq 1$ for any n. Thus Basilica group is generated by a bounded automaton.

Similarly, the activity growth of all the states generating $\mathcal{G}$ is bounded.

On the other hand, the activity growth of every nontrivial state generating $\mathcal{H}^{(4)}$ is exponential (the two loops at each non-trivial state in the automaton in Figure 9 provide exponentially many paths leading to an active state).

Theorem 7.22 (Bartholdi, Kaimanovich, Nekrashevych, Virág [5]) *All groups generated by bounded automata are amenable.*

The proof of the above result is based on self-similar random walks techniques developed by Bartholdi and Virág in [16], "Münchausen trick" of Kaimanovich [58] and embedding techniques of Brunner, Nekrashevych and Sidki reducing the problem, for each arity k, to a specific self-similar automaton group, called the *mother group*.

Greenleaf [34] asked if a non-amenable groups can have amenable actions. An action of G on X is amenable if X admits a G-invariant mean. A group has the property ANA if it is non-amenable but admits a faithful transitive amenable action. It is shown in [86] that F_2 has the property ANA. On the other hand, it is known that groups with Kazhdan Property (T) never have the property ANA.

Theorem 7.23 (Grigorchuk, Nekrashevych, [35]) *Every finitely generated, residually finite, non-amenable group embeds into a finitely generated, residually finite, non-amenable group that has faithful, transitive, amenable actions.*

Groups with the property ANA and other related properties were also studied by Monod and Glasner in [32], where many new amenable actions of non-amenable groups are introduced.

7.5 Problems

Problem 16 ([19]) (a) What is the range of Tarski numbers?

  (b) Give an example of a non-amenable group with explicitly determined Tarski number different from 4.

Problem 17 Is Pol_k amenable?

Problem 18 Are there hereditarily just infinite groups that are amenable but not elementary amenable?

Problem 19 Are there non-amenable branch groups in NF?

Problem 20 Are there non-amenable automaton groups that do not contain the free group F_2 of rank 2?

8   Schreier graphs related to self-similar groups

Let G be a finitely generated spherically transitive group of k-ary automorphisms, ξ be a ray in ∂T and, for $n \geq 0$, ξ_n be the unique level n vertex on ξ. In other words, ξ is an infinite word over X and, for $n \geq 0$, ξ_n is its prefix of length n. Further, let P_n be the stabilizer $P_n = P_{\xi_n} = \mathsf{St}_G(\xi_n)$ and P be the stabilizer $P = P_\xi = \mathsf{St}_G(\xi)$. The sequence of subgroups $\{P_n\}_{n=0}^{\infty}$ is decreasing to $P = \bigcap_{n=0}^{\infty} P_n$.

For a fixed symmetric generating set S the *Schreier graph* $\Gamma_n = \Gamma_n(G, P_n, S)$ is the graph whose vertices are the left cosets of P_n and in which, for each pair of a coset gP_n and a generator s in S, there is an edge from gP_n to sgP_n labeled by s.

Since the action of G is spherically transitive the graph Γ_n, for $n \geq 0$, is a connected graph on k^n vertices that is isomorphic (independently of ξ) to the Schreier graph of the action of G on level n. The vertices of the Schreier graph of the action at level n are the vertices at level n and, for each vertex u and generator s, there is an edge from u to $s(u)$ labeled by s. The isomorphism between the two graphs is given by $gP_{\xi_n} \leftrightarrow u$ if and only if $g(\xi_n) = u$.

On the other hand, since the group G is countable and ∂T is not, the action of G on ∂T is not transitive and the Schreier graph Γ_ξ is isomorphic to the connected component of the Schreier graph of the action of G on ∂T representing the G-orbit of the ray ξ.

The following proposition is rather obvious in the light of the fact that tree automorphisms preserve prefixes.

Proposition 8.1 *For all $n \geq 0$, the map $\Gamma_{n+1} \to \Gamma_n$ given by*

$$wx \mapsto w$$

is a k-fold graph covering. The map $\Gamma \to \Gamma_n$ given by

$$\zeta \mapsto \zeta_n,$$

where ζ is an infinite word in the G orbit of ξ and, for $n \geq 0$, ζ_n is the prefix of ζ of length n, is also a graph covering.

Example 15 (Schreier graphs of $\mathcal{H}^{(3)}$) Let $G = \mathcal{H}^{(3)}$. The Schreier graph $\Gamma_3 = \Gamma_{000}$ corresponding to the action of $\mathcal{H}^{(3)}$ at level 3 is given in Figure 19.

The graphs Γ_n, as n grows, look more and more like the Sierpiński gasket (see Figure 15). We will see that this is not a random phenomenon and the reason for this is that the Sierpiński gasket is the Julia set of the map $f : z \mapsto \bar{z}^2 - \frac{16}{27\bar{z}}$, whose iterated monodromy group is exactly $\mathcal{H}^{(3)}$.

We offer two recursive ways to build the graph Γ_{n+1} from Γ_n. Both are based on the similarity between these graphs.

The graph Γ_0 corresponding to the action at the root is given by

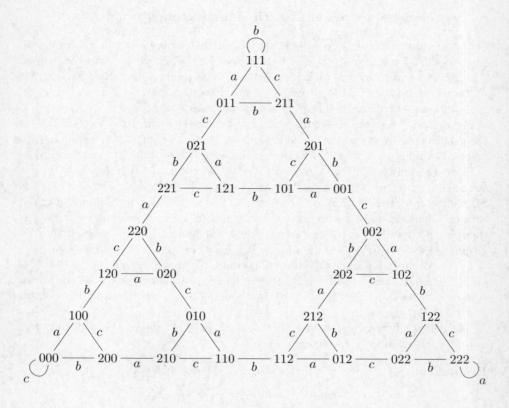

Figure 19. The Schreier graph of $\mathcal{H}^{(3)}$ at level 3

For $n \geq 0$, in order to build Γ_{n+1}, first build 3 copies of Γ_n, denoted by $\Gamma_{n,0}$, $\Gamma_{n,1}$ and $\Gamma_{n,2}$. The copy $\Gamma_{n,x}$ differs from Γ_n only by the fact that each vertex label u in Γ_n is replaced by ux in $\Gamma_{n,x}$. To get Γ_{n+1} delete, for each pair $x, y \in X_3$, $x \neq y$, the loops at $z^n x$ and $z^n y$ in $\Gamma_{n,x}$ and $\Gamma_{n,y}$, respectively, and replace them by a single edge labeled by a_{xy} connecting $z^n x$ and $z^n y$ (here z denotes the third letter in X_3 different from both x and y).

The second recursive way to build Γ_{n+1} from Γ_n is by graph substitution. The axiom is the graph describing Γ_0 corresponding to level 0 and the rules are

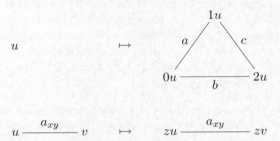

$$u \xrightarrow{\;a_{xy}\;} v \quad \mapsto \quad zu \xrightarrow{\;a_{xy}\;} zv$$

for each vertex and each edge in Γ_n. Given these rules Γ_{n+1} is built from Γ_n by replacing each occurrence of a vertex u in Γ_n by a triangle according to the first rule and replacing each occurrence of a labeled edge in Γ_n by a labeled edge given by the second rule connecting the indicated vertices in Γ_{n+1} (again, here z denotes the third letter in X_3 different from both x and y).

Example 16 (Schreier graphs of $\mathcal{G}$) The Schreier graphs of $\mathcal{G}$ corresponding to the first 3 levels are given in Figure 20

$$b,c,d \;\bigcirc\; 0 \xrightarrow{\;a\;} 1 \;\bigcirc\; b,c,d$$

$$b,c,d \;\bigcirc\; 10 \xrightarrow{\;a\;} 00 \underset{c}{\overset{b}{\rightleftharpoons}} 01 \xrightarrow{\;a\;} 11 \;\bigcirc\; b,c,d$$

$$b,c,d \;\bigcirc\; 110 \xrightarrow{\;a\;} 010 \underset{c}{\overset{b}{\rightleftharpoons}} 000 \xrightarrow{\;a\;} 100 \underset{d}{\overset{b}{\rightleftharpoons}} 101 \xrightarrow{\;a\;} 001 \underset{c}{\overset{b}{\rightleftharpoons}} 011 \xrightarrow{\;a\;} 111 \;\bigcirc\; b,c,d$$

Figure 20. The Schreier graphs of $\mathcal{G}$ corresponding to levels 1, 2 and 3

The graphs Γ_n, $n \geq 1$, can be obtained by graph substitution as follows. The axiom is the graph Γ_1 and the rules are

$$u \xrightarrow{\;a\;} v \quad \mapsto \quad 1u \xrightarrow{\;a\;} 0u \underset{c}{\overset{b}{\rightleftharpoons}} 0v \xrightarrow{\;a\;} 1v$$

$$u \xrightarrow{\;b\;} v \quad \mapsto \quad 1u \xrightarrow{\;d\;} 1v$$

$$u \xrightarrow{\;c\;} v \quad \mapsto \quad 1u \xrightarrow{\;b\;} 1v$$

$$u \xrightarrow{\;d\;} v \quad \mapsto \quad 1u \xrightarrow{\;c\;} 1v$$

Another description of Γ_{n+1} in terms of Γ_n is as follows. Build two copies $\Gamma_{n,0}$ and $\Gamma_{n,1}$ of Γ_n by adding 0 and 1, respectively, on the right of each vertex label in Γ_n. If $n = 3m + 1$ delete the loops labeled by b and c at $1^{n-1}00$ and $1^{n-1}01$ in $\Gamma_{n,0}$ and $\Gamma_{n,1}$ and replace them by two edges labeled by b and c connecting $1^{n-1}00$ and $1^{n-1}01$. In a similar manner, if $n = 3m+2$ replace the loops labeled by b and d by two edges labeled by b and d connecting $1^{n-1}00$ and $1^{n-1}01$ and if $n = 3m$ do the same with the loops labeled by c and d.

Example 17 (The Schreier graphs of $\mathcal{B}$) The Schreier graph Γ_5 of $\mathcal{B}$ is given in Figure 21 (no loops are drawn on any vertex) The resemblance of the Schreier

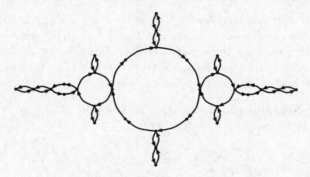

Figure 21. Schreier graphs of Basilica group $\mathcal{B}$ at level 5

graphs of $\mathcal{B}$ to the Julia set of the polynomial $z \mapsto z^2 - 1$ (see Figure 4) is due to the fact that $IMG(z \mapsto z^2 - 1) = \mathcal{B}$.

8.1 Contracting actions and limit spaces

The results in this subsection touch on the phenomenon that Schreier graphs of many self-similar groups exhibit self-similarity features. In particular, when the group in question happens to be the iterated monodromy group of a post-critically finite rational map f on $\hat{\mathbb{C}}$ then the Schreier graphs approximate the Julia set of f.

Definition 27 A self-similar group G, generated by a finite generating set S, is *contracting*, if there exists a constant $C \geq 0$ and integer $n \geq 1$ such that, for all elements g in G with $|g| \geq C$ and all their level n sections g_u, $u \in \mathcal{L}_n$,

$$|g_u| < |g|.$$

We note that all contracting groups are automaton groups. Indeed the above definition implies that the set of sections of every element in G is finite. Therefore one can easily define a finite automaton for each generator in S. Because of the self-similarity, the group generated by all the states in these automata is still just G.

We also note that the question whether a self-similar finitely generated group is contracting or not does not depend on the chosen generating set, i.e., being a contracting self-similar group is property of the group and not of its Cayley graph.

Denote by $X^{-\omega}$ the space of words over X that are infinite to the left. This space is canonically homeomorphic to $\partial\mathcal{T}$.

Definition 28 Let G be a finitely generated self-similar contracting group. Define a relation of *asymptotic equivalence* $\asymp$ on $X^{-\omega}$ by

$$\ldots x_3 x_2 x_1 \asymp \ldots y_3 y_2 y_1$$

if and only if there exists a sequence $\{g_n\}_{n=0}^{\infty}$ of elements in G taking only finitely many different values in G such that, for $n \geq 0$,

$$g_n(x_n \ldots x_1) = y_n \ldots y_1.$$

The *limit space* of G, denoted $\mathcal{J}_G$, is the space $X^{-\omega}/\asymp$.

The following proposition shows that the Schreier graphs $\Gamma_n(G, S)$ approximate the limit space $\mathcal{J}_G$.

Proposition 8.2 *Let G be a finitely generated contracting self-similar group. Then*

$$\ldots x_3 x_2 x_1 \asymp \ldots y_3 y_2 y_1$$

if and only if there exists a constant $C \geq 0$ such that, for all $n \geq 0$, the distance between $x_n \ldots x_1$ and $y_n \ldots y_1$ in $\Gamma_n(G, S)$ is no greater than C.

Recall that the Julia set of a post-critically finite rational map f on $\hat{\mathbb{C}}$ is the closure of the set of repelling cycles of f.

Theorem 8.3 (Nekrashevych) *Let f be a post-critically finite rational map on $\hat{\mathbb{C}}$. Then the action of $IMG(f)$ is contracting and the limit space $\mathcal{J}_{IMG(f)}$ is homeomorphic to the Julia set of f.*

Thus the similarities between Schreier graphs of some self-similar groups and Julia sets of some post-critically finite rational maps that we already observed (compare Figure 15 and Figure 19, as well as Figure 4 and Figure 21) is due to the fact that the group G defining the Schreier graphs is also the iterated monodromy group of the corresponding map.

8.2 Cayley and Schreier spectra

Recall that, given a graph Γ, one can define a *Markov operator* M acting on the Hilbert space $\ell^2(\Gamma, \deg)$ of square integrable functions with weight determined by the vertex degrees by

$$M f(x) = \frac{1}{\deg(x)} \sum_{y \sim x} f(y),$$

where the sum is taken over all neighbors of x. The Markov operator M is a self-adjoint operator of norm ≤ 1. If Γ is m-regular (i.e., all vertex degrees are equal to m) then M is a multiple of the *adjacency operator* (or matrix) usually used in discrete analysis. The operator M corresponds to the simple random walk on Γ in which the random walker is moving from a vertex x to any of its neighbors with equal probability. The *spectrum* of the graph Γ is the spectrum of the corresponding Markov operator M. To each vertex one can associate the spectral measure μ_v, whose moments coincide with the corresponding n-step return probabilities $p_{v,v}^n = \langle M^n \delta_v, \delta_v \rangle = \int_{-1}^{1} \lambda^n d\mu_v(\lambda)$, for $n = 0, 1, 2, \dots$. If Γ is vertex transitive then μ_v does not depend on v. This is the case, for example, when Γ is the Cayley graph of a group with respect to some finite system of generators. When we speak of a spectrum of a group we mean the spectrum of the Cayley graph with respect to some fixed finite symmetric system of generators. By the *Cayley spectrum* of an automaton group $G = G(\mathcal{A})$ we mean the spectrum of the Cayley graph of G with respect to the standard generating set $S \cup S^{-1}$, where S is the set of states of $\mathcal{A}$, and by the *Cayley spectral measure* $\mu_{\mathcal{A}}$ we mean the spectral measure of G.

It is quite a difficult problem to study the spectrum of non-abelian groups because of a lack of well developed theory of representations of such groups. Many fundamental problems of mathematics have interpretations that relate them to particular questions about spectra and spectral properties. For instance, an example of a torsion free group with a gap in the spectrum (for some system of generators) would provide a counterexample to the famous Kadison–Kaplansky Conjecture on Idempotents [59] (and consequently to Baum–Connes [23] and Novikov Conjecture [31]).

Groups generated by finite automata (or realizations of known groups by finite automata) lead to solution of difficult problems through methods and ideas based on self-similarity. The original ideas go back to the paper [4] where the first examples of regular graphs with Cantor spectra were given. The realization of the lamplighter group $L_2 = (\mathbb{Z}/2\mathbb{Z}) \wr \mathbb{Z}$ as automaton group was used in [51] to calculate the spectrum of L_2, which turned out to be the first example of a group with pure point spectrum. It also led to a counterexample to the Strong Atiyah Conjecture on L^2-Betti numbers.

Theorem 8.4 (Grigorchuk, Linell, Schick, Żuk [50]) *There exists a 7-dimensional manifold $\mathcal{M}$ such that all torsion elements in the fundamental group $G = \pi(\mathcal{M})$ have order dividing 2, but for which the second L^2-Betti number is $\frac{7}{3}$.*

Another way to attach a spectrum to a self-similar group G generated by a finite symmetric set S is as follows. Consider the Schreier graph $\Gamma_\xi = \Gamma(G, P_\xi, S)$. The graph Γ_ξ depends on ξ but in case of a spherically transitive action of G on $\mathcal{T}$ all graphs Γ_ξ are locally isomorphic and the spectrum (as a set) does not depend on ξ. In such a situation we call the spectrum of the graph Γ_ξ the *Schreier spectrum* of G and denote it by $\mathsf{spec}(\Gamma)$.

In some cases the Schreier spectrum coincides with the Cayley spectrum and such cases are of special interest since it is usually easier to calculate the Schreier spectrum. For instance this happens whenever P_ξ is trivial or cyclic (as happened for the lamplighter group [51]).

We note that the Schreier graphs Γ_ξ may be far from vertex transitive (and may even have trivial automorphism group as in the case of $\mathcal{G}$). Therefore one has to pay attention to the possibility of having spectral measures μ_v depending on v. There is hope that all μ_v would at least be in the same measure class, but there are no results in this direction. The so called *KNS (Kesten – von Neumann – Serre) spectral measure ν* appears in the study of μ_v, as introduced in [4] and studied further in different situations in [51, 53, 83].

Definition 29 Let G be a self-similar group of k-ary tree automorphisms generated by a symmetric set S and let ξ be a point on the boundary ∂T. For an interval I in $[-1, 1]$ define the *KNS spectral measure* by

$$\nu(I) = \lim_{n \to \infty} \frac{\#_n(I)}{k^n},$$

where $\#_n(I)$ denotes the number of eigenvalues of Γ_n in the interval I.

At the moment there are more complete calculations involving the Schreier spectrum than the Cayley spectrum. We provide the full description in case of $\mathcal{G}$, the Gupta–Sidki 3-group and $\mathcal{H}^{(3)}$. The calculations are based on the following result.

Theorem 8.5 (Bartholdi, Grigorchuk [4]) *Let either the Schreier graph $\Gamma = \Gamma(G, \xi, S)$ or the parabolic subgroup P_ξ be amenable. Then*

$$\mathsf{spec}(\Gamma) = \overline{\bigcup_{n=0}^{\infty} \mathsf{spec}(\Gamma_n)}.$$

Theorem 8.6 (Bartholdi, Grigorchuk [4])

(a) *The n-th level spectrum $\mathsf{spec}(\Gamma_n)$ of $\mathcal{G}$ with respect to $\{a, b, c, d\}$ is, as a set, equal to*

$$\left\{ 1 \pm \sqrt{5 + 4\cos\theta} \ \middle| \ \theta \in \frac{2\pi\mathbb{Z}}{2^n} \right\} \setminus \{0, -2\}.$$

The Schreier spectrum of $\mathcal{G}$ is equal to

$$[-2, 0] \cup [2, 4].$$

(b) *The Schreier spectrum of the Gupta–Sidki 3-group G with respect to $\{a, a^{-1}, b, b^{-1}\}$, given by*

$$a = (012)(1, 1, 1)$$
$$b = \qquad (a, a^{-1}, b)$$

is equal to the closure of the set

$$\left\{ \begin{array}{c} 4, \\ -2, \\ 1, \\ 1 \pm \sqrt{\frac{9 \pm 3}{2}}, \\ 1 \pm \sqrt{\frac{9 \pm \sqrt{45 \pm 4 \cdot 3}}{2}}, \\ 1 \pm \sqrt{\frac{9 \pm \sqrt{45 \pm 4 \cdot \sqrt{45 \pm 4 \cdot 3}}}{2}}, \\ \cdots \end{array} \right\}.$$

The spectrum is a Cantor set symmetric with respect to 1.

Theorem 8.7 (Grigorchuk, Šunić [48]) *The n-th level spectrum of $\mathcal{H}^{(3)}$ with respect to $\{a, b, c\}$, as a set, has $3 \cdot 2^{n-1} - 1$ elements and is equal to*

$$\{3\} \cup \bigcup_{i=0}^{n-1} f^{-i}(0) \cup \bigcup_{j=0}^{n-2} f^{-j}(-2),$$

where f is the polynomial function $f(x) = x^2 - x - 3$.

The multiplicity of the 2^i level n eigenvalues in $f^{-i}(0)$, $i = 0, \ldots, n-1$, is $(3^{n-1-i} + 3)/2$ and the multiplicity of the 2^j eigenvalues in $f^{-j}(-2)$, $j = 0, \ldots, n-2$, is $(3^{n-1-i} - 1)/2$.

The Schreier spectrum of $\mathcal{H}^{(3)}$, as a set, is equal to

$$\{3\} \cup \overline{\bigcup_{i=0}^{\infty} f^{-i}\{0, -2\}} = \overline{\bigcup_{i=0}^{\infty} f^{-i}(0)}.$$

It consists of the set of isolated points $I = \bigcup_{i=0}^{\infty} f^{-i}(0)$ and its set of accumulation points J, which is the Julia set of the polynomial $f(x) = x^2 - x - 3$ and is a Cantor set.

The KNS spectral measure is discrete and concentrated on the the set of eigenvalues in $\bigcup_{i=0}^{\infty} f^{-i}\{0, -2\}$. The KNS measure of the eigenvalues in $f^{-i}\{0, -2\}$ is $1/(6 \cdot 3^i)$, $i = 0, 1, \ldots$.

The spectra of Γ_n are calculated by using operator recursion induced by the self-similarity of the group in question. We illustrate this approach in the case of $\mathcal{H}^{(3)}$.

The action of $\mathcal{H}^{(3)}$ on level n of the ternary tree induces permutational representations of dimension 3^n, recursively defined by

$$a_0 = b_0 = c_0 = [1]$$

$$a_{n+1} = \begin{bmatrix} 0 & 1 & 0 \\ 1 & 0 & 0 \\ 0 & 0 & a_n \end{bmatrix}$$

$$b_{n+1} = \begin{bmatrix} 0 & 0 & 1 \\ 0 & b_n & 0 \\ 1 & 0 & 0 \end{bmatrix}$$

$$c_{n+1} = \begin{bmatrix} c_n & 0 & 0 \\ 0 & 0 & 1 \\ 0 & 1 & 0 \end{bmatrix},$$

where 0 and 1 are the zero and the identity matrix, respectively, of size $3^n \times 3^n$.

The matrix $\Delta_n = a_n + b_n + c_n$ is the adjacency matrix of Γ_n and it satisfies the recursive relation

$$\Delta_0 = [3]$$

$$\Delta_{n+1} = \begin{bmatrix} c_n & 1 & 1 \\ 1 & b_n & 1 \\ 1 & 1 & a_n \end{bmatrix}$$

For $n \geq 1$ and real numbers x and y define $\Delta_n(x,y)$ to be the $3^n \times 3^n$ matrix given by

$$\Delta_n(x,y) = \begin{bmatrix} c-x & y & y \\ y & b-x & y \\ y & y & a-x \end{bmatrix}.$$

Let $D_n(x,y) = \det(\Delta_n(x,y))$. We first find the set of points in the plane for which the matrix $D_n(x,y)$ is not invertible. Call this set the *auxiliary spectrum*. The spectrum we are interested in is the intersection of the auxiliary spectrum with the line $y = 1$.

We first determine a recursive formula for $D_n(x,y)$. The matrix

$$\Delta_n(x,y) = \begin{bmatrix} c-x & 0 & 0 & y & 0 & 0 & y & 0 & 0 \\ 0 & -x & 1 & 0 & y & 0 & 0 & y & 0 \\ 0 & 1 & -x & 0 & 0 & y & 0 & 0 & y \\ y & 0 & 0 & -x & 0 & 1 & y & 0 & 0 \\ 0 & y & 0 & 0 & b-x & 0 & 0 & y & 0 \\ 0 & 0 & y & 1 & 0 & -x & 0 & 0 & y \\ y & 0 & 0 & y & 0 & 0 & -x & 1 & 0 \\ 0 & y & 0 & 0 & y & 0 & 1 & -x & 0 \\ 0 & 0 & y & 0 & 0 & y & 0 & 0 & a-x \end{bmatrix}$$

is row/column equivalent to

$$\begin{bmatrix} c-x' & y' & y' & 0 & 0 & 0 & 0 & 0 & 0 \\ y' & b-x' & y' & 0 & 0 & 0 & 0 & 0 & 0 \\ y' & y' & a-x' & 0 & 0 & 0 & 0 & 0 & 0 \\ 0 & 0 & 0 & P_4 & 0 & 0 & 0 & 0 & 0 \\ 0 & 0 & 0 & 0 & P_5 & 0 & 0 & 0 & 0 \\ 0 & 0 & 0 & 0 & 0 & P_6 & 0 & 0 & 0 \\ 0 & 0 & 0 & 0 & 0 & 0 & y & 0 & 0 \\ 0 & 0 & 0 & 0 & 0 & 0 & 0 & 1 & 0 \\ 0 & 0 & 0 & 0 & 0 & 0 & 0 & 0 & 1 \end{bmatrix},$$

where

$$x' = \frac{x^4 - x^3 - x^2 + x + (x^2 - x^3)y + (2x - 3x^2)y^2 + xy^3 + 2y^4}{(x - 1 - y)(x^2 - 1 + y - y^2)},$$

$$y' = \frac{y^2(x - 1 + y)}{(x - 1 - y)(x^2 - 1 + y - y^2)}.$$

and

$$P_4 P_5 P_6 y = (x^2 - (1+y)^2)^{3^{n-2}} (x^2 - 1 + y - y^2)^{2 \cdot 3^{n-2}}.$$

Direct calculation shows that

$$D_1(x,y) = -(x - 1 - 2y)(x - 1 + y)^2$$

and, for $n \geq 2$,

$$D_n(x,y) = (x^2 - (1+y)^2)^{3^{n-2}} (x^2 - 1 + y - y^2)^{2 \cdot 3^{n-2}} D_{n-1}(F(x,y)),$$

where $F(x,y)$ is the two-dimensional rational map given by $F(x,y) = (x',y')$. Define a transformation $\Psi : \mathbb{R}^2 \to \mathbb{R}$ by

$$\Psi(x,y) = \frac{x^2 - 1 - xy - 2y^2}{y}$$

and a transformation $f : \mathbb{R} \to \mathbb{R}$ by

$$f(x) = x^2 - x - 3.$$

It is easy to check that

$$\Psi(F(x,y)) = f(\Psi(x,y)),$$

i.e., the diagram

$$\begin{array}{ccc} \mathbb{R}^2 & \xrightarrow{F} & \mathbb{R}^2 \\ \Psi \downarrow & & \downarrow \Psi \\ \mathbb{R} & \xrightarrow{f} & \mathbb{R} \end{array}$$

commutes. Thus the two-dimensional rational map F is semi-conjugate, in a non-trivial way, to the one-dimensional map f and this is precisely what makes all calculations possible.

For $n \geq 2$, the auxiliary spectrum consists of 3 lines and $3 \cdot 2^{n-2} - 2$ hyperbolas. For example, for $n = 5$ the auxiliary spectrum is given in Figure 22.

8.3   Problems

Problem 21 ([43])   (a) Does there exist an algorithm that, given a finite automaton $\mathcal{A}$, and a recursively defined ray ξ, decides if the parabolic subgroup P_ξ of $G(\mathcal{A})$ is trivial?

(b) Does there exist an algorithm that, given a finite automaton $\mathcal{A}$, decides if there exists a parabolic subgroup P_ξ in $G(\mathcal{A})$ that is trivial?

(c) Does there exist an algorithm that, given a finite automaton $\mathcal{A}$ and a recursively defined ray ξ, decides if Γ has polynomial growth?

Problem 22 Describe the possible types of growth of the Schreier graph Γ for automaton groups.

Problem 23 (Nekrashevych [71]) Are all contracting groups amenable?

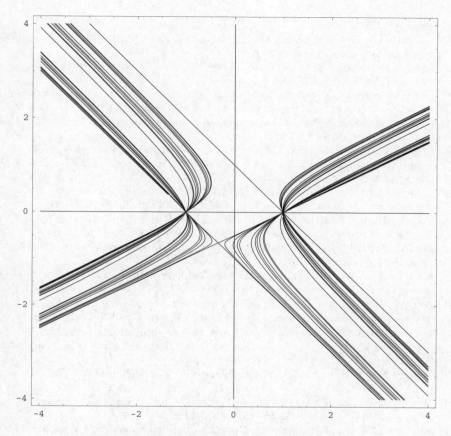

Figure 22. The auxiliary spectrum of Γ_5 for $\mathcal{H}^{(3)}$

Acknowledgments

We are thankful to Volodia Nekrashevych for the willingness to unselfishly share with us his great insight in the subject.

Thanks also to the Organizers of Groups St Andrews 2005, without whose hard work neither the conference nor these proceedings would have been possible.

References

[1] Miklós Abért, Group laws and free subgroups in topological groups, *Bull. London Math. Soc.* **37** (2005), no. 4, 525–534.

[2] Miklós Abért, Representing graphs by the non-commuting relation, *Publ. Math. Debrecen*, to appear, 2005.

[3] S. I. Adian, Random walks on free periodic groups, *Izv. Akad. Nauk SSSR Ser. Mat.* **46** (1982), no. 6, 1139–1149, 1343.

[4] L. Bartholdi and R. I. Grigorchuk, On the spectrum of Hecke type operators related to some fractal groups, *Tr. Mat. Inst. Steklova* **231** (2000), Din. Sist., Avtom. i Beskon.

Gruppy, 5–45.

[5] L. Bartholdi, Vadim Kaimanovich, V. Nekrashevych and Bálint Virág, Amenability of automata groups, preprint, 2005.

[6] Laurent Bartholdi, The growth of Grigorchuk's torsion group, *Internat. Math. Res. Notices* **1998**, no. 20, 1049–1054.

[7] Laurent Bartholdi, Lower bounds on the growth of a group acting on the binary rooted tree, *Internat. J. Algebra Comput.* **11** (2001), no. 1, 73–88.

[8] Laurent Bartholdi, A Wilson group of non-uniformly exponential growth, *C. R. Math. Acad. Sci. Paris* **336** (2003), no. 7, 549–554.

[9] Laurent Bartholdi, Rostislav Grigorchuk and Volodymyr Nekrashevych, From fractal groups to fractal sets, in *Fractals in Graz 2001*, 25–118, Trends Math., Birkhäuser, Basel, 2003.

[10] Laurent Bartholdi and Rostislav I. Grigorchuk, On parabolic subgroups and Hecke algebras of some fractal groups, *Serdica Math. J.* **28** (2002), no. 1, 47–90.

[11] Laurent Bartholdi, Rostislav I. Grigorchuk and Zoran Šunić, Branch groups, in *Handbook of algebra, Vol. 3*, 989–1112, North-Holland, Amsterdam, 2003.

[12] Laurent Bartholdi and Volodymyr Nekrashevych, Thurston equivalence of topological polynomials, preprint, 2005.

[13] Laurent Bartholdi and Said N. Sidki, The automorphism tower of groups acting on rooted trees, *Trans. Amer. Math. Soc.* **358** (2006), no. 1, 329–358 (electronic).

[14] Laurent Bartholdi and Zoran Šunić, On the word and period growth of some groups of tree automorphisms, *Comm. Algebra* **29** (2001), no. 11, 4923–4964.

[15] Laurent Bartholdi and Zoran Šunić, Some solvable automaton groups, in *Topological and Asymptotic Aspects of Group Theory*, 11–30, *Contemp. Math.* **394**, Amer. Math. Soc., Providence, RI, 2006.

[16] Laurent Bartholdi and Bálint Virág, Amenability via random walks, *Duke Math. J.*, to appear, 2003.

[17] Hyman Bass, The degree of polynomial growth of finitely generated nilpotent groups, *Proc. London Math. Soc. (3)* **25** (1972), 603–614.

[18] Kai-Uwe Bux and Rodrigo Perez, Growth of iterated monodromy groups, Contemporary Mathematics, to appear, 2006.

[19] T. Ceccherini-Silberstein, R. I. Grigorchuk and P. de la Harpe. Amenability and paradoxical decompositions for pseudogroups and discrete metric spaces, *Tr. Mat. Inst. Steklova* **224** (1999), *Algebra. Topol. Differ. Uravn. i ikh Prilozh.*, 68–111.

[20] Tullio G. Ceccherini-Silberstein and Rostislav I. Grigorchuk, Amenability and growth of one-relator groups, *Enseign. Math. (2)* **43** (1997), no. 3–4, 337–354.

[21] Bruce Chandler and Wilhelm Magnus, *The history of combinatorial group theory*, Studies in the History of Mathematics and Physical Sciences **9**, Springer-Verlag, New York, 1982.

[22] Ching Chou, Elementary amenable groups, *Illinois J. Math.* **24** (1980), no. 3, 396–407.

[23] Alain Connes, *Noncommutative geometry*, Academic Press Inc., San Diego, CA, 1994.

[24] Mahlon M. Day, Amenable semigroups, *Illinois J. Math.* **1** (1957), 509–544.

[25] Thomas Delzant and Rostislav Grigorchuk, Finiteness properites of branch groups, preprint, 2004.

[26] M. Edjvet and Stephen J. Pride, The concept of "largeness" in group theory, II, in *Groups—Korea 1983 (Kyoungju, 1983)*, Lecture Notes in Math. **1098**, 29–54. Springer, Berlin, 1984.

[27] Anna Erschler, Boundary behavior for groups of subexponential growth, *Ann. of Math. (2)* **160** (2004), no. 3, 1183–1210.

[28] Anna Erschler, Not residually finite groups of intermediate growth, commensurability and non-geometricity, *J. Algebra* **272** (2004), no. 1, 154–172.

[29] Alex Eskin, Shahar Mozes and Hee Oh, On uniform exponential growth for linear groups, *Invent. Math.* **160** (2005), no. 1, 1–30.

[30] Jacek Fabrykowski and Narain Gupta, On groups with sub-exponential growth functions, II, *J. Indian Math. Soc. (N.S.)* **56** (1991), no. 1–4, 217–228.

[31] Steven C. Ferry, Andrew Ranicki and Jonathan Rosenberg, A history and survey of the Novikov conjecture, in *Novikov conjectures, index theorems and rigidity, Vol. 1 (Oberwolfach, 1993)*, London Math. Soc. Lecture Note Ser. **226**, 7–66. Cambridge Univ. Press, Cambridge, 1995.

[32] Yair Glasner and Nicolas Monod, Amenable actions, free products and a fixed point property, preprint, 2006.

[33] Yair Glasner and Shahar Mozes, Automata and square complexes, preprint, 2003.

[34] Frederick P. Greenleaf, *Invariant means on topological groups and their applications.* Van Nostrand Mathematical Studies, No. 16, Van Nostrand Reinhold Co., New York, 1969.

[35] R. Grigorchuk and V. Nekrashevych, Amenable actions of nonamenable groups, *Zap. Nauchn. Sem. S.-Peterburg. Otdel. Mat. Inst. Steklov. (POMI)* **326**, Teor. Predst. Din. Sist. Komb. i Algoritm. Metody. 13) (2005), 85–96, 281.

[36] R. I. Grigorchuk, On Burnside's problem on periodic groups, *Funktsional. Anal. i Prilozhen.* **14** (1980), no. 1, 53–54.

[37] R. I. Grigorchuk, Symmetrical random walks on discrete groups, in *Multicomponent random systems*, Adv. Probab. Related Topics **6**, 285–325, Dekker, New York, 1980.

[38] R. I. Grigorchuk, On the Milnor problem of group growth, *Dokl. Akad. Nauk SSSR* **271** (183), no. 1, 30–33.

[39] R. I. Grigorchuk, Degrees of growth of finitely generated groups and the theory of invariant means, *Izv. Akad. Nauk SSSR Ser. Mat.* **48** (1984), no. 5, 939–985.

[40] R. I. Grigorchuk, Degrees of growth of p-groups and torsion-free groups, *Mat. Sb. (N.S.)* **126 (168)** (1985), no. 2, 194–214, 286.

[41] R. I. Grigorchuk, On the Hilbert-Poincaré series of graded algebras that are associated with groups, *Mat. Sb.* **180** (1989), no. 2, 207–225, 304.

[42] R. I. Grigorchuk and P. de lya Arp, One-relator groups of exponential growth have uniformly exponential growth, *Mat. Zametki* **69** (2001), no. 4, 628–630.

[43] R. I. Grigorchuk, V. V. Nekrashevich and V. I. Sushchanskiĭ, Automata, dynamical systems, and groups, *Tr. Mat. Inst. Steklova* **231** (2000), Din. Sist., Avtom. i Beskon. Gruppy, 134–214.

[44] R. I. Grigorchuk and S. N. Sidki, The group of automorphisms of a 3-generated 2-group of intermediate growth, *Internat. J. Algebra Comput.* **14** (2004), no. 5–6, 667–676.

[45] R. I. Grigorchuk and J. S. Wilson, The uniqueness of the actions of certain branch groups on rooted trees, *Geom. Dedicata* **100** (2003), 103–116.

[46] R. I. Grigorchuk and J. S. Wilson, A minimality property of certain branch groups, in *Groups: topological, combinatorial and arithmetic aspects*, London Math. Soc. Lecture Note Ser. **311**, 297–305, Cambridge Univ. Press, Cambridge, 2004.

[47] R.I. Grigorchuk, Just infinite branch groups, in Markus P. F. du Sautoy, Dan Segal and Aner Shalev, eds., *New horizons in pro-p groups*, 121–179, Birkhäuser Boston, Boston, MA, 2000.

[48] Rostislav Grigorchuk and Zoran Šunik, Hanoi towers groups, preprint, 2006.

[49] Rostislav I. Grigorchuk, On a problem of M. Day on nonelementary amenable groups in the class of finitely presented groups, *Mat. Zametki* **60** (1996), no. 5, 774–775.

[50] Rostislav I. Grigorchuk, Peter Linnell, Thomas Schick and Andrzej Żuk, On a question of Atiyah, *C. R. Acad. Sci. Paris Sér. I Math.* **331** (2000), no. 9, 663–668.

[51] Rostislav I. Grigorchuk and Andrzej Żuk, The lamplighter group as a group generated

by a 2-state automaton, and its spectrum, *Geom. Dedicata* **87** (2001), no. 1–3, 209–244.

[52] Rostislav I. Grigorchuk and Andrzej Żuk, On a torsion-free weakly branch group defined by a three state automaton, *Internat. J. Algebra Comput.* **12** (2002), no. 1–2, 223–246.

[53] Rostislav I. Grigorchuk and Andrzej Żuk, The Ihara zeta function of infinite graphs, the KNS spectral measure and integrable maps, in *Random walks and geometry*, 141–180, Walter de Gruyter GmbH & Co. KG, Berlin, 2004.

[54] Mikhael Gromov, Groups of polynomial growth and expanding maps, *Inst. Hautes Études Sci. Publ. Math.* **53** (1981), 53–73.

[55] Yves Guivarc'h, Croissance polynomiale et périodes des fonctions harmoniques, *Bull. Soc. Math. France* **101** (1973), 333–379.

[56] Narain D. Gupta and Said N. Sidki, On the Burnside problem for periodic groups, *Math. Z.* **182** (1983), no. 3, 385–388.

[57] Andreas M. Hinz, The Tower of Hanoi, *Enseign. Math. (2)* **35** (1989), no. 3–4, 289–321.

[58] Vadim Kaimanovich, "Münchausen trick" and amenability of self-similar groups, preprint, 2006.

[59] Irving Kaplansky, *Bialgebras*, Lecture Notes in Mathematics **218**, Department of Mathematics, University of Chicago, Chicago, Ill., 1975.

[60] Malik Koubi, Croissance uniforme dans les groupes hyperboliques, *Ann. Inst. Fourier (Grenoble)* **48** (1998), no. 5, 1441–1453.

[61] Yaroslav Lavreniuk and Volodymyr Nekrashevych, Rigidity of branch groups acting on rooted trees, *Geom. Dedicata* **89** (2002), 159–179.

[62] Yu. Leonov, On identities of groups of automorphisms of trees, *Visnyk of Kyiv State University of T.G. Shevchenko* **3** (1997), 37–44.

[63] Yu. G. Leonov, On a lower bound for the growth of a 3-generator 2-group, *Mat. Sb.* **192** (2001), no. 11, 77–92.

[64] Alexander Lubotzky and Avinoam Mann, Residually finite groups of finite rank, *Math. Proc. Cambridge Philos. Soc.* **106** (1989), no. 3, 385–388.

[65] Alexander Lubotzky, Avinoam Mann and Dan Segal, Finitely generated groups of polynomial subgroup growth, *Israel J. Math.* **82** (1993), no. 1–3, 363–371.

[66] Alexander Lubotzky, László Pyber and Aner Shalev, Discrete groups of slow subgroup growth, *Israel J. Math.* **96** (1996), part B, 399–418.

[67] John Milnor, Growth of finitely generated solvable groups, *J. Differential Geometry* **2** (1968), 447–449.

[68] John W. Milnor, Problem 5603, *Amer. Math. Monthly* **75** (1968), 685–686.

[69] Roman Muchnik and Igor Pak, On growth of Grigorchuk groups, *Internat. J. Algebra Comput.* **11** (2001), no. 1, 1–17.

[70] V. V. Nekrashevich, Iterated monodromy groups, *Dopov. Nats. Akad. Nauk Ukr. Mat. Prirodozn. Tekh. Nauki* **2003**, no. 4, 18–20.

[71] Volodymyr Nekrashevych, *Self-similar groups*, Mathematical Surveys and Monographs **117**, American Mathematical Society, Providence, RI, 2005.

[72] Peter M. Neumann, Some questions of Edjvet and Pride about infinite groups, *Illinois J. Math.* **30** (1986), no. 2, 301–316.

[73] A. Yu. Ol'šanskiĭ, On the question of the existence of an invariant mean on a group, *Uspekhi Mat. Nauk* **35** (1980), no. 4(214), 199–200.

[74] A. Yu. Ol'shanskii and M. V. Sapir, Non-amenable finitely presented torsion-by-cyclic groups, *Publ. Math. Inst. Hautes Études Sci.* **96** (2002), 43–169 (2003).

[75] A. Yu. Ol'shanskiĭ, Groups of bounded period with subgroups of prime order, *Algebra i Logika* **21** (1982), no. 5, 553–618.

[76] D. V. Osin, Elementary classes of groups, *Mat. Zametki* **72** (2002), no. 1, 84–93.

[77] D. V. Osin, The entropy of solvable groups, *Ergodic Theory Dynam. Systems* **23** (2003), no. 3, 907–918.

[78] Stephen J. Pride, The concept of "largeness" in group theory, in *Word problems, II (Conf. on Decision Problems in Algebra, Oxford, 1976)*, Stud. Logic Foundations Math. **95**, 299–335. North-Holland, Amsterdam, 1980.

[79] Dan Segal, The finite images of finitely generated groups, *Proc. London Math. Soc. (3)* **82** (2001), no. 3, 597–613.

[80] Said Sidki, On a 2-generated infinite 3-group: subgroups and automorphisms, *J. Algebra* **110** (1987), no. 1, 24–55.

[81] Said Sidki, Finite automata of polynomial growth do not generate a free group, *Geom. Dedicata* **108** (2004), 193–204.

[82] P. V. Silva and B. Steinberg, On a class of automata groups generalizing lamplighter groups, *Internat. J. Algebra Comput.*, to appear.

[83] Zoran Šunić, Hausdorff dimension of automaton groups, preprint, 2004.

[84] Jacques Tits, Free subgroups in linear groups, *J. Algebra* **20** (1972), 250–270.

[85] V. A. Ufnarovskiĭ, Criterion for the growth of graphs and algebras given by words, *Mat. Zametki* **31** (1982), no. 3, 465–472, 476.

[86] Eric K. van Douwen, Measures invariant under actions of F_2, *Topology Appl.* **34** (1990), no. 1, 53–68.

[87] John von Neumann, Zur allgemeinen Theorie des Masses, *Fund. Math.* **13** (1929), 73–116 and 333. = *Collected works*, Vol. I, pages 599–643.

[88] M. Vorobets and Y. Vorobets, On a free group of transformations defined by an automaton, *Geom. Dedicata*, to appear, arXiv math.GR/0601231, 2006.

[89] Stan Wagon, *The Banach-Tarski paradox*, Encyclopedia of Mathematics and its Applications **24**, Cambridge University Press, Cambridge, 1985.

[90] J. S. Wilson, Groups with every proper quotient finite, *Proc. Cambridge Philos. Soc.* **69** (1971), 373–391.

[91] John S. Wilson, Further groups that do not have uniformly exponential growth, *J. Algebra* **279** (2004), no. 1, 292–301.

[92] John S. Wilson, On exponential growth and uniformly exponential growth for groups, *Invent. Math.* **155** (2004), no. 2, 287–303.

[93] Joseph A. Wolf, Growth of finitely generated solvable groups and curvature of Riemanniann manifolds, *J. Differential Geometry* **2** (1968), 421–446.

ON SURFACE GROUPS: MOTIVATING EXAMPLES IN COMBINATORIAL GROUP THEORY

PETER ACKERMANN[*], BENJAMIN FINE[†] and GERHARD ROSENBERGER[§]

[*]Peter Ackermann, Fachbereich Mathematik, Universitat Dortmund, 44227 Dortmund, Federal Republic of Germany

[†]Department of Mathematics, Fairfield University, Fairfield, Connecticut 06430, United States
Email: `fine@mail.fairfield.edu`

[§]Fachbereich Mathematik, Universitat Dortmund, 44227 Dortmund, Federal Republic of Germany

Abstract

Surface groups, that is the fundamental groups of compact surfaces, have always played a principal role in the development of combinatorial group theory. It is well known that they have one-relator presentations with quadratic relators. This fact has in essence, from methods of Dehn, motivated the standard decision problems and has led to both small cancellation theory and the theory of hyperbolic groups. Surface groups have been crucial in motivating the concept of test elements and test words. Further surface groups are the prime examples of non-free elementary free groups — that is groups which share the same elementary theory as the class of free groups. We survey all these ideas and then talk about some new results on both tame automorphisms of elementary free groups and on what we call the surface group conjecture.

Contents

1 Introduction

Combinatorial Group Theory refers to the theory of group presentations, that is of groups specified by a set of generators and corresponding defining relations. The theory has its origins and roots in topology and complex analysis and in particular in the theory of the fundamental groups of combinatorial cell complexes. Indeed just as finite group theory deals with the finite groups necessary to study equations over fields and infinite continuous group theory (Lie group theory) deals with those groups necessary to study analysis, combinatorial group theory grew out of the need to study the infinite discrete groups necessary to understand the combinatorial objects in low dimensional topology — specifically originally surface groups. A good source for the history of the development of combinatorial group theory are the books by Chandler and Magnus [18], Collins and Zieschang [20] and Stillwell [119].

Because of its nature and origins, combinatorial group theory comes into contact with, and uses, many diverse areas of mathematics. Clearly algebra and topology as mentioned above, but also hyperbolic geometry via the study of the Cayley graph; pure mathematical logic through the study of various decision problems; and computer science through the study of computational aspects of group theory and through the study of rewriting systems all come into play. These last three areas have been of intense interest and the subject of intense study over the past decade. The ties with hyperbolic geometry have evolved into geometric group theory — especially the theory of the hyperbolic groups — by considering the groups themselves as geometric objects via the Cayley graph. The ties with computation and computer science have evolved into the theory of automatic groups where rewriting and computability properties are paramount. Finally there have been extensive studies on the nature of solutions of the various decision problems and on the logical foundation of the whole theory. Relative to the latter there has been a great deal of work on the Tarski Problem and its ties to purely group theoretical properties. Motivating many if not all of these areas are **surface groups**.

Recall that a **surface group** is the fundamental group of a compact orientable or non-orientable surface. If the genus of the surface is g then we say that the corresponding surface group also has genus g. It can be shown (see [41] and the references there) that an orientable surface group S_g of genus $g \geq 2$ has a one-relator presentation of the form

$$S_g = \langle a_1, b_1, \ldots, a_g, b_g \; ; \; [a_1, b_1] \ldots [a_g, b_g] = 1 \rangle \qquad (1.1)$$

while a non-orientable surface group T_g of genus $g \geq 2$ also has a one-relator presentation — now of the form

$$T_g = \langle a_1, a_2, \ldots, a_g \; ; \; a_1^2 a_2^2 \ldots a_g^2 = 1 \rangle. \qquad (1.2)$$

Much of combinatorial group theory arose originally out of the theory of one-relator groups and the concepts and ideas surrounding the Freiheitssatz or Independence Theorem of Magnus (see Section 2). Going backwards the ideas of the Freiheitssatz were motivated by the topological propeties of surface groups. The purpose of this paper is to examine how surface groups have motivated a great many of the areas of exploration in combinatorial group theory and infinite discrete group theory. This surface group motivation comes from the rich interplay surface groups provide among group theory, topology, hyperbolic geometry and computer science. From topology, a surface group inherits many of its properties from topological properties of the surface for which it is the fundamental group. This raises the questions of which of these properties are actually algebraic, that is dependent on the group theoretic structure and/or the presentational form and independent of the topology. Further surface groups also admit faithful Fuchsian representations in $PSL_2(\mathbb{R})$, that is they can be represented faithfully as discrete subgroups of $PSL_2(\mathbb{R})$. This has two consequences. First a surface group is linear and hence inherits all properties of linear groups. This raises questions, related to the one described above of when a group with a one-relator presentation is actually linear. Secondly a Fuchsian group describes through the upper half-plane model of hyperbolic geometry a discrete group of isometries of the hyperbolic plane. It follows that surface groups have many properties related to this geometric interpretation and as before it raises the purely group theoretic question of which of these properties can be deduced purely from the presentation. Further the method used to determine if an element of a surface group can be trivial has led to small cancellation theory. This in recent years has been closely tied to computer science via the concept of an automatic group. We will also discuss some new results relating surface groups to the elementary theory of groups and in particular the solution of the Tarski problem. We will use this relationships to motivate a generalization of a result on tame automorphism of surface groups to general fully residually free groups (see Section 6) and to make some progress towards what we call the **surface group conjecture**.

The outline of this paper is as follows. In Section 2 we consider the subgroup structure of surface groups and show how this structure has been extended and generalized via the Freiheitssatz and group amalgams. In Section 3 we examine the n-free property. This is the property that any n-generator subgroup of a group for some n must be a free group. 1-free is of course torsion-free. This property holds in surface groups. A surface group of genus g is $(2g - 1)$-free in the orientable case and $(g - 1)$-free in the non-orientable case. In Section 4 we look at finite extensions of surface groups and discuss the Nielsen Realization Problem. In Section 5 we look at the ties between surface groups and the elementary theory of free groups. From the solution of the Tarski Problem by Kharlampovich and Myasnikov [68] and independently by Sela all nonabelian free groups have the same elementary theory. The proofs of this result lead to a characterization of the elementary free groups — that is groups which share the same elementary theory as the class of free groups. The surface groups provide the simplest example of non-free elementary free groups. In Section 6 we look at how this tie to elementary theory can be used to generalize a result on tame automorphism of surface groups. We further use this to make some progress on what we call the surface group conjecture. This is the following: If G is a non-free one-relator group with the property that every subgroup of finite index is a one-relator group and every subgroup of infinite index is free, must G be a surface group? In Section 7 we look at small cancellation theory and how it was motivated by the solution of the word problem in surface groups. In Section 8 we examine test elements in free groups and surface groups and their relation to the surface group relator. Finally in Section 9 we briefly show how the theory of surface groups, as we examined it, fits into a larger program generalizing algebraic properties of discrete groups (see [41]).

2 Subgroup Structure and the Freiheitssatz

Suppose that G is an orientable surface group of genus g, that is $G \cong S_g$. By abelianizing the one-relator presentation (1.1) it is clear that the rank (minimum number of necessary generators) of G is $2g$. Now suppose that H is a proper subgroup of G. It follows then from covering space theory that $H = \pi_1(S)$ where S is a cover of a compact orientable surface of genus g. If $|G : H| < \infty$, then S must be homotopic to another orientable surface of genus $g_1 \geq g$ and hence $H = S_{g_1}$. If H has infinite index in S_g then homotopically S is a wedge of circles and H is a free group. An analogous argument works in the non-orientable case. We thus have the following theorem which completely describes the subgroups of surface groups.

Theorem 2.1 *Let G be a surface group. Then any subgroup of finite index is again a surface group (of higher or equal genus) and any subgroup of infinite index is free.*

This result in its basic form was probably known to Poincaré and definitely known to Dehn. However it was not proved algebraically until 1971–1972 by Hoare, Karrass and Solitar [53, 54] using Reidemeister–Schreier rewriting. They actually proved a stronger result for the class of F-groups. Recall that a **Fuchsian group** is a non-elementary, discrete subgroup (and thus discontinuous) of $PSL_2(\mathbb{R})$ or a conjugate of such a group in $PSL_2(\mathbb{C})$. Since the real axis can be mapped on any given circle by a $T \in PSL_2(\mathbb{C})$, we can equivalently define a Fuchsian group as a non-elementary, discontinuous subgroup of $PSL_2(\mathbb{C})$ which fixes a circle C and maps the interior of C on itself. C would be called the **fixed circle** of the group. A subgroup G of $PSL_2(\mathbb{C})$ is called **elementary** if for $g, h \in G$ with infinite order their commutator $[g, h]$ has trace two, that is $\text{tr}([g, h]) = 2$, or equivalently, if any two elements of infinite order in G have a common fixed point (considered as linear fractional transformations). A finitely generated Fuchsian group always admits a presentation of the following form

$$G = \langle p_1, \ldots, p_t, e_1, \ldots, e_s, a_1, b_1, \ldots, a_g, b_g \; ; \; e_1^{m_1} = \cdots = e_s^{m_s} = R = 1 \rangle \quad (2.1)$$

where

$$R = p_1 \ldots p_t e_1 \ldots e_s [a_1, b_1] \ldots [a_g, b_g].$$

This presentation is called a **Poincaré presentation for G**.

The sequence $(g; m_1, \ldots, m_s; t)$ is called the **signature** of G. The real number

$$\mu(G) = 2\pi \left(2g - 2 + t + \sum_{i=1}^{s} (1 - 1/m_i) \right) \tag{2.2}$$

measures the hyperbolic area of a fundamental domain for G in the hyperbolic plane.

A group G with a presentation of the form (2.1) is called an **F-group** and from a theorem of Poincaré an F-group with signature $(g; m_1, \ldots, m_s; t)$ with $m_i \geq 2$ is a presentation of a non-elementary Fuchsian group with that signature only if

$$2\pi \left(2g - 2 + t + \sum_{i=1}^{s} (1 - 1/m_i) \right) > 0.$$

The integer g is called the genus of G. In particular if $t = s = 0$ then G is a surface group and hence surface groups admit faithful representations as Fuchsian groups.

What Hoare, Karrass and Solitar actually proved was Theorem 2.1 relative to F-groups.

Theorem 2.2 ([53, 54]) *Let G be an F-group with $\mu(G) > 0$. Then any subgroup of finite index is again an F-group of the same form with higher or equal genus and any subgroup of infinite index is a free product of cyclics.*

This theorem can be generalized in a number of ways, all of which are part of the general question of what group theoretical properties of surface groups can be deduced solely from a group having a presentation similar to the standard surface group presentation.

Historically the beginning point for questions of this type was the Freiheitssatz or Independence Theorem of Magnus. If G is an orientable surface group of genus g then its rank is $2g$. From Theorem 2.1 if H is a subgroup of rank $< 2g$ it cannot be of finite index. It follows that H must be a free group. If G has the standard presentation (1.1) this implies that any proper subset of the generators generates a free group. According to Magnus [18] this fact was known to Dehn who presented to Magnus the problem of proving the general result for one-relator groups. This is the basis for the Freiheitssatz.

Theorem 2.3 (Freiheitssatz) *Let $G = \langle x_1, \ldots, x_n \; ; \; R = 1 \rangle$ where R is a cyclically reduced word which involves all the generators. Then the subgroup generated by $x_1, \ldots, x_{n-1}$ is free on these generators.*

In proving the Freiheitssatz, Magnus developed a general method, now called the Magnus Method, to handle one-relator groups. This involved using group amalgams coupled with induction on the length of the relator. These techniques have become standard in combinatorial group theory (see [81], [79], [5], [41]).

The Freiheitssatz itself has been generalized in many directions. In a more general context the Freiheitssatz can be described as follows. Let X, Y be disjoint sets of generators and suppose that the group A has the presentation $A = \langle X \; ; \mathrm{Rel}(X) \rangle$ and that the group G has the presentation

$$G = \langle X, Y \; ; \mathrm{Rel}(X), \mathrm{Rel}(X, Y) \rangle.$$

Then we say that G satisfies a **Freiheitssatz** which we abbreviate by **FHS** {relative to A} if $\langle X \rangle_G \cong A$. In other words the subgroup of G generated by X is isomorphic to A. In simpler language this says that the complete set of relations on X in G is the "obvious" one from the presentation of G. An alternative way to look at this is that A injects into G under the obvious map taking X to X. In this language Magnus' original FHS can be phrased as a one-relator group satisfies a FHS relative to the free group on any proper subset of the generators. In the setting above we say that the group A is a **FHS factor** of G.

From this point of view, for any group amalgam — free product with amalgamation or HNN groups — an amalgam factor is a FHS factor. Thus any factor in a free product with amalgamation and the base in an HNN group embed as a FHS factors in the resulting groups. This becomes the basic idea in Magnus' method. The method is to embed the group into an amalgam in such a way that the proposed FHS factor embeds into an amalgam factor which in turn contains the proposed FHS factor as a FHS factor. The result can then be obtained by applying the FHS for amalgams.

The most extensive work on extending the Freiheitssatz has been to **one-relator products**. Let $\{A_i\}$, i in some index set I, be a family of groups. Then a **one-relator product** is the quotient, $G = A/N(R)$, of the free product $A = *_i A_i$ by the normal closure $N(R)$ of a single non-trivial word R in the free product. We assume that R is cyclically reduced and of syllable length at least two. The groups A_i are called the **factors** while R is the **relator**. In analogy with the one-relator group case we say R **involves** A_i if R has a non-trivial syllable from A_i. If $R = S^m$ with S a non-trivial cyclically reduced word in the free product and $m \geq 2$, then R is a **proper power**. We then also call S a relator.

In this context a one-relator group is just a one-relator product of free groups. From the Freiheitssatz a one-relator group with at least two generators in the given presentation is never trivial. On the other hand a one-relator product of non-trivial groups may completely collapse. For example, consider $A = \langle a \rangle$ and $B = \langle b \rangle$ to be finite cyclic groups of relatively prime order. Then the one-relator product $G = (A * B)/N(ab)$ is a trivial group. Because of examples such as this, a natural question to ask is under what conditions the factors actually inject into a one-relator product. We say that a **Freiheitssatz** holds for a one-relator product G if each factor injects into G via the identity map. In the general framework described above a Freiheitssatz holds for a one-relator product if each factor is a FHS factor.

As with one-relator groups, the starting off point for a study of one-relator products is to determine a Freiheitssatz. The example above shows that there is no such result in general and therefore some restrictions must be imposed. There are two approaches. The first is to impose conditions on the factors while the second is to impose conditions on the relator.

A group H is **locally indicable** if every nontrivial finitely generated subgroup has an infinite cyclic quotient. Local indicability of the factors is a strong enough condition to allow most of the results on one-relator groups to be carried over to one-relator products. In particular it is clear from the subgroup theorem on surface groups that surface groups are locally indicable.

The following theorem was discovered independently by Brodskii [13, 14], J. Howie [55] and H. Short [115]. It is interesting that all three proofs are entirely different. Another proof mimicking Magnus's original proof was given by B. Baumslag [2].

Theorem 2.4 *A one-relator product of locally indicable factors satisfies a Freiheitssatz. That is, if $G = *_i A_i/N(R)$, where each A_i is locally indicable, and R is a cyclically reduced word in the free product $*_i A_i$ of syllable length at least two, then each A_i injects into G under the identity map, i.e., is a FHS factor.*

It has been conjectured that the Freiheitssatz holds for one-relator products of torsion-free factors.

The second approach is to impose restrictions on the relator. The most common relator condition is that R is a proper power of suitably high order, that is $R = S^m$ with $m \geq 2$. We will discuss this approach in Section 5.3. If $m \geq 7$ then the relator satisfies the small cancellation condition $C'(1/6)$ and a FHS can be deduced from small cancellation theory (see [41]). A FHS does hold in the cases $m = 4, 5, 6$ ($m = 6$ due to Gonzalez-Acuna and Short [46], $m = 4, 5$ due to Howie [57, 58]) but the proofs are tremendously difficult. The cases $m = 2, 3$ are still open in general although specific cases where a FHS does hold have been proved. In particular if the factors admit representations into a suitable linear group, a FHS can be given. The technique for handling these proper power situations is combinatorial geometric and closely tied to small cancellation diagrams. A complete treatement of the geometric techniques can be found in the excellent survey articles by Howie [56] and Duncan and Howie [24] as well as the original papers. We summarize the results.

Theorem 2.5 ([57, 58], [25]) *Suppose $G = (* A_i)/N(S^m)$ is a one-relator product where S is a cyclically reduced word in the free product $(* A_i)$ of syllable length at least 2 and suppose $m \geq 4$. Then the FHS holds, that is, each factor A_i naturally injects into G. Further if $m = 3$ and the relator S contains no letters of order 2, then the FHS holds.*

Theorem 2.6 ([35]) *Suppose $G = (* A_i)/N(S^m)$ is a one-relator product where S is a cyclically reduced word in the free product $(* A_i)$ of syllable length at least 2 and suppose $m \geq 2$. Then, if each A_i admits a faithful representation into $PSL_2(\mathbb{C})$ the FHS holds, that is, each factor A_i naturally injects into G.*

One-relator products and the Freiheitssatz are closely tied to the **solution of equations over groups**, also called the **adjunction problem**. Basically the adjunction problem is the following. Let G be a group and $W(x_1, \ldots, x_n)$ be an equation (or system of equations) with coefficients in G. This question is whether or not this can be solved in some overgroup of G. In the case of a single equation

$W(x) = 1$ in a single unknown x a result of F. Levin [74] shows that there exists a solution over G if x appears in W only with positive exponents (see [41]).

A complete discussion of many of the aspects of the Freiheitssatz can be found in [40] and [41].

Since a surface group admits a faithful representation as a Fuchsian group it is of course linear. The basic properties of surface groups and especially their linearity properties are best generalized via the concept of a cyclically pinched one-relator group. Consider the standard one-relator presentation (1.1) for an orientable surface group S_g of genus $g \geq 2$:

$$S_g = \langle a_1, b_1, \ldots, a_g, b_g \; ; [a_1, b_1] \ldots [a_g, b_g] = 1 \rangle.$$

If we let $U = [a_1, b_1] \ldots [a_{g-1}, b_{g-1}]$, $V = [a_g, b_g]$ then S_g decomposes as the free product of the free groups F_1 on $a_1, b_1, \ldots, a_{g_1}, b_{g-1}$ and F_2 on a_g, b_g amalgamated over the maximal cyclic subgroups generated by U in F_1 and V in F_2. Hence $S_g = F_1 \star_{\langle U \rangle = \langle V^{-1} \rangle} F_2$.

The algebraic generalization of this construction leads to **cyclically pinched one-relator groups**. These groups have the general form of a surface group and have proved to be quite amenable to study. In particular a **cyclically pinched one-relator group** is a one-relator group of the following form

$$G = \langle a_1, \ldots, a_p, a_{p+1}, \ldots, a_n \; ; U = V \rangle \qquad (2.3)$$

where $1 \neq U = U(a_1, \ldots, a_p)$ is a cyclically reduced, non-primitive (not part of a free basis) word in the free group F_1 on $a_1, \ldots, a_p$ and $1 \neq V = V(a_{p+1}, \ldots, a_n)$ is a cyclically reduced, non-primitive word in the free group F_2 on $a_{p+1}, \ldots, a_n$.

Clearly such a group is the free product of the free groups on $a_1, \ldots, a_p$ and $a_{p+1}, \ldots, a_n$ respectively amalgamated over the cyclic subgroups generated by U and V.

Cyclically pinched one-relator groups have been shown to be extremely similar to surface groups. G. Baumslag [4] has shown that such a group is residually finite. A group G is **conjugacy separable** if for given elements g, h in G either g is conjugate to h or there exists a finite quotient where they are not conjugate. J. Dyer [27] has proved the conjugacy separability of cyclically pinched one-relator groups. Note that conjugacy separability in turn implies residual finiteness. S. Lipschutz proved that cyclically pinched one-relator groups have solvable conjugacy problem [75].

A group G is **subgroup separable** or **LERF** if given any finitely generated subgroup H of G and an element $g \in G$, with $g \notin H$, then there exists a finite quotient G^* of G such that image of g lies outside the image of H. P. Scott [106] proved that surface groups are subgroup separable and then Brunner, Burns and Solitar [16] showed that in general, cyclically pinched one-relator groups are subgroup separable. Cyclically pinched one-relator groups with neither U nor V (in the presentation (2.1)) proper powers were shown to be linear by Wehrfritz [123] while using a result of Shalen [114], Fine and Rosenberger (see [41]) showed that any cyclically pinched one-relator group with neither U nor V proper powers has a faithful representation in $PSL_2(\mathbb{C})$. Again note the similarity to surface groups.

Along these same lines where neither U nor V is a proper power Bestvinna and Feign [10], Juhasz and Rosenberger [60] and Kharlampovich and Myasnikov [66] all independently have proved that cyclically pinched one-relator groups are hyperbolic. Juhasz and Rosenberger [60] extended this result to the case that at most one of U or V is a proper power and also to more general groups of F-type. We summarize all these results.

Theorem 2.7 *Let G be a cyclically pinched one-relator group. Then*
 (1) *G is residually finite [3].*
 (2) *G has a solvable conjugacy problem [75] and is conjugacy separable [27].*
 (3) *G is subgroup separable [16].*
 (4) *If neither U nor V is a proper power then G has a faithful representation over some commutative field [123].*
 (5) *If neither U nor V is a proper power then G has a faithful representation in $PSL_2(\mathbb{C})$ [41].*
 (6) *If neither U nor V is a proper power then G is hyperbolic ([10], [60], [66])*

Rosenberger [101], using Nielsen cancellation, has given a positive solution to the isomorphism problem for cyclically pinched one-relator groups, that is, he has given an algorithm to determine if an arbitrary one-relator group is isomorphic or not to a given cyclically pinched one-relator group.

Theorem 2.8 *The isomorphism problem for any cyclically pinched one-relator group is solvable; given a cyclically pinched one-relator group G there is an algorithm to decide in finitely many steps whether an arbitrary one-relator group is isomorphic or not to G.*

More specifically let G be a non-free cyclically pinched one-relator group with presentation (2.3) such that at most one of U and V is a power of a primitive element in F_1 respectively F_2. Suppose that $x_1, \ldots, x_{p+q}$ is a generating system for G. Then one of the following two cases occurs:
 (1) *There is a Nielsen transformation from $\{x_1, \ldots, x_{p+q}\}$ to a system $\{a_1, \ldots, a_p, y_1, \ldots, y_q\}$ with $y_1, \ldots, y_q \in F_2$ and $F_2 = \langle V, y_1, \ldots, y_q \rangle$.*
 (2) *There is a Nielsen transformation from $\{x_1, \ldots, x_{p+q}\}$ to a system $\{y_1, \ldots, y_p, b_1, \ldots, b_q\}$ with $y_1, \ldots, y_p \in F_1$ and $F_1 = \langle U, y_1, \ldots, y_p \rangle$.*
For $x_1, \ldots, x_{p+q}$ there is a presentation of G with one-relator. Further G has only finitely many Nielsen equivalence classes of minimal generating systems.

The HNN analogs of cyclically pinched one-relator groups are called **conjugacy pinched one-relator groups** and are also motivated by the structure of orientable surface groups. In particular suppose

$$S_g = \langle a_1, b_1, \ldots, a_g, b_g \, ; \, [a_1, b_1] \ldots [a_g, b_g] = 1 \rangle.$$

Let $b_g = t$ then S_g is an HNN group of the form

$$S_g = \langle a_1, b_1, \ldots, a_g, t \, ; \, tUt^{-1} = V \rangle.$$

where $U = a_g$ and $V = [a_1, b_1] \ldots [a_{g-1}, b_{g-1}]a_g$. Generalizing this we say that a **conjugacy pinched one-relator group** is a one-relator group of the form

$$G = \langle a_1, \ldots, a_n, t \; ; \; tUt^{-1} = V \rangle$$

where $1 \neq U = U(a_1, \ldots, a_n)$ and $1 \neq V = V(a_{p+1}, \ldots, a_n)$ are cyclically reduced in the free group F on $a_1, \ldots, a_n$.

Structurally such a group is an HNN extension of the free group F on $a_1, \ldots, a_n$ with cyclic associated subgroups generated by U and V and is hence the HNN analog of a cyclically pinched one-relator group.

Groups of this type arise in many different contexts and share many of the general properties of the cyclically pinched case. However many of the proofs become tremendously more complicated in the conjugacy pinched case than in the cyclically pinched case. Further in most cases additional conditions on the associated elements U and V are necessary. To illustrate this we state a result ([39], see [41]) which gives a partial solution to the isomorphism problem for conjugacy pinched one-relator groups.

Theorem 2.9 *Let $G = \langle a_1, \ldots, a_n, t \; ; \; tUt^{-1} = V \rangle$ be a conjugacy pinched one-relator group and suppose that neither U nor V is a proper power in the free group on $a_1, \ldots, a_n$. Suppose further that there is no Nielsen transformation from $\{a_1, \ldots, a_n\}$ to a system $\{b_1, \ldots, b_n\}$ with $U \in \{b_1, \ldots, b_{n-1}\}$ and that there is no Nielsen transformation from $\{a_1, \ldots, a_n\}$ to a system $\{c_1, \ldots, c_n\}$ with $V \in \{c_1, \ldots, c_{n-1}\}$. Then:*

(1) *G has rank $n + 1$ and for any minimal generating system for G there is a one-relator presentation.*

(2) *The isomorphism problem for G is solvable, that is, it can be decided algorithmically in finitely many steps whether an arbitrary given one-relator group is isomorphic to G.*

(3) *G is Hopfian.*

For more results on cyclically pinched one-relator groups and conjugacy pinched one-relator groups see [41] and [44].

3   The n-Free Property

If G is an orientable surface group of genus $g \geq 2$ then it is clear from the topological arguments given in the last section that any subgroup H of G generated by less than or equal $2g - 1$ elements must be a free group. Similarly if G is a nonorientable surface group of genus $g \geq 2$ then any subgroup generated by less than or equal to $g - 1$ elements must be free. Motivated by surface groups we generalize this property.

A group G is **n-free** for a positive integer n if any subgroup of G generated by n or fewer elements must be a free group. G is ω-free or **locally free** if it is n-free for every n. In this language we would say that any orientable surface group of genus $g \geq 2$ is $(2g - 1)$-free while a non-orientable surface group of genus $g \geq 2$

is $(g-1)$-free. G. Baumslag [3] first generalized this to certain cyclically pinched one-relator groups.

Theorem 3.1 ([3]) *Let G be a cyclically pinched one-relator group with the property that U and V are not proper powers in the respective free groups on the generators which they involve. Then G is 2-free.*

Using Nielsen and r-stable Nielsen reduction in free products with amalgamation (see the article [43] for terminology), G. Rosenberger [99] was then able to give a complete classification of the subgroups of rank ≤ 4 of such cyclically pinched one-relator groups.

Theorem 3.2 ([99]) *Let G be a cyclically pinched one-relator group with the property that U and V are not proper powers in the respective free groups on the generators which they involve. Then*

(1) *G is 3-free.*

(2) *Let $H \subset G$ be a subgroup of rank 4. Then one of the following two cases occurs:*

   (i) *H is free of rank 4.*

   (ii) *If $\{x_1, \ldots, x_4\}$ is a generating system for H then there is a Nielsen transformation from $\{x_1, \ldots, x_4\}$ to $\{y_1, \ldots, y_4\}$ with $y_1, y_2 \in zF_1z^{-1}$, $y_3, y_4 \in zF_2z^{-1}$ for a suitable $z \in G$. Further there is a one-relator presentation for H on $\{x_1, \ldots, x_4\}$.*

We note that the 3-free part of the above theorem was reproven in a different manner by G. Baumslag and P. Shalen [8].

In conjunction with a study on the universal theory of non-abelian free groups (see Section 5) the freeness part of the above results was extended in the following manner by Fine, Gaglione, Rosenberger and Spellman [34], again using Nielsen reduction techniques.

Theorem 3.3 ([34]) *Let $B_1, \ldots, B_n$ with $n \geq 2$ be pairwise disjoint sets of generators, each of size ≥ 2 and for $i = 1, \ldots, n$ let $W_i = W_i(B_i)$ be a non-trivial word in the free group on B_i, neither a proper power nor a primitive element. Let*

$$G = \langle B_1, \ldots, B_n \; ; \; W_1 W_2 \ldots W_n = 1 \rangle.$$

Then G is n-free.

A similar result can be obtained if the words W_i are proper powers.

Theorem 3.4 ([34]) *Let $B_1, \ldots, B_n$ with $n \geq 2$ be pairwise disjoint non-empty sets of generators, and for $i = 1, \ldots, n$ let $W_i = W_i(B_i)$ be a non-trivial word in the free group on B_i. Let*

$$G = \langle B_1, \ldots, B_n \; ; \; W_1^{t_1} W_2^{t_1} \ldots W_n^{t_n} = 1 \rangle$$

with $t_i \geq 1$. Then G is $(n-1)$-free.

This result is the best possible since a non-orientable surface group of genus $g \geq 2$ is $(g-1)$-free but not g-free.

Motivated by work of Alperin and Bass on group actions on Λ-trees (see [41]), the HNN analogs of these results were obtained. The above 2-free and 3-free results for cyclically pinched one-relator groups do carry over with modifications to conjugacy pinched one-relator groups. The results for cyclically pinched one-relator groups used Nielsen reduction in free products with amalgamation as developed by Zieschang [126], Collins and Zieschang [20], Rosenberger [94, 95, 96, 97] and others (see [43]). The corresponding theory of Nielsen reduction for HNN groups was developed by Peczynski and Reiwer [88] and is used in the analysis of conjugacy pinched one-relator groups. Important for applications of Peczynski and Reiwer's results is the case where the associated subgroups are malnormal in the base. Recall that $H \subset G$ is malnormal if $xHx^{-1} \cap H = \{1\}$ if $x \notin H$. For a cyclic subgroup $\langle U \rangle$ of a free group F this requires that U is not a proper power in F. Using this, Fine, Roehl and Rosenberger proved the following 2-free result.

Theorem 3.5 ([38]) *Let $G = \langle a_1, \ldots, a_n, t \; ; \; tUt^{-1} = V \rangle$ be a conjugacy pinched one-relator group. Suppose that neither U nor V are proper powers in the free group on $a_1, \ldots, a_n$. If $\langle x, y \rangle$ is a two-generator subgroup of G then one of the following holds:*

(1) $\langle x, y \rangle$ is free of rank two;

(2) $\langle x, y \rangle$ is abelian;

(3) $\langle x, y \rangle$ has a presentation $\langle a, b \; ; \; aba^{-1} = b^{-1} \rangle$.

As a direct consequence of the proof we also obtain the following.

Corollary 3.1 *Let G be as in Theorem 3.5 and suppose that U is not conjugate to V^{-1} in the free group on $a_1, \ldots, a_n$. Then any two-generator subgroup of G is either free or abelian.*

The extension of these theorems to a 3-free result proved to be quite difficult and required some further modifications. A two-generator subgroup N of a group G is **maximal** if rank $N = 2$ and if $N \subset M$ for another two-generator subgroup M of G then $N = M$. A maximal two-generator subgroup $N = \langle U, V \rangle$ is **strongly maximal** if for each $X \in G$ there is a $Y \in G$ such that $\langle U, XVX^{-1} \rangle \subset \langle U, YVY^{-1} \rangle$ and $\langle U, YVY^{-1} \rangle$ is maximal. Building upon and extending the theory of Peczynski and Reiwer the following is obtained.

Theorem 3.6 ([39]) *Let $G = \langle a_1, \ldots, a_n, t \; ; \; tUt^{-1} = V \rangle$ be a conjugacy pinched one-relator group. Suppose that $\langle U, V \rangle$ is a strongly maximal subgroup of the free group on $a_1, \ldots, a_n$. Then G is 3-free.*

If $\langle U, V \rangle$ is not strongly maximal we can further obtain that a subgroup of rank 3 is either free or has a one-relator presentation.

Theorem 3.7 ([39]) *Let $G = \langle a_1, \ldots, a_n, t \; ; \; tUt^{-1} = V \rangle$ be a conjugacy pinched one-relator group. Suppose that neither U nor V is a proper power in the free group* on $a_1, \ldots, a_n$ and that in this free group U is not conjugate to either V or V^{-1}. Let $H = \langle x_1, x_2, x_3 \rangle \subset G$. Then H is free or has a one-relator presentation on $\langle x_1, x_2, x_3 \rangle$.

The n-free property has played a role in the elementary theory of groups. We mention one striking result that was used in the classification of fully residually free groups of low rank [33]. Recall that a group G is **residually free** if for each non-trivial $g \in G$ there is a free group F_g and an epimorphism $h_g : G \to F_g$ such that $h_g(g) \neq 1$ and is **fully residually free** provided to every finite set $S \subset G \setminus \{1\}$ of non-trivial elements of G there is a free group F_S and an epimorphism $h_S : G \to F_S$ such that $h_S(g) \neq 1$ for all $g \in S$. Finitely generated fully residually free groups are also known as **limit groups**. In this guise they were studied by Sela (see [107, 108, 109, 110, 111, 112, 113] and [11]) in terms of studying homomorphisms of general groups into free groups. We will say more about this in the next section.

Theorem 3.8 ([33]) *Let G be a 2-free fully residually free group. Then G is 3-free.*

4 The Extension Property

We consider finite torsion-free extensions of surface groups. This was motivated originally not by surface groups but by free groups. However this question as related to surface groups spurred a tremendous amount of research surrounding the Nielsen Realization Problem (see below).

Recall that if P is a group property, then a group G is **virtually P** if G has a subgroup of finite index satisfying P. G is **P-by-finite** if G has a normal subgroup of finite index satisfying P. If P is a subgroup inherited property, such as torsion-freeness, freeness, or solvability then virtually P and P-by-finite are equivalent.

The structure of virtually-free groups {free-by-finite groups} is well understood and generalizes the structure of free groups in expected ways. For virtually free groups there is a very detailed structure theorem due to Karrass, Pietrowski and Solitar [62]; Stallings, relying on his theory of ends, also proved its main portion.

Theorem 4.1 ([62]) *A finitely generated group G is virtually free if and only if G is an HNN group of the form*

$$G = \langle t_1, \ldots, t_n, K \; ; \; t_1 L_1 t_1^{-1} = M_1, \ldots, t_n L_n t_n^{-1} = M_n \rangle$$

where K is a tree product of finitely many finite groups {the vertices of K} and each associated subgroup L_i, M_i is a subgroup of a vertex of K.

In particular, if there is no torsion in G then the base K is trivial and G is a free group.

Corollary 4.1 ([118]) *A torsion-free finite extension of a free group is free.*

Using the structure theorem Theorem 4.1, Karrass, Pietrowski and Solitar were able to prove several other nice results on virtually free groups. Stallings using his theory of ends proved many others. For a general discussion see [9]. A very surprising characterization of virtually free groups in terms of formal languages was developed by Muller and Schupp [84].

Motivated by the free group result Karrass asked whether torsion-free finite extensions of surface groups must also be surface groups. Zieschang originally gave the following positive answer to this question.

Theorem 4.2 ([127], [128]) *A torsion-free finite extension of a surface group is also a surface group.*

Note the analogy in the above statement with Stallings' theorem on free-by-finite groups.

A planar discontinuous group is a non-elementary discontinuous group of isometries of the hyperbolic plane. Here we say that a planar discontinuous group is non-elementary if the subgroup of orientation preserving elements is non-elementary. If a planar discontinuous group has only orientation preserving mappings it is a Fuchsian group. It can be proven algebraically as well as geometrically {see [79] and the references there and [53, 54]} that planar discontinuous groups contain surface groups as subgroups of finite index. It is natural then to ask the question: under what conditions are finite extensions of surface groups isomorphic to planar discontinuous groups. This is the basis of the **Nielsen Realization Problem**. The results we outline here are described in detail in [129].

Given any group H we say that H is **geometrically realized** if H is isomorphic to a planar discontinuous group. For a planar discontinuous group G, the centralizer $C_K(G)$ of any surface group K contained in G must be trivial. Therefore to properly pose the Nielsen Realization Problem this fact must be included as a necessary condition. To this end we say that G is an **effective extension** of a group K if

$$\text{whenever } g^{-1}xg = x \text{ for all } x \in K \text{ then } g \in K \qquad (4.1)$$

If K is a surface group this means $g = 1$ since K is centerless. We are now ready to pose:

Nielsen Realization Problem *Suppose G is a finite effective extension of a surface group K. Under what conditions is G isomorphic to a planar discontinuous group — in other words, is G geometrically realized?*

The Nielsen Realization Problem has an equivalent topological interpretation. Nielsen [86] proved that any automorphism of the fundamental group of a closed surface S is induced by a homeomorphism of S. The **mapping class group** or **homeotopy group** of S is $\mathrm{Meom}(S)/\mathrm{Meot}(S)$, that is, the group of homeomorphisms of S modulo the isotopy group of S. Theorems of Baer and Nielsen {see [129]} say that for a closed surface S, the mapping class group is isomorphic to $\mathrm{Aut}\Pi_1(S)/\mathrm{Inn}\Pi_1(S)$ — the outer automorphism group of $\Pi_1(S)$, where $\Pi_1(S)$ is the fundamental group of S.

In this context the Nielsen Realization Problem corresponds to determining the finite subgroups of the mapping class groups or equivalently determining which finite groups can be viewed as finite groups of mapping classes of S.

Nielsen Realization Problem *Suppose G is a finite effective extension of a surface group $\Pi_1(S)$ with $A_0 = G/\Pi_1(S)$. Can the group A_0 be realized as a group of homeomorphisms of S? That is for each class $a \in A_0$, does there exist a homeomorphism $f_a : S \to S$ such that $\{f_a \; ; \; a \in A_0\}$ form a group of mappings of S.*

In [128] Zieschang stated the following {a complete proof is in [129]}:

Theorem 4.3 ([128]) *Let G be a finite effective extension of a surface group K. Suppose that G satisfies the following condition*

> *Whenever $x^a = y^b = (xy)^c$, with $x, y \in G$, $a, b, c \geq 2$, then x, y generate a cyclic subgroup.*

Then G is isomorphic to a finitely generated planar discontinuous group.

Using a somewhat different approach Fenchel proved:

Theorem 4.4 ([31, 32]) *Let G be a finite effective extension of a surface group K. Suppose that G/K is a solvable group. Then G is isomorphic to a finitely generated planar discontinuous group.*

From a different viewpoint Eckman and Muller [29] considered Poincaré duality groups {see [129]}. They then proved:

Theorem 4.5 ([29]) *Let G be a finite extension of a surface group K such that the first Betti number of G is positive. Then G is isomorphic to a planar discontinuous group.*

Consider the triangle group $D^\star$ with presentation $\langle u, v \; ; \; u^a = v^b = (uv)^c = 1\rangle$. Then a **semi-triangle group** is a two-generator, non-triangle and non-cyclic group $D = \langle x, y\rangle$ of isometries of the hyperbolic plane such that x, y, xy have finite orders $a, b, c \geq 2$ respectively and there is an epimorphism $\phi : D^\star \to D$ with torsion-free kernel. Now suppose K is a Fuchsian group and G is a finite effective extension represented as isometries of the hyperbolic plane. Zieschang [129] then gives the following extension of Theorem 4.5.

Theorem 4.6 ([129]) *Suppose G and K are as above.*
  (a) *If all elements of G preserve orientation then the following is a complete list of possibilities.*
   (a1) *G can be realized by a Fuchsian group.*
   (a2) *G is a semi-triangle group. Every proper normal subgroup of G is torsion-free. Moreover, either G contains a triangle group or else G does not contain any semi-triangle group as proper subgroup.*

(b) *If G contains orientation reversing elements then either G is a planar discontinuous group or else the index 2 subgroup of orientation preserving elements is of the form described in* (a2).

The complete solution of the Nielsen Realization Problem was finally given by Eckman and Muller [28, 29] and Kerchoff [63]. (See also [129].)

We close this section with some results of Schneebeli on extensions of surface groups. Consider a class **C** of groups containing the trivial group that is closed with respect to isomorphisms and subgroups of finite index. Recall that a group is **poly-C** if there exists a finite normal series

$$G = G_0 \supset G_1 \supset \cdots \supset G_n = 1$$

with $G_i/G_{i+1} \in \mathbf{C}$ for each i. Schneebeli proved the following.

Theorem 4.7 ([105]) *The class of virtually poly-{surface groups} is extension closed.*

This theorem has the following immediate consequences.

Corollary 4.2 *Any extension of a surface group is virtually a poly-surface group.*

5   The Elementary Free Property and the Tarski Problem

Surface groups and the surface group relators play a large role in the solution of the Tarski problem. We briefly describe this and in the next section present some newer related results.

The **elementary theory** of a group G consists of all the **first-order** or **elementary sentences** which are true in G. We refer the reader to [33] for the formal defintions. Although this is a concept which originated in formal logic, in particular model theory, it arises independently from the theory of equations **within** groups. Recall that an **equation** in a group G is a word $W(x_1, \ldots, x_n, g_1, \ldots, g_k)$ in free variables $x_1, \ldots, x_n$ and constants $g_1, \ldots, g_k$ which are elements of G. A **solution** consists of an n-tuple $(h_1, \ldots, h_n)$ of elements from G which upon substitution for $x_1, \ldots, x_n$ make the word trivial in G. Hence an equation is a first-order sentence in the language $L[G]$ consisting of the elementary language of group theory (again see Section 2) augmented by allowing constants from the group G.

Equations within groups play an important role in many areas of both algebra and formal logic (see [79] or [49]) and the development of the theory of equations within groups has a long history. The greatest progress has been with the theory of equations within free groups. Fundamental work was done by Makanin [82], [83] and Razborov [90] in finding both an algorithm to determine if a system of equations in a free group has a solution, and if it does, to describe a solution. Another breakthrough result in this area was obtained by O. Kharlampovich and A. Myasnikov in the series of two papers, [64], [65]. They proved that every finite system of equations $S(X) = 1$ over a free group can be effectively transformed into

a finite number of systems of a very particular type, so-called non-degenerate triangular quasi-quadratic systems, thus reducing the system $S = 1$ into finitely many
quadratic equations over groups which are pretty close to being free (universally free
groups). This work in part was motivated by several long-standing problems known
as the **Tarski Conjectures**. The first of these is that all non-abelian free groups
have the same elementary theory while the second is that the elementary theory
of these free groups is decidable. While significant work on these conjectures was
done they resisted complete solution until the development (by Baumslag, Myasnikov and Remesslennikov [6, 7] of an algebraic geometry over groups mirroring
classical algebraic geometry over fields. Applying this algebraic geometry a positive
solution to the Tarski problems was given by Kharlampovich and Myasnikov (see
[64, 65, 67, 66, 69], [68]) and independently by Sela [108, 109, 110, 111, 112, 113].
Sela independently also developed most of the ideas surrounding the algebraic geometry over groups. The notes by Bestvina and Feighn [11] are the first in a series
to give an exposition of the work of both Sela and Kharlampovich–Myasnikov.

Concurrently with the work on the Tarski conjectures there has also been a great
deal of work tying logical concepts such as universal freeness with algebraic concepts such as free discrimination, commutative transitivity, equationally Noetherian
property, theory of quasi-varieties, etc. This has led to further work in this area,
including the development of discriminating and squarelike groups (see [33]).

We call a group an **elementary free group** if it has the same elementary theory
as the nonabelian free groups. Prior to the solution of the Tarski problem it was
asked whether there exist finitely generated non-free elementary free groups. The
answer is yes and prominent among them are the surface groups. As part of the
solution of the Tarski problems by Kharlampovich, Myasnikov and Sela a complete
characterization of the elementary free groups is obtained. The class of elementary
free groups extends beyond the class of solely the free groups and in the terminology
of Kharlampovich and Myasnikov consist precisely of the *NTQ-groups.* These are
the coordinate groups of regular NTQ systems of equations over free groups (see
[64, 65, 67, 66, 69] or [68] for relevant definitions).

Theorem 5.1 *An orientable surface group of genus $g \geq 2$ is an elementary free
group.*

However the connection of surface groups and especially the surface group relators to the solution of the Tarski problems goes much deeper than this. Part
of the solution depends on the fact that a finitely generated fully residually free
group admits a non-trivial cyclic JSJ-decomposition unless it is free, abelian or
a surface group. The **universal theory** of a group G consists of all universal
sentences (sentences with only universal quantifiers) that are true in G (see [33]).
All nonabelian finitely generated free groups have the same universal theory. A
universally free group is a group that has the same universal theory as a non-
abelian finitely generated free group. An elementary free group must be universally
free and hence fully residually free from a theorem of Gaglione and Spellman [45]
and Remesslennikov [91].

JSJ-decompositions were introduced by Rips and Sela [92] (see [68], [11] or

[107, 108, 109, 110, 111, 112, 113]) and have played a fundamental role in the study of both hyperbolic groups and fully residually free groups. Roughly a JSJ-decomposition of a group G is a splitting of G as a graph of groups with abelian edges which is canonical in the sense that it encodes all other such abelian splittings. If each edge group is cyclic it is called a **cyclic JSJ-decomposition**. For a formal definition we refer to [68]. The relevant fact for fully residually free groups is the following.

Theorem 5.2 ([68], [108, 109, 110, 111, 112, 113])
 (a) *A finitely generated fully residually free group which is indecomposable relative to JSJ-decompositions is either the fundamental group of a closed surface, a free group or a free abelian group.*
 (b) *A finitely generated fully residually free group admits a non-trivial cyclic JSJ-decomposition if it is not abelian or a surface group.*

Further in the JSJ decomposition very important components are what are called QH-vertices. These are vertices which are free groups that have specific "surface-group"-like presentations. QH stands for **quadratically hanging** and a QH-vertex is a free group that geometrically is the fundamental group of a punctured surface. Specifically a QH-vertex is a vertex group P in a splitting of a group G as a graph of groups that is isomorphic to the fundamental group of a compact surface S with boundary and such that the boundary components correspond to the incident edge groups. Hence a QH-vertex admits one of the following presentations
 (1) $\langle p_1, \ldots, p_m, a_1, b_1, \ldots, a_g, b_g ; \prod_{k=1}^{m} p_k \prod_{i=1}^{g} [a_i, b_i] \rangle$
 (2) $\langle p_1, \ldots, p_m, v_1, \ldots, v_g ; \prod_{k=1}^{m} p_k \prod_{i=1}^{g} v_i^2 \rangle$.
The theory of fully residually free groups is also in some sense motivated by surface groups. Magnus posed the question (see [33]) as to whether the surface groups S_g with $g \geq 2$ are residually free. G. Baumslag answered this affirmatively in the following manner. He observed that each S_g embeds in S_2 and residual freeness is inherited by subgroups so it suffices to show that S_2 is residually free. He actually showed more. If F is a nonabelian free group and $u \in F$ is a nontrivial element which is neither primitive nor a proper power then the group K given by

$$K = \langle F \star \overline{F} ; u = \overline{u} \rangle$$

where $\overline{F}$ is an identical copy of F and $\overline{u}$ is the corresponding element to u in $\overline{F}$, is residually free. A one-relator group of this form is called a **Baumslag double**. Baumslag proceeded by embedding K in the group

$$H = \langle F, t ; t^{-1}ut = u \rangle$$

by

$$K = \langle F, t^{-1}Ft \rangle.$$

The group H is then residually free and hence K is residually free. Therefore every Baumslag double is residually free. The group

$$S_2 = \langle a_1, b_1, a_2, b_2 ; [a_1, b_1] = [a_2, b_2] \rangle$$

is a Baumslag double answering the original question.

The group H is called a free rank one extension of centralizers of the free group F. Extensions of centralizers subsequently have played a large role in the structure theory of fully residually free groups and hence of elementary free groups (see [33]).

The theory of fully residually free groups is now very well developed (see [68]). Recently Kharlampovich, Myasnikov, Remeslennikov and Serbin [70] were able to translate the method of Stallings foldings to fully residually free groups by doing this they were able to algorithmically solve many problems in fully residually free groups mirroring the algorithmic solutions in absolutely free groups (see [70]).

Fully residually free groups also appear prominently in Sela's solution to the Tarski problem (see [108, 109, 110, 111, 112, 113]. In Sela's approach the finitely generated fully residually free groups are identified with **limit groups**. If G is a finitely generated group and F is a fixed nonabelian free group then a sequence $\{f_i\}$ in $\mathrm{Hom}(G, F)$ is **stable** if for all $g \in G$ the sequence $\{f_i(g)\}$ is eventually always 1 or eventually never 1. The **stable kernel** of $\{f_i\}$ denoted by $\underrightarrow{\mathrm{Ker}} f_i$ is the set

$$\{g \in G \; ; \; f_i(g) = 1 \text{ for almost all } i\}.$$

A finitely generated group Γ is a **limit group** is there is a finitely generated group G such that

$$\Gamma = G/\underrightarrow{\mathrm{Ker}} f_i$$

for some stable sequence

$$\{f_i\} \subset \mathrm{Hom}(G, F).$$

Sela proves that a finitely generated group Γ is a limit group if and only if it is fully residually free (see [108] and [11]).

6 Tame Automorphisms and the Surface Group Conjecture

The fact that surface groups are elementary free raises the question as to which properties of surface groups are actually first-order, that is shared by all elementary free groups. We explore this relative to automorphisms.

Suppose that $G = \langle F \mid R \rangle$ is a finitely presented group with F a finitely generated free group. An automorphsim $\alpha : G \rightarrow G$ is **tame** if it is induced by or lifts to an automorphism on F. If each automorphism of G is tame we say that the automorphism group $\mathrm{Aut}(G)$ is **tame**. In [116] Shpilrain gives a survey of some of the known general results on tame automorphisms and tame automorphism groups. If G is a surface group a result of Zieschang [125] and improved upon by Rosenberger [93, 94] shows that G has only one Nielsen class of minimal generating systems. An easy consequence of this is that $\mathrm{Aut}(G)$ is tame. Rosenberger uses the term **almost quasifree** for a group which has a tame automorphism group. If $G = \langle F \mid R \rangle$ is almost quasifree and in addition each automorphism of F induces an automorphism of G, G is called **quasifree**. Rosenberger observed that a non-free, non-cyclic one-relator group is quasifree only if it has a presentation $\langle a, b \; ; \; [a, b]^n = 1 \rangle$ for $n \geq 1$. This is a Fuchsian group if $n \geq 2$ or isomorphic to a free abelian group of rank 2 if

$n = 1$. S. Pride [89] proved that a 2-generator one-relator group with torsion, not a free product of cyclics, is almost quasifree. Moreover the isomorphism problem is solvable for such groups. This is a consequence of the fact that in such groups there is only one Nielsen class of generating pairs. Further from Theorem 2.8 it is clear that a cyclically pinched one-relator group is almost quasifree (see also [98, 100]).

Bumagin, Kharlampovich and Myasnikov [17] were able to prove that the isomorphism problem is solvable for finitely generated fully residually free groups. In doing this they developed a great deal of information about automorphisms of fully residually free groups. This can be applied to the study of the automorphism group of an elementary free group. Recently Fine, Kharlampovich, Myasnikov and Remeslennikov [36] proved the following.

Theorem 6.1 *The automorphism group* $\mathrm{Aut}(G)$ *of a finitely generated freely indecomposable fully residually free group* G *is tame.*

In particular since elementary free groups, that is groups which share the same elementary theory as the class of free groups, are fully residuallly free we obtain:

Corollary 6.1 *The automorphism group of a finitely generated freely indecomposable elementary free group* G *is tame.*

The proofs of these results depend upon an analysis of the automorphisms of the JSJ decomposition of a fully residually free group.

The methods used in considering tame automorphisms have also been used to make some progress on what we call the **surface group conjecture**. In the Kourovka [72] notebook Melnikov proposed the following problem.

Surface Group Conjecture A *Suppose that* G *is a non-free residually finite one-relator group such that every subgroup of finite index is again a one-relator group. Then* G *is a surface group.*

Since subgroups of infinite index in surface groups must be free groups this conjecture was modified to:

Surface Group Conjecture B *Suppose that* G *is a non-free one-relator group such that every subgroup of finite index is again a one-relator group and every subgroup of infinite index is a free group. Then* G *is a surface group.*

Using the structure theorem for fully residually free groups in terms of its JSJ decomposition Fine, Kharlampovich, Myasnikov, Remeslennikov and Rosenberger [37] have made some progress on these conjectures. We concentrate on the property that subgroups of infinite index must be free.

We say that a group G satisfies **Property IF** if every subgroup of infinite index is free. We obtain first.

Lemma 6.1 *If a one-relator group has Property IF and is freely decomposable then it is a free group.*

Theorem 6.2 ([37]) *Suppose that G is a finitely generated freely indecomposable fully residually free group with property IF. Then G is either a cyclically pinched one relator group or a conjugacy pinched one relator group.*

The proof uses the fact that a finitely generated freely indecomposable fully residually free group has a canonical cyclic JSJ decomposition.

Lemma 6.2 *Suppose that G is a finitely generated freely decomposable fully residually free group with property IF. Then G is a free group.*

Corollary 6.2 *Suppose that G is a finitely generated fully residually free group with property IF. Then G is either free or every subgroup of finite index is freely indecomposable and hence a one-relator group.*

If the surface group conjecture is true then a group satisfying the conditions of the conjecture must be hyperbolic or free abelian of rank 2. In [37] the following is proved.

Theorem 6.3 ([37]) *Let G be a finitely generated fully residually free group with property IF. Then either G is hyperbolic or G is free abelian of rank 2.*

In light of these results we give a modified version of the surface group conjecture.

Surface Group Conjecture C *Suppose that G is a finitely generated freely indecomposable fully residually free group with property IF. Then G is a surface group or a free abelian group of rank 2. We call a free abelian group of rank 2 a surface group of genus $g = 1$.*

We note that Surface group Conjecture C is true under either of the following two conditions:
(1) The original relator is quadratic;
(2) There is only one QH-vertex in the JSJ decomposition for G.
That is:

Theorem 6.4 ([37]) *Suppose that G is a finitely generated freely indecomposable fully residually free group with property IF. If either*
(1) *G is a one-relator group with a quadratic relator, or*
(2) *there is only one QH-vertex in the JSJ decomposition for G,*
then G is a surface group of genus $g \geq 1$.

7 Small Cancellation Theory

Dehn in 1912 [22] provided a solution to the word problem for a finitely generated orientable surface group. In doing so he ultimately paved the way for both the development of small cancellation theory and the theory of hyperbolic groups.

Dehn proved that in a surface group S_g, with $g \geq 2$, any non-empty word w in the generators which represents the identity must contain at least half of the original relator $R = [a_1, b_1] \ldots [a_g, b_g]$ where $a_1, b_1, \ldots, a_g, b_g$ are the generators. That is, if $w = 1$ in S_g, then $w = bcd$ where for some cyclic permutation R' of $R^{\pm 1}$, $R' = ct$ with $|t| < |c|$ where $|\ |$ represents free group length. It follows then that $w = bt^{-1}d$ in S_g and this word representation of w has shorter length than the original. Given an arbitrary w in S_g we can apply this type of reduction process to obtain shorter words. After a finite number of steps we will either arrive at 1 showing that $w = 1$ or at a word that cannot be shortened in which case $w \neq 1$. This procedure solves the word problem for S_g and is known as **Dehn's Algorithm** for a surface group. Dehn's original approach was geometric and relied on an analysis of the tessellation of the hyperbolic plane provided by a surface group.

The idea of a Dehn algorithm can be generalized in the following manner. Suppose G has a finite presentation $\langle X ; R \rangle$ (R here is a set of words in X). Let F be the free group on X and N the normal closure in F of R, $N = N_F(R)$, so that $G = F/N$. G, or more precisely the finite presentation $\langle X ; R \rangle$, has a **Dehn Algorithm** if there exists a finite set of words $D \subset N$ such that any non-empty word w in N can be shortened by applying a relator in D. That is, given any non-empty w in N, w has a factorization $w = ubv$ where there is a cyclic permutation r' of an element r of $D^{\pm 1}$ with $r' = bc$ and $|c| < |b|$. Then applying bc to w we have $w = uc^{-1}v$ in G where $|uc^{-1}v| < |ubv|$. By the same argument as in the surface group case the existence of a Dehn Algorithm leads to a solution of the word problem. Further Dehn also presented an algorithm based on the word problem algorithm to solve the conjugacy problem in surface groups.

The general idea of a Dehn algorithm is clearly that there is "not much cancellation possible in multiplying relators". Although Dehn's approach was geometric the idea can be phrased purely algebraically. This is the basic notion of **small cancellation theory**. This theory was initiated in 1947 by Tartakovskii [120] who showed, using purely algebraic methods, that certain groups, besides surface groups, also satisfy a Dehn Algorithm. His conditions were that in these groups again there is not much cancellation in multiplying relators. Greendlinger [47], [48], Schiek [104] and Britton [12] introduced other "small cancellation conditions" and also obtained Dehn Algorithms, and thus greatly expanded the class of groups with solvable word problem. Lyndon [76], [77], [78] in the mid 1960's, placed the whole theory in a geometric context and thus returned to Dehn's original approach. Lyndon also used this geometric approach to reprove the Freiheitssatz (see [78] or [79]). The geometric constructions used by Lyndon, now called Lyndon–Van Kampen diagrams, have been extended and modified for use in many areas of infinite group theory. A complete and readable account of small cancellation theory can be found in Chapter 5 of Lyndon and Schupp's book [79]. The proofs, both algebraic and geometric, are quite complex. We describe the small cancellation conditions.

Suppose F is free on a set of generators X. Let R be a *symmetrized* set of words in F. By this we mean that all elements of R are cyclically reduced and for each r in R all cyclically reduced conjugates of both r and r^{-1} are in R. If r_1 and r_2 are distinct elements of R with $r_1 = bc_1$ and $r_2 = bc_2$, then b is called a **piece**. Pieces represent those subwords of elements of R which can be cancelled by multiplying two non-inverse elements of R. The small cancellation hypotheses state that pieces must be relatively small parts of elements of R. We define three small cancellation conditions. The most common is a metric condition denoted $C'(\lambda)$ where λ is a positive real number. This condition asserts that if r is an element of R with $r = bc$ and b a piece, then $|b| < \lambda|c|$. If G is a group with a presentation $\langle X \; ; \; R \rangle$ where R is symmetrized and satisfies $C'(\lambda)$, then G is called a λ-*group*. So for example, if $\lambda = 1/6$, G is a **sixth group**, etc.

If p is a natural number, the second small cancellation condition is a non-metric one denoted $C(p)$. This asserts that no element of R is a product of fewer than p pieces. Notice that $C'(\lambda)$ implies $C(p)$ for $\lambda \le 1/(p-1)$.

The final small cancellation condition is also a non-metric condition denoted $T(q)$ for q a natural number. This asserts the following: Suppose $r_1, \ldots, r_h$ with $3 \le h < q$ are elements of R with no successive pair inverses. Then at least one of the products $r_1 r_2, \ldots, r_{h-1} r_h, r_h r_1$ is reduced without cancellation. $T(q)$ is dual to $C(p)$ for $(1/p) + (1/q) = 1/2$ in a suitable geometric context [79].

Greendlinger [47] proved purely algebraically that sixth-groups satisfy a Dehn Algorithm while Schiek showed the same for fourth groups also satisfying $T(4)$ [104]. Lyndon [76], [77], [78] placed the study of small cancellation theory in a geometric context and this is the way it is most often looked at. Essentially the small cancellation conditions lead to geometric tesselations and the solutions to the various decision problems are when these tesselations are non-Euclidean. This led Gromov to define negatively curved or hyperbolic groups.

In hyperbolic geometry there is a universal constant A such that triangles are **A-thin**. By this we mean that if XYZ is any geodesic triangle then any point on one side is at a distance less than A from some point on one of the other two sides. Now suppose G is a finitely generated group with fixed finite symmetric generating set X. We say that a generating system X for a group is symmetric if whenever $a \in X$ we also have $a^{-1} \in X$. Let Γ be the Cayley graph of G relative to this symmetric generating set X equipped with the word metric. A **geodesic** in the Cayley graph is a path between two points with minimal length relative to the word metric. A geodesic triangle is a triangle with geodesic sides. A geodesic triangle in Γ is **δ-thin** if any point on one side is at a distance less than δ from some point on one of the other two sides. Γ is **δ-hyperbolic** if every geodesic triangle is δ-thin. Finally G is **word-hyperbolic** or just **hyperbolic** if G is δ-hyperbolic with respect to some symmetric generating set X and some fixed $\delta \ge 0$. Gromov further showed that being hyperbolic is independent of the generating set although the δ may change, that is if G is hyperbolic with respect to one finite symmetric generating set it is hyperbolic with respect to all finite symmetric generating sets. For a full account of hyperbolic groups see the original paper of Gromov [49] or the notes [1]. Further suppose G is a finitely generated group with finite presentation $\langle X \mid R \rangle$. If W is a freely reduced word in the finitely generated free group $F(X)$ on X of length $L(W)$ and $\overline{W} = 1$ in G then there are words $P_i \in F(X)$ and relators $R_i \in R$ such that

$$W = \prod_{i=1}^{N} P_i R_i^{\epsilon_i} P_i^{-1} \quad \text{in } F(X)$$

where $\epsilon_i = \pm 1$ for each i. G then satisfies a **linear isoperimetric inequality** if there exists a constant K such that for all words W such that $W = 1$ we have $N < KL(W)$. A summary of results of Gromov [49] and independently of Lysenok [80] and Shapiro (see [1]) ties all these ideas together

Theorem 7.1 ([49],[80]) *The following conditions on a finitely presented group are equivalent:*

(1) *G is hyperbolic;*

(2) *G satisfies a linear isoperimetric inequality;*

(3) *G has a Dehn algorithm.*

If G is hyperbolic it must have a Dehn algorithm. Further a group that is finitely presented and has a Dehn algorithm must be hyperbolic. It follows that all the orientable surface groups are hyperbolic returning us to the genesis of the whole idea.

Theorem 7.2 *An orientable surface group is hyperbolic.*

Recall (see Section 2) that an orientable surface group is a cyclically pinched one-relator group. Juhasz and Rosenberger [60], and independently Bestvina and Feighn [10] and Kharlampovich and Myasnikov [67] proved the generalization of Theorem 7.2.

Theorem 7.3 *Let G be a cyclically pinched one-relator group with presentation*

$$G = \langle a_1, \ldots, a_p, a_{p+1}, \ldots, a_n \, ; U = V \rangle$$

If neither U nor V is a proper power in the free group on the generators which they involve then G is hyperbolic.

Further Gromov showed that all fundamental groups of closed compact hyperbolic manifolds are hyperbolic. From the existence of the Dehn algorithm we get the following immediate results again generalizing the results for surface groups.

Theorem 7.4 *A hyperbolic group G is finitely presented.*

Theorem 7.5 *A hyperbolic group G has a solvable word problem.*

In fact if G is δ-hyperbolic with symmetric generating set X and if we define

$$R = \{W \in F(X) \, ; L(W) \leq 8\delta \text{ and } \overline{W} = 1 \text{ in } G\}$$

then $\langle X \mid R \rangle$ is a Dehn presentation for G, that is has a Dehn algorithm.

Gromov [49] further has proved that hyperbolic groups have solvable conjugacy problem while Sela [107] has shown that the isomorphism problem is solvable for the class of torsion-free hyperbolic groups.

8   Test Words and Generic Elements

A **test element** in a group G is an element g with the property that if $f(g) = g$ for an endomorphism f from G to G then f must be an automorphism. A test element in a free group is called a **test word**. There has been a great deal of reseach on test words and related elements of free groups. As is the theme of this article much of this work was motivated by surface groups and especially the surface group relators.

Nielsen [86] gave the first non-trivial example of a test word by showing that in the free group on x, y the commutator $[x, y]$ satisfies this property. Other examples of test words have been given by Zieschang [124], [125], Rosenberger [93, 94, 95] Kalia and Rosenberger [61], Hill and Pride [52] and Durnev [26]. These are related to the surface group relator word. Gupta and Shpilrain [50] have studied the question as to whether the commutator $[x, y]$ is a test element in various quotients of the free group on x, y.

Test elements in finitely generated free groups were completely characterized by the **retraction theorem** of Turner [87]. Recall that a subgroup H of a group G is a **retract** if there exists a homomorphism $f : G \to H$ which is the identity on H. Clearly in a free group F any free factor is a retract. However there do exist retracts in free groups which are not free factors. Turner [87] characterized test words as those elements of a free group which do not lie in any proper retract.

Theorem 8.1 (Retraction Theorem [87]) *Let F be a free group. The test words in F are precisely those elements that do not lie in any proper retract.*

Using this characterization Turner was able to give several straightforward criteria to determine if a given element of a free group is a test word. Using these criteria, Comerford [21] proved that it is effectively decidable whether elements of free groups are test words. Since free factors are retracts, Turner's result implies that no test word can fall in a proper free factor. Therefore being a test word is a very strong form of non-primitivity. Shpilrain [117] defined the **rank** of an element w in a free group F as the smallest rank of a free factor containing w. Clearly in a free group of rank n a test word has maximal rank n. Shpilrain conjectured that the converse was also true but Turner gave an example showing this to be false. However Turner also proved that Shpilrain's conjecture is true if only test words for monomorphisms are considered.

As a direct consequence of the characterization Turner obtains the following result [121] (Example 5) which shows that there is a fairly extensive collection of test words in a free group of rank two.

Lemma 8.1 ([121]) *In a free group of rank two any non-trivial element of the commutator subgroup is a test word.*

It is fairly straightforward to show that the retraction theorem also holds in non-abelian surface groups.

Theorem 8.2 *Let S_g be an orientable surface group. The test elements in S_g are precisely those elements that do not lie in any proper retract.*

Examples of test elements in surface groups were constructed by Konieczny, Rosenberger and Wolny [71]. D. Hennig [51] constructed a specific example of a test element in each nonelementary Fuchsian group. In particular relative to surface groups:

Theorem 8.3 *Let S_g be a surface group of genus $g \geq 2$*

$$S_g = \langle a_1, b_1, \ldots, a_g, b_g \; ; \; [a_1, b_1] \ldots [a_g, b_g] = 1 \rangle.$$

Then if $p \geq 2$ is prime the element $a_1^p b_1^p \ldots a_g^p b_g^p$ is a test element.

For triangle groups Hennig proved:

Theorem 8.4 (1) *Let G be a (p, q, r) triangle group so that G has the presentation*

$$G = \langle a, b \; ; \; a^p = b^q = (ab)^r = 1 \rangle$$

with $2 \leq p, q, r < \infty$ and $\frac{1}{p} + \frac{1}{q} + \frac{1}{r} < 1$. Then the element $[a, b]$ is a test element.

(2) *If G has the presentation $G = \langle a, b \; ; \; [a, b]^n = 1 \rangle$ with $n \geq 2$ then $[a, b]$ is a test element.*

(3) *If G has the presentation*

$$G = \langle s_1, s_2, s_3, s_4 \; ; \; s_1^2 = s_2^2 = s_3^2 = s_4^q = s_1 s_2 s_3 s_4 = 1 \rangle$$

with $q = 2k + 1 \geq 3$ then $[s_1 s_2, s_3 s_4]$ is a test element.

O'Neill and Turner later proved [87] that the retraction theorem also holds in certain torsion-free hyperbolic groups. Whether it holds in all is still an open question.

Fine, Rosenberger, Spellman and Stille [42] considered the relationships between test words and certain additional types of elements in free groups, specifically generic elements and almost primitive elements (APE's). There are strong ties to the surface group relators.

An **almost primitive element** — (APE) — is an element of a free group F which is not primitive in F but which is primitive in any proper subgroup of F containing it. This can be extended to arbitrary groups in the following manner. An element $g \in G$ is **primitive** in G if g generates an infinite cyclic free factor of G, that is, g has infinite order and $G = \langle g \rangle \star G_1$ for some $G_1 \subset G$. g is then an APE if it is not primitive in G but primitive in any proper subgroup containing it. Rosenberger [93] proved that in the free group $F = F(x_i, y_i, z_j)$, $1 \leq i \leq m$, $1 \leq j \leq n$, of rank $2m + n$ the element

$$[x_1, y_1] \ldots [x_m, y_m] z_1^{p_1} \ldots z_n^{p_n}$$

where the p_i are not necessarily distinct primes, is an APE in F. Rosenberger [93] proved, in a different setting, that if A, B are arbitrary groups containing APE's a, b respectively, then the product ab is either primitive or an APE in the free product $A \star B$. This was reproved by Brunner, Burns and Oates-Williams [15] who also considered the more difficult problem of determining when ab is a tame APE in $A \star B$ where $a \in A$ and $b \in B$. An APE w in a group G is a **tame APE** if whenever $w^\alpha \in H \subset G$ with $\alpha \geq 1$ minimal, then either w^α is primitive in H or the index $[G : H]$ is α. It follows easily that the surface group word $[a_1, b_1] \ldots [a_g, b_g]$, $g \geq 1$, is a tame APE in the free group on $a_1, b_1, \ldots, a_g, b_g$ (see [95]). Further $a_1^p \ldots a_g^p$ with $g \geq 2$ and p prime is a tame APE if and only if $p = 2$.

Let $\mathcal{U}$ be a variety defined by a set of laws $\mathcal{V}$. (We refer to the book of H. Neumann [85] for relevant terminology). For a group G we let $\mathcal{V}(G)$ denote the verbal subgroup of G defined by $\mathcal{V}$. An element $g \in G$ is $\mathcal{U}$-**generic** in G if $g \in \mathcal{V}(G)$ and whenever H is a group, $f : H \to G$ a homomorphism and $g = f(u)$ for some $u \in \mathcal{V}(H)$, it follows that f is surjective. Equivalently $g \in G$ is $\mathcal{U}$-generic in G if $g \in \mathcal{V}(G) \subset G$ but $g \notin \mathcal{V}(K)$ for every proper subgroup K of G [118]. An element is **generic** if it is $\mathcal{U}$-generic for some variety $\mathcal{U}$. Let $\mathcal{U}_n$ be the variety defined by the set of laws $\mathcal{V}_n = \{[x, y], z^n\}$. For $n = 0$ we have $\mathcal{U}_n = \mathcal{A}$ the abelian variety. Stallings [118] and Dold [23] have given sufficient conditions for an element of a free group to be $\mathcal{U}_n$-generic. Using this it can be shown that $x_1^n x_2^n \ldots x_m^n$ is $\mathcal{U}_n$-generic in the free group on $x_1, \ldots, x_m$ for all $n \geq 2$ and if m is even $[x_1, x_2] \ldots [x_{m-1}, x_m]$ is $\mathcal{U}_n$-generic in the free group on $x_1, \ldots, x_m$ for $n = 0$ and for all $n \geq 2$. These facts are also consequences of a result of Rosenberger [94, 95].

Comerford [21] points out that if G is Hopfian, which is the case if G is free, then being generic implies being a test word. Thus for free groups we have

$$\text{generic} \longrightarrow \text{test word}.$$

Comerford also shows that there is no converse. In particular he shows that in a free group of rank 3 on x, y, z the word $w = x^2[y^2, z]$ is a test word but is not generic. We can also show that in general, generic does not imply APE. Suppose $F = F(x, y)$ is the free group of rank two on x, y and let $w = x^4 y^4$. Then w is $\mathcal{U}_4$-generic but w is not an APE since $w \in \langle x^2, y^2 \rangle$ and is not primitive in this subgroup while this subgroup is not all of F.

Further, in general it is not true that being an APE implies being a test word. Again let $F = F(x, y)$ and let $w = x^2 y x^{-1} y^{-1}$. Brunner, Burns and Oates-Williams show that w is an APE. However Turner shows that w is not a test word. Since generic elements are test words in a Hopfian group this example shows further that APE does not imply generic in general. This is really to be expected since test words are strongly non-primitive. However the following result shows that many APE's are indeed generic and therefore test words.

Recall that a variety $\mathcal{U}$ defined by the set of laws $\mathcal{V}$ is a non-trivial variety if it consists of more than just the trivial group. In this case $\mathcal{V}(F) \neq F$ for any free group F.

Theorem 8.5 *Let F be a free group and $\mathcal{B}$ a non-trivial variety defined by the set* *of laws* $\mathcal{V}$. Let $w \in \mathcal{V}(F)$. *If w is an APE then w is $\mathcal{B}$-generic. In particular w is a test word.*

More results on these elements can be found in [42].

Suppose that

$$S_g = \langle a_1, b_1, \ldots, a_g, b_g \; ; [a_1, b_1] \ldots [a_g, b_g] = 1 \rangle$$

with $g \geq 2$ is an orientable surface group. It is difficult to work in this group directly with the above definitions of APE and tame APE. If we modify the definition slightly we have more success. An element $w \in S_g$ is m-primitive if it is a member of a minimal generating system for S_g. With this definition we may define, analogously as in a free group, the properties m-APE and tame m-APE. In this sense Konieczny, Rosenberger and Wolny [71] showed that the element $a_1^p b_1^p \ldots a_g^p b_g^p$ with $p \geq 2$ a prime is an m-APE in S_g if and only if $p = 2$.

An extension of the test element concept is to the idea of a **test rank**. If g is a test element of a group G then it is straightforward to see that this is equivalent to the fact that if $f(g) = \alpha(g)$ for some endomorphism f of G and some automorphism α of G then f must also be an automorphism. A **test set** in a group G consists of a set of elements $\{g_i\}$ with the property that if f is an endomorphism of G and $f(g_i) = \alpha(g_i)$ for some automorphism α of G and for all i then f must also be an automorphism. Any set of generators for G is a test set, and if G possesses a test element then this is a singleton test set. The **test rank** of a group is the minimal size of a test set. Clearly the test rank of any finitely generated group is finite and bounded above by the rank and below by 1. Further the test rank of any free group of finite rank is 1 since these contain test elements. For a free abelian group of rank n the test rank is precisely n. Turner and Rocca [122] have developed a method to find the test rank of any abelian group and are able to prove that given any m, n with $1 \leq m \leq n$ there exists a group with rank n and test rank m.

9 Algebraic Generalizations of Discrete Groups

As we have seen orientable surface groups have faithful representations as Fuchsian groups and hence are linear. The questions motivated by surface groups are part of a more general program motivated by both the properties of surface groups and the linear properties of discrete groups.

Recall that a **Fuchsian group** F is a non-elementary discrete subgroup of $PSL_2(\mathbb{R})$ or a conjugate of such a group in $PSL_2(\mathbb{C})$ (see Section 2). If F is finitely generated then F has a presentation, called the **Poincaré presentation**, of the form

$$F = \langle e_1, \ldots, e_p, h_1, \ldots, h_t, a_1, b_1, \ldots, a_g, b_g \; ; e_i^{m_i} = 1, \; i = 1, \ldots, p, \; R = 1 \rangle \quad (9.1)$$

where $R = e_1 \ldots e_p h_1 \ldots h_t [a_1, b_1] \ldots [a_g, b_g]$ and $p \geq 0$, $t \geq 0$, $g \geq 0$, $p + t + g > 0$, and $m_i \geq 2$ for $i = 1, \ldots, p$.

The **Euler Characteristic** of F is given by $\chi(F) = -\mu(F)$ where

$$\mu(F) = 2g - 2 + t + \sum_{i=1}^{p} (1 - 1/m_i).$$

An **F-group** G is a group with a presentation of the form (9.1). An F-group G with $\mu(G) > 0$ has a faithful representation as a Fuchsian group and $2\pi\mu(G)$ represents the hyperbolic area of a fundamental polygon for G.

From the Poincaré presentation, Fuchsian groups fall into the class of one-relator products of cyclics. The basic question then is which properties of Fuchsian groups are shared by one-relator products of cyclics.

Fuchsian groups are of course linear and therefore satisfy many "linearity" properties. One-relator groups are also known to satisfy many of these same linearity properties even though they may or may not be linear. The specific questions explored are:

(1) Which properties of Fuchsian groups are shared by all one-relator products of cyclics?

(2) If a property of a Fuchsian group does not hold in all one-relator products of cyclics, then is there a subclass — specifically a special form of the relator — in which it does hold?

A research program was developed and worked on at various times by Fine, Rosenberger, Howie, Gaglione, Spellman, Levin, Thomas, Tang, Kim, Allenby, Stille, Scheer and others to then study the following general question.

What linear properties of discrete groups are shared by all one-relator products of cyclics?

The specific properties that received the most attention were:

(1) Faithful representations into $PSL_2(\mathbb{C})$,

(2) Virtual torsion-free property,

(3) Tits Alternative,

(4) The Classification of the Finite Generalized Triangle Groups,

(5) SQ-universality,

(6) Subgroup structure.

Many of the results are summarized in the book *Algebraic Generalizations of Discrete Groups*, Fine–Rosenberger, Marcel-Dekker (1999). Subsequent to the book there has been more work done on the classification of the finite generalized triangle groups and the finite generalized tetrahedron groups. We refer to the papers [30], [59], [73], [102] and the thesis of Scheer [103] for more on this.

References

[1] J. Alonzo, T. Brady, D. Cooper, T. Delzant, V. Ferlini, M. Lustig, M. Mihalik, M. Shapiro and H. Short, *Notes on Negatively Curved Groups*, MSRI preprint 1989.

[2] B. Baumslag, Free products of locally indicable groups with a single relator, *Bull. Austral. Math. Soc.* **29** (1984), 401–404.

[3] G. Baumslag, On generalized free products, *Math. Z.* **78** (1962), 423–438.

[4] G. Baumslag, A survey of groups with a single relation, in *Groups St Andrews 1985*, London Math. Soc. Lecture Notes Series **121**, 30–58, 1987.

[5] G. Baumslag, *Topics in Combinatorial Group Theory*, Birkhäuser, 1993.

[6] G. Baumslag, A. G. Myasnikov and V. N. Remeslennikov, Algebraic geometry over groups, I, *J. Algebra* **219** (1999), 16–79.

[7] G. Baumslag, A. G. Myasnikov and V. N. Remeslennikov, Discriminating and co-discriminating groups, *J. Group Theory* **3** (2000), 467–479.

[8] G. Baumslag and P. Shalen, Groups whose three generator subgroups are free, *Bull. Austral. Math. Soc.* **40** (1989), 163–174.

[9] K. Bencsath and B. Fine, Reflections on virtually one-relator groups, in *Groups Galway/St Andrews 1993*, 37–57, London Math. Soc. Lecture Notes Series **211**, 1994.

[10] M. Bestvina and M. Feighn, A combination theorem for negatively curved groups, *J. Differential Geom.* **35** (1992), 85–101.

[11] M. Bestvina and M. Feighn, Notes on Sela's work: Limit groups and Makanin–Razborov diagrams, preprint.

[12] J. L. Britton, Solution of the word problem for certain types of groups I, II, *Proc. Glasgow Math. Assoc.* **3** (1956), 45–54; **3** (1957) 68–90.

[13] S. D. Brodskii, Equations over groups and groups with a single defining relation, (Russian), *Uspekhi Mat. Nauk* **35** (1980), no. 4, 183.

[14] S. D. Brodskii, Anomalous products of locally indicable groups, (Russian), *Algebraicheskie Sistemy, pub. Ivanovo University*, 51–77, 1981.

[15] A. M. Brunner, R. G. Burns and S. Oates-Williams, On almost primitive elements of free groups with an application to Fuchsian groups, *Canad. J. Math.* **45** (1993), 225–254.

[16] A. M. Brunner, R. G. Burns and D. Solitar, The subgroup separability of free products of two free groups with cyclic amalgamation, *Contemp. Math.* **33** (1984), 90–115.

[17] I. Bumagin, O. Kharlampovich, A. Myasnikov, Isomorphism problem for finitely generated fully residually free groups, in progress.

[18] B. Chandler and W. Magnus, *The History of Combinatorial Group Theory*, Springer-Verlag, 1982.

[19] D. Cohen, *Groups of Cohomological Dimension One*, Lecture Notes in Math. **245**, Springer-Verlag, 1972.

[20] D. J. Collins and H. Zieschang, *Combinatorial Group Theory and Fundamental Groups*, Algebra VII, 3–166, Springer-Verlag, 1993.

[21] L. P. Comerford, Generic Elements of Free Groups, *Arch. Math.*, to appear.

[22] M. Dehn, Transformation der Kurve auf zweiseitigen Flachen, *Math. Ann.* **72** (1912), 413–420.

[23] A. Dold, Nullhomologous words in free groups which are not nullhomologous in any proper subgroup, *Arch. Math.* **50** (1988), 564–569.

[24] A. J. Duncan and J. Howie, One relator products with high-powered relators, in *Proceedings of the Geometric Group Theory Symposium*, University of Sussex, 1991.

[25] A. J. Duncan and J. Howie, Weinbaum's Conjecture on unique subwords of nonperiodic words, *Proc. Amer. Math. Soc.* **115** (1992), no. 4, 947–954.

[26] V. G. Durnev, The Mal'cev–Nielsen equation in a free metabelian group of rank two, *Math. Notes* **6**4 (1989), 927–929.

[27] J. L. Dyer, Separating conjugates in amalgamated free porducts and HNN extensions *J. Austral. Math. Soc. Ser. A* **29** (1980), 35–51.

[28] B. Eckman and H. Müller, Virtual Surface Groups, ETH preprint, 1980.

[29] B. Eckman and H. Müller, Plane motion groups and virtual Poincaré duality of dimension 2, *Invent. Math.* **69** (1982), 293–310.

[30] M. Edjvet, G. Rosenberger, M. Stille and R. Thomas, On certain finite gneralized

tetrahedron groups, in *Computational and Geometric Aspects of Modern Algebra*, London Math. Soc. Lecture Notes Series **275**, 54–65, 2000.

[31] W. Fenchel, Estensioni gruppi descontinui e transformazioni periodiche delle surficie, *Rend. Acc. Naz. Lincei* **5** (1948), 326–329.

[32] W. Fenchel, Bemarkogen om endliche gruppen af abbildungsklasser, *Mat. Tidskrift* B (1950), 90–95.

[33] B. Fine, A. Gaglione, A. Myasnikov, G. Rosenberger and D. Spellman, A classification of fully residually free groups of rank three or less, *J. Algebra* **200** (1998), 571–605.

[34] B. Fine, A. Gaglione, G. Rosenberger and D. Spellman, Free Groups and Questions About Universally Free Groups, in *Groups Galway/St Andrews 1993*, London Math. Soc. Lecture Notes Series **211**, 191–204, 1994.

[35] B. Fine, J. Howie and G. Rosenberger, One-relator quotients and free products of cyclics, *Proc. Amer. Math. Soc.* **102** (1988), 1–5.

[36] B. Fine, O. Kharlampovich, A. Myasnikov and V. Remesslennikov, Tame automorphisms of elementary free groups, to appear.

[37] B. Fine, O. Kharlampovich, A. Myasnikov, V. Remesslennikov and G. Rosenberger, On the surface group conjecture, to appear.

[38] B. Fine, F. Roehl and G. Rosenberger, Two generator subgroups of certain HNN groups, *Contemp. Math.* **109** (1990), 19–25.

[39] B. Fine, F. Roehl and G. Rosenberger, A three-free theorem for certain HNN groups, in *Infinite Groups and Group Rings*, 13–37, World Scientific, 1993.

[40] B. Fine and G. Rosenberger, The Freiheitssatz and its extensions, *Contemp. Math.* **169** (1994), 213–252.

[41] B. Fine and G. Rosenberger, *Algebraic Generalizations of Discrete Groups*, Marcel-Dekker, 1999.

[42] B. Fine, G. Rosenberger, D. Spellman and M. Stille, Test words, generic elements and almost primitivity, *Pacific J. Math.* **190** (1999), no. 2, 277–297.

[43] B. Fine, G. Rosenberger and M. Stille, Nielsen transformations and applications: a survey, in *Groups Korea 1994* (eds. Kim, Johnson), 69–105, DeGruyter, 1995.

[44] B. Fine, G. Rosenberger and M. Stille, Conjugacy pinched and cyclically pinched one-relator groups, *Revista Math. Madrid* **10** (1997), no. 2, 207–227.

[45] A. Gaglione and D. Spellman, Some model theory of free groups and free algebras, *Houston J. Math.* **19** (1993), 327–356.

[46] F. Gonzalez-Acuna and H. Short, Knot surgery and primeness, *Math. Proc. Cambridge Philos. Soc.* **99** (1986), 89–102.

[47] M. Greendlinger, Dehn's algorithm for the word problem, *Comm. Pure Appl. Math.* **13** (1960), 67–83.

[48] M. Greendlinger, On Dehn's algorithm for conjugacy and word problems with applications, *Comm. Pure Appl. Math.* **13** (1960), 641–677.

[49] M. Gromov, Hyperbolic Groups, in *Essays in Group Theory* (S. Gersten ed.), MSRI Publication 8, Springer-Verlag, 1987.

[50] N. Gupta and V. Shpilrain, Nielsen's commutator test for two-generator groups, *Arch. Math.* **44** (1985), 1–14.

[51] D. Hennig, Test Elements in Fuchsian Groups, thesis, Univ. of Dortmund.

[52] P. Hill and S. Pride, Commutators, generators and conjugacy equations in groups, *Math. Proc. Cambridge Philos. Soc.* **114** (1993), 295–301.

[53] A. Hoare, A. Karrass and D. Solitar, Subgroups of finite index of Fuchsian groups, *Math. Z.* **120** (1971), 289–298.

[54] A. Hoare, A. Karrass and D. Solitar, Subgroups of infinite index in Fuchsian groups, *Math. Z.* **125** (1972), 59–69.

[55] J. Howie, On pairs of 2-complexes and systems of equations over groups, *J. Reine Angew. Math.* **324** (1981), 165–174.

[56] J. Howie, How to generalize one-relator group theory, *Ann. of Math. Stud.* **111** (1987), 53–78.

[57] J. Howie, The quotient of a free product of groups by a single high-powered relator, I. Pictures. Fifth and higher powers, *Proc. London Math. Soc.* **59** (1989), 507–540.

[58] J. Howie, The quotient of a free product of groups by a single high-powered relator, II. Fourth powers, *Proc. London Math. Soc.* **61** (1990), 33–62.

[59] J. Howie, V. Metafsis and R. Thomas, Finite generalized triangle groups, *Trans. Amer. Math. Soc.* **347** (1995), 3613–3623.

[60] A. Juhasz and G. Rosenberger, On the combinatorial curvature of groups of F-type and other one-relator products of cyclics, *Contemp. Math.* **169** (1994), 373-384.

[61] R. N. Kalia and G.Rosenberger, Automorphisms of the Fuchsian groups of type $(0, 2, 2, 2, q : 0)$, *Comm. Algebra* **6** (1978), no. 11, 115–129.

[62] A. Karrass, A. Pietrowski and D. Solitar, Finite and infinite cyclic extensions of free groups, *J. Austral. Math. Soc.* **16** (1972), 458–466.

[63] S. P. Kerchoff, The Nielsen realization problem, *Ann. of Math.* **117** (1983), 235–265.

[64] O. Kharlampovich and A. Myasnikov, Irreducible affine varieties over a free group, I. Irreducibility of quadratic equations and Nullstellensatz, *J. Algebra* **200** (1998), 472–516.

[65] O. Kharlampovich and A. Myasnikov, Irreducible affine varieties over a free group, II. Systems in triangular quasi-quadratic form and a description of residually free groups, *J. Algebra* **200** (1998), 517–569.

[66] O. Kharlampovich and A. Myasnikov, Hyperbolic groups and free constructions, *Trans. Amer. Math. Soc.* **350** (1998), no. 2, 571–613.

[67] O. Kharlampovich and A. Myasnikov, Description of fully residually free groups and irreducible affine varieties over free groups, in *Summer school in Group Theory in Banff, 1996*, CRM Proceedings and Lecture Notes **17**, 71–81, 1999.

[68] O. Kharlampovich and A.Myasnikov, Algebraic geometry over free groups, to appear.

[69] O.Kharlampovich and A.Myasnikov, Solution of the Tarski Problem, to appear.

[70] O. Kharlampovich, A. Myasnikov, V. Remeslennikov and D. Serbin, Subgroups of fully residually free groups: algorithmic problems, *Contemp. Math.* **360** (2004).

[71] J. Konieczny, G. Rosenberger, J. Wolny, Tame almost primitive elements, *Results Math.* **38** (2000), 116–129.

[72] Kourovka Notebook — Unsolved Problems in Group Theory (ed. Y. I. Merzlyakov).

[73] L. Levai, G. Rosenberger and B. Souvignier, All finite generalized triangle groups, *Trans. Amer. Math. Soc.* **347** (1995), 3625–3627.

[74] F. Levin, Solutions of equations over groups, *Bull. Amer. Math. Soc.* **68** (1962), 603–604.

[75] S. Lipschutz, The conjugacy problem and cyclic amalgamation, *Bull. Amer. Math. Soc.* **81** (1975), 114–116.

[76] R. C. Lyndon, A maximum principle for graphs, *J. Combinatorial Theory* **3** (1967), 34–37.

[77] R. C. Lyndon, On the Freiheitssatz, *J. London Math. Soc.* **5** (1972), 95–101.

[78] R. C. Lyndon, Equations in groups, *Bol. Soc. Bras. Math.* **11** (1980), 79–102.

[79] R. C. Lyndon and P. E. Schupp, *Combinatorial Group Theory*, Springer-Verlag, 1977.

[80] I. G. Lysenok, On some algorithmic properties of hyperbolic groups, *Math. USSR Izv.* **35** (1990), 145–163.

[81] W. Magnus, A. Karrass, and D. Solitar, *Combinatorial Group Theory*, Wiley, 1966; Second Edition, Dover, New York 1976.

[82] G. S. Makanin, Equations in a free group (Russian), *Izv. Akad. Nauk SSSR, Ser.*

Mat. **46** (1982), 1199–1273; transl. in *Math. USSR Izv.* **21** (1983).

[83] G. S. Makanin, Decidability of the universal and positive theories of a free group, *Math. USSR Izv.* **25** (1985), no. 1, 75–88.

[84] D. E. Muller and P. E. Schupp, Groups, the theory of ends and context-free languages, *J. Comput. System Sci.* **26** (1983), 295–310.

[85] H. Neumann, *Varieties of Groups*, Springer-Verlag, 1967.

[86] J. Nielsen, Die Automorphisem der algemeinen unendlichen Gruppe mit zwei Erzeugenden, *Math. Ann.* **78** (1918), 385-397.

[87] J. C. O'Neill and E. C. Turner, Test elements and the retract theorem in hyperbolic groups, *New York J. Math.* **6** (2000), 107–117.

[88] N. Peczynski and W. Reiwer, On cancellations in HNN groups, *Math. Z.* **158** (1978), 79–86.

[89] S. J. Pride, The isomorphism problem for two generator one-relator groups with torsion is solvable, *Trans. Amer. Math. Soc.* **227** (1977), 109–139.

[90] A. A. Razborov, On systems of equations in free groups *Izv. Akad. Nauk SSSR* **48** (1984), 779–832; English transl.: *Math. USSR Izv.* **25**, 115–162.

[91] V. N. Remeslennikov, ∃-free groups, *Siberian Math. J.* **30** (1989), 998–1001.

[92] E. Rips and Z. Sela, Cyclic splittings of finitely presented groups and the canonical JSJ decomposition, *Ann. of Math. (2)* **146** (1997), no. 1, 53–109.

[93] G. Rosenberger, Zum Rang und Isomorphieproblem für freie Produkte mit Amalgam, Habilitationsschrift, Hamburg 1974.

[94] G. Rosenberger, Zum Isomorphieproblem für Gruppen mit einer definierenden Relation, *Illinois J. Math.* **20** (1976), 614–621.

[95] G. Rosenberger, On Discrete Free Subgroups of Linear Groups, *J. London Math. Soc. (2)* **17** (1978), 79–85.

[96] G. Rosenberger, Alternierende Produkte in freien Gruppen, *Pacific J. Math.* **78** (1978), 243–250.

[97] G. Rosenberger, Applications of Nielsen's reduction method in the solution of combinatorial problems in group theory, in *London Math. Soc. Lecture Notes* **36**, 339–358, 1979.

[98] G. Rosenberger, Gleichungen in freien Produkten mit Amalgam, *Math. Z.* **173** (1980), 1–12.

[99] G. Rosenberger, On one-relator groups that are free products of two free groups with cyclic amalgamation, in *Groups St Andrews 1981*, 328–344, Cambridge University Press, 1982.

[100] G. Rosenberger, Bemerkungen zu einer Arbeit von R. C. Lyndon, *Archiv. Math.* **40** (1983), 200–207.

[101] G. Rosenberger, The isomorphism problem for cyclically pinched one-relator groups, *J. Pure Appl. Algebra* **95** (1994), 75–86.

[102] G. Rosenberger and M. Scheer, Classification of the finite generalized tetrahedron groups, *Contemp. Math.* **296** (2002), 207–229.

[103] M. Scheer, Classification of the Finite Generalized Tetrahedron Groups, Thesis, Univ. Dortmund, 2002.

[104] H. Schiek, Ahnlichkeitsanalyse von Gruppenrelationen, *Acta Math.* **96** (1956), 157–252.

[105] H. R. Schneebeli, On virtual properties and group extensions, *Math. Z.* **159** (1978), no. 2, 159–167.

[106] G. P. Scott, Subgroups of surface groups are almost geometric, *J. London Math. Soc.* **17** (1978), no. 2, 555–565.

[107] Z. Sela, The isomorphism problem for hyperbolic groups I, *Ann. of Math.* **141** (1995), no. 2, 217–283.

[108] Z. Sela, Diophantine geometry over groups I: Makanin–Razborov diagrams, *Publ. Math. Inst. Hautes Études Sci.* **93** (2001), 31–105.

[109] Z. Sela, Diophantine geometry over groups II: Completions, closures and formal solutions, *Israel J. Math.* **104** (2003), 173–254.

[110] Z. Sela, Diophantine geometry over groups III: Rigid and solid solutions, *Israel J. Math.* **147** (2005), 1–73.

[111] Z. Sela, Diophantine geometry over groups IV: An iterative procedure for validation of a sentence, *Israel J. Math.* **143** (2004), 1–130.

[112] Z. Sela, Diophantine geometry over groups V: Quantifier elimination, *Israel J. Math.* **150** (2005), 1–97.

[113] Z. Sela, Diophantine geometry over groups VI: The elementary theory of a free group, to appear.

[114] P. Shalen, Linear representations of certain amalgamated products, *J. Pure Appl. Algebra* **15** (1979), 187–197.

[115] H. Short, Topological Methods in Group Theory; the Adjunction Problem, Ph.D. Thesis, University of Warwick, 1984.

[116] V. Shpilrain, Recognizing automorphisms of the free groups, *Arch. Math.* **62** (1994), 385–392.

[117] V. Shpilrain, Test elements for endomorphisms of free groups and algebras, preprint.

[118] J. Stallings, Problems about free quotients of groups, preprint.

[119] J. Stillwell, *Classical Topology and Combinatorial Group Theory*, Springer, 1980.

[120] V. A. Tartakovskii, Solution of the word problem for groups with a k reduced basis for $k > 6$, *Izv. Akad. Nauk SSSR Ser. Math.* **13** (1949), 483–494.

[121] E. C. Turner, Test words for automorphisms of the free groups, *Bull. London Math. Soc.* **28** (1996), 255–263.

[122] E. C. Turner and C. Rocca, Test ranks of abelian groups, to appear.

[123] B. A. F. Wehrfritz, Generalized free products of linear groups, *Proc. London Math. Soc.* **273** (1973), 402–424.

[124] H. Zieschang, Alternierende Produkte in Freien Gruppen, *Abh. Math. Sem. Univ. Hamburg* **27** (1964), 12–31.

[125] H. Zieschang, Automorphismen ebener discontinuerlicher Gruppen *Math. Ann.* **166** (1966), 148–167.

[126] H. Zieschang, Uber die Nielsensche Kurzungsmethode in freien Produkten mit Amalgam, *Invent. Math.* **10** (1970), 4–37.

[127] H. Zieschang, On extensions of fundamental groups of surfaces and related groups, *Bull. Amer. Math. Soc.* **77** (1971), 1116–1119.

[128] H. Zieschang, Addendum to: On extensions of fundamental groups of surfaces and related groups, *Bull. Amer. Math. Soc.* **80** (1974), 366–367.

[129] H. Zieschang, *Finite Groups of Mapping Classes of Surfaces*, Lecture Notes in Math. **875**, Springer-Verlag, 1980.

NILPOTENT p-ALGEBRAS AND FACTORIZED p-GROUPS

BERNHARD AMBERG[*] and LEV KAZARIN[†1]

[*]Institut für Mathematik, Johannes-Gutenberg-Universität, 55099 Mainz, Germany
Email: Amberg@Mathematik.Uni-Mainz.de
[†]Department of Mathematics, Yaroslavl State University, 150000 Yaroslavl, Russia

Abstract

An associative algebra R over a field of characteristic p is called a p-algebra. Every nilpotent p-algebra forms a p-group under the operation $x \circ y = x + y + xy$ for every two elements $x, y \in R$. This group is called the adjoint group of R. The structure of a finite p-algebra and the relation to its adjoint group is discussed. This is used to study finite factorized groups, for instance to obtain information on the Prüfer rank of a finite group which is the product of two subgroups.

AMS Classification: 16N40

1 Introduction

An associative algebra R is called *nilpotent* with *nilpotency class* $n = n(R)$ if there exists a natural number n such that $R^n \neq 0$ and $R^{n+1} = 0$. Nilpotent algebras have been studied in several papers; see for instance the monographs by R. Kruse and D. Price [34] and D. Suprunenko and R. Tyshkevich [49]. Every nilpotent algebra R forms a group under the "circle operation" $x \circ y = x + y + xy$ for every two elements $x, y \in R$. This group is called the *adjoint group* R° of R (see for instance [34] or [1]) .

An associative algebra R over a field F of characteristic p for the prime p is called a *p-algebra*. It is easy to see that the adjoint group of a nilpotent p-algebra is a p-group. This raises the question which finite p-groups occur as the adjoint group of some finite nilpotent p-algebra.

Examples of nilpotent algebras are for instance the subalgebras of the algebra of all upper triangular matrices with zero diagonal entries and with dimension n over a field F. The augmentation ideal of a group algebra of a finite p-group over a field of characteristic p is a nilpotent p-algebra. Therefore the study of nilpotent algebras can be used to obtain information about finite p-groups and their group algebras. If a finite p-group G occurs as the adjoint group of some nilpotent p-algebra, then it is complemented in the group of units of the corresponding group algebra (see for instance [17]).

It is clear that every elementary abelian p-group is the adjoint group of the corresponding null-algebra (i.e., the algebra with zero multiplication). On the

---

[1]The second author likes to thank the Deutsche Forschungsgemeinschaft and RFBR for financial support and the Department of Mathematics of the University of Mainz for its excellent hospitality during the preparation of this paper.

other hand, the cyclic p-group of order at least p^2 can be the adjoint group of some nilpotent p-algebra only if it has order 4. In general there will be many nilpotent algebras with isomorphic adjoint groups. For instance, it is observed in [34], Chapter V, that there are at least 100000 pairwise non-isomorphic nilpotent rings and more than 35000 nilpotent 2-algebras of order 2^6, but there exist only 267 groups of order 2^6. Note also that it was proved in [49] that there are infinitely many non-isomorphic nilpotent algebras of dimension 6 over the field of real numbers.

The well-known construction by E. Golod of a non-nilpotent nil algebra R whose adjoint group contains an infinite finitely generated residually finite p-subgroup gives for each prime p a series of finite p-algebras R/R^k (see [22] and [42]).

We will also discuss a connection between nilpotent p-algebras and their adjoint groups with the structure of finite p-groups that can be represented as the product of two its proper subgroups.

There is also a relation between commutative nilpotent algebras with some enumerating problems described in [48] and based on F. Macaulay [37].

All group-theoretical and ring-theoretical notation is standard. In particular we note the following.

The n-th power of an algebra R is the subalgebra R^n of R generated by the set of elements of the form $x_1 x_2 \ldots x_k$ with $k \geq n$, where $x_1, x_2, \ldots, x_k \in R$. In particular $R^1 = R$. The subalgebra L of an algebra R generated by the set of elements $x_1, x_2, \ldots, x_k$ will be denoted by $\langle\langle x_1, x_2, \ldots, x_k \rangle\rangle$, whereas the subspace of the algebra R generated by these elements is $\langle x_1, x_2, \ldots, x_k \rangle$. An algebra $R = \langle\langle a \rangle\rangle$ over the field F is said to be *one-generated* if there exists an element a in R such that every element of R can be expressed as $af(a)$ for some polynomial $f \in F[x]$.

For every natural number i consider the subalgebra $R^{(i)} = \langle\, a^{p^i} \mid a \in R \,\rangle$ and the ideal $R_{(i)} = \langle\, a \in R \mid a^{p^i} = 0 \,\rangle$ of the commutative algebra R. Then clearly $R/R_{(i)} \simeq R^{(i)}$. Furthermore $d(R)$ is the minimal number of generators of an algebra R and $d(R^\circ)$ is the minimal number of generators of its adjoint group.

2   Some open problems

There is a close relationship between the structure of the adjoint group of a nilpotent p-algebra and certain factorized groups. We begin with an open problem concerning such groups.

A group G has *finite Prüfer rank* $r = r(G)$ if every finitely generated subgroup of G can be generated by r elements and r is the least such number.

Problem 1 Let the finite p-group $G = AB$ be the product of two subgroups A and B, whose Prüfer ranks do not exceed the nonnegative number r. Does there exist a positive constant c such that the Prüfer rank of G satisfies $r(G) \leq cr$?

It can be shown that this problem can be solved at least for abelian factors A and B provided the following question has a positive answer (see Section 6 below).

Problem 2 Let R be a finite nilpotent p-algebra. Does there exist a positive constant e such that
$$r(R^\circ) \geq e \dim R?$$

Clearly this question can be formulated for a commutative nilpotent algebra R in terms of the subalgebra $R^{(1)} = \{a^p \mid a \in R\}$ as follows (see [3]).

Problem 3 Let R be a commutative nilpotent p-algebra. Does there exist a constant $f > 2$ such that
$$\dim R \geq f \dim R^{(1)}?$$

It was shown in [3], Proposition 5.2, that if this question has a positive answer for every commutative nilpotent p-algebra satisfying the identity $a^{p^2} = 0$, then
$$r(R^\circ) \geq \frac{f-2}{f-1} \dim R,$$

so that also Problem 2 has a positive answer.

N. Eggert proved in [20] the following interesting result.

Theorem 2.1 *Let R be a commutative nilpotent algebra over a perfect field of characteristic p. If $\dim R^{(1)} = 2$, then $\dim R \geq 2p$.*

This lead Eggert to the following conjecture.

Problem 4 (Eggert's conjecture) Let R be a commutative nilpotent p-algebra. Is it true that
$$\dim R \geq p \dim R^{(1)}?$$

It is clear that Problem 3 is a weak form of this conjecture.

If Eggert's conjecture could be proved, this has a number of important consequences. For instance it was already noted in [20], Theorem 2, that the structure of the adjoint group of a commutative nilpotent algebra R could then easily be described.

More precisely, then there exist one-generated algebras $L_1, L_2, \ldots, L_k$ for some $k \geq 1$ such that $R^\circ \simeq L_1{}^\circ \times L_2{}^\circ \times \cdots \times L_k{}^\circ$. Note that these algebras L_i need not be subalgebras of R with the property $L_1 + L_2 + \cdots + L_k = R$ and $L_i \cap L_j = 0$ for each $0 \leq i < j \leq k$.

As an application we would have an answer to the following problem.

Problem 5 Determine the structure of the group of units of a finite associative commutative algebra.

This question is connected to the following problem of Suprunenko [49].

Problem 6 What is the structure of a maximal abelian p-subgroup of the group $GL_n(F)$ over a finite field F of characteristic p?

W. Brown solved this problem in [18] for $n \leq 5$. However, in general only some bounds for the orders of the maximal abelian p-subgroups are known. By classical results due to I. Schur, W. Gustafson and T. Laffey ([45], [24], [35]) the dimension of the maximal commutative nilpotent subalgebra R of the matrix algebra $M_n(F)$ over the field F satisfies the following inequalities

$$(2n)^{2/3} \leq \dim R \leq \lceil n^2/4 \rceil.$$

In these investigations the nilpotency class $n(R)$ of a commutative nilpotent algebra R plays a role, so that also the following question is natural.

Problem 7 What is the structure of a nilpotent algebra R with relatively high nilpotency class $n(R)$? In particular does this imply the existence of a large commutative subalgebra of R?

In general the minimal number of generators of a nilpotent algebra and that of its adjoint group do not coincide. Therefore the following may be asked.

Problem 8 Let R be a nilpotent algebra. Is the dimension of R bounded in terms of the minimal number $d(R^\circ)$ of generators of its adjoint group?

More generally, we ask the following.

Problem 9 Is every nil algebra R nilpotent if its adjoint group has only finitely many generators?

Note that Golod's group is only a subgroup of the adjoint group of a certain nil algebra, but not necessarily the whole adjoint group. On the other hand, J. Isbell has proved in [29], that every commutative algebra whose multiplicative semigroup has finitely many generators, is finite.

A very useful concept in the theory of finite p-groups is the notion of a *powerful* subgroup introduced by A. Lubotzky and A. Mann in [36]. It is not difficult to show that every nilpotent p-algebra R contains a subalgebra L of relatively small codimension such that its adjoint group is powerful. What can be said about the structure of this subalgebra L? For instance, the following question is natural.

Problem 10 Let R be a nilpotent algebra whose adjoint group is powerful. Is $r(R^\circ) \geq h \dim R$ for some positive constant h depending only on p?

3 The adjoint group of a finite nilpotent algebra

In this section we will discuss which finite p-groups occur as the full adjoint group of some (finite) nilpotent ring or even nilpotent algebra.

It should be noted that every finite p-group is isomorphic with a subgroup of the adjoint group of some nilpotent p-algebra. To see this, observe that every finite p-group is isomorphic to a subgroup of a Sylow p-subgroup of a general linear group $GL(n, q)$ for some positive integer n over a field with $q = p^n$ elements. It is well-known that a Sylow p-subgroup of $GL(n, q)$ can be expressed as a group of all upper triangular matrices over $GF(q)$ of size n with entries "1" in the main diagonal, and it is obvious that this group is the adjoint group of the nilpotent p-algebra of all upper triangular matrices over $GF(q)$ of size n with entries "0" in the main diagonal.

L. Fischer and K. Eldridge have classified in [21] all finite rings with cyclic adjoint groups. There are only two types of nilpotent p-algebras with this property: the null algebra with p elements and a 2-algebra generated by an element u such that $u^2 \neq 0 = u^3$.

Metacyclic groups that occur as the adjoint groups of a finite nilpotent p-algebra are described by B. Gorlov in [23]. The main result is as follows.

Theorem 3.1 *If R is a finite nilpotent p-algebra whose adjoint group is metacyclic, then R° is either an elementary abelian p-group of order at most p^2, or $p = 3$ and $R^\circ \simeq \mathbb{Z}_9 \times \mathbb{Z}_3$, or $p = 2$ and R° is one of the following groups: $\mathbb{Z}_4, \mathbb{Z}_2 \times \mathbb{Z}_4, \mathbb{Z}_4 \times \mathbb{Z}_4, \mathbb{Z}_2 \times \mathbb{Z}_8, \mathbb{Z}_4 \rtimes \mathbb{Z}_4, D_8, Q_8$.*

The following two results from [4] generalize Theorem 3.1.

Theorem 3.2 *Let R be a nilpotent noncommutative finite 2-algebra whose adjoint group G has no elementary abelian subgroups of rank 3. If G is not metacyclic, then one of the following holds.*

(i) *$G \simeq Q_8 \times Q_8$, where $G' \simeq \mathbb{Z}_2 \times \mathbb{Z}_2$ or $G \simeq Q_8 * Q_8$ is a central product, or G is isomorphic with a subgroup of these groups;*

(ii) *$G \simeq \langle a, b, c \rangle$ with the relations $a^4 = b^4 = c^4 = 1; a^2 = c^2, (a, c) = b^2, (b, c) = c^2, (a, b) = 1$;*

(iii) *G is isomorphic to a Sylow 2-subgroup of a group of type $U_3(4)$: $G = \langle a, b, c, d \rangle$ with relations $a^4 = b^4 = c^4 = d^4 = 1, d^2 = a^2 \circ b^2, c^2 = b^2, (a, c) = a^2, (a, d) = (b, c) = a^2 \circ b^2, (b, d) = b^2, (a, b) = (c, d) = 1$.*

Theorem 3.3 *Let R be a nilpotent p-algebra whose dimension is at least $p(p + 1)/2$ if $p > 2$ and $\dim R \geq 7$ if $p = 2$. Then R° contains an elementary abelian p-subgroup of rank p.*

It is much easier to describe commutative nilpotent p-algebras R with rank less than the prime p. For these we have $\dim R \leq p + 1$ if $p > 2$ and $\dim R \leq 4$, if $p = 2$. This result depends on the description of a commutative nilpotent algebra which is either one-generated or has a one-generated subalgebra with codimension one (see [4]).

The structure of the adjoint group of a nilpotent one-generated p-algebra $R = \langle\langle a \rangle\rangle$ can easily be described. Suppose that $\dim(R) = n$, i.e., $a^{n+1} = 0 \neq a^n$. Then the nility of a is $\nu(a) = n$ and the rank of the adjoint group of R is $r(R^\circ) = r = n - [n/p]$ (see [4]).

For the p-algebra R define a matrix $M(R) = (\alpha_{ij})$ of size $l \times r$, where $l = 1 + [\log_p n]$, as follows.

The elements $\alpha_{11}, \alpha_{12}, \ldots, \alpha_{1l}$ of the first row of $M(R)$ are integers between n and $n - [n/p] + 1$, namely $\alpha_{11} = n$ and $\alpha_{1i+1} = \alpha_{1i} - 1$ for $i = 1, 2, \ldots, l - 1$. The elements in each column of $M(R)$ are as follows: if α_{1j} is the first element in the j-th column then we put $\alpha_{2j} = \alpha_{1j}/p$ if p divides α_{1j} and $\alpha_{2j} = 0$ otherwise. Define the elements α_{kj} inductively by $\alpha_{k+1j} = \alpha_{kj}/p$ if p divides α_{kj} and $\alpha_{k+1j} = 0$ otherwise for each $k = 1, 2, \ldots, l$.

For illustration consider the following example: $R = \langle\langle a \rangle\rangle, n = \dim R = 17, p = 2$. Then $r = 17 - [17/2] = 9$ and $l = [\log_2 17] + 1 = 5$. Hence

$$M(R) = \begin{pmatrix} 17 & 16 & 15 & 14 & 13 & 12 & 11 & 10 & 9 \\ 0 & 8 & 0 & 7 & 0 & 6 & 0 & 5 & 0 \\ 0 & 4 & 0 & 0 & 0 & 3 & 0 & 0 & 0 \\ 0 & 2 & 0 & 0 & 0 & 0 & 0 & 0 & 0 \\ 0 & 1 & 0 & 0 & 0 & 0 & 0 & 0 & 0 \end{pmatrix}$$

The matrix $M(R)$ is of type $(m_1, m_2, \ldots, m_l)$ if it has m_1 columns with exactly 1 non-zero element, m_2 columns with 2 non-zero elements, $\ldots$, m_l columns with l non-zero elements. Note that $\sum_{i=1}^{l} i m_i = n$. In particular in the above example $M(R)$ is of type $(5, 2, 1, 0, 1)$ and $1 \cdot 5 + 2 \cdot 2 + 3 \cdot 0 + 4 \cdot 1 = 17$.

The following result appears in [4].

Theorem 3.4 *Let $R = \langle\langle a \rangle\rangle$ be a one-generated nilpotent p-algebra of dimension n whose matrix $M(R)$ is of type $(m_1, m_2, \ldots, m_l)$. Then the adjoint group of R is isomorphic to a group*

$$G = \mathbb{Z}_p{}^{m_1} \times \mathbb{Z}_{p^2}{}^{m_2} \times \cdots \times \mathbb{Z}_{p^l}{}^{m_l}.$$

The description of nilpotent algebras over a field of characteristic zero having a one-generated subalgebra of codimension 1 can be found in [49]. For the description of the adjoint group of nilpotent p-algebras of this type see [4].

L. Kaloujnine has shown in [30] that for each odd prime p every finite p-group of nilpotency class 2 is isomorphic with the adjoint group of some nilpotent ring. In fact, all groups of order p, p^2 and p^3 occur as the adjoint group of some nilpotent ring, but a group of order p^4, if and only if, it is nilpotent of class ≤ 2 (see [34], Chapter I, Section 6).

L. Moran and R. Tench [41] and independently A. Bovdi [16] proved that every finite p-group of nilpotency class 2 and exponent p and every finite 2-group of class 2 and exponent 4 is the adjoint group of some nilpotent p-algebra (respectively 2-algebra) (see also [4]).

The results above allow a description in terms of generators and defining relations of all p-groups of order at most p^5 which occur as the adjoint group of some nilpotent p-algebra. The p-groups for $p > 2$ of order at most p^5 which occur as the adjoint groups of nilpotent rings were described by K. Tahara and A. Hosomi [51]. The comparison of their list with that in [4] shows that not every group that is the adjoint group of a nilpotent ring is the adjoint group of a nilpotent algebra. On the other hand, the proof of the corresponding result in [4] is substantially shorter and covers even the more difficult case $p = 2$.

The p-groups of order p^3 and p^4 which are the adjoint group of some nilpotent p-algebra can be found in [4].

Theorem 3.5 *Let G be the adjoint group of some nilpotent p-algebra R of dimension ≤ 4. Then one of the following holds.*

(i) *$|G| = p$ and G is cyclic of order p;*

(ii) *$|G| = p^2$ and G is either an elementary abelian group of order p^2 or $p = 2$ and G is a cyclic group of order 4;*

(iii) *$|G| = p^3$ and G is either an elementary abelian p-group, or $p \geq 3$ and G is a p-group of class 2 with exponent p, or $p = 3$ and $G \simeq \mathbb{Z}_9 \times \mathbb{Z}_3$, or $p = 2$ and G is a non-cyclic group of order 8;*

(iv) *$|G| = p^4$ and G is either an elementary abelian p-group, or $p \geq 5$ and G is a group of exponent p with commutator subgroup of order p, or G is a group of exponent ≤ 9 and commutator subgroup of order 3, or $G \simeq \mathbb{Z}_9 \times \mathbb{Z}_3 \times \mathbb{Z}_3$, or G has class 2 and exponent 4, or $G \simeq \mathbb{Z}_8 \times \mathbb{Z}_2$.*

The following theorem describes all nonabelian p-groups of order p^5 which occur as the adjoint group of some nilpotent p-algebra.

Theorem 3.6 *Let G be a non-abelian adjoint group of some nilpotent p-algebra R of dimension 5. Then one of the following holds.*

(i) *G is p-group of class 2 and of exponent p;*

(ii) *G is one of the 5 groups of class 3;*

(iii) *G is a p-group for $p \leq 3$ of class at most 2 and of exponent $\leq p^2$ such that G' is group of order $\leq p^2$ and $\Phi(G) \leq Z(G)$ is an elementary abelian p-group.*

Kruse proved in [34] that the adjoint group of a nilpotent algebra of dimension n has nilpotency class at most $(n + 1)/2$. Hence a p-group of maximal class occurs as the adjoint group of some nilpotent p-algebra only if $n \leq 3$. In [4] the finite p-algebras for odd primes p whose adjoint group has at most two generators are classified. The dimension of these algebras is at most 3 (see [4], Theorem 3.6). This is not the case for $p = 2$, since there exists a finite nilpotent 2-algebra with dimension 5 whose adjoint group has two generators.

Nevertheless the following holds (see [5]).

Theorem 3.7 *Let R be a nilpotent p-algebra whose adjoint group has only two generators. Then $\dim R \leq 5$.*

If the adjoint group of a nilpotent p-algebra R has at most 3 generators there is the following result (see [32]).

Theorem 3.8 *Let R be a nilpotent p-algebra with $p > 3$, whose adjoint group has at most 3 generators. If R has two generators, then $\dim R \leq 9$.*

The two preceding theorems show that Golod's group, which is infinite, is not the whole adjoint group of the corresponding nil algebra.

The following theorem from [9] excludes some other classes of p-groups that cannot occur as the adjoint group of a finite nilpotent algebra. If G is a finite p-group, we consider the natural numbers $\mu_i = \mu_i(G)$ determined by $|\gamma_i(G)/\gamma_{i+1}(G)| = p^{\mu_i}$; here $\gamma_i(G)$ denotes the i-th term of the lower central series of G.

Theorem 3.9 *Let G be a finite p-group such that the numbers $\mu_i(G) = \mu_i$ satisfy $\sum_{\mu_i \geq 3} \mu_i \leq k$. If G is the adjoint group of some nilpotent p-algebra, then $|G| \leq f(p, k)$ for some function f depending only on p and k.*

Some further information on the relations between the structure of a nilpotent algebra and its adjoint group can for instance be found in the survey articles [12] and [17].

4   The nilpotency class of a nilpotent algebra and connections with finite p-groups

It is well-known that the structure theory of finite nilpotent algebras is closely related with the theory of finite p-groups (see for instance [1]). But although there are many papers about finite p-groups, the theory of finite nilpotent algebra is not very well developed. One reason for this may be that even in the case of a commutative nilpotent p-algebra one has to deal with non-abelian p-groups. For instance, the product ab of two elements a and b of a commutative nilpotent algebra R corresponds to a commutator $[c, d]$ of two elements in a nilpotent group G. Hence the appropriate analog to the Engel condition in group theory is an identity of the form $ax^n = 0$ in the algebra R.

It is natural to ask for the nilpotency class of a commutative nilpotent algebra satisfying the identity $x^n = 0$ for all $x \in R$. A classical result of Frobenius (see [49]) says that if R is such a commutative algebra over a field of characteristic zero, then the nilpotency class of R is at most $n - 1$. This approach fails however for commutative algebras over a field of prime characteristic. The bound for the nilpotency class for these algebras satisfying an identity of the form $x^n = 0$ was determined by A. Belov [15] and improved by A. Klein [33]. However it is difficult to use corresponding bounds when we intend to find relations between the dimension of a commutative nilpotent algebra and the rank of its adjoint group.

What relations exist between the nilpotency class of an algebra, the rank of its adjoint group and its dimension? In these investigations the existence of a "large" one-generated subalgebra plays an important role. It is also of some interest to find analogs for algebras of some recent results about the small co-class problem for finite p-groups.

If G is a finite p-group of order p^n with nilpotency class m, then the co-class of G is $n - m$. Hence we may define the co-class of a nilpotent algebra R of dimension n and nilpotency class $n(R)$ as $n - n(R)$. Clearly, the p-groups with co-class 1 are the groups of maximal class. Nilpotent algebras with co-class at most 2 over a field of characteristic zero were studied in [49].

It was proved by Stack [46] that if R is a nilpotent p-algebra with $d_i(R) = \dim R^i/R^{i+1} = 1$, then there exists a one-generated subalgebra L of R such that $R^i = L^i$. This indicates that the case when the parameters $d_i(R)$ are relatively small plays a special role.

The following general result was proved in [4] and solves the "small co-class problem" for nilpotent algebras. Recall that the minimal number of generators of the algebra R is denoted by $d(R)$.

Theorem 4.1 *Let R be a nilpotent algebra over an arbitrary field F. Then there exists a one-generator subalgebra L in R such that $\dim L = n(R)$ provided that one of the following conditions holds*

   (i) $\dim R^i/R^{i+1} = 1$ *for some integer* $0 < i < n(R)$;

  (ii) $\dim R \leq 2n(R) + d(R) - 3$;

 (iii) $2k + 2 \leq \dim R \leq n(R) + k$ *for some natural number* k;

 (iv) $n(R) > \lceil \frac{1}{2}(\dim R + 1) \rceil$.

Thus if the co-class k of a nilpotent algebra R is less than $\lceil \frac{1}{2}(\dim R - 1) \rceil$, then there exists a commutative subalgebra with codimension depending only on k. The situation when the nilpotency class $n(R)$ of the algebra R is close to $\frac{1}{2} \dim R$ is discussed in [4].

In their classical paper [40] Miller and Moreno considered finite non-abelian groups, in which every proper subgroup is abelian. A complete list of these groups was later given in [43]. These results suggested to consider similar classifications for other group classes. In [38] MacDonald studied finite nilpotent groups, in which every proper subgroup or every m-th maximal subgroup is nilpotent of a given class. In particular he proved that a finite p-group of nilpotency class $2n$ in which every proper subgroup has nilpotency class n, has at most two generators and if G is metabelian, then $n = 1$ ([38], Theorem 1). He showed also that if G is a group of order p^{r+m} in which every subgroup of order p^r has nilpotency class n, whereas G is not of class n, then the following holds.

   (i) G has a minimal generating set of $m + n$ generators;

  (ii) The class of G is bounded by some function $f(m, n)$;

 (iii) The order of the $(n + 1)$-th term of the lower central series $\gamma_{n+1}(G)$ divides $p^{g(m,n)}$;

 (iv) The order of the factor group $G/\zeta_n(G)$ divides $p^{h(m,n)}$.

Here f, g, h are certain functions of m and n, which do not dependent on r ([38], Theorem 2). If G is a p-group of order p^{r+2} in which every subgroup of order p^r has class 2, then the nilpotency class of G is at most 4 ([38], Theorem 3).

In the following we list corresponding results [8] for finite-dimensional nilpotent algebras over an arbitrary field F, which are even more precise than those for finite p-groups. The proofs of these depend heavily on the above-mentioned results concerning algebras having "large" one-generator subalgebras.

A well-known theorem of Fitting says that the product of two nilpotent normal subgroups of nilpotency classes m and n of a group G is likewise a nilpotent normal subgroup of G with class at most $m + n$. This theorem has an obvious analog for associative algebras. If a nilpotent algebra R is the sum of two ideals of nilpotency classes m and n respectively, then the nilpotency class of R is at most $m + n$. Note that this bound is sharp, as can be seen from the nilpotent Grassman algebra R with n generators, which is the sum $R = A + B$ of two ideals A and B such that $n = n(R) = n(A) + n(B)$.

In particular a nilpotent algebra with at least two maximal subalgebras of class n has nilpotency class at most $2n$. The following theorem shows that this bound is only rarely attained.

Theorem 4.2 *Let R be a nilpotent algebra over the field F such that all proper subalgebras of R have nilpotency class at most n, whereas the nilpotency class of R is at least $2n$. Then R either is a one-generator algebra with $\dim R = 2n$ or $2n+1$, or $n = 1$ and $\dim R = 3$.*

The next result corresponds to a group theoretical result in [38], Theorem 2.

Theorem 4.3 *Let R be a nilpotent algebra over the field F such that every subalgebra of R with codimension m has nilpotency class at most n, whereas the nilpotency class of R is at least $n + 1$. Then the following holds*

(i) *R has a generating set of at most $m + n$ elements;*

(ii) *The nilpotency class of R is bounded by a function $f(m, n)$;*

(iii) *The dimension of R is bounded by a function $g(m, n)$, where*
$$g(1, n) < 2(n + 1)^{n+1} \quad and \quad g(m, n) \le m + g(1, n) - 1;$$

(iv) *If there exists a nilpotent algebra A with nilpotency class $n(A) = k > n$ and minimal number of generators $d(A) = d$, all of whose proper subalgebras have nilpotency class at most n, then there is a nilpotent algebra R such that the dimension of R is at least $(d^{k+1} - 1)/(d - 1)$.*

Here the functions f and g depend only on m and n and not on the field F. The function $f(m, n)$ may be chosen as $f(m, n) = n \cdot 2^m$ (as in the case of finite p-groups in [38]). However, in most cases a better bound is given by the function $f(m, n) = m(n + 1) - 1$. In the case of an algebra which is not one-generated we may improve this bound by $f(m, n) = (mn + m + 2n)/2$ if m is even and $f(m, n) = (mn + m + n - 1)/2$ if m is odd and $m > 1$. In all cases $f(1, n) = 2n + 1$.

The function $g(m, n)$ satisfies $g(m, n) \le m + g(1, n) - 1$. It follows from condition (iv) that there always exists a nilpotent algebra R with nilpotency class $n(R) = n + 1$, all of whose proper subalgebras have nilpotency class at most n such that $\dim R = ((n + 1)^{n+1} - 1)/n$. This bound is in a sense best possible, since the nilpotent Grassman algebra with $n + 1$ generators has all its subalgebras of class at most n.

Finally we mention the following result.

Theorem 4.4 *Let R be a nilpotent algebra over the field F such that every proper subalgebra of R has nilpotency class at most 2, whereas the nilpotency class of R is at least 3. Then $d(R) \le 3$ and $n(R) \le 13$.*

5 Hilbert functions of commutative nilpotent algebras

The n-th power of an associative algebra R over the field F is the subalgebra R^n of R generated by the set of elements of the form $x_1 x_2 \dots x_k$ with $k \ge n$, where $x_1, x_2 \dots, x_k \in R$. The descending series $R \ge R^1 \ge R^2 \cdots \ge R^n \ge \dots$ is called the *power series* of R. The dimensions $d_i = d_i(R) = \dim R^i/R^{i+1}$ of the factors of the power series of R are also called the *Hilbert functions* of R, usually denoted by $H(R, i)$.

In this section we are interested in the Hilbert functions of a nilpotent algebra R. It turns out that if d_i is relatively small and $i > 1$, then $d_j \leq d_i$ for each $j \geq i$. This property is very useful in the investigation of the structure of a finite nilpotent algebra (see for instance [2], [3]).

Let R be Noetherian commutative ring with unit element, graded by the set of nonnegative integers N, then R has a direct-sum decomposition

$$R = R_0 \oplus R_1 \oplus R_2 \oplus \ldots,$$

where $R_i R_j \subseteq R_{i+j}$ and $1 \in R_0$. If in addition $R_0 = F$ is a field, so that R is an F-algebra, we will say that R is a *G-algebra* in the sense of [48]. The assumption that R is Noetherian implies that R is finitely generated and that each R_n is a finite-dimensional vector space over F. In particular we have $d_0 = 1$ and $d = d_1$ is the minimal number of generators of R.

If R is a commutative nilpotent algebra over the field F and $Q = R \otimes_F K$ for some extension field K of F, then the dimensions and the Hilbert functions of R and Q coincide. Furthermore each linearly independent subset of R remains linearly independent in Q. These simple observations are used frequently (see for instance [48]). From this it follows in particular that we may assume in our considerations that the ground field of the algebra R is algebraically closed.

Let $A = F[y_1, y_2, \ldots, y_d]$ be a polynomial ring over the field F with d independent variables. There is a canonical surjection p from A to R defined by $p(y_i) = x_i$. A non-empty subset T of A consisting of monomials of the form $y_1^{a_1} y_2^{a_2} \ldots y_d^{a_d}$ in the variables $y_1, y_2, \ldots, y_d$ is said to be an *order ideal of monomials* if whenever $t \in T$ and t' divides t, then $t' \in T$.

The following result is due to Macaulay (see [48], Theorem 2.1).

Theorem 5.1 *There exists in A an order ideal T of monomials in variables y_1, y_2, $\ldots$, y_d such that the elements $p(t)$, $t \in T$, form a basis of R.*

It follows from Theorem 2.2 in [48] that there exists a commutative nilpotent algebra R with given Hilbert functions $d_i(R)$, if and only if there is an ideal of monomials T in A such that $d_i(R) = |\{u \in T \mid \deg u = i\}|$ for each $i \geq 1$.

As a corollary of Theorem 5.1 Stanley showed that if $d_m(R) \leq m$ for some $m \geq 2$, then $d_{n+1}(R) \leq d_n(R)$ for all $n \geq m$. This important observation was used by K. R. McLean [39] in his proof of some generalizations of Eggert's theorem in a class of graded algebras.

The following theorem of [11] gives some information about the Hilbert functions of a commutative nilpotent algebra.

Theorem 5.2 *Let R be a commutative nilpotent algebra over the field F with a minimal generating set X with d elements. Then the following holds.*

(i) *If $d_k \leq k$ for some integer $k \geq 2$, then $d_j \leq d_k$ for each $j \geq k$. Moreover, if the field F is large enough, then there exists an element $x \in R \setminus R^2$ such that $R^j = R^{j-1} x + R^{j+1}$;*

(ii) *If there exists an integer $m \geq 2$ such that for every non-zero element $x \in \langle X \rangle$ we have $x^{m+1} \in R^{m+1} \setminus R^{m+2}$, then $d_m \geq d$. Moreover, if $|F| \geq d$ and $d_m < d$, then there exists some element $y \in R \setminus R^2$ such that $R^{m-1}y^2 \subseteq R^{m+2}$.*

The proof of Theorem 5.2 depends heavily on the following result.

Lemma 5.3 *Let $\{X_{i,j} \mid 1 \leq i \leq m, \ 1 \leq j \leq n\}$ be a set of matrices with sizes $a_i \times k$ and entries in F such that for each $i \leq m$ there exists a number $s = s(i)$ such that the matrix $X_{i,s}$ has rank a_i. If the field F is large enough, then there exists a sequence $\{\lambda_1, \lambda_2, \ldots, \lambda_n\}$ of elements in F such that for each i the linear combination $\sum_{j=1}^{n} \lambda_j X_{i,j}$ has rank a_i.*

The case $d_i(R) \leq 3$ for a nilpotent algebra R and $i \geq 2$ was discussed in [10].

The following theorem (see [5]) is similar to the above-mentioned result of Stanley for noncommutative algebras.

Theorem 5.4 *Let R be a nilpotent algebra over an arbitrary field F and $d_i(R) \leq 2$ for some integer $i > 1$. Then $d_j(R) \leq 2$ for each $j \geq i$.*

6 Products of finite p-groups and their ranks

Let the finite group $G = AB$ be the product of two its subgroups A and B. It is well-known that for each prime p dividing the order of G there exist Sylow p-subgroups A_p and B_p of A and B, respectively such that $A_p B_p = B_p A_p = G_p$ is a Sylow p-subgroup of G (see for instance [1], Corollary 1.3.3). This fact allows to reduce many problems concerning the structure of a finite factorized group G to factorized p-groups.

If the properties of the subgroups A and B are inherited by subgroups and epimorphic images, then we may apply induction arguments and consider minimal counterexamples. In many cases such a minimal counterexample is a finite p-group $G = AB = AM = BM$ for three subgroups A, B and M of G, where M is normal in G. If in addition $A \cap M = M \cap B = 1$, then G is called *triply factorized*. It was observed by Ya. Sysak in [50] that such triply factorized groups may be constructed from finite nilpotent p-algebras (see for instance [1], Chapter 6).

More precisely, let R be any finite nilpotent p-algebra. Then the adjoint group R° of R operates on its additive group $R^+ = M$ mapping the element r of R onto the element $r + ra$ for every a in R°. The corresponding semidirect product $G(R) = M \rtimes A = M \rtimes B = AB$ is a triply factorized p-group (the so-called *associated group* of R), where the normal subgroup M is isomorphic with R^+ and A and B are subgroups of $G(R)$ isomorphic with R°.

Conversely, Sysak also proved in [50] the following theorem.

Theorem 6.1 *Let the finite p-group $G = AB = AM = BM$ be the product of three abelian subgroups A, M, B with $A \cap B = A \cap M = M \cap B = 1$, where M is normal in G. Then there exists a commutative nilpotent p-algebra R such that its associated subgroup $G(R)$ is isomorphic with G and $A \simeq R^\circ \simeq B$.*

It should be noted that in the above construction the normal subgroup M of the triply factorized group G is always abelian. Using a modification of these ideas with a near-ring in the place of the nilpotent algebra R, P. Hubert constructed also triply factorized groups, where the normal subgroup M of G is not necessarily abelian, and showed that Theorem 6.1 can be generalized widely (see [26] and [27]).

If the finite p-group $G = AB$ is the product of two subgroups A and B, whose Prüfer ranks are bounded by the real number r, then it was proved by D. Zaitsev [52] and L. Kazarin [31] that the Prüfer rank of G is bounded by a polynomial function of r (see [1], proof of Theorem 4.3.5).

This result was considerably improved in [2], where a bound for the normal rank $r_n(G)$ is given. Recall that the *normal rank* $r_n(G)$ of G is the maximum of the minimal number of generators of each normal subgroup of G.

It is still not known whether there exists even a linear bound with this properties (see Problem 1 above).

The results from [2] were further improved in [3] as follows.

Theorem 6.2 *Let the finite p-group $G = AB$ be the product of two subgroups A and B such that $r(A) \leq r(B)$. If $p \geq 3$, then the normal rank $r_n(G)$ satisfies the following inequality:*

$$r_n(G) \leq 2r(A)(\lceil \log_2 r(A) \rceil + 1) + r(A) + r(B).$$

As a consequence of Theorem 6.2 a bound for the Prüfer rank of G is obtained in terms of the Prüfer ranks of A and B which is almost linear.

Corollary 6.3 *Let the finite p-group $G = AB$ be the product of two subgroups A and B, whose Prüfer ranks do not exceed the real number r. If $p \geq 3$, then the Prüfer rank of G satisfies the following inequality:*

$$r(G) \leq 4r(\lceil \log_2 r \rceil + 2)^2.$$

Note that the bound given in Corollary 6.3 is better than a bound of the form $r(G) \leq cr^\alpha$ with $\alpha > 1$ and $c > 0$ for almost all r.

The case when the ranks of the two factors A and B of the factorized p-group $G = AB$ are small is of special interest. B. Huppert showed that for $p > 2$ every finite p-group, which is the product of two cyclic subgroups, is metacyclic; see for instance [28], Theorem III.(11.5). A similar result holds for finite p-groups which are the product of two metacyclic subgroups.

Theorem 6.4 *Let the finite p-group $G = AB$ with $p > 3$ be the product of two subgroups A and B. If A is metacyclic, then the normal rank of G does not exceed $r(B) + 2$. In particular, if A and B are both metacyclic, then the Prüfer rank of G is at most 4.*

It also follows from [3] that for every positive integer k there are only finitely many primes p such that a finite p-group $G = AB$, which is the product of two subgroups with Prüfer ranks at most k, has Prüfer rank more than $2k$. It is enough to see that this property holds if $2k < p$. Note that in the case $p = 3$ there exists a finite 3-group $G = AB$ with Prüfer rank 5, which is the product of two metacyclic abelian subgroups (see [3]).

Recently M. Conder and M. Isaacs [19] proved the following result.

Theorem 6.5 *Let the finite p-group $G = AB$ be the product of an abelian subgroup A and cyclic subgroup B. Then $G'/(G' \cap A)$ is cyclic.*

The proofs of the preceding four theorems are group-theoretical. As a consequence of Theorems 6.2 and 6.4 we obtain the following information about the dimension of a nilpotent p-algebra (see [3]).

Theorem 6.6 *Let R be a nilpotent p-algebra for $p > 2$. Then*

$$\dim R \leq 2r(R^\circ)(\lceil \log_2 r(R^\circ) \rceil + 1).$$

If R is commutative, then

$$\dim R \leq r(R^\circ)(\lceil \log_p(p/(p-1)r(R^\circ) \rceil + 1).$$

It is surprising that the difference between the bounds in the commutative and in the non-commutative cases is not very large.

7   On Eggert's conjecture

Eggert's conjecture concerning commutative nilpotent algebras has influence on other algebraic problems, in particular on questions about factorized groups (see for instance [3]). As an example we note the following.

Lemma 7.1 *Let the finite p-group $G = AB$ be the product of two abelian subgroups A and B. If Eggert's conjecture is true, then the Prüfer rank $r(G)$ of G satisfies the inequality*

$$r(G) \leq 2(r(A) + r(B)) + \frac{p}{(p-1)} \min\{r(A), r(B)\}.$$

As was already noted above, there exists a linear bound for the Prüfer rank of a product of two abelian p-groups, if there exists a constant $c > 2$ such that $\dim R \geq c \dim R^{(1)}$. In general there is only the trivial bound $\dim R \geq 2 \dim R^{(1)}$, which holds for all nilpotent p-algebras of nilpotency class $\geq p$.

It seems difficult to decide whether Eggert's conjecture holds in general. Here we confirm it for commutative nilpotent p-algebras with certain restrictions and discuss some related results.

The conjecture of Eggert was proved 1976 by R. Bautista ([14], see also [47]) for the case $\dim R^{(1)} \leq 3$. In 1996 C. Stack [46] showed that if a nilpotent algebra over a field of characteristic p contains two elements a and b such that a^p and b^p are linearly independent, then $\dim R \geq 2p$.

Note that it is not assumed here that R is commutative. The question of Stack whether Eggert's conjecture is true also in the noncommutative case, was answered in the negative in [4], where a counterexample is given.

The above-mentioned result of Stack is generalized in [6] as follows.

Theorem 7.2 *Let R be a nilpotent algebra over an arbitrary field F containing two elements x, y such that the system $\{x^k, y^k\}$ is linearly independent for some integer $k \geq 1$. If for any two elements a and b in R such that $a^k = x^k$, $b^k = y^k$, the subalgebra generated by these elements cannot be generated by one element, then $\dim R \geq 2k$.*

The bound given in Theorem 7.2 is for instance attained by an algebra of dimension $2k$ generated by an element x where $y = x^2$, and also by an algebra $R = L_1 \oplus L_2$ which is the direct sum of two one-generator subalgebras L_1 and L_2 of dimension k generated by elements x and y respectively. Also, it follows for instance from Theorem 7.2 that the adjoint group of a finite commutative nilpotent 2-algebra containing a subgroup isomorphic to the direct product of two cyclic groups of order 32 has order at least 2^{32}.

The next theorem of [6] contains the main results of Stack, Bautista and Eggert as special cases and holds for every field of prime characteristic.

Theorem 7.3 *Let R be a commutative nilpotent p-algebra containing three elements x, y, z such that the system $\{x^{p^m}, y^{p^m}, z^{p^m}\}$ is linearly independent for some integer $m \geq 1$. Then $\dim R \geq 3p^m$.*

The following theorem is contained in [7].

Theorem 7.4 *Let R be a commutative nilpotent p-algebra for some prime p. If $\dim R^{(1)} \leq 4$, then Eggert's conjecture is true.*

In his Diplomarbeit (Mainz, 2002) J. Zahn showed that if a finite noncommutative p-algebra has a one-generated subalgebra with codimension at most $p - 1$, then the inequality $\dim R \geq p \dim R^{(1)}$ still holds.

As a direct group-theoretic consequence of Theorem 7.4 we obtain the following.

Corollary 7.5 *Let the finite group $G = AB$ be the product of two p-subgroups A and B for the prime p such that B is abelian and either $|A| \leq p^{4p}$ or $r(A) < 2p$. Then the Prüfer rank $r(G)$ of G satisfies the inequality*

$$r(G) \leq 2(r(A) + r(B)) + \frac{p}{(p-1)} \min\{r(A), r(B)\}.$$

If the minimal number of generators $d(R)$ of the commutative nilpotent p-algebra R is less than its characteristic p we have proved that $\dim R \leq er(R^\circ)$, where e is the Euler number. This unpublished result of the authors has the following immediate consequence.

Theorem 7.6 *Let $G = AB$ be a product of an abelian p-group A and a group B, which is a direct product of an elementary abelian p-group and a p-group with Prüfer at most $p - 1$. Then the Prüfer rank $r(G)$ satisfies the inequality*

$$r(G) \leq 5(r(A) + r(B)).$$

Using results similar to Theorem 5.2 for graded commutative nilpotent p-algebras, McLean [39] proved the following.

Theorem 7.7 *Let R be a graded commutative nilpotent p-algebra. If $d(R) = 2$ or $(R^{(1)})^2 = 0$, then Eggert's conjecture is true.*

The following result of [11] holds for all commutative nilpotent p-algebras with nilpotency class p.

Theorem 7.8 *Let R be a commutative nilpotent p-algebra with nilpotency class p. Then Eggert's conjecture is true.*

Final Remark   The paper [25] claims to give a full proof of Eggert's conjecture, but contains several severe gaps.

The essential part of the proof of the Theorem in [25] requires to have a basis B of the p-algebra $R = \langle\langle x, y_1, y_2 \ldots y_k\rangle\rangle$ over a field F consisting of monomials with the following properties:

1. The basis $B^{(1)}$ of the algebra $R^{(1)}$ has the form $B^{(1)} = C_1 \cup C_2$, where C_1 is a basis of the algebra $(Ax)^{(1)}$ and C_2 is a basis of the algebra $D^{(1)}$, where $D = \langle\langle y_1, y_2 \ldots y_k\rangle\rangle$ and $A = R + 1.F$.

2. The basis $B^{(1)}$ is contained in the basis B and C_2 is contained in a basis of the algebra D, which is a subset of B.

Note that properties 1. and 2. are equivalent to the statement that the vector space R can be expressed as $R = Ax \oplus D$ for some element $x \in R$.

If we could find a basis with the above properties, then the proof of the Theorem in [25] is indeed trivial.

However, the following example of a commutative nilpotent algebra R shows that the construction of a basis B with the required properties 1. and 2. is impossible for any choice of the generators of R.

Let $R = \langle\langle x, y\rangle\rangle$ be a commutative nilpotent 2-algebra with defining relations $x^6 = 0 \neq x^5$, $y^4 = 0 \neq y^3$ and $xy = 0$, $y^3 = x^5$. Clearly, R has a basis $B = \{x, x^2, x^3, x^4, x^5, y, y^2\}$, satisfying the conditions of Lemma 3 in [25]. Then $\{x^2, x^4, y^2\}$ is a basis of the subalgebra $R^{(1)}$. But obviously it is impossible to complete this to a basis of R satisfying properties 1. and 2.

We remark also that the proof of Lemma 3 in [25] is incomplete, but it could be saved by Theorem 5.1 above.

Although the given example is not a counterexample to Eggert's conjecture, but it shows that the arguments in [25] fail. Thus, the conjecture of Eggert remains open.

References

[1] B. Amberg, S. Franciosi, and F. de Giovanni, *Products of groups*, Clarendon Press, Oxford (1992).

[2] B. Amberg and L. Kazarin, On the rank of a finite product of two p-groups, in *Groups—Korea 1994*, W. de Gruyter, Berlin (1995), 1–8.

[3] B. Amberg and L. Kazarin, On the rank of the product of two finite p-groups and nilpotent p-algebras, *Comm. Algebra* **27** (1999), 3895–3907.

[4] B. Amberg and L. Kazarin, On the adjoint group of a finite nilpotent p-algebra, Algebra 13., *J. Math. Sci. (New York)* **102** (2000) 3979–3997.

[5] B. Amberg and L. Kazarin, The dimension of nilpotent 2-algebras with two generators, *Proc. F. Scorina Gomel State Univ.* **3** (2000), 76–79.

[6] B. Amberg and L. Kazarin, On the dimension of a nilpotent algebra, *Mat. Zametki (Math. Notes)* **70** (2001), 133–144.

[7] B. Amberg and L. Kazarin, Commutative nilpotent p-algebras with small dimension, *Quaderni di matematica (Napoli)* **8** (2001), 1–20.

[8] B. Amberg and L. Kazarin, On certain minimal nilpotent algebras of a given nilpotency class, *Proc. Steklov. Inst. Math., Suppl.* **2** (2001), 16–25.

[9] B. Amberg and L. Kazarin, On the central series of the adjoint group of a nilpotent p-algebra, *Publ. Math. Debrecen* **63** (2003), 473–482.

[10] B. Amberg and L. Kazarin, On the powers of a commutative nilpotent algebra, *Advances in Algebra*, World Sci. Publishing, River Edge, NJ, USA (2003), 1–12.

[11] B. Amberg and L. Kazarin, On the power series of a nilpotent algebra, in preparation.

[12] B. Amberg and Ya. P. Sysak, Radical rings and products of groups, in *Groups St Andrews 1997 in Bath, I*, London Math. Soc. Lecture Note Series **260** (1999), 1–19.

[13] J. C. Ault and J. F. Watters, Circle groups of multiplicative rings, *Amer. Math. Monthly* **80** (1973), 48–52.

[14] R. Bautista, Units of finite algebras, *Ann. Inst. Mat. Univ. Nac. Autónoma, México* **16**, no. 2 (1976), 1–78 (in Spanish).

[15] A. Ya. Belov, Some estimations for nilpotence of nil-algebras over a field of an arbitrary characteristic and height theorem, *Comm. Algebra* **20** (1992), 2919–2922.

[16] A. A. Bovdi, On circle groups of nilpotent rings of characteristic 2, *Periodica Math. Hungarica* **32** (1996), 31–34.

[17] A. A. Bovdi, The group of units of a group algebra of characteristic p, *Publ. Math. Debrecen* **52** (1998), 193–244.

[18] W. C. Brown, Constructing maximal commutative subalgebras of matrix rings in small dimensions, *Comm. Algebra* **25** (1997), 3923–3946.

[19] M. D. E. Conder and I. M. Isaacs, Derived subgroups of products of an abelian and cyclic subgroup, *J. London Math. Soc. (2)* **69** (2004), 333–348.

[20] N. H. Eggert, Quasi-regular groups of finite commutative nilpotent algebras, *Pacific J. Math.* **36** (1971), 631–634.

[21] L. Fischer and K. Eldridge, Artinian rings with cyclic quasiregular group, *Duke Math. J.* **36** (1969), 43–47.

[22] E. S. Golod, On nil-algebras and finitely approximable p-groups, *Izv. Akad. Nauk SSSR* **28** (1964), 273–276.

[23] B. O. Gorlov, Finite nilpotent algebras with metacyclic adjoint group, *Ukrain. Math. Z.* **47** (1995), 1426–1431.

[24] W. H. Gustafson, On maximal commutative algebras of linear transformations, *J. Algebra* **42** (1976), 557–563.

[25] L. Hammoudi, Eggert's conjecture on the dimensions of nilpotent algebras, *Pacific J. Math.* **202** (2002), 93–97.

[26] P. Hubert, Local near-rings and triply factorized groups, *Comm. Algebra* **32** (2004),

1229–1235.

[27] P. Hubert, Triply factorized groups and nearrings, in *Groups St Andrews 2005*.

[28] B. Huppert, *Endliche Gruppen I*, Springer, Berlin (1967).

[29] J. R. Isbell, On the multiplicative semigroup of commutative, *Proc. Amer. Math. Soc.* **10** (1959), 908–909.

[30] L. Kaloujnine, Zum Problem der Klassifikation der endlichen metabelschen p-Gruppen, *Wiss. Z. Humboldt-Univ. Berlin, Math.-Nat. Reihe* **4** (1954/55), 1–7.

[31] L. Kazarin, On factorizable groups, *Dokl. Akad. Nauk SSSR* **256** (1981), 26–29.

[32] L. Kazarin and P. Soules, On the adjoint group of a nilpotent algebra with two generators, *JP J. Pure Appl. Algebra* **4** (2004), 108–118.

[33] A. A. Klein, Bounds for Indices of nilpotency and nility, *Arch. Math.* **74** (2000), 6–10.

[34] R. L. Kruse and D. T. Price, *Nilpotent rings*, Gordon and Breach, New York (1967).

[35] T. J. Laffey, The minimal dimension of maximal commutative subalgebra of full matrix algebra, *Linear Algebra Appl.* **71** (1985), 199–212.

[36] A. Lubotzky and A. Mann, Powerful p-groups, I. Finite groups, *J. Algebra* **105** (1987), 485–505.

[37] F. S. Macaulay, Some properties of enumeration in the theory of modular systems, *Proc. London Math. Soc.* **26** (1927), 531–555.

[38] I. D. MacDonald, Generalizations of a classical theorem about nilpotent groups, *Illinois J. Math.* **8** (1964), 556–570.

[39] K. R. McLean, Eggert's conjecture on nilpotent algebras, *Comm. Algebra* **32** (2004), 997–1006.

[40] G. A. Miller and H. C. Moreno, Non-abelian groups in which every proper subgroup is abelian, *Trans. Amer. Math. Soc.* **4** (1903), 398–404.

[41] L. E. Moran and R. N. Tench, Normal complements in mod p envelopes, *Israel J. Math.* **27** (1977), 331–338.

[42] A. Yu. Ol'shanskii, A simplification of Golod's example, in *Groups—Korea 1994*, W. de Gruyter, Berlin (1995), 263–265.

[43] L. Redei, Das schiefe Produkt in der Gruppentheorie, *Comment. Math. Helv.* **20** (1947), 225–264.

[44] A. Shalev, The structure of finite p-groups: effective proof of the coclass conjectures, *Invent. Math.* **115** (1994), 315–345.

[45] I. Schur, Zur Theorie der vertauschbaren Matrizen, *J. Reine Angew. Math.* **130** (1905), 66–76.

[46] C. Stack, Dimensions of nilpotent algebras over fields of prime characteristic, *Pacific J. Math.* **176** (1996), 263–266.

[47] C. Stack, Some results on the structure of finite nilpotent algebras over fields of prime characteristic, *J. Combinat. Math. Combin. Comput.* **28** (1998), 327–335.

[48] R. P. Stanley, Hilbert functions of graded algebras, *Advances in Math.* **28** (1978), 57–83.

[49] D. A. Suprunenko and R. I. Tyshkevich, *Commutative matrices*, Academic Press, New York (1968).

[50] Ya. P. Sysak, Some examples of factorized groups and their relation to group theory, in *Infinite Groups 1994*, W. de Gruyter, Berlin, (1995), 263–265.

[51] K.-I. Tahara and A. Hosomi, On the circle group of finite nilpotent rings, in *Groups—Korea 1983*, Lecture Notes Math. **1098**, Springer, Berlin (1984), 161–179.

[52] D.-I. Zaitsev, Factorizations of polycyclic groups, *Mat. Zametki (Math. Notes)* **29** (1981), 481–490.

CLASSIFICATION OF FINITE GROUPS BY THE NUMBER OF ELEMENT CENTRALIZERS

ALI REZA ASHRAFI* and BIJAN TAERI†

*Department of Mathematics, University of Kashan, Kashan, Iran
E-mail: ashrafi@kashanu.ac.ir
†Department of Mathematics, Isfahan University of Technology, Isfahan, Iran

Abstract

For a finite group G, $\#\mathrm{Cent}(G)$ denotes the number of centralizers of its elements. A group G is called n-centralizer if $\#\mathrm{Cent}(G) = n$, and primitive n-centralizer if $\#\mathrm{Cent}(G) = \#\mathrm{Cent}(G/Z(G)) = n$.

In some research papers the problem is posed to find the structure of finite n-centralizer and primitive n-centralizer groups, for a given positive integer n. In this paper, we report on recent work on this problem and prove some new results.

AMS Classification: 20D99, 20E07.
Keywords: Finite group, n-centralizer group, primitive n-centralizer group, simple group.

1 Introduction

In this section we describe some notation which will be used throughout. Following B. H. Neumann [18], we shall say that a group G is covered by a family of cosets or subgroups if G is simply the set-theoretic union of the family. Throughout this paper we consider covering by subgroups. We shall say that a covering of G is irredundant if none of the subgroups X_i can be omitted; that is, $X_i \not\subseteq \bigcup_{j \neq i} X_j$ for each i. Finally, a group G is called *capable* if there exists a group H such that $G \cong H/Z(H)$.

The maximum value of $|G : \bigcap_{i=1}^{n} X_i|$ in a group G with an irredundant covering by n subgroups will be denoted by $f_2(n)$. In [23], Tomkinson has proved that $f_2(3) = 4$ and $f_2(4) = 9$. He also announced that, by a detailed investigation of the different situations which can arise when $n = 5$, he has been able to show that $f_2(5) = 16$. The unpublished proof of that result was a lengthy calculation using the known structure of groups covered by three or four subgroups but gave no structural information about the groups covered by five subgroups. In [7], Bryce, Fedri and Serena presented a short proof of the above mentioned result. Let $\mathrm{Cent}(G)$ be the set of all element centralizers of a finite group G, $\#\mathrm{Cent}(G) = |\mathrm{Cent}(G)|$ and $\mathrm{PrCent}(G) = \#\mathrm{Cent}(G)/|G|$.

Definition 1.1 A group G is called *n-centralizer* if $\#\mathrm{Cent}(G) = n$, and *primitive n-centralizer* if $\#\mathrm{Cent}(G) = \#\mathrm{Cent}(G/Z(G)) = n$.

It is obvious that G is 1-centralizer if and only if G is abelian. In [6], Belcastro and Sherman proved that there is no n-centralizer group for $n = 2, 3$, and that G is 4-centralizer if and only if $G/Z(G) \cong Z_2 \times Z_2$. Furthermore, they proved that G is 5-centralizer if and only if $G/Z(G) \cong Z_3 \times Z_3$ or S_3, the symmetric group on three letters. By these results one can see that there is no primitive 4-centralizer group and that G is a primitive 5-centralizer group if and only if $G/Z(G) \cong S_3$.

In this connection one might ask about the structure of primitive n-centralizer groups, for $n \geq 6$. In this paper, we report on this problem. Throughout this paper all groups mentioned are assumed to be finite. Our notation is standard and taken mainly from [16], [21] and [22].

2   Distinct Centralizers of some Finite Groups

In this section we present some examples which show that the number of distinct centralizers of a finite group can be arbitrarily large.

Example 2.1 Consider the generalized quaternion group Q_{4m} of order $4m$, $m \geq 2$. This group is defined by

$$Q_{4m} = \langle a, b \mid a^{2m} = 1,\ b^2 = a^m,\ bab^{-1} = a^{-1} \rangle.$$

It is well known that $Z(Q_{4m}) = \langle b^2 \rangle$, and we can see that

$$C_{Q_{4m}}(a^i b) = \{1, a^i b, a^m, a^{m+i} b\}, \quad 0 \leq i \leq 2m - 1.$$

Now since $C_{Q_{4m}}(a) = \langle a \rangle$ and $C_{Q_{4m}}(a^i b) = C_{Q_{4m}}(a^{m+i} b)$, we have $\#\mathrm{Cent}(Q_{4m}) = m + 2$.

By the previous example, if n is an arbitrary positive integer different from 2 and 3 then $\#\mathrm{Cent}(Q_{4n-8}) = n$. This shows that for any positive integer $n \neq 2, 3$, there exists a finite group G such that $\#\mathrm{Cent}(G) = n$. In fact, we prove the following result:

Proposition 2.2 *For every positive integer $n \neq 2, 3$, there exists an n-centralizer group.*

In [6] it is proved that there is no primitive 4-centralizer group. Now it is natural to ask that for which positive integers n there exists a primitive n-centralizer group. A partial answer is provided by the following example:

Example 2.3 Consider the dihedral group D_{2n} of order $2n$. This group can be presented in the form

$$D_{2n} = \langle x, y \mid x^n = y^2 = 1,\ y^{-1} x y = x^{-1} \rangle.$$

The elements of D_{2n} are then $1, x, \cdots, x^{n-1}$ together with $y, xy, \cdots, x^{n-1}y$. Also we have $yx^i = x^{-i}y$, for all integers i, and that $C_{D_{2n}}(x) = \langle x \rangle$. Assume that $n \geq 3$ is an odd integer, then $|Z(D_{2n})| = 1$ and so $C_{D_{2n}}(x^i y) = \{1, x^i y\}$, $1 \leq i \leq n$. This implies that $\#\mathrm{Cent}(D_{2n}) = n+2$. Therefore, for all odd $n > 3$, $\#\mathrm{Cent}(D_{2n-4}) = n$.

By the previous example, if $n > 3$ is an arbitrary odd positive integer then $\#\mathrm{Cent}(D_{2n-4}) = n$. This shows that for any odd positive integer $n > 3$, there exists a primitive n-centralizer finite group. On the other hand, every abelian group is primitive 1-centralizer. Therefore, we prove the following result:

Proposition 2.4 *Let $n \neq 3$ is an odd positive integer. Then there exists a primitive n-centralizer group.*

In general, we can calculate the number of element centralizers of dihedral groups D_{2n}, for any positive integer n. We have:

$$\#\mathrm{Cent}(D_{2n}) = \begin{cases} n+2 & \text{if } 2 \nmid n \\ \frac{n}{2} + 2 & \text{if } 2 \mid n. \end{cases}$$

Our computations with the computational group theory system GAP [20], in investigating the finite groups of small order, suggests the following conjecture:

Conjecture 2.5 *For any natural number n, there is no primitive 2^n-centralizer group.*

In [6], Belcastro and Sherman asked whether or not there exists a finite group G other than Q_8 and D_{2p}, p a prime, such that $\mathrm{PrCent}(G) \geq \frac{1}{2}$. We found several examples to answer this question in [3]. In this paper we conjectured that if $\mathrm{PrCent}(G) \geq \frac{2}{3}$ then G is isomorphic to the group S_3, $S_3 \times S_3$, or a dihedral group of order 10, which is still open.

Example 2.6 Suppose that $F = GF(2^n)$ and θ is an automorphism of F. Following A. Hanaki [12], we shall use the notation $A(n, \theta)$ to denote the set of all matrices of the form $U(a,b) = \begin{bmatrix} 1 & 0 & 0 \\ a & 1 & 0 \\ b & a\theta & 1 \end{bmatrix}$ with $a, b \in F$. The multiplication is defined as matrix multiplication. Then $A(n, \theta)$ is a 2^n-centralizer group of order 2^{2n}. But this group is not primitive, since $A(n, \theta)/Z(A(n, \theta)) \cong F$. On the other hand, assume that $F = GF(q)$, $q = p^n$ a prime power, and $A(n, p)$ is the set of all matrices of the form $V(a,b,c) = \begin{bmatrix} 1 & 0 & 0 \\ a & 1 & 0 \\ b & c & 1 \end{bmatrix}$ with $a, b, c \in F$. Then $A(n, p)$, with matrix multiplication, is a $(q+2)$-centralizer group of order p^{3n}. Similarly, since $A(n, p)/Z(A(n, p)) \cong F \times F$, the group $A(n, p)$ is not primitive.

We now assume that p is a prime number and $N(p)$ denotes the set of all integers n such that $n = \#\mathrm{Cent}(G)$ for some p-group G. In Example 2.6, we show that for every positive integer m, $p^m + 2 \in N(p)$ and $2^m, 2^m + 2 \in N(2)$. Now it is natural to ask about $N(p)$, p a prime. For $p = 2$, by [6, Theorem 4], $5 \notin N(2)$ and so $N(2) \neq \mathbb{N} - \{2, 3\}$. The following lemma proves that for any prime number p, $N(p) \neq \mathbb{N} - \{2, 3\}$.

Lemma 2.7 ([3, Lemma 2.7]) *If G is a non-abelian p-group, then $\#\mathrm{Cent}(G) \geq p + 2$, with equality if and only if $G/Z(G) \cong Z_p \times Z_p$.*

Set $N(\pi) = \bigcup\limits_{p \text{ prime}} N(p)$. It it interesting to study $N(\pi)$. However, the investigation of $N(\pi)$ does not seem to be simple.

Question 2.8 Is it true that $N(\pi) = \mathbb{N} - \{2, 3\}$?

We now assume that G is a finite group in which every non-trivial conjugacy class has length p or p^2, where p is prime. To end this section, we state two lemmas to obtain the number of centralizers of such a group:

Lemma 2.9 ([4, Lemma 5]) *Let G be an n-centralizer group, $n - 2 = p$, p a prime, and $X_1, X_2, \cdots, X_n$ be distinct centralizers of elements of G such that $X_1 = G$. If $|G : X_2| = |G : X_3| = \cdots = |G : X_n| = n - 2$ then $G/Z(G) \cong Z_p \times Z_p$.*

Theorem 2.10 ([3, Theorem 3.3.]) *Let G be an n-centralizer group, $n - 2 = p^2$, p a prime, and $X_1, X_2, \cdots, X_n$ be distinct centralizers of elements of G such that $X_1 = G$. If $|G : X_2| = |G : X_3| = \cdots = |G : X_n| = n - 2$ and two of X_i's are normal in G, then $G/Z(G) \cong Z_p \times Z_p \times Z_p \times Z_p$.*

3   On Primitive 6- and 7-Centralizers Groups

In this section we study the structure of finite groups with exactly six and seven distinct centralizers.

In [23], Tomkinson gives bounds $16 \leq f_2(5) \leq 54$. He has also announced that, by a detailed investigation of the different situations which can arise when $n = 5$, he has been able to show that $f_2(5) = 16$. Next, in [7], Bryce, Fedri and Serena presented a short proof for this result.

We now ready to state an important result on the characterization problem of primitive 6-centralizer groups (see [4, Theorem 1] and [3, Theorem 3.6]).

Theorem 3.1 *If G is a 6-centralizer group then $G/Z(G) \cong D_8$, A_4, $Z_2 \times Z_2 \times Z_2$ or $Z_2 \times Z_2 \times Z_2 \times Z_2$. Moreover, if $G/Z(G) \cong A_4$ then $\#\mathrm{Cent}(G) = 6$ or 8, if $G/Z(G) \cong D_8$ then $\#\mathrm{Cent}(G) = 6$ and if $G/Z(G) \cong Z_2 \times Z_2 \times Z_2$ then $\#\mathrm{Cent}(G) = 6$ or 8.*

It is natural to ask about the converse of the previous theorem. The following example shows that there is an 8-centralizer group such that $G/Z(G) \cong A_4$.

Example 3.2 Assume that $H = \langle x \rangle$ is the cyclic group of order 3 and $N = Q_8$ is the quaternion group of order 8. Then $\mathrm{Aut}(N)$ has a unique conjugacy type of automorphism of order 3. Consider the semidirect product $G = H \times_\theta N$, where $\theta(x)$ is an automorphism of order 3 of the group Q_8. Then G is a group of order 24, $G/Z(G) \cong A_4$ and $\#\mathrm{Cent}(G) = 8$.

In what follows, we investigate some special cases where group structure is determined by the number of distinct centralizers. We need these results for the characterization of finite groups with exactly seven element centralizers.

First, we investigate the number of element centralizers of finite groups G such that $G/Z(G) \cong Z_2 \times Z_2 \times Z_2$.

Lemma 3.3 *Let G be a finite group and $G/Z(G) \cong Z_2 \times Z_2 \times Z_2$. Then $\#\mathrm{Cent}(G) = 6$ or 8.*

Proof It is clear that $\#\mathrm{Cent}(G) \leq |G : Z(G)| = 8$. Suppose $Z = Z(G)$, $G/Z(G) = Z \cup x_1 Z \cup x_2 Z \cup \cdots \cup x_7 Z$ and $\#\mathrm{Cent}(G) < 8$. Then there are i and j such that $1 \leq i \neq j \leq 7$ and $C_G(x_i) = C_G(x_j)$. It is an easy fact that $C_G(x_i)/Z = C_G(x_j)/Z = \{Z, x_i Z, x_j Z, x_i x_j Z\} = C_G(x_i x_j)/Z$. This shows that $\#\mathrm{Cent}(G) \leq 6$ and, since G is not abelian, $\#\mathrm{Cent}(G) \geq 4$. Now by Theorems 2 and 4 of [6], $\#\mathrm{Cent}(G) = 6$, proving the lemma. □

Belcastro and Sherman, in [6, Theorem 5], proved that if $G/Z(G) \cong D_{2p}$, for p an odd prime, then $\#\mathrm{Cent}(G) = p + 2$. In the following lemma we generalize this result to the case where p is an arbitrary positive integer.

Lemma 3.4 *Let G be a finite group and $G/Z(G) \cong D_{2n}$, $n \geq 2$. Then $\#\mathrm{Cent}(G) = n + 2$.*

Proof Set $Z = Z(G)$ and suppose that $G/Z(G)$ has the following presentation:

$$\langle\, xZ, yZ \mid x^n Z = y^2 Z = Z,\ yZ\,xZ\,yZ = x^{-1}Z \,\rangle.$$

Then it is obvious that $\{x^i y^j \mid 0 \leq i \leq n - 1,\ 0 \leq j \leq 1\}$ is a left transversal for $Z = Z(G)$ in G. It is enough to investigate the element centralizers $C_G(x^i y^j)$, for $0 \leq i \leq n - 1$, $0 \leq j \leq 1$. Suppose $aZ \in \langle xZ \rangle/Z$ then there exists i, $0 \leq i \leq n - 1$, such that $a \in x^i Z$. Therefore, $C_G(a) = C_G(x^i) = \langle x \rangle$ or G. If n is odd, then $C_G(x^i y) = Z \cup x^i y Z$, $0 \leq i \leq n - 1$ and $\#\mathrm{Cent}(G) = n + 2$. We now assume that $n = 2k$. Using a simple calculation we can see that $C_G(x^i y)/Z \subseteq C_{G/Z}(x^i yZ) = \{Z, x^i yZ, x^k Z, x^{i+k} yZ\}$. Suppose $C_G(x^i y)/Z = C_{G/Z}(x^i yZ)$, for some i, $0 \leq i \leq n - 1$. Then

$$C_G(x^{i+k} y)/Z \subseteq C_{G/Z}(x^{i+k} yZ) = C_{G/Z}(x^i yZ) = C_G(x^i y)/Z.$$

Thus, $C_G(x^{i+k} y) \subseteq C_G(x^i y)$. So $x^k y = y x^k$. This means that $x^k \in Z(G)$, which is a contradiction. Hence, $|C_G(x^i y)/Z| = 2$ and $C_G(x^i y) = Z \cup x^i yZ$, for every i, $0 \leq i \leq n - 1$. Therefore,

$$\mathrm{Cent}(G) = \{G, C_G(x), C_G(y), C_G(xy), \cdots, C_G(x^{n-1} y)\}.$$

This completes the proof. □

It is an elementary fact that D_{12}, T and A_4 are the only non-abelian groups of order 12, where the group T is presented by:

$$\langle\, a, b \mid a^6 = 1,\ a^3 = b^2,\ ba = a^{-1}b \,\rangle.$$

If $G/Z(G) \cong A_4$ then by Theorem 3.6 of [3], $\#\mathrm{Cent}(G) = 6$ or 8. Also, by the previous lemma if $G/Z(G) \cong D_{12}$ then $\#\mathrm{Cent}(G) = 8$. In the following lemma we compute $\#\mathrm{Cent}(G)$, when $G/Z(G) \cong T$.

Lemma 3.5 *Let G be a finite group and $G/Z(G) \cong T$. Then $\#\mathrm{Cent}(G) = 8$.*

Proof First of all, we can assume that $G/Z(G)$ has the following presentation:

$$\langle\, aZ, bZ \mid a^6 Z = Z,\ a^3 Z = b^2 Z,\ bZ\, aZ = a^{-1}Z\, bZ \,\rangle.$$

It is obvious that $\langle aZ \rangle \subseteq C_G(a^i)/Z$, $0 \le i \le 5$. If for some i, $0 \le i \le 5$, $\langle aZ \rangle \ne C_G(a^i)/Z$ then $a^i \in Z(G)$, a contradiction. Thus for every i, $0 \le i \le 5$, we have:

$$C_G(a^i) = Z \cup aZ \cup a^2 Z \cup a^3 Z \cup a^4 Z \cup a^5 Z.$$

Using a similar argument to that in Lemma 3.4, we can see that the element centralizers $C_G(a^i b)$, $0 \le i \le 5$, are distinct and so $\#\mathrm{Cent}(G) = 8$, proving the lemma. $\qquad\square$

Suppose $r(G)$ denotes the maximal size of a subset of pairwise non-commuting elements of G. In the following lemma, under a certain condition on $r(G)$, we prove that every element centralizer of G is abelian. In fact, we prove that:

Lemma 3.6 *Let G be a finite n-centralizer group and $r(G) = n - 1$. Then every proper element centralizer of G is abelian. Moreover, for every two non-central elements x and y of G, $C_G(x) = C_G(y)$ or $C_G(x) \cap C_G(y) = Z(G)$.*

Proof Let $\{x_1, x_2, \cdots, x_{n-1}\}$ be a set of pairwise non-commuting elements of G having maximal size. Then we have

$$\mathrm{Cent}(G) = \{G, C_G(x_1), \cdots, C_G(x_{n-1})\}.$$

Suppose for some i, $1 \le i \le n-1$, $X = C_G(x_i)$ is non-abelian. Choose elements $a, b \in X$ such that $ab \ne ba$. It is an easy fact that $C_G(a) \ne G$, $C_G(b) \ne G$ and $C_G(a) \ne C_G(b)$. Without loss of generality, we can assume that $C_G(a) = C_G(x_j)$, for some $j \ne i$. So $x_i \in C_G(a) = C_G(x_j)$, which is a contradiction. We now assume that x and y are non-central and $C_G(x) \ne C_G(y)$. Set $T = C_G(x) \cap C_G(y)$. Choose, if possible, an element $a \in T - Z(G)$. It is obvious that $C_G(a) \ne G$ and $C_G(x) \cup C_G(y) \subseteq C_G(a)$, which contradicts by our assumption. This proves the lemma. $\qquad\square$

Let R be a group of order 20 presented by

$$R = \langle a, b \mid a^5 = b^4 = 1,\ b^{-1}ab = a^2 \rangle.$$

It is easy to see that R has trivial center and it is an extension of a cyclic group of order 5 by a cyclic group of order 4 acting faithfully. Let R_1 be the unique Sylow 5-subgroup and $R_2, R_3, \cdots, R_6$ be the Sylow 2-subgroups of R. We can see that $\mathrm{Cent}(R) = \{R, R_1, \cdots, R_6\}$ and so R is 7-centralizer. In the following lemma, we investigate the number of element centralizers of the group G, when $G/Z(G) \cong R$.

Lemma 3.7 *Let G be a finite group and $G/Z(G) \cong R$. Then G is 7-centralizer.*

Proof Suppose $Z = Z(G)$. By assumption, we have

$$G/Z = \langle aZ, bZ \mid a^5 Z = b^4 Z = Z,\ b^{-1}abZ = a^2 Z \rangle.$$

It is clear that $\{a^i b^j \mid 0 \leq i \leq 4,\ 0 \leq j \leq 3\}$ is a left transversal to $Z = Z(G)$ in G. So, it is enough to investigate the element centralizers $C_G(a^i b^j)$, for $0 \leq i \leq 4$, $0 \leq j \leq 3$. Since $\langle aZ \rangle$ is the unique Sylow 5-subgroup of G/Z, $C_{G/Z}(aZ) = \langle aZ \rangle$. On the other hand, $C_G(a)/Z \subseteq C_{G/Z}(aZ) = \langle aZ \rangle$ and so

$$C_G(a) = C_G(a^2) = C_G(a^3) = C_G(a^4) = Z \cup aZ \cup a^2 Z \cup a^3 Z \cup a^4 Z.$$

We now compute the element centralizers $C_G(ab)$, $C_G(a^2 b)$, $C_G(a^2 b^3)$, $C_G(b)$ and $C_G(a^3 b)$. To do this, we note that $\langle abZ \rangle \subseteq C_G(ab)/Z \subseteq C_{G/Z}(abZ) = \langle abZ \rangle$. On the other hand,

$$Y_1 = \langle abZ \rangle = \{Z, abZ, a^4 b^2 Z, a^3 b^3 Z\}$$
$$Y_2 = \langle a^2 bZ \rangle = \{Z, a^2 bZ, a^3 b^2 Z, ab^3 Z\}$$
$$Y_3 = \langle a^2 b^3 Z \rangle = \{Z, a^2 b^3 Z, ab^2 Z, a^4 bZ\}$$
$$Y_4 = \langle bZ \rangle = \{Z, bZ, b^2 Z, b^3 Z\}$$
$$Y_5 = \langle a^3 bZ \rangle = \{Z, a^3 bZ, a^2 b^2 Z, a^4 b^3 Z\}.$$

Thus, $C_G(ab) = Z \cup abZ \cup a^4 b^2 Z \cup a^3 b^3 Z$. We now prove that $C_G(ab) = C_G(a^4 b^2) = C_G(a^3 b^3)$. Since

$$\langle abZ \rangle = \langle a^3 b^3 Z \rangle \subseteq C_G(a^3 b^3)/Z \subseteq C_{G/Z}(a^3 b^3 Z) = \langle a^3 b^3 Z \rangle = \langle abZ \rangle,$$

$C_G(a^3 b^3) = C_G(ab)$. Suppose $C_G(a^4 b^2) = Z \cup a^4 b^2 Z$. Then $C_G(a^4 b^2)$ is abelian and $C_G(a^4 b^2) \subseteq C_G(ab)$. On the other hand, $C_G(ab)/Z$ is cyclic and so $C_G(ab)$ is abelian, a contradiction. Thus $C_G(a^4 b^2) = C_G(ab)$. This shows that $C_G(ab) = C_G(a^4 b^2) = C_G(a^3 b^3)$.

Using a similar argument to the above, we can prove that for every non-identity elements $uZ, vZ \in Y_i$, $1 \leq i \leq 5$, $C_G(u) = C_G(v)$. This completes the proof. □

Now we are ready to prove an important result for classifying primitive 7-centralizers group.

Theorem 3.8 *A group G is primitive 7-centralizer if and only if $G/Z(G) \cong D_{10}$ or R.*

Proof Suppose $G/Z(G) \cong D_{10}$ or R. Then by Lemmas 3.4 and 3.7, G is 7-centralizer. On the other hand, by Lemma 2.4 of [6], D_{10} is 7-centralizer, and by the paragraph before Lemma 3.7, R is 7-centralizer. Hence, G is primitive 7-centralizer. We now assume that G is a primitive 7-centralizer finite group. We also suppose that $\{x_1, \cdots, x_r\}$ is a set of pairwise non-commuting elements of G having maximal size. Set $X_i = C_G(x_i)$, $1 \le i \le r$. By Theorem 5.2 of [23], $3 \le r \le 6$. Thus, it is enough to investigate four cases which we now treat separately.

Case 1: $r = 3$. In this case by Corollary 5.2 of Tomkinson [23], $|G/Z(G)| \le f_2(3) = 4$. So $G/Z(G) \cong Z_2 \times Z_2$, a contradiction.

Case 2: $r = 4$. Using Corollary 5.2 of Tomkinson [23], we can see that $|G/Z(G)| \le f_2(4) = 9$. So $G/Z(G)$ is isomorphic to S_3, D_8 or Q_8. But, Q_8 is not capable and if $G/Z(G) \cong S_3$ then by Theorem 4 of [6], G is 5-centralizer, a contradiction. Finally, if $G/Z(G) \cong D_8$ then, by Lemma 3.4, G is 6-centralizer, which is a contradiction.

Case 3: $r = 5$. In this case $|G/Z(G)| \le f_2(5)$ and by a result of Bryce, Fedri and Serena [7], $f_2(5) = 16$. So $|G/Z(G)| \le 16$ and we can assume that $|G : Z(G)| = 10$, $12, 14, 16$. Our computation together with the computational group theory system GAP [20] in investigating groups of order 16 shows that there is no 7-centralizer group of order 16. Suppose $|G/Z(G)| = 14$ then $G/Z(G) \cong D_{14}$ and by Lemma 3.4, G is 9-centralizer, a contradiction. We now assume that $|G/Z(G)| = 12$. Then $G/Z(G) \cong D_{12}$, A_4 or T. If $G/Z(G) \cong D_{12}$ or T then by Lemmas 3.4 and 3.5, G is 8-centralizer. On the other hand, if $G/Z(G) \cong A_4$ then by Theorem 3.6 of Ashrafi [3], $\#\mathrm{Cent}(G) = 6$ or 8. So $|G/Z(G)| = 10$ and $G/Z(G) \cong D_{10}$.

Case 4: $r = 6$. Set $\beta_i = |G : X_i|$, $\beta_1 \le \beta_2 \le \cdots \le \beta_6$. By Lemma 3.3 of Tomkinson [23], $\beta_2 \le 5$ and by Lemma 3.6, $X_i \cap X_j = Z(G)$, for every i, j such that $1 \le i \ne j \le 6$. If $\beta_2 = 2$ then $\beta_1 = 2$ and $G/Z(G) \cong Z_2 \times Z_2$, a contradiction. Suppose $\beta_2 = 3$. If $\beta_1 = 3$ and one of the element centralizers of index 3 is normal in G then G is 5-centralizer, which is impossible. If $\beta_1 = 2$ or $\beta_1 = 3$ and G does not have a normal element centralizer then $|G : Z(G)| = 6$, which is a contradiction. We now assume that $\beta_2 = 4$. Then $\beta_1 = 2$ implies that $|G : Z(G)| = 8$ and $\beta_1 = 3$ implies that $|G : Z(G)| = 12$. Thus $\beta_1 = 4$ and $|G : Z(G)| \le 16$. But these cases cannot occur. Therefore $\beta_2 = 5$ and by Lemma 3.3 of Tomkinson [23], $\beta_3 = \cdots = \beta_6 = 5$.

If $\beta_1 = 5$ then by Ito's Theorem [14], $G \cong A \times P$, where A is an abelian group and P is a 5-group. Since $G/Z(G) \cong P/Z(P)$ and $\#\mathrm{Cent}(G/Z(G)) = 7$, by Lemma 2.7 of [3], $G/Z(G) \cong Z_5 \times Z_5$, a contradiction. Thus $\beta_1 < 5$. Now by a result of Haber and Rosenfeld [11], since G is an irredundant union of $X_1, \cdots, X_6$, $\bigcap_{i=1}^{6} X_i = \bigcap_{i=2}^{6} X_i = Z(G)$. If $X_2 \trianglelefteq G$ then the X_i's, $2 \le i \le 6$, are normal in G and so $G/Z(G) \cong Z_5 \times Z_5$, which is impossible. So X_2 is a non-normal subgroup of G and $Z(G) = \mathrm{Core}_G X_2$. This shows that $G/Z(G)$ is isomorphic to a subgroup of S_5. Using a simple GAP program, we can see that $\#\mathrm{Cent}(S_5) > 7$ and $\#\mathrm{Cent}(A_5) > 7$. Thus $G/Z(G)$ is a solvable subgroup of S_5. On the other hand, S_5 does not have a subgroup of index 3 or 4. Hence, $|G/Z(G)| = 10$ or 20. If $|G/Z(G)| = 10$, then $G/Z(G) \cong D_{10}$, as desired. Suppose $|G/Z(G)| = 20$. But there are three non-abelian groups of order 20, $SmallGroup(20, 1)$, $SmallGroup(20, 3)$ and $SmallGroup(20, 4)$, in GAP notation. Using a GAP program, we can see that these groups are 7-centralizer. But $SmallGroup(20, 1)$ and $SmallGroup(20, 4)$ have an abelian element centralizer of order 10 and S_5 does not have an element of order 10, which is a contradiction. Therefore, $G/Z(G) \cong SmallGroup(20, 3) \cong R$. This completes the proof.   □

To end this paper we state a result on classifying the alternating group A_5 by the number of element centralizers.

Theorem 3.9 *If G is a finite group and $G/Z(G) \simeq A_5$, then $\#\mathrm{Cent}(G) = 22$ or 32. Moreover, every n-centralizer finite group with $n \leq 21$ is solvable.*

Proof Use the classification of finite simple groups, for details see [4].   □

In what follows, we show that there exists a finite group G such that $G/Z(G) \cong A_5$ and $\#\mathrm{Cent}(G) = 32$. To do this, we assume that G is a group generated by two permutations x and y such that

$$x = (1, 20, 17, 5, 12)(2, 3, 9, 19, 10)(4, 14, 22, 11, 6)(7, 8, 15, 13, 16),$$
$$y = (2, 18)(5, 1)(6, 21)(7, 24)(9, 17)(10, 16)(12, 23)(13, 20)(14, 19)(15, 22).$$

Then $|G| = 240$, $G/Z(G) \cong A_5$ and $\#\mathrm{Cent}(G) = 32$.

It may appear that if G and H are finite simple groups and $|G| \leq |H|$, then $\#\mathrm{Cent}(G) \leq \#\mathrm{Cent}(H)$. But this is not true. To see this, we notice that

$$\#\mathrm{Cent}(PSL(2, 7)) = 79 > 74 = \#\mathrm{Cent}(PSL(2, 8)).$$

The following question arises naturally.

Question 3.10 Let G and H be finite simple groups. Is it true that if $\#\mathrm{Cent}(G) = \#\mathrm{Cent}(H)$, then G is isomorphic to H?

References

[1] M. Aschbacher, *Finite group theory*, Cambridge Univ. Press, 1986.

[2] A. R. Ashrafi, On the n-sum groups, $n = 6, 7$, *Southeast Asian Bull. Math.* **2** (1998), 111–114.

[3] A. R. Ashrafi, On finite groups with a given number of centralizers, *Algebra Colloq.* **7** (2000), no. 2, 139–146.

[4] A. R. Ashrafi, Counting the centralizers of some finite groups, *Korean J. Comput. & Appl. Math.* **7** (2000), no. 1, 115–124.

[5] A. R. Ashrafi and B. Taeri, On finite groups with a certain number of centralizers, *J. Appl. Math. Comput.* **17** (2005), no. 1–2, 217–227.

[6] S. M. Belcastro and G. J. Sherman, Counting centralizers in finite groups, *Math. Mag.* **5** (1994), 366–374.

[7] R. A. Bryce, V. Fedri and L. Serena, Covering groups with subgroups, *Bull. Austral. Math. Soc.* **55** (1997), no. 3, 469–476.

[8] J. H. E. Cohn, On n-sum groups, *Math. Scand.* **75** (1994), 44–58.

[9] J. H. Conway, R. T. Curtis, S. P. Norton, R. A. Parker and R.A. Wilson, *Atlas of finite groups*, Oxford Univ. Press (Clarendon), Oxford, 1985.

[10] D. Gorenstien, *Finite simple groups*, Plenum Press, New York, 1982.

[11] S. Haber and A. Rosenfeld, Groups as unions of proper subgroups, *Amer. Math. Monthly* **66** (1959), 491–494.

[12] A. Hanaki, A condition of lengths of conjugacy classes and character degree, *Osaka J. Math.* **33** (1996), 207–216.

[13] B. Huppert, *Endliche Gruppen I*, Springer-Verlag, Berlin, 1967.

[14] N. Ito, On the degrees of irreducible representations of a finite group, *Nagoya Math. J.* **3** (1951), 5–6.

[15] G. D. James and M. W. Liebeck, *Representations and characters of groups*, Cambridge University Press, 1993.

[16] G. Karpilovsky, *Group Representations, Volume 2*, North-Holland Mathematical Studies, Vol. 177, Amesterdam–New York–Oxford–Tokyo.

[17] P. B. Kleidman and R. A. Wilson, Sparodic simple subgroups of finite exceptional groups of Lie type, *J. Algebra* **157** (1993), 316–330.

[18] B. H. Neumann, Groups covered by permutable subsets, *J. London Math. Soc.* **29** (1954), 236–248.

[19] D. J. S. Robinson, *A course in the theory of groups, 2nd edn.*, Springer-Verlag, Berlin, 1996.

[20] M. Schonert et al., *GAP: Groups, algorithms, and programming*, Lehrstuhl D fur Mathematik, RWTH Aachen, 2002.

[21] M. Suzuki, *Group Theory I*, Springer-Verlag, New York, 1982.

[22] M. Suzuki, *Group Theory II*, Springer-Verlag, New York, 1986.

[23] M. J. Tomkinson, Groups covered by finitely many cosets or subgroups, *Comm. Algebra* **15** (1987), 845–859.

ALGORITHMIC USE OF THE MAL'CEV CORRESPONDENCE

BJÖRN ASSMANN[1]

Centre for Interdisciplinary Research in Computational Algebra (CIRCA),
University of St Andrews, North Haugh, St Andrews, KY16 9SS Fife, Scotland
Email: bjoern@mcs.st-and.ac.uk

Abstract

Mal'cev showed in the 1950s that there is a correspondence between radicable torsion-free nilpotent groups and rational nilpotent Lie algebras. In this paper we show how to establish the connection between the radicable hull of a finitely generated torsion-free nilpotent group and its corresponding Lie algebra algorithmically. We apply it to fast multiplication of elements of polycyclically presented groups.

1 Introduction

The connection between groups and Lie rings, respectively Lie algebras, is a well-known and mathematically very useful concept. For example, a typical way to solve a problem in a Lie group is to transfer the problem to the Lie algebra of the group, study it there with the help of tools from linear algebra and transfer the result back into the Lie group.

Mechanisms of this kind have also been shown to be useful for algorithmic applications. For instance, Vaughan-Lee and O'Brien used Lie ring techniques to construct a consistent polycyclic presentation of $R(2, 7)$, the largest 2-generator group of exponent 7 [16].

In this paper we demonstrate the algorithmic usefulness of the Mal'cev correspondence between torsion-free radicable nilpotent groups and rational nilpotent Lie algebras. We use it to develop an algorithm for fast multiplication in polycyclic groups. To be more precise, for a infinite polycyclic group G, given by a polycyclic presentation, we show how to determine a subgroup H of finite index in G, and construct an algorithm which carries out fast multiplication in H.

For the setup of this algorithm we determine a finitely generated torsion-free nilpotent subgroup N of H and compute a matrix representation of N. Using the latter we calculate a Lie algebra $\mathcal{L}$ which corresponds to the radicable hull of N. Multiplication in H is first reduced to some calculations with automorphisms of N, which correspond to automorphisms of $\mathcal{L}$, which, being linear, are much easier to handle.

An alternative approach would be to compute a matrix representation for the whole group G. That this is possible for any polycyclic group was shown by Auslander [1]. Lo and Ostheimer constructed an algorithm for this task [12]. Such a representation could then be applied to carry out multiplications in G. However

---

[1]I gratefully acknowledge support from the "Gottlieb Daimler und Karl Benz-Stiftung" and the EPSRC.

it seems to be much harder to compute a representation for G then for N. Also, the author is not aware of a practical implementation of the algorithm of Lo and Ostheimer.

2 Mal'cev correspondence

A group G is called *radicable* if for all $g \in G$ and for all $m \in \mathbb{N}$ there exists an $h \in G$ such that $h^m = g$. In this section we recall some well known facts about the connection between radicable torsion-free nilpotent groups and nilpotent Lie algebras, the Mal'cev correspondence, discovered by Mal'cev in 1951 [13].

We denote by $\mathrm{Tr}_1(n, \mathbb{Q})$ the group of all upper unitriangular matrices over $\mathbb{Q}$ and by $\mathrm{Tr}_0(n, \mathbb{Q})$ the set of all upper triangular rational matrices with zeros on the diagonal. Although this can be done in a more general way, we will restrict ourself in this section to the description of the connection between subgroups of $\mathrm{Tr}_1(n, \mathbb{Q})$ and Lie subalgebras of $\mathrm{Tr}_0(n, \mathbb{Q})$. For proofs of the stated results and more background see [17, Chapter 6].

Let $g \in \mathrm{Tr}_1(n, \mathbb{Q})$. The *logarithm* of $g = 1 + u$ is

$$\log g = u - \frac{1}{2}u^2 + \cdots + \frac{(-1)^n}{(n-1)}u^{n-1}$$

and for $x \in \mathrm{Tr}_0(n, \mathbb{Q})$, the *exponential* of x is

$$\exp x = 1 + x + \frac{1}{2}x^2 + \cdots + \frac{1}{(n-1)!}x^{n-1}.$$

Note that this coincides with the definition of log and exp on the complex numbers, since $u^n = x^n = 0$. These mappings are mutually inverse bijections and moreover, for commuting matrices $x, y \in \mathrm{Tr}_0(n, \mathbb{Q})$, we have $(\exp x)(\exp y) = \exp(x + y)$.

The vector space $\mathrm{Tr}_0(n, \mathbb{Q})$ with the Lie bracket $[x, y] = xy - yx$ has the structure of a Lie algebra. For longer Lie brackets we use the left norm convention, i.e., $[x_1, \ldots, x_r] = [[x_1, \ldots, x_{r-1}], x_r]$, and for a vector of positive integers $e = (e_1, \ldots, e_r)$ we define

$$[x, y]_e = [x, \underbrace{y, \ldots, y}_{e_1}, \underbrace{x, \ldots, x}_{e_2}, \ldots].$$

Since $[x_1, \ldots, x_n] = 0$ for all $x_i \in \mathrm{Tr}_0(n, \mathbb{Q})$, the latter is a nilpotent Lie algebra of class $n - 1$.

Theorem 2.1 (Baker–Campbell–Hausdorff formula) *There exist universal constants $q_e \in \mathbb{Q}$ not depending on n such that for all $x, y \in \mathrm{Tr}_0(n, \mathbb{Q})$ and $z(x, y) = \log((\exp x)(\exp y))$ we have*

$$z(x, y) = x + y + \sum_e q_e[x, y]_e,$$

where we take the sum over all vectors $e = (e_1, \ldots, e_r)$, with positive integer entries, such that $e_1 + \cdots + e_r < n - 1$. In particular this means that $z(x, y)$ is an element of the Lie subalgebra of $\mathrm{Tr}_0(n, \mathbb{Q})$ generated by x and y.

A similar formula holds for $\log([\exp(x), \exp(x)]) = [x, y] + \sum_e p_e[x, y]_e$, where the $p_e \in \mathbb{Q}$ are again universal constants. Using the BCH-formula we can define a group multiplication $x * y = z(x, y)$ on $\mathrm{Tr}_0(n, \mathbb{Q})$. The exponential map is then an isomorphism of groups between $(\mathrm{Tr}_0(n, \mathbb{Q}), *)$ and $\mathrm{Tr}_1(n, \mathbb{Q})$. The following theorem explains the interplay of subgroups of $\mathrm{Tr}_1(n, \mathbb{Q})$ and Lie subalgebras of $\mathrm{Tr}_0(n, \mathbb{Q})$ via this mechanism.

Theorem 2.2 *Let* $G \leq \mathrm{Tr}_1(n, \mathbb{Q})$ *and let* $\mathcal{L}(G) = \mathbb{Q} \log G$ *be the* $\mathbb{Q}$-*vector space spanned by* $\log G = \{\log g \mid g \in G\}$. *Let* L *be a Lie subalgebra of* $\mathrm{Tr}_0(n, \mathbb{Q})$. *Then the following holds:*

- $\exp L$ *is a radicable torsion-free nilpotent subgroup of* $\mathrm{Tr}_1(n, \mathbb{Q})$.
- $\mathcal{L}(G)$ *is a Lie subalgebra of* $\mathrm{Tr}_0(n, \mathbb{Q})$.
- $G \leq \exp \mathcal{L}(G)$ *and every element of* $\exp \mathcal{L}(G)$ *has some positive power lying in* G.

A group N is called a $\mathcal{T}$-*group* if it is finitely generated torsion-free nilpotent. It is well-known that every $\mathcal{T}$-group can be embedded in $\mathrm{Tr}_1(n, \mathbb{Q})$ for some $n \in \mathbb{N}$ [17, Chapter 5]. Therefore, given a faithful representation $\beta : N \to \mathrm{Tr}_1(n, \mathbb{Q})$, we can construct the Lie algebra $\mathcal{L}(N\beta)$ of $N\beta$. In the following we will denote by $\mathcal{L}(N)$ the Lie algebra $\mathcal{L}(N\beta)$ and for $g \in N$ by $\mathrm{Log}(g)$ the element $\log(g\beta)$, i.e., we identify the groups N and $N\beta$. This is justified by the fact that for two embeddings β_1, β_2 the Lie algebras $\mathcal{L}(N\beta_1)$ and $\mathcal{L}(N\beta_2)$ are isomorphic, see [17, Chapter 6].

A group $\hat{N}$ is said to be a *radicable hull* of a $\mathcal{T}$-group N if it is a radicable torsion-free nilpotent group that contains N and has the property that every element in $\hat{N}$ has some positive power lying in N. It can be shown that $\hat{N}$ is unique up to isomorphism [17, Chapter 6]. Theorem 2.2 shows that $\exp \mathcal{L}(G)$ is a radicable hull of a finitely generated subgroup $G \leq \mathrm{Tr}_1(n, \mathbb{Q})$. Using a faithful representation $\beta : N \to \mathrm{Tr}_1(n, \mathbb{Q})$ and identifying N and $N\beta$ we can construct the radicable hull of any $\mathcal{T}$-group N.

Theorem 2.3 *Let* $G \leq \mathrm{Tr}_1(n, \mathbb{Q})$ *and let* Γ *be the group of automorphisms of the Lie algebra* $\mathcal{L}(G)$ *which map* $\log G$ *onto itself. Then the function* $\phi : \mathrm{Aut}(G) \to \Gamma$, $\alpha \mapsto \exp \circ \alpha \circ \log$ *is an isomorphism.*

$$
\begin{array}{ccc}
G & \xrightarrow{\ \alpha\ } & G \\
{\scriptstyle \exp} \uparrow & & \downarrow {\scriptstyle \log} \\
\mathcal{L}(G) & \dashrightarrow & \mathcal{L}(G)
\end{array}
$$

3 Polycyclic presentations

Let G be a polycyclic group. A *polycyclic sequence* of G is a sequence of elements $\mathcal{G} = (g_1, \ldots, g_n)$ of G such that the subgroups $G_i = \langle g_i, \ldots, g_n \rangle$ form a subnormal series $G = G_1 > \ldots > G_n > G_{n+1} = \{1\}$ with non-trivial cyclic factors.

Let $r_i = [G_i : G_{i+1}]$ and $I = \{i \in \{1, \ldots, n\} \mid r_i < \infty\}$. Then each element $g \in G$ has a unique normal form with respect to the polycyclic sequence:

$\text{nf}(g) = g_1^{e_1} \cdots g_n^{e_n}$ with $e_i \in \mathbb{Z}$ and $0 \leq e_i < r_i$ for $i \in I$. Thus g can be represented by the exponent vector $(e_1, \ldots, e_n)$ with respect to $\mathcal{G}$.

Each polycyclic sequence $(g_1, \ldots, g_n)$ of G defines a presentation of G on the generators $g_1, \ldots, g_n$ with relations of the form

$$g_i^{g_j} = g_{j+1}^{e(i,j,j+1)} \cdots g_n^{e(i,j,n)} \qquad \text{for } 1 \leq j < i \leq n,$$

$$g_i^{g_j^{-1}} = g_{j+1}^{f(i,j,j+1)} \cdots g_n^{f(i,j,n)} \qquad \text{for } 1 \leq j < i \leq n,$$

$$g_i^{r_i} = g_{i+1}^{\ell(i,i+1)} \cdots g_n^{\ell(i,n)} \qquad \text{for } i \in I,$$

where the right hand sides in these relations are the normal forms of the elements on the left hand sides. It is well-known that these relations define G. Such a presentation is called a *consistent polycyclic presentation* for G. The term consistent refers to the fact that every element in this finitely presented group on the abstract generators $g_1, \ldots, g_n$ has a unique normal form $g_1^{e_1} \cdots g_n^{e_n}$ with $e_i \in \mathbb{Z}$ and $0 \leq e_i < r_i$ for $i \in I$. Throughout this paper all polycyclic presentations are consistent.

Let H be a $\mathcal{T}$-group, i.e., a finitely generated torsion-free nilpotent group. A polycyclic sequence $(h_1, \ldots, h_n)$ is called a *Mal'cev basis* for H if the subgroups $\langle h_i, \ldots, h_n \rangle$ form a central series with infinite cyclic factors. Since the upper central series of a $\mathcal{T}$-group has torsion-free factors, every $\mathcal{T}$-group has a Mal'cev basis.

4 Algorithmic realization of the Mal'cev correspondence

The aim of this section is to show how to realize algorithmically the correspondence between elements of a $\mathcal{T}$-group N, represented by their exponent vectors with respect to a certain polycyclic sequence, and their counterparts in the Lie algebra $\mathcal{L}(N)$, represented as coefficient vectors with respect to a certain basis.

Lemma 4.1 *Let $(g_1, \ldots, g_l)$ be a Mal'cev basis for a $\mathcal{T}$-group N. Then $\mathcal{B} = \{\text{Log}(g_1), \ldots, \text{Log}(g_l)\}$ is a basis for the Lie algebra $\mathcal{L}(N)$. In particular, the dimension of $\mathcal{L}(N)$ is equal to the Hirsch length of N.*

Proof Let $g = g_1^{a_1} \cdots g_l^{a_l} \in N$. Then $\text{Log}(g) = a_1 \text{Log}(g_1) * \cdots * a_l \text{Log}(g_l)$. Thus it is sufficient to show that $\mathcal{B}$ is basis for the Lie algebra L generated by $\mathcal{B}$. We will show this via induction over l, the Hirsch length of N.

If $N = \langle g_1 \rangle$ then $\{\text{Log}(g_1)\}$ is a basis for $\mathcal{L}(N)$. Assume that the lemma is true for all $\mathcal{T}$-groups of Hirsch length $l - 1$. First we show that $\mathcal{B}$ is a generating set for the $\mathbb{Q}$-vector space $\mathcal{L}(N)$. By assumption the vector spaces $\langle \text{Log}(g_2), \ldots, \text{Log}(g_l) \rangle_{\mathbb{Q}}$ and $\langle \text{Log}(g_1), \text{Log}(g_3), \ldots, \text{Log}(g_l) \rangle_{\mathbb{Q}}$ are closed under taking Lie brackets. Thus we have to show that $[\text{Log}(g_1), \text{Log}(g_2)] \in \langle \mathcal{B} \rangle_{\mathbb{Q}}$. By Corollary 3 of [17, Chapter 6] we have that $[\text{Log}(g_1), \text{Log}(g_2)] = \text{Log}([g_1, g_2]) + \sum_i \alpha_i \text{Log}(\chi_i(g_1, g_2))$, where $\alpha_i \in \mathbb{Q}$ and χ_i is a repeated group theoretic commutator in g_1, g_2 of length ≥ 3. Since $[g_1, g_2], \chi_i(g_1, g_2) \in \langle g_3, \ldots, g_l \rangle$ the right hand side of the last equation is contained in $\mathcal{L}(\langle g_3, \ldots, g_l \rangle)$ and thus in the $\mathbb{Q}$-vector space spanned by $\mathcal{B}$.

It remains to show that the elements of $\mathcal{B}$ are linearly independent. So assume that $\mathrm{Log}(g_1) \in \langle \mathrm{Log}(g_2), \ldots, \mathrm{Log}(g_l) \rangle_{\mathbb{Q}} = L_2$. Therefore $g_1 \in \exp(L_2)$ (recall that we identify N and $N\beta \leq \mathrm{Tr}_1(n, \mathbb{Q})$) and, according to Theorem 2.2, there must be an $m \in \mathbb{N}$ such that $g_1^m \in \langle g_2, \ldots, g_l \rangle$. Since $(g_1, \ldots, g_l)$ is a Mal'cev basis this is a contradiction. $\qquad\square$

Lo/Ostheimer [12] and de Graaf/Nickel [2] describe practical algorithms to compute a faithful representation of a $\mathcal{T}$-group in $\mathrm{Tr}_1(n, \mathbb{Q})$. The latter has been implemented in GAP [6] and is part of the Polycyclic package [5]. We use this implementation for the computation of representations of $\mathcal{T}$-groups.

A polycyclic sequence $\mathcal{M} = (M_1, \ldots, M_l)$ for a group $H \leq \mathrm{Tr}_1(n, \mathbb{Q})$ is called *constructive* if there exists a practical algorithm, which, given any $h \in H$, determines the normal form $\mathrm{nf}(h)$ with respect to $\mathcal{M}$. It is well-known how to compute a constructive polycyclic sequence for a given finitely generated group $H \leq \mathrm{Tr}_1(n, \mathbb{Q})$ which is also a Mal'cev basis of H, see for example [18, Chapter 9].

We now summarize the algorithms which set up the Mal'cev correspondence between a $\mathcal{T}$-group N and the Lie algebra $\mathcal{L}(N)$.

SetupMalcevCorrespondence(N)
1: compute a faithful representation $\beta : N \to \mathrm{Tr}_1(n, \mathbb{Q})$.
2: compute a constructive pc-sequence $\mathcal{M} = (M_1, \ldots, M_l)$ for $N\beta$.
3: determine the basis $\mathcal{B} = \{\log M_1, \ldots, \log M_l\}$ of $\mathcal{L}(N)$.
4: compute a polycyclic presentation for N with respect to $\beta^{-1}(\mathcal{M})$.

Given $g \in N$, the following algorithm computes the corresponding element in $\mathcal{L}(N)$. We represent elements in N by their normal form with respect to the polycyclic sequence $\beta^{-1}(\mathcal{M}) = (n_1, \ldots, n_l)$ of N, and the elements of $\mathcal{L}(N)$ as coordinate vectors with respect to the canonical basis $\log(\mathcal{M})$ of Lemma 4.1.

Logarithm(g)
1: let $(e_1, \ldots, e_l)$ be the exponent vector of g with respect to $\beta^{-1}(\mathcal{M})$.
2: determine $\beta(g) = M_1^{e_1} \cdots M_l^{e_l}$.
3: determine $\log \beta(g)$.
4: determine $\gamma = (\gamma_1, \ldots, \gamma_l)$ such that $\log \beta(g) = \sum \gamma_i \log M_i$.
5: return γ.

Finally, given an element in $\log(N\beta)$, represented as coordinate vector with respect to $\log(\mathcal{M})$, the following algorithm computes its counterpart in N.

Exponential($(\gamma_1, \ldots, \gamma_l)$)
1: determine $N = \sum \gamma_i \log M_i$.
2: determine $\exp N$.
3: compute $(e_1, \ldots, e_l)$ such that $M_1^{e_1} \cdots M_l^{e_l} = \exp N$.
4: return $n_1^{e_1} \cdots n_l^{e_l}$.

5   Collection in polycyclic groups

Let G be an infinite polycyclically presented group. In this section we show that the Mal'cev correspondence can be used for a efficient multiplication algorithm in a certain subgroup H of finite index in G.

A polycyclic presentation (pcp) is a good computer representation of a polycyclic group. Algorithms for finite solvable groups given by a pcp were developed in the 1980s by Laue, Neubüser and Schoenwaelder [9]. More recently, calculations with infinite polycyclically presented groups have been shown to be practical, see for example [4, 18].

The efficiency of calculations with polycyclic presentations depends on the ability to compute quickly the normal form of the product of two elements given in normal form. The original methods for doing this were based on rewriting using pcp and called *collection*. Nowadays, collection is used as a general term for an algorithm which computes the normal form of an element of a polycyclically presented group. In the recent past, various strategies for collection algorithms have been discussed [7, 10, 19]. "Collection from the left" is the current state of the art.

The collection process in $\mathcal{T}$-groups is much better understood than in the more general class of polycyclic groups. The following theorem states that collection with respect to a Mal'cev basis can be done symbolically. For a proof see [8].

Theorem 5.1 (Hall) *Let G be a $\mathcal{T}$-group with Mal'cev basis $(g_1, \ldots, g_l)$. Define the functions $\zeta_1, \ldots, \zeta_l$ in $2l$ integer variables $x_1, \ldots, x_l, y_1, \ldots, y_l$ such that*

$$g_1^{x_1} \cdots g_l^{x_l} g_1^{y_1} \cdots g_l^{y_l} = g_1^{\zeta_1} \cdots g_l^{\zeta_l}.$$

Then the functions ζ_i are rational polynomials.

This result can be used for computational applications. In the 90s Leedham-Green and Soicher developed the algorithm "Deep Thought" [11], which computes these polynomials and uses them for collection in $\mathcal{T}$-groups. An implementation of "Deep Thought" by Merkwitz [14] is part of the GAP system.

Having said that collection in $\mathcal{T}$-groups is easier than in the general case, it is interesting to note that $\mathcal{T}$-groups can be regarded as main ingredients of polycyclic groups. Namely, if G is a polycyclic group, then it has a normal series $G \geq K \geq N \geq 1$, where N is a $\mathcal{T}$-group, K/N is free-abelian and G/K is finite. Further the next theorem shows that, up to a finite index, the group G is made out of two $\mathcal{T}$-groups. A proof can be found in [17, Chapter 3].

Theorem 5.2 (Newell) *Let K be a polycyclic group and N a $\mathcal{T}$-group which is normal in K and has the property that K/N is nilpotent. Then there exists a $\mathcal{T}$-group C such that CN is of finite index in K.*

The group C from Theorem 5.2 is said to be a *nilpotent almost-supplement* for N in K. 'Almost' because C and N generate a subgroup of finite index, and 'supplement' because C may intersect N non-trivially. Since $[G : K] < \infty$ the group C is also a nilpotent almost-supplement for N in G. This structure of polycyclic groups can also be explored algorithmically. In [4, Chapter 9] Eick describes a practical algorithm to compute a nilpotent-by-abelian-by-finite series $G \geq K \geq N \geq 1$ and methods to determine a nilpotent almost-supplement C for N in G.

Therefore, up to a finite index, collection in infinite polycyclic groups can be reduced to the case of a polycyclic group H generated by two $\mathcal{T}$-groups C and N such that $N \lhd H$ and H/N is free-abelian. We explain how collection in H can be realized efficiently. Let $\mathcal{N} = (n_1, \ldots, n_l)$ be a Mal'cev basis of N and let $(c_1 N, \ldots, c_k N)$ be a basis for the free abelian group CN/N. Then $\mathcal{H} = (c_1, \ldots, c_k, n_1, \ldots, n_l)$ is a polycyclic sequence for H.

Lemma 5.3 *The list $(c_1, \ldots, c_k)$ can be extended to a Mal'cev basis*

$$\mathcal{C} = (c_1, \ldots, c_k, c_{k+1}, \ldots, c_{k+m})$$

of the $\mathcal{T}$-group C.

Proof The upper central series of $C \cap N$ has torsion-free factors and is invariant under the action of $(c_1, \ldots, c_k)$. This series can be refined to a central series with torsion-free factors which are centralized by $(c_1, \ldots, c_k)$. To see this, let M be one of the torsion-free factors. Denote by A the centralizer of $(c_1, \ldots, c_k)$ in M. Then A is non-trivial, since C acts nilpotently on M, and furthermore M/A is torsion-free, since for $m \in M, z \in \mathbb{N}$, the equality $zm = (zm)^{c_i} = z(m^{c_i})$ implies $m^{c_i} = m$. Thus, by induction on the dimension of M, we get a strictly ascending series of submodules of M with torsion-free factors, which are by construction centralized by $(c_1, \ldots, c_k)$. $\square$

Denote by $c^x n^{\bar{x}}$ the element in H given by the exponent vector $(x_1, \ldots, x_k, \bar{x}_1, \ldots, \bar{x}_l)$ with respect to $\mathcal{H}$. For two elements $c^x n^{\bar{x}}, c^y n^{\bar{y}} \in H$ we have

$$c^x n^{\bar{x}} c^y n^{\bar{y}} = c^x c^y (n^{\bar{x}})^{(c^y)} n^{\bar{y}}.$$

Since CN/N is free abelian, the normal form of $c^x c^y$ with respect to $\mathcal{C}$ is of the form $c^{x+y} c_{k+1}^{z_{k+1}} \cdots c_{k+m}^{z_{k+m}}$. The computation of the tail $t = c_{k+1}^{z_{k+1}} \cdots c_{k+m}^{z_{k+m}}$ is easy in the sense that we can apply standard methods for $\mathcal{T}$-groups, such as Deep Thought. We can also compute the normal form of the tail $t \in C \cap N$ with respect to $\mathcal{N}$. The efficient computation of the normal form of

$$(n^{\bar{x}})^{(c^y)} = (n^{\bar{x}})^{(c_1^{y_1} \cdots c_k^{y_k})}$$

is the crucial step of our method. For this, we will describe an algorithm which applies powers of automorphisms using the Mal'cev correspondence. Finally, the multiplication of $\mathrm{nf}(t)$, $\mathrm{nf}((n^{\bar{x}})^{(c^y)})$ and $n^{\bar{y}}$ in N can be done by standard techniques.

It remains to describe an algorithm for the computation of the normal form of

$$n^{(\varphi^q)},$$

where $n \in N$, $\varphi \in \mathrm{Aut}(N)$ and $q \in \mathbb{Z}$. As described in Section 4, let $\beta : N \to \mathrm{Tr}_1(m, \mathbb{Q})$ be a faithful representation of N and $\mathcal{M} = (M_1, \ldots, M_l)$ be a constructive polycyclic sequence for $N\beta$, which is also a Mal'cev basis for $N\beta$. Denote by n_i the element $\beta^{-1}(M_i) \in N$. We define a $l \times l$ matrix Φ by

$$\mathrm{Log}(n_i^{\varphi}) = \sum_{j=1}^{l} \Phi_{ij} \mathrm{Log}(n_i).$$

Then by Theorem 2.3 the matrix Φ is a representation of the Lie algebra isomorphism $\exp \circ \varphi \circ \log$, with respect to the basis $\{\log M_1, \ldots, \log M_l\}$. This yields the following algorithm.

ApplyPowerOfAutomorphism(n, φ, q)
1: determine $\gamma = \mathbf{Logarithm}(\ n\ )$.
2: compute $\bar{\gamma} = \gamma \cdot \Phi^q$.
3: compute $g = \mathbf{Exponential}(\ \bar{\gamma}\ )$.
4: return g.

If we want to apply several powers of automorphisms, as in the computation of the normal form of $(n^{\bar{x}})^{(c^y)}$, we switch only once from $n^{\bar{x}}$ to the corresponding element γ in the Lie algebra, then multiply γ with $\Phi(c_1)^{y_1} \cdots \Phi(c_k)^{y_k}$, where $\Phi(c_i)$ is the matrix representation of the Lie algebra isomorphism corresponding to the conjugation with c_i, and then switch back to the representation with respect to $(n_1, \ldots, n_l)$.

Given that this algorithm works for arbitrary automorphisms, it can also be used as a partial step for collection in arbitrary polycyclic groups. If G is a polycyclic group, N a $\mathcal{T}$-group which is normal in G and C a nilpotent almost-supplement for N in G, then G has a polycyclic sequence of the form

$$\mathcal{G} = (g_1, \ldots, g_j, c_1, \ldots, c_k, n_1, \ldots, n_l),$$

where $\langle g_1 CN, \ldots, g_j CN \rangle = G/CN$ and $\mathcal{H} = (c_1, \ldots, c_k, n_1, \ldots, n_l)$ is as before. The action of $g^z = g_1^{z_1} \cdots g_j^{z_j}$ on an element $n^{\bar{x}} \in N$ can be computed in the same way as the action of c^y on $n^{\bar{x}}$.

The development of this algorithm was motivated by a recent result of du Sautoy. In [3] he proves that symbolic collection in so-called *splittable* polycyclic groups is possible. For definition and background of a semi-simple splitting of a polycyclic group see [17, Chapter 7]. Every polycyclic group has a splittable subgroup of finite index, which can be constructed as a subgroup of finite index of the group CN mentioned in the last paragraph. However, for practical collection algorithms it seems to be better to avoid this concept. In our approach we avoid having to pass to a subgroup of finite index of CN, and we also avoid computations with elements of finite extension of $\mathbb{Q}$, which are used by du Sautoy to describe symbolic collection in splittable groups.

Group	Hirsch length	Hl(N)	Class N	Dimension $N\beta$	Time
$G(\mathrm{Tr}_3(\mathcal{O}_1))$	9	6	2	6	00:00.229
$G(\mathrm{Tr}_4(\mathcal{O}_1))$	16	12	3	16	00:03.752
$G(\mathrm{Tr}_5(\mathcal{O}_1))$	25	20	4	46	05:46.230
$G(\mathrm{Tr}_3(\mathcal{O}_2))$	12	9	2	9	00:00.669
$G(\mathrm{Tr}_4(\mathcal{O}_2))$	22	18	3	27	00:32.664
$G(\langle\bar{\varphi}_1\rangle \rtimes F_{2,4})$	9	8	4	10	00:00.801
$G(\langle\bar{\varphi}_1\rangle \rtimes F_{2,5})$	15	14	5	20	00:34.531
$G(\langle\bar{\varphi}_2\rangle \rtimes F_{3,4})$	33	32	4	43	12:22.355

Table 1. Setup of the Mal'cev correspondence: The third column displays the Hirsch Length of the $\mathcal{T}$-group N for which the Lie algebra $\mathcal{L}(N)$ is computed. In the fifth column the dimension m of the vector space on which $N\beta \leq \mathrm{Tr}_1(m, \mathbb{Q})$ acts is indicated. The sixth column contains the time which is needed by the algorithm SetupMalcevCorrespondence(N) of Section 4.

6 Runtimes

In this section we compare the performance of the collection algorithm presented in Section 5 with the classical "Collection from the left" method, as implemented in the GAP package Polycyclic [5] Version 1.1. We also specify the time required to set up the Mal'cev correspondence.

We construct our first class of examples of polycyclically presented groups with the help of matrices over algebraic integers. Let $\mathbb{Q}(\theta)$ be an algebraic extension of $\mathbb{Q}$ and $\mathcal{O}$ its maximal order. We denote by $\mathrm{Tr}_n(\mathcal{O})$ the group of upper-triangular matrices in $GL_n(\mathcal{O})$, by $\mathrm{Tr}_1(n, \mathcal{O})$ the subgroup of matrices in $\mathrm{Tr}_n(\mathcal{O})$ with 1s on the diagonal and by $D_n(\mathcal{O})$ the group of diagonal matrices in $GL_n(\mathcal{O})$. Note that every polycyclic group has a subgroup of finite index which can be embedded in some $\mathrm{Tr}_n(\mathcal{O})$ [17, page 132].

Let $U(\mathcal{O})$ be the group of units of the maximal order $\mathcal{O}$. By Dirichlet's Units Theorem, $U(\mathcal{O})$ is polycyclic and therefore also $D_n(\mathcal{O})$. Using the torsion unit and fundamental units of $U(\mathcal{O})$, it is straightforward to obtain a polycyclic presentation for $D_n(\mathcal{O})$. In a similar way to $\mathrm{Tr}_1(n, \mathbb{Q})$, we can compute a constructive polycyclic sequence for $\mathrm{Tr}_1(n, \mathcal{O})$, which then yields a polycyclic presentation for $\mathrm{Tr}_1(n, \mathcal{O})$. Using the fact that $\mathrm{Tr}_n(\mathcal{O}) = D_n(\mathcal{O}) \rtimes \mathrm{Tr}_1(n, \mathcal{O})$, we obtain a polycyclically presented group $G(\mathrm{Tr}_n(\mathcal{O}))$ which is isomorphic to $\mathrm{Tr}_n(\mathcal{O})$.

We use the irreducible polynomials $p_1(x) = x^2 - 3$ and $p_2(x) = x^3 - x^2 + 4$ for our examples. By $\mathcal{O}_i$ we denote the maximal order of $\mathbb{Q}(\theta_i)$ where θ_i is a zero of the polynomial p_i.

The second class of examples is constructed as follows: Let F_n be the free group on n generators $f_1, \ldots, f_n$. Then $F_{n,c} = F_n/\gamma_{c+1}(F_n)$ is the free nilpotent group on n generators of class c. It is a $\mathcal{T}$-group and we use the nilpotent quotient algorithm in the GAP package NQ [15] to compute a polycyclic presentation for it. An automorphism φ of F_n naturally induces an automorphism $\bar{\varphi}$ of $F_{n,c}$.

Group	range = 2		range = 4		range = 8	
	Cftl	Malcev	Cftl	Malcev	Cftl	Malcev
$G(\mathrm{Tr}_3(\mathcal{O}_1))$	0.001	0.002	0.951	0.003	*	0.004
$G(\mathrm{Tr}_4(\mathcal{O}_1))$	0.011	0.013	22:24.238	0.017	*	0.025
$G(\mathrm{Tr}_5(\mathcal{O}_1))$	0.035	0.148	*	0.174	*	0.223
$G(\mathrm{Tr}_3(\mathcal{O}_2))$	0.002	0.004	00:31.017	0.006	*	0.009
$G(\mathrm{Tr}_4(\mathcal{O}_2))$	5.808	0.049	*	0.061	*	0.088
$G(\langle\bar{\varphi}_1\rangle \rtimes F_{2,4})$	0.001	0.004	0.001	0.005	7.099	0.008
$G(\langle\bar{\varphi}_1\rangle \rtimes F_{2,5})$	0.001	0.020	0.221	0.027	*	0.043
$G(\langle\bar{\varphi}_2\rangle \rtimes F_{3,4})$	0.001	0.179	0.531	0.221	*	0.304

Table 2. Runtimes for the multiplication of two random elements: This table specifies the average runtime of 100 computations of the normal form of gh where g, h are radomly chosen group elements of range r. The two compared methods are Cftl, i.e., 'Collection from the left' as implemented in the GAP package Polycyclic [5] Version 1.1, and Malcev, i.e., the algorithm explained in Section 5. The symbol * means that the average runtime is more than 1 hour.

We use the automorphism φ_1 of F_2 which maps f_1 to f_2^{-1} and f_2 to $f_1 f_2^3$ and the automorphism φ_2 of F_3 mapping f_1 to f_2^{-1}, f_2 to f_3^{-1} and f_3 to $f_2^{-3} f_1^{-1}$ for our examples.

An automorphism ψ of $F_{n,c}$ and can be used to construct a polycyclically presented group $G(\langle\psi\rangle \rtimes F_{n,c})$ which is isomorphic to $\langle\psi\rangle \rtimes F_{n,c}$.

The collection algorithm of Section 5 can be applied to these examples by setting $N = \mathrm{Tr}_1(n, \mathcal{O})$ (respectively $N = F_{n,c}$) and $C = D_n(\mathcal{O})$ (respectively $C = \langle\psi\rangle$). All calculation in N were carried out via the matrix representation $\beta : N \to \mathrm{Tr}_1(m, \mathbb{Q})$.

In Table 1 we give an overview of the example groups and specify the time required to set up the Mal'cev correspondence between the normal nilpotent $\mathcal{T}$-group N and the Lie algebra $\mathcal{L}(N)$, as described in Section 4. Most of the time for this set up is used to construct a faithful matrix representation for N.

In Table 2 we indicate the average runtime for the multiplication of two random elements. We generate random elements of a group G with polycyclic sequence $(g_1, \ldots, g_k)$ in the following way. Let r be a natural number. A random element $g \in G$ of *range* r is of the form $g = g_1^{e_1} \cdots g_k^{e_k}$, where e_i is a randomly chosen integer in $[-r, \ldots, r]$.

All computations where carried out in GAP Version 4.4.5 on a Pentium 4 machine with 3.2 gigahertz and 1 gigabyte of memory. The display format for runtimes is minutes:seconds.milliseconds.

Discussion of the results: The runtimes displayed in Table 2 show that Collection from the left (Cftl) is faster than the collection algorithm of Section 5 (Malcev) only for the multiplication of group elements with very small exponents, like integers between -2 and 2. In these cases little rewriting has to be done and so Cftl is quicker.

For slightly bigger exponents Malcev outperforms Cftl. For example the multiplication of two random elements of range 5 in the group $G(\mathrm{Tr}_5(\mathcal{O}_1))$ using Cftl took at least 1 hour on average. In fact, we found frequently examples of random elements $g, h \in G(\mathrm{Tr}_5(\mathcal{O}_1))$ of range 5, whose multiplication using Cftl did not terminate after 24 hours. The average runtime for the same kind of computation using Malcev was 189 milliseconds.

The reason why Cftl fails in these situations is that the exponents arising in the normal form of the product become too big. For instance, the size of the exponents of $\mathrm{nf}(gh)$ for two random elements $g, h \in G(\mathrm{Tr}_5(\mathcal{O}_1))$ of range 5 can easily be bigger than 10^{10}. Malcev can handle these cases, since the algorithm is based on arithmetic with the exponents rather than classical rewriting.

It is interesting to note that the size of the exponents of $\mathrm{nf}(gh)$ varies a lot for random elements g, h of same range. As a consequence the runtime of Cftl for several multiplications differs considerably. For example, the average runtime for 100 computations of $\mathrm{nf}(gh)$, where $g, h \in G(\mathrm{Tr}_4(\mathcal{O}_1))$ were random elements of range 3, was 4.296 seconds with a standard deviation of 29.901 seconds. Malcev showed a stable behavior in this respect in all experiments. In the cited example the average runtime was 0.014 seconds with a standard deviation of 0.009 seconds.

Moreover, Malcev is able to deal with multiplications of random elements of much higher range. For example in the group $G(\mathrm{Tr}_3(\mathcal{O}_2))$ it is possible to multiply two random elements of range 1000 on average in 466 milliseconds. It seems that the average runtime of Malcev as function in the range of the input elements is linear, see Figure 1.

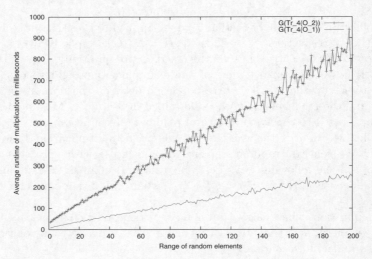

Figure 1. For the groups $G(\mathrm{Tr}_4(\mathcal{O}_1))$ and $G(\mathrm{Tr}_4(\mathcal{O}_2))$ the average runtime of the collection algorithm Malcev as a function in the range of the multiplied random elements is displayed. The graph suggests a linear relationship. For comparison, Collection from the left is not able to carry out multiplications of random elements of range 5 within 1 hour in either group.

The results make it clear that it is worthwhile to set up the Mal'cev correspondence and to use the collection algorithm Malcev, if at least one non-trivial multiplication is going to be carried out. However for more complicated groups it can be very hard to set up this correspondence. The most time consuming step is the calculation of the matrix representation $\beta : N \to \mathrm{Tr}_1(m, \mathbb{Q})$. For example, we were not able to compute a matrix representation of $F_{2,7}$ in less than 1 hour. Future work will explore alternatives to this calculation.

Acknowledgement I would like to thank Steve Linton for various inspiring discussions and Colva Roney-Dougal for helpful comments on an earlier version of this article.

References

[1] L. Auslander, On a problem of Philip Hall, *Ann. of Math. (2)* **86** (1967), 112–116.
[2] W. de Graaf and W. Nickel, Constructing faithful representations of finitely-generated torsion-free nilpotent groups, *J. Symbolic Comput.* **33** (2002), 31–41.
[3] M. du Sautoy, Polycyclic groups, analytic groups and algebraic groups, *Proc. London Math. Soc. (3)* **85** (2002), 62–92.
[4] B. Eick, Algorithms for polycyclic groups, Habilitationsschrift, Universität Kassel, 2001.
[5] B. Eick and W. Nickel, *Polycyclic — computing with polycyclic groups*, 2000. A refereed GAP 4 package, see [6].
[6] The GAP Group, *GAP — Groups, Algorithms and Programming*, http://www.gap-system.org, 2005.
[7] V. Gebhardt, Efficient collection in infinite polycyclic groups, *J. Symbolic Comput.* **34** (2002), no. 3, 213–228.
[8] P. Hall, Nilpotent groups, in *The collected works of Philip Hall* (Clarendon Press, Oxford, 1988), 415–462. Notes of lectures given at the Canadian Mathematical Congress 1957 Summer Seminar.
[9] R. Laue, J. Neubüser, and U. Schoenwaelder, Algorithms for finite soluble groups and the SOGOS system, in *Computational Group Theory (Durham, 1982)* (Academic Press, London, New York, 1984), 105–135.
[10] C. R. Leedham-Green and L. H. Soicher, Collection from the left and other strategies. *J. Symbolic Comput.* **9** (1990), 665–675.
[11] C. R. Leedham-Green and L. H. Soicher, Symbolic collection using Deep Thought. *LMS J. Comput. Math.* **1** (1998), 9–24.
[12] E. H. Lo and G. Ostheimer, A practical algorithm for finding matrix representations for polycyclic groups. *J. Symbolic Comput.* **28** (1999), 339–360.
[13] A. J. Mal'cev, On certain classes of infinite soluble groups, *Mat. Sb.* **28** (1951), 567–588.
[14] W. Merkwitz, Symbolische Multiplikation in nilpotenten Gruppen mit Deep Thought, Diplomarbeit, RWTH Aachen, 1997.
[15] W. Nickel, *NQ*, 1998. A refereed GAP 4 package, see [6].
[16] E. O'Brien and M. Vaughan-Lee, The 2-generator restricted burnside group of exponent 7, *Internat. J. Algebra Comput.* **12** (2002), 575–592.
[17] D. Segal, *Polycyclic Groups* (Cambridge University Press, Cambridge, 1983).
[18] C. C. Sims, *Computation with finitely presented groups*, (Cambridge University Press, Cambridge, 1994).
[19] M. Vaughan-Lee, Collection from the left, *J. Symbolic Comput.* **9** (1990), 725–733.

MINIMAL BUT INEFFICIENT PRESENTATIONS FOR SEMI-DIRECT PRODUCTS OF FINITE CYCLIC MONOIDS

FIRAT ATEŞ and A. SINAN ÇEVIK

Balikesir Universitesi, Fen-Edebiyat Fakultesi, Matematik Bolumu, 10100 Balikesir/Turkey
Email: firat@balikesir.edu.tr, scevik@balikesir.edu.tr

Abstract

Let A and K be arbitrary two monoids. For any connecting monoid homomorphism $\theta : A \longrightarrow \mathrm{End}(K)$, let $M = K \rtimes_\theta A$ be the corresponding monoid semi-direct product. In [2], Cevik discussed necessary and sufficient conditions for the standard presentation of M to be efficient (or, equivalently, p-Cockcroft for any prime p or 0), and then, as an application of this, he showed the efficiency for the presentation, say $\mathcal{P}_M$, of the semi-direct product of any two finite cyclic monoids. As a main result of this paper, we give sufficient conditions for $\mathcal{P}_M$ to be *minimal* but not *efficient*. To do that we will use the same method as given in [3].

AMS Classification: 20L05, 20M05, 20M15, 20M50, 20M99.
Keywords: Minimality, Efficiency, p-Cockcroft property, Finite cyclic monoids.

1 Introduction

Let $\mathcal{P} = [X \; ; \; \mathbf{r}]$ be a monoid presentation where a typical element $R \in \mathbf{r}$ has the form $R_+ = R_-$. Here R_+, R_- are words on X (that is, elements of the free monoid $F(X)$ on X). The *monoid defined by* $[X \; ; \; \mathbf{r}]$ is the quotient of $F(X)$ by the smallest congruence generated by $\mathbf{r}$.

We have a (Squier) graph $\Gamma = \Gamma(X; \mathbf{r})$ associated with $[X \; ; \; \mathbf{r}]$, where the vertices are the elements of $F(X)$ and the edges are the 4-tuples $e = (U, R, \varepsilon, V)$ where $U, V \in F(X)$, $R \in \mathbf{r}$ and $\varepsilon = \pm 1$. The initial, terminal and inversion functions for an edge e as given above are defined by $\iota(e) = UR_\varepsilon V$, $\tau(e) = UR_{-\varepsilon} V$ and $e^{-1} = (U, R, -\varepsilon, V)$.

Two paths π and π' in a 2-complex are equivalent if there is a finite sequence of paths $\pi = \pi_0, \pi_1, \cdots, \pi_m = \pi'$ where for $1 \leq i \leq m$ the path π_i is obtained from π_{i-1} either by inserting or deleting a pair ee^{-1} of inverse edges or else by inserting or deleting a defining path for one of the 2-cells of the complex. There is an equivalence relation, $\sim$, on paths in Γ which is generated by $(e_1.\iota(e_2))(\tau(e_1).e_2) \sim (\iota(e_1).e_2)(e_1.\tau(e_2))$ for any edges e_1 and e_2 of Γ. This corresponds to requiring that the closed paths $(e_1.\iota(e_2))(\tau(e_1).e_2)(e_1^{-1}.\tau(e_2))(\iota(e_1).e_2^{-1})$ at the vertex $\iota(e_1)\iota(e_2)$ are the defining paths for the 2-cells of a 2-complex having Γ as its 1-skeleton. This 2-complex is called the Squier complex of $\mathcal{P}$ and denoted by $\mathcal{D}(\mathcal{P})$ (see, for example, [7], [12], [13], [15]). The paths in $\mathcal{D}(\mathcal{P})$ can be represented by geometric configurations, called *monoid pictures*. Monoid pictures and group pictures have been used in several papers by S. Pride and other authors. We assume here that the reader is familiar with monoid pictures (see [7, Section 4], [12, Section 1] or [13,

Section 2]). Typically, we will use blackboard bold, e.g., $\mathbb{A}, \mathbb{B}, \mathbb{C}, \mathbb{P}$, as notation for monoid pictures. Atomic monoid pictures are pictures which correspond to paths of length 1. Write $[[U, R, \varepsilon, V]]$ for the atomic picture which corresponds to the edge (U, R, ε, V) of the Squier complex. Whenever we can concatenate two paths π and π' in Γ to form the path $\pi\pi'$, then we can concatenate the corresponding monoid pictures $\mathbb{P}$ and $\mathbb{P}'$ to form a monoid picture $\mathbb{P}\mathbb{P}'$ corresponding to $\pi\pi'$. The equivalence of paths in the Squier complex corresponds to an equivalence of monoid pictures. That is, two monoid pictures $\mathbb{P}$ and $\mathbb{P}'$ are equivalent if there is a finite sequence of monoid pictures $\mathbb{P} = \mathbb{P}_0, \mathbb{P}_1, \cdots, \mathbb{P}_m = \mathbb{P}'$ where, for $1 \leq i \leq m$, the monoid picture $\mathbb{P}_i$ is obtained from the picture $\mathbb{P}_{i-1}$ either by inserting or deleting a subpicture $\mathbb{A}\mathbb{A}^{-1}$ where $\mathbb{A}$ is an atomic monoid picture or else by replacing a subpicture $(\mathbb{A}.\iota(\mathbb{B}))(\tau(\mathbb{A}).\mathbb{B})$ by $(\iota(\mathbb{A}).\mathbb{B})(\mathbb{A}.\tau(\mathbb{B}))$ or vice versa, where $\mathbb{A}$ and $\mathbb{B}$ are atomic monoid pictures.

A monoid picture is called a *spherical* monoid picture when the corresponding path in the Squier complex is a closed path. Suppose $\mathbf{Y}$ is a collection of spherical monoid pictures over $\mathcal{P}$. Two monoid pictures $\mathbb{P}$ and $\mathbb{P}'$ are *equivalent relative to* $\mathbf{Y}$ if there is a finite sequence of monoid pictures $\mathbb{P} = \mathbb{P}_0, \mathbb{P}_1, \cdots, \mathbb{P}_m = \mathbb{P}'$ where, for $1 \leq i \leq m$, the monoid picture $\mathbb{P}_i$ is obtained from the picture $\mathbb{P}_{i-1}$ either by the insertion, deletion and replacement operations of the previous paragraph or else by inserting or deleting a subpicture of the form $W.\mathbb{Y}.V$ or of the form $W.\mathbb{Y}^{-1}.V$ where $W, V \in F(X)$ and $\mathbb{Y} \in \mathbf{Y}$. By definition, a set $\mathbf{Y}$ of spherical monoid pictures over $\mathcal{P}$ is a *trivializer of* $\mathcal{D}(\mathcal{P})$ if every spherical monoid picture is equivalent to an empty picture relative to $\mathbf{Y}$. By [13, Theorem 5.1], if $\mathbf{Y}$ is a trivializer for the Squier complex, then the elements of $\mathbf{Y}$ generate the first homology group of the Squier complex. The trivializer is also called a set of generating pictures. Some examples and more details of the trivializer can be found in [4], [2], [3], [9], [12], [13], [15] and [17].

For any monoid picture $\mathbb{P}$ over $\mathcal{P}$ and for any $R \in \mathbf{r}$, $\exp_R(\mathbb{P})$ denotes the *exponent sum* of R in $\mathbb{P}$ which is the number of positive discs labelled by R_+ minus the number of negative discs labelled by R_-. For a non-negative integer n, $\mathcal{P}$ is said to be *n-Cockcroft* if $\exp_R(\mathbb{P}) \equiv 0 \,(\mathrm{mod}\; n)$ (where congruence (mod 0) is taken to be equality) for all $R \in \mathbf{r}$ and for all spherical pictures $\mathbb{P}$ over $\mathcal{P}$. Then a monoid $\mathcal{M}$ is said to be *n-Cockcroft* if it admits an n-Cockcroft presentation. In fact to verify the n-Cockcroft property, it is enough to check for pictures $\mathbb{P} \in \mathbf{Y}$, where $\mathbf{Y}$ is a trivializer (see [12], [13]). The 0-Cockcroft property is usually just called Cockcroft. In general we take n to be equal to 0 or a prime p. Examples of monoid presentations with Cockcroft and p-Cockcroft properties can be found in [2]. We note that the group case of these properties can be found in [10].

In group theory, the homological concept of efficiency has been widely studied. In [1], the authors defined efficiency for *finite* semigroups and hence for *finite* monoids. The following definition for not necessarily finite monoids follows [2] and is equivalent to the definition in [1] when the monoids are finite. For an abelian group G, $rk_{\mathbb{Z}}(G)$ denotes the $\mathbb{Z}$-rank of the torsion free part of G and $d(G)$ means the minimal number of generators of G. Suppose that $\mathcal{P} = [X; \mathbf{r}]$ is a finite presentation for a monoid $\mathcal{M}$. Then the *Euler characteristic*, $\chi(\mathcal{P})$, is defined by

$\chi(\mathcal{P}) = 1 - |X| + |\mathbf{r}|$ and $\delta(\mathcal{M})$ is defined by $\delta(\mathcal{M}) = 1 - rk_{\mathbb{Z}}(H_1(\mathcal{M})) + d(H_2(\mathcal{M}))$. In unpublished work, S. Pride has shown that $\chi(\mathcal{P}) \geq \delta(\mathcal{M})$. With this background, we define the finite monoid presentation $\mathcal{P}$ to be *efficient* if $\chi(\mathcal{P}) = \delta(\mathcal{M})$ and we define the monoid $\mathcal{M}$ to be *efficient* if it has an efficient presentation. Moreover a presentation $\mathcal{P}_0$ for $\mathcal{M}$ is called *minimal* if $\chi(\mathcal{P}_0) \leq \chi(\mathcal{P})$ for all presentations $\mathcal{P}$ of $\mathcal{M}$. There is also interest in finding *inefficient* finitely presented monoids since if we can find a minimal presentation $\mathcal{P}_0$ for a monoid $\mathcal{M}$ such that $\mathcal{P}_0$ is not efficient then we have $\chi(\mathcal{P}') \geq \chi(\mathcal{P}_0) > \delta(\mathcal{M})$, for all presentations $\mathcal{P}'$ defining the same monoid $\mathcal{M}$. Thus there is no efficient presentation for $\mathcal{M}$, that is, $\mathcal{M}$ is not an efficient monoid.

The following result is also an unpublished result by S. Pride. We will use this result rather than making more direct computations of homology for monoids. Kilgour and Pride prove the analogous result for groups in [10] and credit an earlier proof by Epstein, [6].

Theorem 1.1 *Let $\mathcal{P}$ be a monoid presentation. Then $\mathcal{P}$ is efficient if and only if it is p-Cockcroft for some prime p.*

The definition and a standard presentation for the semi-direct product of two monoids can be found in [2], [14], [16] or [17]. Let A and K be arbitrary monoids with associated presentations $\mathcal{P}_A = [X ; \mathbf{r}]$ and $\mathcal{P}_K = [Y ; \mathbf{s}]$, respectively. Let $M = K \rtimes_\theta A$ be the corresponding semi-direct product of these two monoids where θ is a monoid homomorphism from A to $\mathrm{End}(K)$. (We note that the reader can find some examples of monoid endomorphisms in [5].) The elements of M can be regarded as ordered pairs (a, k) where $a \in A$, $k \in K$ with multiplication given by $(a, k)(a', k') = (aa', (k\theta_{a'})k')$. The monoids A and K are identified with the submonoids of M having elements $(a, 1)$ and $(1, k)$, respectively. We want to define standard presentations for M. For every $x \in X$ and $y \in Y$, choose a word, which we denote by $y\theta_x$, on Y such that $[y\theta_x] = [y]\theta_{[x]}$ as an element of K. To establish notation, let us denote the relation $yx = x(y\theta_x)$ on $X \cup Y$ by T_{yx} and write $\mathbf{t}$ for the set of relations T_{yx}. Then, for any choice of the words $y\theta_x$,

$$\mathcal{P}_M = [Y, X ; \mathbf{s}, \mathbf{r}, \mathbf{t}] \tag{1}$$

is a standard monoid presentation for the semi-direct product M.

In [17], Wang constructed a finite trivializer set for the standard presentation $\mathcal{P}_M$, as given in (1), for the semi-direct product M. We will essentially follow [2] in describing this trivializer set using spherical pictures and certain non-spherical subpictures of these.

If $W = y_1 y_2 \cdots y_m$ is a positive word on Y, then for any $x \in X$, we denote the word $(y_1\theta_x)(y_2\theta_x) \cdots (y_m\theta_x)$ by $W\theta_x$. If $U = x_1 x_2 \cdots x_n$ is a positive word on X, then for any $y \in Y$, we denote the word $(\cdots ((y\theta_{x_1})\theta_{x_2})\theta_{x_3} \cdots)\theta_{x_n})$ by $y\theta_U$ and this can be represented by a monoid picture, say $\mathbb{A}_{U,y}$, as in Figure 1 (see also [2]). For $y \in Y$ and the relation $R_+ = R_-$ in the relation set $\mathbf{r}$, we have two important special cases, $\mathbb{A}_{R_+,y}$ and $\mathbb{A}_{R_-,y}$, of this consideration. We should note that these non-spherical pictures consist of only T_{yx}-discs ($x \in X$).

$\mathbb{A}_{U,y}$

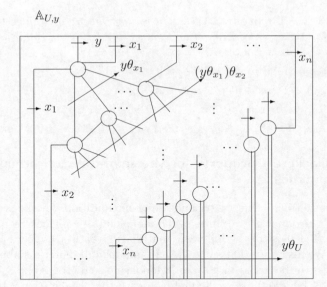

Figure 1.

Let $S \in \mathbf{s}$, $x \in X$. Since $[S_+\theta_x]_{\mathcal{P}_K} = [S_-\theta_x]_{\mathcal{P}_K}$, there is a non-spherical picture, say $\mathbb{B}_{S,x}$, over $\mathcal{P}_K$ with $\iota(\mathbb{B}_{S,x}) = S_+\theta_x$ and $\tau(\mathbb{B}_{S,x}) = S_-\theta_x$.

Let $R_+ = R_-$ be a relation $R \in \mathbf{r}$ and $y \in Y$. Since θ is a homomorphism, by our definition for $y\theta_U$, we have that $y\theta_{R_+}$ and $y\theta_{R_-}$ must represent the same element of the monoid K. That is, $[y\theta_{R_+}]_{\mathcal{P}_K} = [y\theta_{R_-}]_{\mathcal{P}_K}$. Hence there is a non-spherical picture over $\mathcal{P}_K$ which we denote by $\mathbb{C}_{y,\theta_R}$ with $\iota(\mathbb{C}_{y,\theta_R}) = y\theta_{R_+}$ and $\tau(\mathbb{C}_{y,\theta_R}) = y\theta_{R_-}$.

In fact there may be many different ways to construct the pictures $\mathbb{B}_{S,x}$ and $\mathbb{C}_{y,\theta_R}$. These pictures must exist, but they are not unique. On the other hand the picture $\mathbb{A}_{U,y}$ will depend upon our choices for words $y\theta_x$, but this is unique once these choices are made.

After all, for $x \in X$, $y \in Y$, $R \in \mathbf{r}$ and $S \in \mathbf{s}$, one can construct spherical monoid pictures, say $\mathbb{P}_{S,x}$ and $\mathbb{P}_{R,y}$, by using the non-spherical pictures $\mathbb{B}_{S,x}$, $\mathbb{A}_{R_+,y}$, $\mathbb{A}_{R_-,y}$ and $\mathbb{C}_{y,\theta_R}$ (see [2, Figure 2] for the details).

Let $\mathbf{X_A}$ and $\mathbf{X_K}$ be trivializer sets of $\mathcal{D}(\mathcal{P}_A)$ and $\mathcal{D}(\mathcal{P}_K)$, respectively. Also, let

$$\mathbf{C_1} = \{\mathbb{P}_{S,x} : S \in \mathbf{s}, \ x \in X\} \quad \text{and} \quad \mathbf{C_2} = \{\mathbb{P}_{R,y} : R \in \mathbf{r}, \ y \in Y\}.$$

The proof of the following lemma can be found in [17].

Lemma 1.2 *Suppose that $M = K \rtimes_\theta A$ is a semi-direct product with associated presentation $\mathcal{P}_M$, as in (1). Then a trivializer set of $\mathcal{D}(\mathcal{P}_M)$, say $\mathbf{X_M}$, is*

$$\mathbf{X_A} \cup \mathbf{X_K} \cup \mathbf{C_1} \cup \mathbf{C_2}.$$

Now, by using Lemma 1.2, we get the following main result in [2].

Theorem 1.3 [2, Theorem 3.1] *Let p be a prime or 0. Then the presentation $\mathcal{P}_M$, as in (1), is p-Cockcroft if and only if the following conditions hold.*

  (i) *$\mathcal{P}_A$ and $\mathcal{P}_K$ are p-Cockcroft,*

 (ii) *$\exp_y(S) \equiv 0 \pmod{p}$ for all $S \in \mathbf{s}$, $y \in Y$,*

(iii) *$\exp_S(\mathbb{B}_{S,x}) \equiv 1 \pmod{p}$ for all $S \in \mathbf{s}$, $x \in X$,*

(iv) *$\exp_S(\mathbb{C}_{y,\theta_R}) \equiv 0 \pmod{p}$ for all $S \in \mathbf{s}$, $y \in Y$, $R \in \mathbf{r}$,*

 (v) *$\exp_{T_{yx}}(\mathbb{A}_{R_+,y}) \equiv \exp_{T_{yx}}(\mathbb{A}_{R_-,y}) \pmod{p}$ for all $R \in \mathbf{r}$, $y \in Y$ and $x \in X$.*

2   The p-Cockcroft property for the semi-direct products of finite cyclic monoids

In this section we will give necessary and sufficient conditions for the presentation of the semi-direct product of two finite cyclic monoids to be p-Cockcroft (p a prime). We first note that most of the following materials given in this section can also be found in [2]. We also note that some of the fundamental materials about the finite cyclic monoids can be found in [8] (see "Monogenic semigroups").

Let A and K be two finite cyclic monoids with the presentations (by [2], [8])

$$\mathcal{P}_A = [x\,;\, x^\mu = x^\lambda] \quad \text{and} \quad \mathcal{P}_K = [y\,;\, y^k = y^l] \tag{2}$$

respectively, where $l, k, \lambda, \mu \in \mathbb{Z}^+$ such that $l < k$ and $\lambda < \mu$.

We can give the following lemma for a trivializer set of the finite cyclic monoids (see [2]).

Lemma 2.1 *Let K be the finite cyclic monoid with a presentation $\mathcal{P}_K$. Then a trivializer set $\mathbf{X_K}$ of the Squier complex $\mathcal{D}(\mathcal{P}_K)$ is given by the pictures $\mathbb{P}_{k,l}^m$ $(1 \leq m \leq k-1)$, as in Figure 2.*

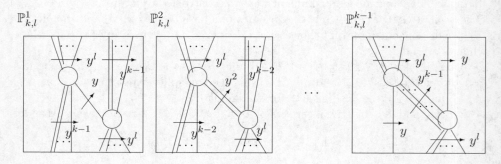

Figure 2.

Let ψ_i $(0 \leq i \leq k-1)$ be an endomorphism of K. Then we have a mapping

$$x \longrightarrow \mathrm{End}(K), \quad x \longmapsto \psi_i.$$

In fact this induces a homomorphism $\theta : A \longrightarrow \text{End}(K)$, $x \longmapsto \psi_i$ if and only if $\psi_i^\mu = \psi_i^\lambda$. Since ψ_i^μ and ψ_i^λ are equal if and only if they agree on the generator y of K, we must have

$$[y^{i^\mu}] = [y^{i^\lambda}]. \tag{3}$$

We then have the semi-direct product $M = K \rtimes_\theta A$ and, by [2], have a standard presentation

$$\mathcal{P}_M = [y, x \, ; \, S, R, T_{yx}], \tag{4}$$

as in (1), for the monoid M where

$$S : y^k = y^l, \quad R : x^\mu = x^\lambda \quad \text{and} \quad T_{yx} : yx = xy^i.$$

At the rest of the paper we will assume that the equality (3) holds when we talk about the semi-direct product M of K by A.

The subpicture $\mathbb{B}_{S,x}$ can be drawn as in Figure 3-(a) and in fact, by considering this subpicture, we have the following lemma .

$\mathbb{B}_{S,x}$

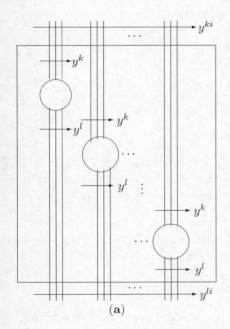

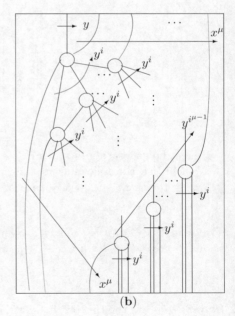

(a)   (b)

Figure 3.

Lemma 2.2 $\exp_S(\mathbb{B}_{S,x}) = i$.

As it seen in Figure 3-(b), we also have the subpicture $\mathbb{A}_{R_+,y}$ (and similarly $\mathbb{A}_{R_-,y}$) with

$$\exp_{T_{yx}}(\mathbb{A}_{R_+,y}) = 1 + i + i^2 + \cdots + i^{\mu-1} = \frac{i^\mu - 1}{i - 1}$$

and

$$\exp_{T_{yx}}(\mathbb{A}_{R_-,y}) = 1 + i + i^2 + \cdots + i^{\lambda-1} = \frac{i^\lambda - 1}{i - 1}.$$

By equality (3), we must have $[y^{i^\mu}] = [y^{i^\lambda}]$. Hence, by [2], the subpicture $\mathbb{C}_{y,\theta_R}$ with

$$\iota(\mathbb{C}_{y,\theta_R}) = y^{i^\mu}, \quad \tau(\mathbb{C}_{y,\theta_R}) = y^{i^\lambda} \quad \text{and} \quad \exp_S(\mathbb{C}_{y,\theta_R}) = \frac{i^\mu - i^\lambda}{k - l}$$

can be depicted as in Figure 4.

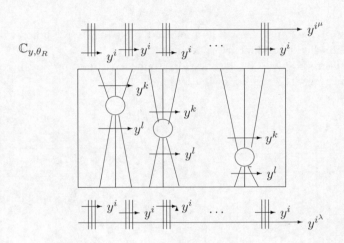

Figure 4.

Hence the generating pictures $\mathbb{P}_{S,x}$ and $\mathbb{P}_{R,y}$ can be depicted as in Figure 5.

We will use the following special case of the main result in [2]. In fact this following theorem is a consequence of Theorem 1.3. Let $d = k - l$, $n = i - 1$, $t = i^\mu - i^\lambda$.

Theorem 2.3 *Let p be a prime. Suppose that $K \rtimes_\theta A$ is a monoid with the associated monoid presentation $\mathcal{P}_M$, as in (4). Then $\mathcal{P}_M$ is p-Cockcroft if and only if*

$$p \mid d, \quad p \mid n, \quad p \mid (t/d), \quad p \mid (t/n).$$

Proof To prove the this result we will check the conditions of Theorem 1.3 hold.

   (i) By Lemma 2.1, trivializer sets $\mathbf{X_A}$ and $\mathbf{X_K}$ of the Squier complexes $\mathcal{D}(\mathcal{P}_A)$ and $\mathcal{D}(\mathcal{P}_K)$ respectively, can be given as in Figure 2. Thus it can be seen that $\mathcal{P}_A$ and $\mathcal{P}_K$ are p-Cockcroft (in fact Cockcroft), and then the condition (i) holds.

  (ii) $\exp_y(S) = k - l$ so, for (ii) to hold, we must have $p \mid k - l$.

 (iii) To make (iii) hold, we need $i \equiv 1 \pmod{p}$, so that $p \mid i - 1$.

 (iv) For the subpicture $\mathbb{C}_{y,\theta_R}$, we must have $p \mid \frac{i^\mu - i^\lambda}{k - l}$, to make (iv) hold.

$\mathbb{P}_{S,x}$ $\mathbb{P}_{R,y}$

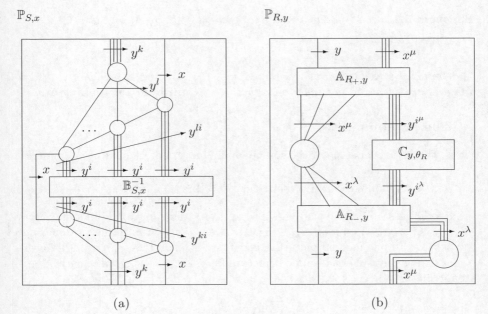

(a)   (b)

Figure 5.

(v) Also, to make (v) hold, we need $\frac{i^{\mu}-1}{i-1} \equiv \frac{i^{\lambda}-1}{i-1}$ (mod p), or equivalently, $\frac{i^{\mu}-i^{\lambda}}{i-1} \equiv 0$ (mod p) since

$$\exp_{T_{yx}}(\mathbb{A}_{R,y}) = \exp_{T_{yx}}(\mathbb{A}_{R_+,y}) - \exp_{T_{yx}}(\mathbb{A}_{R_-,y}).$$

Conversely suppose that the conditions of the theorem hold. Then, by the meaning of the trivializer set $\mathbf{X_M}$, it is easy to see that $\mathcal{P}_M$ is p-Cockcroft where p is a prime.

Hence the result.   □

The examples and consequences of Theorem 2.3 can be found in [2].

3   The main theorem

By Theorems 1.3 and 2.3, one can say that the monoid presentation $\mathcal{P}_M$, as in (4), is efficient if and only if there is a prime p such that

$$\exp_y(S) \equiv 0 \;(\text{mod } p), \qquad \exp_S(\mathbb{B}_{S,x}) \equiv 1 \;(\text{mod } p),$$
$$\exp_{T_{yx}}(\mathbb{C}_{y,\theta_R}) \equiv 0 \;(\text{mod } p), \qquad \exp_{T_{yx}}(\mathbb{A}_{R,y}) \equiv 0 \;(\text{mod } p).$$

In particular, if we choose $\exp_S(\mathbb{B}_{S,x}) = 0$ or 2 (or, by Lemma 2.2, $i = 0$ or $i = 2$) then $\mathcal{P}_M$ will be inefficient.

Let $d = \exp_y(S) = k - l$. We note that, by the meaning of finite cyclic monoids, the value of d cannot be equal to 0.

The main result of this paper is the following.

Theorem 3.1 *Let M be the semi-direct product of K by A, and let*

$$\mathcal{P}_M = [y, x \; ; \; y^k = y^l, \, x^\mu = x^\lambda, \, yx = xy^i]$$

*be a standard presentation of M where $l, k, \lambda, \mu, i \in \mathbb{Z}^+$ and $l < k$, $\lambda < \mu$. If $i = 2$ and d is not even and not equal to 1, then $\mathcal{P}_M$ is **minimal** but **inefficient**.*

4   The preliminaries

We should note that some of the similar preliminary material can also be found in [3].

Let M be a monoid with a presentation $\mathcal{P} = [Y \; ; \; \mathbf{s}]$, and let $P^{(l)} = \bigoplus_{S \in \mathbf{s}} \mathbb{Z}M e_S$ be the free left $\mathbb{Z}M$-module with basis $\{e_S : S \in \mathbf{s}\}$. For an atomic monoid picture, say $\mathbb{A} = (U, S, \varepsilon, V)$ where $U, V \in F(\mathbf{y})$, $S \in \mathbf{s}$, $\varepsilon = \pm 1$, the left evaluation of the positive atomic monoid picture $\mathbb{A}$ is defined by $eval^{(l)}(\mathbb{A}) = \varepsilon \overline{U} e_S \in P^{(l)}$ where $\overline{U} \in M$. For any spherical monoid picture $\mathbb{P} = \mathbb{A}_1 \mathbb{A}_2 \cdots \mathbb{A}_n$, where each $\mathbb{A}_i$ is an atomic picture for $i = 1, 2, \cdots, n$, we then define

$$eval^{(l)}(\mathbb{P}) \;=\; \sum_{i=1}^{n} eval^{(l)}(\mathbb{A}_i) \in P^{(l)}.$$

We let $\delta_{\mathbb{P},S}$ be the coefficient of e_S in $eval^{(l)}(\mathbb{P})$, so we can write

$$eval^{(l)}(\mathbb{P}) = \sum_{S \in \mathbf{s}} \delta_{\mathbb{P},S} e_S \in P^{(l)}.$$

Let $I_2^{(l)}(\mathcal{P})$ be the 2-sided ideal of $\mathbb{Z}M$ generated by the set

$$\{\delta_{\mathbb{P},S} : \mathbb{P} \text{ is a spherical monoid picture}, S \in \mathbf{s}\}.$$

Then this ideal is called the *second Fox ideal* of $\mathcal{P}$.

The fact of the following lemma has been also discussed in [3].

Lemma 4.1 *If $\mathbf{Y}$ is a trivializer of $\mathcal{D}(\mathcal{P})$ then second Fox ideal is generated by the set $\{\delta_{\mathbb{P},S} : \mathbb{P} \in \mathbf{Y}, S \in \mathbf{s}\}$.*

The concept of the second Fox ideals will be needed for the following theorem which can be thought as a *test of minimality of monoid presentations* and has been proved in an unpublished work by S. Pride (as depicted also in [3]). We remark that the case of the group presentation version of this result has been firstly proved by M. Lustig in [11].

Theorem 4.2 *Let $\mathbf{Y}$ be a trivializer of $\mathcal{D}(\mathcal{P})$ and let ψ be a ring homomorphism from $\mathbb{Z}M$ into the ring of all $k \times k$ matrices over a commutative ring L with 1, for some $k \geq 1$, and suppose $\psi(1) = I_{k \times k}$. If $\psi(\delta_{\mathbb{P},S}) = 0$ for all $\mathbb{P} \in \mathbf{Y}$, $S \in \mathbf{s}$ then $\mathcal{P}$ is minimal.*

Theorem 4.2 can be restated by *if there is a ring homomorphism ψ as above such that $I_2^{(l)}(\mathcal{P})$ is contained in the kernel of ψ, then $\mathcal{P}$ is minimal.*

Let us suppose that both K and A be the finite cyclic monoids with presentations $\mathcal{P}_K$ and $\mathcal{P}_A$ as in (2), and let M be the semi-direct product of K by A with a presentation as in (4).

Let us consider the picture $\mathbb{P}_{S,x}$, as in Figure 5-(a).

For the generator y, let us assume that $\dfrac{\partial}{\partial y}$ denotes the Fox derivation with respect to y, and let $\dfrac{\partial^M}{\partial y}$ be the composition

$$\mathbb{Z}F(y) \xrightarrow{\frac{\partial}{\partial y}} \mathbb{Z}F(y) \longrightarrow \mathbb{Z}M,$$

where $F(y)$ is the free monoid on y. Also, for the relator $S : y^k = y^l$, let us define $\dfrac{\partial^M S}{\partial y}$ to be

$$\frac{\partial^M S_+}{\partial y} - \frac{\partial^M S_-}{\partial y}.$$

In fact, for this y, the left evaluations of the positive atomic monoid pictures in $\mathbb{P}_{S,x}$ containing a T_{yx} disc are

$$\overline{1}e_{T_{yx}}, \ \overline{y}e_{T_{yx}}, \ \overline{y}^2 e_{T_{yx}}, \ \cdots, \overline{y}^{l-1}e_{T_{yx}},$$

and the left evaluations of the negative atomic monoid pictures in $\mathbb{P}_{S,x}$ containing a T_{yx} disc are

$$-\overline{1}e_{T_{yx}}, \ -\overline{y}e_{T_{yx}}, \ -\overline{y}^2 e_{T_{yx}}, \ \cdots, -\overline{y}^{k-1}e_{T_{yx}}.$$

Thus the coefficient of $e_{T_{yx}}$ in $eval^{(l)}(\mathbb{P}_{S,x})$ is

$$\overline{1} + \overline{y} + \overline{y}^2 + \cdots + \overline{y}^{l-1} - (\overline{1} + \overline{y} + \overline{y}^2 + \cdots + \overline{y}^{k-1}) = \frac{\partial^M S}{\partial y}. \tag{5}$$

Proposition 4.3 *The second Fox ideal $I_2^{(l)}(\mathcal{P}_M)$ of $\mathcal{P}_M$ is generated by the elements*

$$1 - \bar{x}(eval^{(l)}(\mathbb{B}_{S,x})), \qquad\qquad \frac{\partial^M S}{\partial y},$$

$$eval^{(l)}(\mathbb{A}_{R_+,x}) - eval^{(l)}(\mathbb{A}_{R_-,x}), \qquad eval^{(l)}(\mathbb{C}_{y,\theta_R}),$$

$$1 - \overline{y}^{k-1}, \ 1 - \overline{y}^{k-2}, \cdots, 1 - \overline{y} \quad and \quad 1 - \overline{x}^{\mu-1}, \ 1 - \overline{x}^{\mu-2}, \cdots, 1 - \overline{x}.$$

Proof In the proof, for simplicity, we shall not use the characters with "bar" in the evaluations.

By Lemma 1.2, since $\mathcal{D}(\mathcal{P}_M)$ has a trivializer $\mathbf{X}_M$ consisting of the sets $\mathbf{X}_A$, $\mathbf{X}_K$, $\mathbf{C}_1$ and $\mathbf{C}_2$ where $\mathbf{X}_A$, $\mathbf{X}_K$ are the trivializer sets of $\mathcal{D}(\mathcal{P}_A)$ and $\mathcal{D}(\mathcal{P}_K)$ respectively and $\mathbf{C}_1$, $\mathbf{C}_2$ consist of the single pictures $\mathbb{P}_{S,x}$, $\mathbb{P}_{R,y}$ respectively (see

Figures 2 and 5), we need to calculate $eval^{(l)}(\mathbb{P}_{S,x})$, $eval^{(l)}(\mathbb{P}_{R,y})$, $eval^{(l)}(\mathbb{P}_{k,l}^m)$ $(1 \leq m \leq k-1)$ and $eval^{(l)}(\mathbb{P}_{\mu,\lambda}^n)$ $(1 \leq n \leq \mu - 1)$. So we have

$$
\begin{aligned}
eval^{(l)}(\mathbb{P}_{S,x}) &= \delta_{\mathbb{P}_{S,x},S}e_S + \delta_{\mathbb{P}_{S,x},T_{yx}}e_{T_{yx}} \\
&= (1 - x(eval^{(l)}(\mathbb{B}_{S,x})))e_S + \left(\frac{\partial^M S}{\partial y}\right)e_{T_{yx}} \quad \text{by (5)} \\
eval^{(l)}(\mathbb{P}_{R,y}) &= \delta_{\mathbb{P}_{R,y},T_{yx}}e_{T_{yx}} + \delta_{\mathbb{P}_{R,y},R}e_R + \delta_{\mathbb{P}_{R,y},S}e_S \\
&= (eval^{(l)}(\mathbb{A}_{R_+,y}) - eval^{(l)}(\mathbb{A}_{R_-,y}))e_{T_{yx}} + (1 - y)e_R \\
&\qquad\qquad + (eval^{(l)}(\mathbb{C}_{y,\theta_R}))e_S.
\end{aligned}
$$

Also, for each $1 \leq m \leq k-1$ and $1 \leq n \leq \mu - 1$,

$$
eval^{(l)}(\mathbb{P}_{k,l}^m) = \delta_{\mathbb{P}_{k,l}^m,S}e_S \quad \text{and} \quad eval^{(l)}(\mathbb{P}_{\mu,\lambda}^n) = \delta_{\mathbb{P}_{\mu,\lambda}^n,R}e_R,
$$

where $\delta_{\mathbb{P}_{k,l}^m,S} = 1 - y^{k-m}$ and $\delta_{\mathbb{P}_{\mu,\lambda}^n,R} = 1 - x^{\mu-n}$.

Thus, by Lemma 4.1, we get the result as required. $\qquad\square$

Let

$$
aug : \mathbb{Z}M \longrightarrow \mathbb{Z}, \quad b \longmapsto 1
$$

be the augmentation map. We then have the following lemma.

Lemma 4.4 *We have the following equalities.*
(1) $aug(eval^{(l)}(\mathbb{B}_{S,x})) = \exp_S(\mathbb{B}_{S,x})$.

(2) $aug\left(\frac{\partial^M S}{\partial y}\right) = \exp_y(S) = k - l$.

(3) $aug(eval^{(l)}(\mathbb{A}_{R_+,y}) - eval^{(l)}(\mathbb{A}_{R_-,y})) = \exp_{T_{yx}}(\mathbb{P}_{R,y}) = \frac{i^\mu - i^\lambda}{i-1}$.

(4) $aug(eval^{(l)}(\mathbb{C}_{y,\theta_R})) = \exp_S(\mathbb{P}_{R,y}) = \frac{i^\mu - i^\lambda}{k-l}$.

(5) $aug(eval^{(l)}(\mathbb{P}_{k,l}^m)) = 0$ and $aug(eval^{(l)}(\mathbb{P}_{\mu,\lambda}^n)) = 0$, *for each* $1 \leq m \leq k-1$ *and* $1 \leq n \leq \mu - 1$.

Proof The proofs of the equalities given in (1) and (2) can be found in [3].

Proof of (3): We can write

$$
\begin{aligned}
eval^{(l)}(\mathbb{A}_{R_+,y}) &= \varepsilon_1 \overline{W_1}e_{T_{yx}} + \varepsilon_2 \overline{W_2}e_{T_{yx}} + \cdots + \varepsilon_j \overline{W_j}e_{T_{yx}}, \\
eval^{(l)}(\mathbb{A}_{R_-,y}) &= \gamma_1 \overline{U_1}e_{T_{yx}} + \gamma_2 \overline{U_2}e_{T_{yx}} + \cdots + \gamma_q \overline{U_q}e_{T_{yx}}
\end{aligned}
$$

where, for $1 \leq a \leq j$, $1 \leq b \leq q$, each $\varepsilon_a = 1$, $\gamma_b = -1$ and each of W_a, U_b is a certain word on the set $\{x,y\}$. In the right hand side of the above equalities, each term $\varepsilon_a \overline{W_a}e_{T_{yx}}$ and $\gamma_q \overline{U_q}e_{T_{yx}}$ corresponds to a single T_{yx}-disc and, in fact, the value of each ε_a and γ_b gives the sign of this single T_{yx}-disc. Therefore the sum of the ε_a's and γ_b's, that is, $aug(eval^{(l)}(\mathbb{A}_{R_+,y}) - eval^{(l)}(\mathbb{A}_{R_-,y}))$ must give the exponent sum of the T_{yx}-discs in the picture $\mathbb{P}_{R,y}$, as required since the T_{yx}-discs can only be occured in the subpictures $\mathbb{A}_{R_+,y}$ and $\mathbb{A}_{R_-,y}$.

Proof of (4): We have just the S-discs in the subpicture $\mathbb{C}_{y,\theta_R}$ (see Figure 4) in $\mathbb{P}_{R,y}$. Then, by writing

$$eval^{(l)}(\mathbb{C}_{y,\theta_R}) = \overline{x}^\mu(\epsilon_1\overline{V_1}e_S + \epsilon_2\overline{V_2}e_S + \cdots + \epsilon_g\overline{V_g}e_S)$$

and adapting the proof of (3) into this case, we get the result.

Proof of (5): For each $1 \le m \le k-1$ and $1 \le n \le \mu-1$, since each of $\mathbb{P}_{k,l}^m$ and $\mathbb{P}_{\mu,\lambda}^n$ contains just two S-discs and R-discs (which one is positive and the other is negative) respectively, we write

$$eval^{(l)}(\mathbb{P}_{k,l}^m) = -\overline{W_1^m}e_S + \overline{W_2^m}e_S \quad \text{and} \quad eval^{(l)}(\mathbb{P}_{\mu,\lambda}^n) = -\overline{U_1^n}e_R + \overline{U_2^n}e_R,$$

where W_i^m's are words on y and U_j^n's are words on x ($1 \le i,j \le 2$). Again as in the previous cases, by considering the each term in the above equalities, we get the sign of this single S-disc and single R-disc. Then the sum of the these signs, in other words, augmentation of the evaluation of each picture must give the exponent sum of S and R-discs. That is,

$$aug(eval^{(l)}(\mathbb{P}_{k,l}^m)) = \exp_S(\mathbb{P}_{k,l}^m) = -1+1 = 0 = \exp_S(\mathbb{P}_{\mu,\lambda}^n) = aug(eval^{(l)}(\mathbb{P}_{\mu,\lambda}^n)),$$

as required.

Hence the result.   □

5 Proof of the main Theorem

Suppose that $d = k - l$ is not equal to 1 and $2n$ ($n \in \mathbb{Z}^+$). Let $\mathbb{Z}_d$ denote $\mathbb{Z}$ (mod d) while $d \ne 0$. (Recall that d cannot be equal to 0.) Suppose also that $\exp_S(\mathbb{B}_{S,x}) = 2$ (or, equivalently, $i = 2$ by Lemma 2.2).

Let $M_{\mu,\lambda}$ denote the finite cyclic monoid generated by x. Let us consider the homomorphism from M onto $M_{\mu,\lambda}$ defined by

$$y \longmapsto 1, \quad x \longmapsto x.$$

This induces a ring homomorphism

$$\gamma : \mathbb{Z}M \longrightarrow M_{\mu,\lambda}[x].$$

We note that the restriction of γ to the subring $\mathbb{Z}K$ of $\mathbb{Z}M$ is just the augmentation map $aug : \mathbb{Z}K \longrightarrow \mathbb{Z}$. Thus, by Lemma 4.4, the image of $I_2^{(l)}(\mathcal{P}_M)$ under γ is the ideal of $M_{\mu,\lambda}[x]$ generated by

$$1 - \overline{x}(\exp_S(\mathbb{B}_{S,x})) = 1 - 2\overline{x}, \ \exp_y(S), \ \exp_{T_{yx}}(\mathbb{P}_{R,y}) \text{ and } \exp_S(\mathbb{P}_{R,y}).$$

Let η be the composition of γ and the mapping

$$M_{\mu,\lambda}[x] \longrightarrow \mathbb{Z}_d[x], \quad x \longmapsto x, \ n \longmapsto \overline{n} \ (n \in \mathbb{Z}),$$

where $\bar{n}$ is n (mod d). Then, since $\exp_y(S) = d \equiv 0$ (mod d), $\exp_{T_{yx}}(\mathbb{P}_{R,y}) \equiv 0$ (mod d), $\exp_S(\mathbb{P}_{R,y}) \equiv 0$ (mod d), we get

$$\eta(I_2^{(l)}(\mathcal{P}_M)) = \langle 1 - \overline{2}\bar{x} \rangle$$
$$= I, \text{ say.}$$

A quite similar proof for the following lemma can be found in [3].

Lemma 5.1 $I \neq \mathbb{Z}_d[x]$.

Remark 5.2 In the proof of Lemma 5.1 one can see that if $d = 1$ then $I = \mathbb{Z}_d[x]$ (see [3, Lemma 4.6]).

Let ψ be the composition

$$\mathbb{Z}M \xrightarrow{\eta} \mathbb{Z}_d[x] \xrightarrow{\phi} \mathbb{Z}_d[x]/I,$$

where ϕ is the natural epimorphism. Then ψ sends $I_2^{(l)}(\mathcal{P}_M)$ to 0, and $\psi(1) = 1$. In other words, the images of the generators of $I_2(\mathcal{P}_M)$ are all 0 under ψ. That is,

$$\psi(1 - \bar{x}(eval^{(l)}(\mathbb{B}_{S,x}))) = \phi\eta(1 - \bar{x}(eval^{(l)}(\mathbb{B}_{S,x})))$$
$$= \phi(1 - \bar{x}(\overline{\exp_S(\mathbb{B}_{S,x})}))$$

since η is a ring homomorphism
and by Lemma 4.4-(1)

$$= \phi(1 - \bar{x}\overline{2}) \quad \text{since } \exp_S(\mathbb{B}_{S,x}) = 2$$
$$= 0,$$

$$\psi\left(\frac{\partial^M S}{\partial y}\right) = \phi\eta\left(\frac{\partial^M S}{\partial y}\right)$$
$$= \phi(\overline{\exp_y(S)})$$

since η is a ring homomorphism
and by Lemma 4.4-(2)

$$= \phi(0) \quad \text{since } \exp_y(S) = d \equiv 0 \pmod{d}$$
$$= 0,$$

$$\psi(eval^{(l)}(\mathbb{A}_{R_+,y}) - eval^{(l)}(\mathbb{A}_{R_-,y}))$$
$$= \phi\eta(eval^{(l)}(\mathbb{A}_{R_+,y}) - eval^{(l)}(\mathbb{A}_{R_-,y}))$$
$$= \phi(\overline{\exp_{T_{yx}}(\mathbb{P}_{R,y})})$$

since η is a ring homomorphism and by Lemma 4.4-(3)

$$= \phi\left(\frac{i^\mu - i^\lambda}{i - 1}\right)$$

$$= \phi(0) \qquad \text{since the equation (3) implies}$$

$$i^\mu - i^\lambda \equiv 0 \pmod{p}, \text{ and so } \frac{i^\mu - i^\lambda}{i - 1} \equiv 0 \pmod{d}$$

$$= 0,$$

$$\phi(eval^{(l)}(\mathbb{C}_{y,\theta_R})) = \phi\eta(eval^{(l)}(\mathbb{C}_{y,\theta_R}))$$

$$= \phi(\overline{\exp_S(\mathbb{P}_{R,y})}) \quad \text{since } \eta \text{ is a ring homomorphism}$$

$$\text{and by Lemma 4.4-(4)}$$

$$= \phi\left(\overline{\frac{i^\mu - i^\lambda}{k - l}}\right)$$

$$= \phi(0) \quad \text{by the equation (3)}$$

$$= 0.$$

So, by Theorem 4.2 (Pride), $\mathcal{P}_M$ is minimal and so, by the definition, M is a minimal but inefficient monoid.

These above processes complete the proof of Theorem 3.1.   □

Lemma 5.3 *Suppose that* $d = 2n$ $(n \in \mathbb{Z}^+)$. *Then* $I = \mathbb{Z}_d[x]$.

Proof For simplicity, let us replace $\bar{x}$ by x and $\bar{2}$ by 2.

In the proof it is enough to show $2 \in I = \langle 1 - 2x \rangle$. Because we certainly have $1 - 2x \in I$ and if $2 \in I$ then we must have $1 \in I$.

Thus let us take $1 - 2x \in I$. Then, by the meaning of $\langle 1 - 2x \rangle$, we can write

$$2(n - 1)(1 - 2x) \in I \Rightarrow 2(n - 1) - 4(n - 1)x \in I \Rightarrow$$
$$2(n - 1) - 4nx + 4x \in I \Rightarrow 2(n - 1) \in I$$
$$\text{since } 4nx = 0 \text{ and } 4x = 0 \text{ in } \mathbb{Z}_d[x].$$

Then,

$$2(n - 2)(1 - 2x) \in I \Rightarrow 2(n - 2) - 4(n - 2)x \in I \Rightarrow$$
$$2(n - 2) - 4(n - 1 - 1)x \in I \Rightarrow 2(n - 2) - (4(n - 1)x - 4x) \in I \Rightarrow$$
$$2(n - 2) - 2(2(n - 1)x - 2x) \in I \Rightarrow 2(n - 2) \in I$$
$$\text{since, by the above calculation, } 2(n - 1) \in I \Rightarrow 2(n - 1)x \in I$$
$$\text{and } 2x = 0 \text{ in } \mathbb{Z}_d[x]$$

$$\Rightarrow \qquad \cdots \text{ by iterating this procedure, we get } \cdots \Rightarrow$$
$$2 \in I \Rightarrow 1 \in I,$$

as required.   □

Remark 5.4 Suppose that $i = 0$. Then the presentation $\mathcal{P}_M$, as in (4), can be written

$$\mathcal{P}_M = [y, x \; ; \; y^k = y^l, \; x^\mu = x^\lambda, \; yx = x]. \tag{6}$$

Then it is easy to see that there will not be any $\mathbb{B}_{S,x}$ and $\mathbb{C}_{y,\theta_R}$ subpictures. Also there is no need any restriction on $d = k - l$ since $i = 0$ and so the equality (3) always holds. However, by using same progress as in the proof of Theorem 3.1, we have

$$\eta(I_2^{(l)}(\mathcal{P}_M)) = \langle 1 \rangle = I.$$

That means the minimality test (Theorem *4.2*) used in this paper cannot work for this case since $1 \in I$ and so $I = \mathbb{Z}_d[x]$. Therefore it remains a conjecture whether the presentation $\mathcal{P}_M$ given in (6) is minimal.

6   Some examples

In this section we will give two applications of Theorem 3.1.

Example 6.1 Let us take $i = 2$, $k = 4t$, $l = t$, $\mu = 3t$ and $\lambda = t$ where t is an odd positive integer. Clearly the equality in (3) holds and so we obtain a presentation

$$\mathcal{P}_M = [y, x \; ; \; y^{4t} = y^t, \, x^{3t} = x^t, \, yx = xy^2], \tag{7}$$

as in (4), for the monoid $M = K \rtimes_\theta A$. By Theorem 2.3, $\mathcal{P}_M$ is an inefficient presentation. Moreover $d = 3t \neq 2n$ $(n \in \mathbb{Z}^+)$.

Thus as a consequence of Theorem 3.1, we have

Corollary 6.2 *For every positive odd integer t, the presentation $\mathcal{P}_M$ given in (7) is minimal but inefficient.*

Example 6.3 Let $i = 2$, $k = 2t+1$, $l = 2s$ $(s < t, \, t, s \in \mathbb{Z}^+)$, $\mu = k - l$, $\lambda = 1$. Therefore equality (3) satisfies and then we have a semi-direct product M with the presentation

$$\mathcal{P}_M = [y, x \; ; \; y^{2t+1} = y^{2s}, \, x^{k-l} = x, \, yx = xy^2]. \tag{8}$$

It is clear that $d = 2(t - s) + 1 \neq 2n$ $(n \in \mathbb{Z}^+)$. Also, by Theorem 2.3, $\mathcal{P}_M$ is an inefficient presentation.

As an application of Theorem 3.1, we have

Corollary 6.4 *For all $t, s \in \mathbb{Z}^+$ such that $s < t$, the presentation $\mathcal{P}_M$, as in (8), is minimal but inefficient.*

References

[1] H. Ayık, C. M. Campbell, J. J. O'Connor and N. Ruškuc, Minimal presentations and efficiency of semigroups, *Semigroup Forum* **60** (2000), 231–242.

[2] A. S. Çevik, The p-Cockcroft property of the semi-direct products of monoids, *Internat. J. Algebra Comput.* **13** (2003), no. 1, 1–16.

[3] A. S. Çevik, Minimal but inefficient presentations of the semi-direct products of some monoids, *Semigroup Forum* **66** (2003), 1–17.

[4] R. Cremanns and F. Otto, Finite derivation type implies the homological finiteness condition FP_3, *J. Symbolic Comput.* **18** (1994), 91–112.

[5] V. Dlab and B. H. Neumann, Semigroups with few endomorphisms, *J. Austral. Math. Soc. Ser. A* **10** (1969), 162–168.

[6] D. B. A. Epstein, Finite presentations of group and 3-manifolds, *Quart. J. Math. Oxford* **12** (1961), 205–212.

[7] V. Guba and M. Sapir, *Diagram groups*, American Mathematical Society, Providence, 1997.

[8] J. M. Howie, *Fundamentals of Semigroup Theory*, Oxford University Press, 1995.

[9] S. V. Ivanov, Relation modules and relation bimodules of groups, semigroups and associative algebras, *Internat. J. Algebra Comput.* **1** (1991), 89–114.

[10] C. W. Kilgour and S. J. Pride, Cockcroft presentations, *J. Pure Appl. Algebra* **106** (1996), no. 3, 275–295.

[11] M. Lustig, Fox ideals, $\mathcal{N}$-torsion and applications to groups and 3-monifolds, in *Two-dimensional homotopy and combinatorial group theory* (C. Hog-Angeloni, W. Metzler and A. J. Sieradski, editors), 219–250, Cambridge University Press, 1993.

[12] S. J. Pride, Geometric methods in combinatorial semigroup theory, in *Semigroups, Formal Languages and Groups*, (J. Fountain, editor), 215–232, Kluwer Academic Publishers, 1995.

[13] S. J. Pride, Low-dimensional homotopy theory for monoids, *Internat. J. Algebra Comput.* **5** (1995), no. 6, 631-649.

[14] N. Ruskuc, *Semigroup Presentations*, Ph.D. Thesis, University of St Andrews, 1996.

[15] C. C. Squier, Word problems and a homological finiteness condition for monoids, *J. Pure Appl. Algebra* **49** (1987), 201–216.

[16] T. Saito, Orthodox semidirect products and wreath products of monoids, *Semigroup Forum* **38** (1989), 347–354.

[17] J. Wang, Finite derivation type for semi-direct products of monoids, *Theoret. Comput. Sci.* **191** (1998), no. 1–2, 219–228.

THE MODULAR ISOMORPHISM PROBLEM FOR FINITE p-GROUPS WITH A CYCLIC SUBGROUP OF INDEX p^2

CZESŁAW BAGIŃSKI* and ALEXANDER KONOVALOV†[1]

*Institute of Computer Science, Technical University of Białystok, Wiejska 45A, 15-351 Białystok, Poland

†Department of Mathematics, Zaporozhye National University, Zhukovskogo street, 66, Zaporozhye, 69063, Ukraine

Email: alexander.konovalov@gmail.com

Abstract

Let p be a prime number, G be a finite p-group and K be a field of characteristic p. The Modular Isomorphism Problem (MIP) asks whether the group algebra KG determines the group G. Dealing with MIP, we investigated a question whether the nilpotency class of a finite p-group is determined by its modular group algebra over the field of p elements. We give a positive answer to this question provided one of the following conditions holds: (i) $\exp G = p$; (ii) $\operatorname{cl}(G) = 2$; (iii) G' is cyclic; (iv) G is a group of maximal class and contains an abelian subgroup of index p.

As a consequence, the positive solution of MIP for all p-groups containing a cyclic subgroup of index p^2 was obtained.

1 Introduction

Though the Modular Isomorphism Problem is known for more than 50 years, up to now it remains open. It was solved only for some classes of p-groups, in particular:

- abelian p-groups (Deskins [10]; alternate proof by Coleman [9]);
- p-groups of class 2 with elementary abelian commutator subgroup (Sandling, theorem 6.25 in [22]);
- metacyclic p-groups (for $p > 3$ by Bagiński [1]; completed by Sandling [24]);
- 2-groups of maximal class (Carlson [8]; alternate proof by Bagiński [2]);
- p-groups of maximal class, $p \neq 2$, when $|G| \leq p^{p+1}$ and G contains an abelian maximal subgroup (Caranti and Bagiński [4]);
- elementary abelian-by-cyclic groups (Bagiński [3]);
- p-groups with the center of index p^2 (Drensky [11]),

where the results for abelian case, 2-groups of maximal class and p-groups with the center of index p^2 are valid for arbitrary fields of characteristic p. Also it was solved for a number of groups of small orders and a field of p elements, in particular:

- groups of order not greater then p^4 (Passman [18]);

---

[1]The second author wishes to thank Adalbert and Victor Bovdi for drawing attention to the classification [16] and their warm hospitality during his stay at the University of Debrecen in March–April 2002, and also the NATO Science Fellowship Programme for the support of this visit.

- groups of order 2^5 (Makasikis [14] with remarks by Sandling [22]; alternate proof by Michler, Newman and O'Brien [15]);
- groups of order p^5 (Kovacs and Newman, due to Sandling's remark in [23]; alternate proof by Salim and Sandling [20, 21]);
- groups of order 2^6 (Wursthorn, using computer [29, 30]);
- groups of order 2^7 (Wursthorn, using computer [7]).

Besides this, a lot of MIP invariants (i.e., group properties which are determined by the group algebra) are known, and they are very useful for research in MIP. In the Theorem 1 we summarize some of them for further usage.

Theorem 1 *Let G be a finite p-group, and let F be a field of characteristic p. Then the following properties of G are determined by the group algebra FG:*

(i) *the exponent of the group G ([12]; see also [24]);*

(ii) *the isomorphism type of the center of the group G ([25, 28]);*

(iii) *the isomorphism type of the factor group G/G' ([28]; see also [18, 22]);*

(iv) *the minimal number of generators $d(G')$ of the commutator subgroup G' (follows immediately from Prop. III.1.15(ii) of [26]);*

(v) *the length of the Brauer–Jennings–Zassenhaus $\mathcal{M}$-series of the group G, that is $\mathcal{M}_1(G) = G$, $\mathcal{M}_{n+i}(G) = (\mathcal{M}_n(G), G)\mathcal{M}_i(G)^p$, where i is the smallest integer such that $ip > n$, as well as the isomorphism type of their factors $\mathcal{M}_i(G)/\mathcal{M}_{i+1}(G)$, $\mathcal{M}_i(G)/\mathcal{M}_{i+2}(G)$, $\mathcal{M}_i(G)/\mathcal{M}_{2i+1}(G)$ ([17, 19]).*

It could be useful for further research in MIP to extend the above list of invariants determined by the modular group algebra by adding to it at least the nilpotency class of a group. We are able to it in several cases, listed below.

Theorem 2 *Let G be a p-group and let F be a field of characteristic p. Then $\mathrm{cl}(G)$ is determined by the group algebra FG provided one of the following conditions holds:*

(i) $\exp G = p$;

(ii) $\mathrm{cl}(G) = 2$;

(iii) G' *is cyclic;*

(iv) G *is a group of maximal class and contains an abelian subgroup of index p.*

In addition, we give an application of Theorem 2, solving MIP for finite non-abelian p-groups containing a cyclic subgroup of index p^2 (these groups were classified by Ninomia in [16]):

Theorem 3 *Let G be a p-group containing a cyclic subgroup of index p^2, and let F be the field of p elements. If for a group H we have $FG \cong FH$, then $G \cong H$.*

Our notations are standard. $\Delta = \Delta_K(G)$ denotes the augmentation ideal of the modular group algebra KG. C_{2^m} denotes the cyclic group of order 2^m. We will also use the following notations for 2-groups of order 2^m and exponent p^{m-1}:

- the dihedral group $D_m = \langle a, b \mid a^{2^{m-1}} = 1, b^2 = 1, b^{-1}ab = a^{-1} \rangle$, $m \geq 3$;

- the generalized quaternion group $Q_m = \langle a, b \mid a^{2^{m-1}} = 1, b^2 = a^{2^{m-2}}, b^{-1}ab = a^{-1} \rangle$, $m \geq 3$;
- the semidihedral group $S_m = \langle a, b \mid a^{2^{m-1}} = 1, b^2 = 1, b^{-1}ab = a^{-1+2^{m-2}} \rangle$, $m \geq 4$;
- the quasi-dihedral group $M_m(2) = \langle a, b \mid a^{2^{m-1}} = 1, b^2 = 1, b^{-1}ab = a^{1+2^{m-2}} \rangle$, $m \geq 4$;

2 Determination of the nilpotency class of a p-group

In this section we give the proof of the Theorem 2.

Proof (i) Let $\mathcal{M}_i(G)$ be the i-th term of the Brauer–Jennings–Zassenhaus $\mathcal{M}$-series of G, that is

$$\mathcal{M}_1(G) = G, \quad \mathcal{M}_{n+i}(G) = (\mathcal{M}_n(G), G)\mathcal{M}_i(G)^p,$$

where i is the smallest integer such that $ip > n$. It is clear that if $\exp G = p$, then for all positive integers i we have $\mathcal{M}_i(G) = \gamma_i(G)$. Hence, by Theorem 3(i) of [17] we have $\gamma_i(G)/\gamma_{i+1}(G) = \gamma_i(H)/\gamma_{i+1}(H)$ for all $i \geqslant 1$. This means that $\mathrm{cl}(G) = \mathrm{cl}(H)$.

(ii) It is well known that if x is a non-central element of G and C_x is the conjugacy class of x in G then $\hat{C}_x = \sum_{x \in C_x} x$ lies in the subspace $[FG, FG]$. Moreover the ideal $\langle [FG, FG] \rangle$ of FG is equal to $\Delta(G')FG$. Hence the ideal of FG generated by all central elements of $\Delta(G)$ and the subspace $[FG, FG]$ is equal to $\Delta(Z(G)G')FG$. In particular, the order $|Z(G)G'|$ is determined by FG. But the orders $|Z(G)|$ and $|G'|$ are determined, so is $|Z(G) \cap G'| = |Z(G)||G'|/|Z(G)G'|$. Since $\mathrm{cl}(G) = 2$ if and only if $G' = Z(G) \cap G'$, one can recognize it from the structure of FG.

(iii) Let G be a p-group with G' cyclic of order p^m. Let H be a group such that $FG \cong FH$. It follows immediately from Prop. III.1.15(ii) of [25] that

$$d(G') = \dim_F(\Delta(G')FG/\Delta(G')\Delta(G))$$

is determined by FG. Thus, since G/G' and $d(G')$ are determined, we obtain that H' is also cyclic of order p^m. Let $G' = \langle g \mid g^{p^m} = 1 \rangle$, and $H' = \langle h \mid h^{p^m} = 1 \rangle$. To prove the theorem, we will use induction on the nilpotency class of G. For G abelian the statement is obviously true. Let $\mathrm{cl}(G) = c > 1$ and assume that the theorem is proved for all groups with the nilpotency class less than c. Since $|Z(G) \cap G'|$ is determined, we may assume that $Z(G) \cap G' = \langle g^{p^k} \rangle$, $Z(H) \cap H' = \langle h^{p^k} \rangle$. Consider the ideal $[FG, FG]FG = \Delta(G')FG = (g-1)FG$. Then $\Delta(G')^{p^k}FG = (g^{p^k}-1)FG$, and $FG/\Delta(G')^{p^k}FG \cong F[G/\langle g^{p^k} \rangle]$. Repeating the same conclusions for FH, we get $F[G/\langle g^{p^k} \rangle] \cong F[H/\langle h^{p^k} \rangle]$, or $F\overline{G} \cong F\overline{H}$, where $\overline{G} = G/\langle g^{p^k} \rangle, \overline{H} = H/\langle h^{p^k} \rangle$. Since $\mathrm{cl}(\overline{G}) = \mathrm{cl}(G) - 1$, $\mathrm{cl}(\overline{H}) = \mathrm{cl}(H) - 1$, we get by induction that $\mathrm{cl}(\overline{G}) = \mathrm{cl}(\overline{H})$ and then $\mathrm{cl}(G) = \mathrm{cl}(H)$, which proves the theorem.

(iv) See [4], Theorem 3.2. □

For $p > 2$ one can give another proof of (ii). Namely, it can be noted that in this case the ideal generated by all central elements from the ideal $\Delta(G')FG$ is equal to $\Delta(G')FG$ if and only if $\mathrm{cl}(G) = 2$. The example of 2-groups of maximal class shows that it is not the case when $p = 2$.

It is worth to note that the nilpotency class of the group of units of FG does not depend on the nilpotency class of G. As it was shown in Theorem B of [27], if G' is cyclic, then the nilpotency class of $U(FG)$ is equal to $|G'|$.

3 Presentations of finite 2-groups with a cyclic subgroup of index 4

Finite nonabelian p-groups of order p^m and exponent p^{m-2} are classified in [16]. For the case $p = 2$, they are given by the following presentations:

(a) $m \geq 4$:

$G_1 = \langle a, b \mid a^{2^{m-2}} = 1, b^4 = 1, b^{-1}ab = a^{1+2^{m-3}} \rangle$;

$G_2 = Q_{m-1} \times C_2 = \langle a, b, c \mid a^{2^{m-2}} = 1, b^2 = a^{2^{m-3}}, c^2 = 1, b^{-1}ab = a^{-1},$
$\qquad\qquad\qquad\qquad\qquad\qquad\qquad ac = ca, bc = cb \rangle$;

$G_3 = D_{m-1} \times C_2 = \langle a, b, c \mid a^{2^{m-2}} = 1, b^2 = 1, c^2 = 1, b^{-1}ab = a^{-1},$
$\qquad\qquad\qquad\qquad\qquad\qquad\qquad ac = ca, bc = cb \rangle$;

$G_4 = \langle a, b, c \mid a^{2^{m-2}} = 1, b^2 = 1, c^2 = 1, ab = ba, ac = ca, c^{-1}bc = a^{2^{m-3}}b \rangle$;

$G_5 = \langle a, b, c \mid a^{2^{m-2}} = 1, b^2 = 1, c^2 = 1, ab = ba, c^{-1}ac = ab, bc = cb \rangle$;

(b) $m \geq 5$:

$G_6 = \langle a, b \mid a^{2^{m-2}} = 1, b^4 = 1, b^{-1}ab = a^{-1} \rangle$;

$G_7 = \langle a, b \mid a^{2^{m-2}} = 1, b^4 = 1, b^{-1}ab = a^{-1+2^{m-3}} \rangle$;

$G_8 = \langle a, b \mid a^{2^{m-2}} = 1, b^4 = a^{2^{m-3}}, b^{-1}ab = a^{-1} \rangle$;

$G_9 = \langle a, b \mid a^{2^{m-2}} = 1, b^4 = 1, a^{-1}ba = b^{-1} \rangle$;

$G_{10} = M_{m-1}(2) \times C_2 = \langle a, b, c \mid a^{2^{m-2}} = 1, b^2 = 1, c^2 = 1, b^{-1}ab = a^{1+2^{m-3}},$
$\qquad\qquad\qquad\qquad\qquad\qquad\qquad ac = ca, bc = cb \rangle$;

$G_{11} = S_{m-1} \times C_2 = \langle a, b, c \mid a^{2^{m-2}} = 1, b^2 = 1, c^2 = 1, b^{-1}ab = a^{-1+2^{m-3}},$
$\qquad\qquad\qquad\qquad\qquad\qquad\qquad ac = ca, bc = cb \rangle$;

$G_{12} = \langle a, b, c \mid a^{2^{m-2}} = 1, b^2 = 1, c^2 = 1, ab = ba, c^{-1}ac = a^{-1}, c^{-1}bc = a^{2^{m-3}}b \rangle$;

$G_{13} = \langle a, b, c \mid a^{2^{m-2}} = 1, b^2 = 1, c^2 = 1, ab = ba, c^{-1}ac = a^{-1}b, bc = cb \rangle$;

$G_{14} = \langle a, b, c \mid a^{2^{m-2}} = 1, b^2 = 1, c^2 = a^{2^{m-3}}, ab = ba, c^{-1}ac = a^{-1}b, bc = cb \rangle$;

$G_{15} = \langle a, b, c \mid a^{2^{m-2}} = 1, b^2 = 1, c^2 = 1, b^{-1}ab = a^{1+2^{m-3}}, c^{-1}ac = a^{-1+2^{m-3}},$
$\qquad\qquad\qquad\qquad\qquad\qquad\qquad\qquad bc = cb \rangle$;

$G_{16} = \langle a, b, c \mid a^{2^{m-2}} = 1, b^2 = 1, c^2 = 1, b^{-1}ab = a^{1+2^{m-3}}, c^{-1}ac = a^{-1+2^{m-3}},$
$\qquad\qquad\qquad\qquad\qquad\qquad\qquad\qquad c^{-1}bc = a^{2^{m-3}}b \rangle$;

$G_{17} = \langle a, b, c \mid a^{2^{m-2}} = 1, b^2 = 1, c^2 = 1, b^{-1}ab = a^{1+2^{m-3}}, c^{-1}ac = ab, bc = cb \rangle$;

$G_{18} = \langle a, b, c \mid a^{2^{m-2}} = 1, b^2 = 1, c^2 = b, b^{-1}ab = a^{1+2^{m-3}}, c^{-1}ac = a^{-1}b \rangle$;

(c) $m \geq 6$:

$G_{19} = \langle a, b \mid a^{2^{m-2}} = 1, b^4 = 1, b^{-1}ab = a^{1+2^{m-4}} \rangle$;

$G_{20} = \langle a, b \mid a^{2^{m-2}} = 1, b^4 = 1, b^{-1}ab = a^{-1+2^{m-4}} \rangle$;

$G_{21} = \langle a, b \mid a^{2^{m-2}} = 1, a^{2^{m-3}} = b^4, a^{-1}ba = b^{-1} \rangle$;

$G_{22} = \langle a, b, c \mid a^{2^{m-2}} = 1, b^2 = 1, c^2 = 1, ab = ba, c^{-1}ac = a^{1+2^{m-4}}b,$
$\qquad\qquad\qquad\qquad\qquad\qquad\qquad c^{-1}bc = a^{2^{m-3}}b \rangle$;

$$G_{23} = \langle a, b, c \mid a^{2^{m-2}} = 1, b^2 = 1, c^2 = 1, ab = ba, c^{-1}ac = a^{-1+2^{m-4}}b,$$
$$c^{-1}bc = a^{2^{m-3}}b \rangle;$$
$$G_{24} = \langle a, b, c \mid a^{2^{m-2}} = 1, b^2 = 1, c^2 = 1, b^{-1}ab = a^{1+2^{m-3}}, c^{-1}ac = a^{-1+2^{m-4}}b,$$
$$bc = cb \rangle;$$
$$G_{25} = \langle a, b, c \mid a^{2^{m-2}} = 1, b^2 = 1, c^2 = a^{2^{m-3}}, b^{-1}ab = a^{1+2^{m-3}},$$
$$c^{-1}ac = a^{-1+2^{m-4}}b, bc = cb \rangle;$$

(d) $m = 5$:
$$G_{26} = \langle a, b, c \mid a^8 = 1, b^2 = 1, c^2 = a^4, b^{-1}ab = a^5, c^{-1}ac = ab, bc = cb \rangle;$$

In the next table we listed some properties of these groups that are important for the proof of the Theorem 3.

n	$\gamma_2(G)$	$Z(G)$	$\mathrm{cl}(G)$
1	$C_2 = \langle a^{2^{m-3}} \rangle$	$C_{2^{m-3}} \times C_2 = \langle a^2, b^2 \rangle$	2
2	$C_{2^{m-3}} = \langle a^2 \rangle$	$C_2 \times C_2 = \langle a^{2^{m-3}}, c \rangle$	$m-2$
3	$C_{2^{m-3}} = \langle a^2 \rangle$	$C_2 \times C_2 = \langle a^{2^{m-3}}, c \rangle$	$m-2$
4	$C_2 = \langle a^{2^{m-3}} \rangle$	$C_{2^{m-2}} = \langle a \rangle$	2
5	$C_2 = \langle b \rangle$	$C_{2^{m-3}} \times C_2 = \langle a^2, b \rangle$	2
6	$C_{2^{m-3}} = \langle a^2 \rangle$	$C_2 \times C_2 = \langle a^{2^{m-3}}, b^2 \rangle$	$m-2$
7	$C_{2^{m-3}} = \langle a^2 \rangle$	$C_2 \times C_2 = \langle a^{2^{m-3}}, b^2 \rangle$	$m-2$
8	$C_{2^{m-3}} = \langle a^2 \rangle$	$C_4 = \langle b^2 \rangle$	$m-2$
9	$C_2 = \langle b^2 \rangle$	$C_{2^{m-3}} \times C_2 = \langle a^2, b^2 \rangle$	2
10	$C_2 = \langle a^{2^{m-3}} \rangle$	$C_{2^{m-3}} \times C_2 = \langle a^2, c \rangle$	2
11	$C_{2^{m-3}} = \langle a^2 \rangle$	$C_2 \times C_2 = \langle a^{2^{m-3}}, c \rangle$	$m-2$
12	$C_{2^{m-3}} = \langle a^2 \rangle$	$C_4 = \langle a^{2^{m-4}}b \rangle$	$m-2$
13	$C_{2^{m-3}} = \langle a^2 b \rangle$	$C_2 \times C_2 = \langle a^{2^{m-3}}, b \rangle$	$m-2$
14	$C_{2^{m-3}} = \langle a^2 b \rangle$	$C_2 \times C_2 = \langle a^{2^{m-3}}, b \rangle$	$m-2$
15	$C_{2^{m-3}} = \langle a^2 b \rangle$	$C_2 = \langle a^{2^{m-3}} \rangle$	$m-2$
16	$C_{2^{m-3}} = \langle a^2 b \rangle$	$C_2 = \langle a^{2^{m-3}} \rangle$	$m-2$
17	$C_2 \times C_2 = \langle a^{2^{m-3}}, b \rangle$	$C_{2^{m-4}} = \langle a^4 \rangle$	3
18	$C_{2^{m-3}} = \langle a^2 b \rangle$	$C_4 = \langle a^2 \rangle, m = 5$ $C_2 = \langle a^{2^{m-3}} \rangle, m > 5$	$m-2$
19	$C_4 = \langle a^{2^{m-4}} \rangle$	$C_{2^{m-4}} = \langle a^4 \rangle$	2
20	$C_{2^{m-3}} = \langle a^2 \rangle$	$C_2 = \langle a^{2^{m-3}} \rangle$	$m-2$
21	$C_4 = \langle b^2 \rangle$	$C_{2^{m-3}} = \langle a^2 \rangle$	3
22	$C_4 = \langle a^{2^{m-4}}b \rangle$	$C_{2^{m-3}} = \langle a^2 b \rangle$	3
23	$C_{2^{m-3}} = \langle a^2 b \rangle$	$C_4 = \langle a^{2^{m-4}}b \rangle$	$m-2$
24	$C_{2^{m-3}} = \langle a^2 b \rangle$	$C_2 = \langle a^{2^{m-3}} \rangle$	$m-2$
25	$C_{2^{m-3}} = \langle a^2 b \rangle$	$C_2 = \langle a^{2^{m-3}} \rangle$	$m-2$
26	$C_2 \times C_2 = \langle a^4, b \rangle$	$C_2 = \langle a^4 \rangle$	3

4   The Modular Isomorphism Problem for finite 2-groups containing a cyclic subgroup of index 4

In this section we give the proof of the Theorem 3 for the case when $p = 2$. It appears that some of 2-groups of order 2^m and exponent 2^{m-2} are either metacyclic groups or 2-groups of almost maximal class, for which the modular isomorphism problem is already solved [5, 24]. Thus, it suffices to show that modular group algebras of remaining 2-groups containing a cyclic subgroup of index 4 are non-isomorphic pairwise.

Proof   We may assume that $m \geq 6$, since for $m < 6$ the modular isomorphism problem is already solved (see the review of known results in the section 1). Note that for $m = 6, 7$ it was also solved using computer (see [7, 29, 30]).

Let H be a finite 2-group of order 2^m and exponent 2^{m-2}, and $KG \cong KH$. Since the exponent of the group is determined by its group algebra [24], it follows that H is also a 2-group of order 2^m and exponent 2^{m-2}. Thus, the family of finite 2-groups of order 2^m and exponent 2^{m-2} is determined, and to complete the proof it remains to show that group algebras of such groups are non-isomorphic pairwise.

First we note that G_n is metacyclic for $n \in \{1, 6, 7, 8, 9, 19, 20, 21\}$. Then these groups are determined by their modular group algebras by [24].

Among the remaining non-metacyclic groups G_n is a 2-group of almost maximal class for $n \in \{2, 3, 11, 12, 13, 14, 15, 16, 18, 23, 24, 25\}$. As it was shown in [5], these groups have different sets of invariants, determined by their modular group algebras, so their modular group algebras are non-isomorphic pairwise.

Thus, it remains to deal with G_n for $n \in \{4, 5, 10, 17, 22\}$.

Indeed, any of these groups can not be isomorphic to any of the 2-groups of almost maximal class for $n \in \{2, 3, 11, 12, 13, 14, 15, 16, 18, 23, 24, 25\}$. The derived subgroups of G_{17} is 2-generated which splits it apart from the mentioned twelve 2-groups of maximal class as well as from groups G_4, G_5, G_{10}, G_{22}. The latter ones have the cyclic commutator subgroup, so their nilpotency class (which is equal to 2 or 3) is determined by the Theorem 2 (iii), and this splits them from the groups of almost maximal class as well.

Now since the isomorphism type of the center $Z(G)$ is determined by [25], we may split the groups G_4 and G_{10}.

Groups G_5 and G_{22} have isomorphic centers, but they have cyclic commutator subgroups. Thus, again we may apply Theorem 2 (iii) and split them since $\text{cl}(G_5) = 2$ while $\text{cl}(G_{22}) = 3$, and this completes the proof for the case $p = 2$.   □

5   The Modular Isomorphism Problem for finite p-groups, $p > 2$, containing a cyclic subgroup of index p^2

The case $p > 2$ is easier and we need only the following property of these groups:

Lemma 5.1 *Let G be a finite nonabelian p-group, $p > 2$, containing a cyclic subgroup of index p^2. Then G satisfies at least one of the following three conditions:*
   (a) *G is metacyclic;*

(b) $|G'| = p$;

(c) G' *is elementary abelian of order* p^2 *and* $d(G) = 2$.

Proof For p-groups of order $\leqslant p^4$ the lemma is clear. So assume that $|G| = p^n$, $n > 4$ and suppose G is not metacyclic. Since $\exp G = p^{n-2}$, G cannot be a p-group of maximal class (see for instance [13] 3.3). Hence by the main theorem of [6], G contains a normal subgroup A of order p^3 and exponent p. Let C be a cyclic subgroup of index p^2 in G. It is clear that $AC = G$ and $|A \cap C| = p$. Now, G/A is cyclic and G/G' is not cyclic, so $|G'|$ divides p^2. If $|G'| = p^2$, then obviously $2 = d(G/G') = d(G)$ and the lemma follows. □

Now we are able to complete the proof of the Theorem 3 for the case $p > 2$.

Proof

Case (a). If G is metacyclic, then it is determined by [24].

Case (b). If G has the commutator subgroup of order p, then it is determined by [23].

Case (c). If $d(G) = 2$ and G' is elementary abelian of order p^2, then G is determined by [3]. □

References

[1] C. Bagiński, The isomorphism question for modular group algebras of metacyclic p-groups, *Proc. Amer. Math. Soc.* **104** (1988), no. 1, 39–42. MR0958039 (89i:20016)

[2] C. Bagiński, Modular group algebras of 2-groups of maximal class, *Comm. Algebra* **20** (1992), no. 5, 1229–1241. MR1157906 (93a:20010)

[3] C. Bagiński, On the isomorphism problem for modular group algebras of elementary abelian-by-cyclic p-groups, *Colloq. Math.* **82** (1999), no. 1, 125–136. MR1736040 (2000j:20005)

[4] C. Bagiński and A. Caranti, The modular group algebras of p-groups of maximal class, *Canad. J. Math.* **40** (1988), no. 6, 1422–1435. MR0990107 (90a:20012)

[5] C. Bagiński and A. Konovalov, On 2-groups of almost maximal class, *Publ. Math. Debrecen* **65** (2004), no. 1–2, 97–131. MR2075257 (2005f:20033)

[6] N. Blackburn, Generalizations of certain elementary theorems on p-groups, *Proc. London Math. Soc. (3)* **11** (1961), 1–22. MR0122876 (23 #A208)

[7] F. M. Bleher et al., Computational aspects of the isomorphism problem, in *Algorithmic algebra and number theory (Heidelberg, 1997)*, 313–329, Springer, Berlin, 1999. MR1672070 (2000c:20007)

[8] J. F. Carlson, Periodic modules over modular group algebras, *J. London Math. Soc. (2)* **15** (1977), no. 3, 431–436. MR0472985 (57 #12664)

[9] D. B. Coleman, On the modular group ring of a p-group, *Proc. Amer. Math. Soc.* **15** (1964), 511–514. MR0165015 (29 #2306)

[10] W. E. Deskins, Finite Abelian groups with isomorphic group algebras, *Duke Math. J.* **23** (1956), 35–40. MR0077535 (17,1052c)

[11] V. Drensky, The isomorphism problem for modular group algebras of groups with large centres, in *Representation theory, group rings, and coding theory*, 145–153, Contemp. Math., 93, Amer. Math. Soc., Providence, RI, 1989. MR1003349 (90e:20006)

[12] B. Külshammer, Bemerkungen über die Gruppenalgebra als symmetrische Algebra. II, *J. Algebra* **75** (1982), no. 1, 59–69. MR0650409 (83j:16017b)

[13] C. R. Leedham-Green, and S. McKay, *The structure of groups of prime power order*, Oxford Univ. Press, Oxford, 2002. MR1918951 (2003f:20028)

[14] A. Makasikis, Sur l'isomorphie d'algèbres de groupes sur un champ modulaire, *Bull. Soc. Math. Belg.* **28** (1976), no. 2, 91–109. MR0561324 (81b:20009)

[15] G. O. Michler, M. F. Newman and E. A. O'Brien, Modular group algebras. Unpublished report, Australian National Univ., Canberra, 1987

[16] Y. Ninomiya, Finite p-groups with cyclic subgroups of index p^2, *Math. J. Okayama Univ.* **36** (1994), 1–21 (1995). MR1349018 (96h:20044)

[17] I. B. S. Passi and S. K. Sehgal, Isomorphism of modular group algebras, *Math. Z.* **129** (1972), 65–73. MR0311752 (47 ♯314)

[18] D. S. Passman, The group algebras of groups of order p^4 over a modular field, *Michigan Math. J.* **12** (1965), 405–415. MR0185022 (32 ♯2492)

[19] J. Ritter and S. Sehgal, Isomorphism of group rings, *Arch. Math. (Basel)* **40** (1983), no. 1, 32–39. MR0720891 (84k:16015)

[20] M. A. M. Salim and R. Sandling, The modular group algebra problem for groups of order p^5, *J. Austral. Math. Soc. Ser. A* **61** (1996), no. 2, 229–237. MR1405536 (97e:16064)

[21] M. A. M. Salim and R. Sandling, The modular group algebra problem for small p-groups of maximal class, *Canad. J. Math.* **48** (1996), no. 5, 1064–1078. MR1414071 (97k:20010)

[22] R. Sandling, The isomorphism problem for group rings: a survey, in *Orders and their applications (Oberwolfach, 1984)*, 256–288, Lecture Notes in Math., 1142, Springer, Berlin, 1985. MR0812504 (87b:20007)

[23] R. Sandling, The modular group algebra of a central-elementary-by-abelian p-group, *Arch. Math. (Basel)* **52** (1989), no. 1, 22–27. MR0980047 (90b:20007)

[24] R. Sandling, The modular group algebra problem for metacyclic p-groups, *Proc. Amer. Math. Soc.* **124** (1996), no. 5, 1347–1350. MR1343723 (96g:20003)

[25] S. K. Sehgal, On the isomorphism of group algebras, *Math. Z.* **95** (1967), 71–75. MR0206125 (34 ♯5950)

[26] S. K. Sehgal, *Topics in group rings*, Dekker, New York, 1978. MR0508515 (80j:16001)

[27] A. Shalev, The nilpotency class of the unit group of a modular group algebra. I, *Israel J. Math.* **70** (1990), no. 3, 257–266. MR1074491 (92a:16029)

[28] H. N. Ward, Some results on the group algebra of a p-group over a prime field, Seminar on Finite Groups and Related Topics, pp.13–19. Mimeographed notes, Harvard Univ.

[29] M. Wursthorn, Die modularen Gruppenringe der Gruppen der Ordnung 2^6. Diplomarbeit, Universität Stuttgart, 1990.

[30] M. Wursthorn, Isomorphisms of modular group algebras: an algorithm and its application to groups of order 2^6, *J. Symbolic Comput.* **15** (1993), no. 2, 211–227. MR1218760 (94h:20008)

ON ONE-GENERATED FORMATIONS

A. BALLESTER-BOLINCHES*, CLARA CALVO*, and
R. ESTEBAN-ROMERO[†][1]

*Departament d'Àlgebra, Universitat de València, Dr. Moliner, 50, 46100 Burjassot
(València), Spain
Email: Clara.Calvo@uv.es

†Departament de Matemàtica Aplicada-IMPA, Universitat Politècnica de València, Camí
de Vera, s/n, 46022 València, Spain

Abstract

In this survey, some results on one-generated Baer-local and $\mathfrak{X}$-local formations of
finite groups are presented, where $\mathfrak{X}$ is a class of simple groups. A summary of
characterisations of $\mathfrak{X}$-local formations via Frattini-like subgroups is also given.

In this paper, all groups considered are supposed to be finite.

A *formation* is a class of groups that is both residually- and quotient-closed.
Given a formation $\mathfrak{F}$ and a group G, the $\mathfrak{F}$-*residual* $G^{\mathfrak{F}}$ of G is the smallest normal
subgroup N of G such that G/N belongs to $\mathfrak{F}$. A formation which is closed under
Frattini extensions is said to be *saturated*.

Gaschütz [6] introduced the concept of *local formation*, which enabled him to
construct a rich family of saturated formations. In fact, the family of local formations coincides with the one of saturated formations. This was proved by Gaschütz
and Lubeseder in the soluble universe and later generalised by Schmid to the general finite universe. This is now known as Gaschütz–Lubeseder–Schmid theorem
(see [4, IV, 4.6]).

Another generalisation of the Gaschütz-Lubeseder theorem to the general finite
universe was given by Baer. He used another definition of local formation, which
pays attention to the simple components of a chief factor to label it, rather than to
the primes dividing its order (see [4, Section IV]). This concept coincides with the
one of local formation in the soluble universe. Baer theorem states that these Baer-
local formations are exactly the formations which are closed under extensions by the
Frattini subgroup of the soluble radical, that is, the *solubly saturated formations*.

A different approach to the notion of local formation is the concept of ω-*local
formation*, where ω is a non-empty set of primes. The ω-local formations can be
characterised as the ones which are closed under extensions by the Hall ω-subgroup
of the Frattini subgroup. They have been studied, for instance, in [3, 10, 12]. They
appear in a natural way when the saturation of formation products is considered.

[1]Supported by Proyecto MTM2004-08219-C02-02, MEC (Spain) and FEDER (European
Union).

Given two classes $\mathfrak{Y}$ and $\mathfrak{Z}$ of groups, a product class can be defined by setting

$$\mathfrak{Y}\mathfrak{Z} = (G \in \mathfrak{E} \mid \text{there is a normal subgroup } N \text{ of } G$$
$$\text{such that } N \in \mathfrak{Y} \text{ and } G/N \in \mathfrak{Z}),$$

where $\mathfrak{E}$ denotes the class of all finite groups. This product class turns out to be useful in the theory of classes of groups, especially when certain formations are considered. However this class product is not in general a formation when $\mathfrak{Y}$ and $\mathfrak{Z}$ are formations. Fortunately, there is a way of modifying the above definition to ensure that the class product of two formations is again a formation. If $\mathfrak{F}$ and $\mathfrak{G}$ are formations, the *formation product* or *Gaschütz product* of $\mathfrak{F}$ and $\mathfrak{G}$ is the class $\mathfrak{F} \circ \mathfrak{G}$ defined by

$$\mathfrak{F} \circ \mathfrak{G} := (X \in \mathfrak{E} \mid X^{\mathfrak{G}} \in \mathfrak{F}).$$

It is known that $\mathfrak{F} \circ \mathfrak{G}$ is again a formation and if $\mathfrak{F}$ is closed under taking subnormal subgroups, then $\mathfrak{F}\mathfrak{G} = \mathfrak{F} \circ \mathfrak{G}$ (see [4, IV, 1.7 and 1.8]).

A formation $\mathfrak{F}$ is said to be a one-generated Baer-local formation if there exists a group G such that $\mathfrak{F}$ is the smallest Baer-local formation containing G. In [8], Skiba posed the following question:

> If $\mathfrak{H} = \mathfrak{F} \circ \mathfrak{G}$ is a one-generated Baer-local formation, where $\mathfrak{F}$ and $\mathfrak{G}$ are non-trivial formations, is $\mathfrak{F}$ a Baer-local formation?

It is announced in the 1999 edition of the same book [9] that Skiba has answered the question negatively. An example can be found in [7, page 224]: Let G be a group with a unique non-abelian normal subgroup $R = O^p(G)$. Let $A = G \wr C_p$, $\mathfrak{F} = \text{form}(A)$ (the smallest formation containing A), and $\mathfrak{G} = \mathfrak{S}_p \text{form}(G)$. By [4, A, 18.5], A has a unique minimal subgroup, the base group $R^\natural$. By [11, 18.2], every simple group in $\mathfrak{F}$ has order p. Hence by [7, 4.5.25], $\mathfrak{H} = \mathfrak{F} \circ \mathfrak{G}$ is a one-generated Baer-local formation and since $\mathfrak{S}_p \not\subseteq \mathfrak{F}$, $\mathfrak{F}$ is not a Baer-local formation.

We note that in the known examples of that situation, the equalities $\mathfrak{H} = \mathfrak{G}$ and $\mathfrak{H} = \mathfrak{S}_p\mathfrak{H}$ for a prime p hold, where $\mathfrak{S}_p$ denotes the class of all p-groups. Consequently the following question arises naturally:

> Assume that $\mathfrak{H} = \mathfrak{F} \circ \mathfrak{G}$ is a Baer formation generated by a group G, where $\mathfrak{F}$ and $\mathfrak{G}$ are non-trivial formations. Is $\mathfrak{F}$ a Baer formation provided that $\mathfrak{H} \neq \mathfrak{G}$ or $\mathfrak{H} \neq \mathfrak{S}_p\mathfrak{H}$ for every prime p?

The authors gave in [1, Theorem 1] an affirmative answer to a more general question by using the notion of $\mathfrak{X}$-local formation introduced by Förster [5]. Here $\mathfrak{X}$ denotes here a class of simple groups such that $\pi(\mathfrak{X}) = \text{char } \mathfrak{X}$, where $\pi(\mathfrak{X}) := \{p \in \mathbb{P} \mid \text{there exists } G \in \mathfrak{X} \text{ such that } p \text{ divides } |G|\}$ and $\text{char } \mathfrak{X} := \{p \in \mathbb{P} \mid C_p \in \mathfrak{X}\}$.

Theorem 1 *Let $\mathfrak{X}$ be a class of simple groups such that $\pi(\mathfrak{X}) = \text{char } \mathfrak{X}$. Let $\mathfrak{H} = \mathfrak{F} \circ \mathfrak{G}$ be an $\mathfrak{X}$-saturated formation generated by a group G. If $\mathfrak{F}$ and $\mathfrak{G}$ are non-trivial and either $\mathfrak{H} \neq \mathfrak{G}$ or $\mathfrak{S}_p\mathfrak{H} \neq \mathfrak{H}$ for all primes $p \in \text{char } \mathfrak{X}$, then $\mathfrak{F}$ is $\mathfrak{X}$-saturated.*

Here we explain the concept of $\mathfrak{X}$-local formation. Let $\mathfrak{J}$ denote the class of all simple groups. For any subclass $\mathfrak{Y}$ of $\mathfrak{J}$, we write $\mathfrak{Y}' := \mathfrak{J} \setminus \mathfrak{Y}$. Denote by $\mathrm{E}\,\mathfrak{Y}$ the class of groups whose composition factors belong to $\mathfrak{Y}$. It is clear that $\mathrm{E}\,\mathfrak{Y}$ is a Fitting class, and so each group G has a largest normal $\mathrm{E}\,\mathfrak{Y}$-subgroup, the $\mathrm{E}\,\mathfrak{Y}$-radical $O_{\mathfrak{Y}}(G)$. A chief factor which belongs to $\mathrm{E}\,\mathfrak{Y}$ is called a $\mathfrak{Y}$-chief factor. If p is a prime, we write $\mathfrak{Y}_p$ to denote the class of all simple groups $S \in \mathfrak{Y}$ such that $p \in \pi(S)$. The class of all π-groups, where π is a set of primes, is denoted by $\mathfrak{E}_\pi$.

Definition 2 ([5]) An $\mathfrak{X}$-*formation function* f associates to each $X \in \mathrm{char}(\mathfrak{X}) \cup \mathfrak{X}'$ a formation $f(X)$ (possibly empty). If f is an $\mathfrak{X}$-formation function, then $\mathrm{LF}_{\mathfrak{X}}(f)$ is the class of all groups G satisfying the following two conditions:

1. If H/K is an $\mathfrak{X}_p$-chief factor of G, then $G/\mathrm{C}_G(H/K) \in f(p)$.

2. If G/L is a monolithic quotient of G such that $\mathrm{Soc}(G/L)$ is an $\mathfrak{X}'$-chief factor of G, then $G/L \in f(E)$, where E is the composition factor of $\mathrm{Soc}(G/L)$.

The class $\mathrm{LF}_{\mathfrak{X}}(f)$ is a formation ([5]). A formation $\mathfrak{F}$ is said to be $\mathfrak{X}$-*local* if there exists an $\mathfrak{X}$-formation function f such that $\mathfrak{F} = \mathrm{LF}_{\mathfrak{X}}(f)$. In this case we say that f is an $\mathfrak{X}$-local definition of $\mathfrak{F}$ or that f defines $\mathfrak{F}$.

If $\mathfrak{X} = \mathfrak{J}$, the class of all simple groups, an $\mathfrak{X}$-formation function is simply a formation function and the $\mathfrak{X}$-local formations are exactly the local formations. If $\mathfrak{X} = \mathbb{P}$, the class of all abelian simple groups, an $\mathfrak{X}$-formation function is a Baer function and the $\mathfrak{X}$-local formations are exactly the Baer-local ones (see [4, IV, 4.9]). Moreover, every formation is $\mathfrak{X}$-local for $\mathfrak{X} = \emptyset$. Förster also introduced in [5] an $\mathfrak{X}$-Frattini subgroup $\Phi_{\mathfrak{X}}^*(G)$ for every group G. He defined $\mathfrak{X}$-saturation in the obvious way and he proved that the $\mathfrak{X}$-saturated formations are exactly the $\mathfrak{X}$-local ones. From this one can deduce at once the theorems of Gaschütz–Lubeseder–Schmid and Baer. However, Förster's definition of $\mathfrak{X}$-saturation is not the natural one if our aim is to generalise the concepts of saturation and soluble saturation. Since $O_{\mathfrak{J}}(G) = G$ and $O_{\mathbb{P}}(G) = G_{\mathfrak{S}}$, we would expect the $\mathfrak{X}$-Frattini subgroup of a group G to be defined as $\Phi(O_{\mathfrak{X}}(G))$. In general $\Phi_{\mathbb{P}}^*(G)$ does not coincide with $\Phi(G_{\mathfrak{S}})$, as we can see in [1, Example 2.4]. Hence the proof of Baer's theorem does not follow immediately from Förster's result.

In [2] another $\mathfrak{X}$-Frattini subgroup $\Phi_{\mathfrak{X}}(G)$ in every group G is introduced. It is smaller than Förster's one.

Definition 3   1. Let p be a prime number. We say that a group G belongs to the class $\mathrm{A}_{\mathfrak{X}_p}(\mathfrak{P}_2)$ provided there exists an elementary abelian normal p-subgroup N of G such that

   (a) $N \leq \Phi(G)$ and G/N is a primitive group with a unique non-abelian minimal normal subgroup, i.e., G/N is a primitive group of type 2,

   (b) $\mathrm{Soc}(G/N) \in \mathrm{E}\,\mathfrak{X} \setminus \mathfrak{E}_{p'}$, and

   (c) $\mathrm{C}_G^h(N) \leq N$, where

$$\mathrm{C}_G^h(N) := \bigcap \{\mathrm{C}_G(H/K) \mid H/K \text{ is a chief factor of } G \text{ below } N\}.$$

2. The $\mathfrak{X}$-Frattini subgroup of a group G is the subgroup $\Phi_{\mathfrak{X}}(G)$ defined as

$$\Phi_{\mathfrak{X}}(G) := \begin{cases} \Phi(O_{\mathfrak{X}}(G)) & \text{if } G \notin A_{\mathfrak{X}_p}(\mathfrak{P}_2) \text{ for all } p \in \operatorname{char} \mathfrak{X}, \\ \Phi(G) & \text{otherwise.} \end{cases}$$

It is proved in [2, Theorem A] that the $\mathfrak{X}$-saturated formations associated with this new $\mathfrak{X}$-Frattini subgroup are exactly the $\mathfrak{X}$-local formations. Moreover, since $\Phi_{\mathbb{P}}(G) = \Phi(G_{\mathfrak{S}})$, Baer's theorem is a direct consequence of this result.

There exist groups G for which $\Phi(O_{\mathfrak{X}}(G))$ is a proper subgroup of $\Phi_{\mathfrak{X}}(G)$, as the following example, suggested by John Cossey, shows (see [2]):

Example 4 Let X be the maximal Frattini extension of the alternating group A_5 of degree 5 corresponding to the prime $p = 5$. Then X has an elementary abelian normal subgroup M of order 5^3 contained in $\Phi(X)$ and X/M is isomorphic to A_5 (see [4, Appendix β]). Let T be a non-abelian group of order 55, and consider $Y = X \wr T$. Let G be a subdirect product of Y and a cyclic group of order 25 with amalgamated factor group isomorphic to C_5. Consider the class $\mathfrak{X} = (A_5, C_2, C_3, C_5)$. Then G belongs to $A_{\mathfrak{X}_5}(\mathfrak{P}_2)$, and so $\Phi_{\mathfrak{X}}(G) = \Phi(G)$, which has order $(5^3)^{55} \cdot 5$. On the other hand, $\Phi(O_{\mathfrak{X}}(G))$ has order $(5^3)^{55}$.

If we force the groups in $A_{\mathfrak{X}_p}(\mathfrak{P}_2)$ to be monolithic, then we get an $\mathfrak{X}$-Frattini subgroup for a group G which is smaller than $\Phi_{\mathfrak{X}}(G)$. With the corresponding $\mathfrak{X}$-saturation, we have that the $\mathfrak{X}$-saturated formations are exactly the $\mathfrak{X}$-local ones. At the moment of writing, it is an open question whether or not there exist monolithic groups in $A_{\mathfrak{X}_p}(\mathfrak{P}_2)$ such that $\Phi(O_{\mathfrak{X}}(G)) \neq \Phi(G)$. Hence the following question remains open:

> Let $\mathfrak{F}$ be a formation such that, for every group G, if $G/\Phi(O_{\mathfrak{X}}(G)) \in \mathfrak{F}$, then $G \in \mathfrak{F}$. Is $\mathfrak{F}$ $\mathfrak{X}$-local?

Note that an analogous result to Theorem 1 was proved by Vishnevskaya in [13] for p-saturated formations. She shows that the p-saturated formation $\mathfrak{H}$ generated by a finite group cannot be the Gaschütz product $\mathfrak{F} \circ \mathfrak{G}$ of two non-p-saturated formations provided $\mathfrak{H} \neq \mathfrak{G}$. Although in general there does not exist a class of simple groups $\mathfrak{X}(\omega)$ such that the $\mathfrak{X}(\omega)$-saturated formations are exactly the ω-saturated formations (see [1, Section 3]), the arguments used in the proof of Theorem 1 still hold for ω-saturated formations. It leads to an alternative proof of Vishnevskaya's result.

References

[1] A. Ballester-Bolinches, Clara Calvo and R. Esteban-Romero, A question from the Kourovka Notebook on formation products, *Bull. Austral. Math. Soc.* **68** (2003), no. 3, 461–470.

[2] A. Ballester-Bolinches, Clara Calvo and R. Esteban-Romero, On $\mathfrak{X}$-saturated formations of finite groups, *Comm. Algebra* **33** (2005), 1053–1064.

[3] A. Ballester-Bolinches and L. A. Shemetkov, On lattices of p-local formations of finite groups, *Math. Nachr.* **186** (1997), 57–65.

[4] K. Doerk and T. Hawkes, *Finite soluble groups*, De Gruyter Expositions in Mathe-
     matics, no. 4, Walter de Gruyter, Berlin, New York, 1992.
[5] P. Förster, Projektive Klassen endlicher Gruppen. IIa. Gesättigte Formationen: ein
     allgemeiner Satz von Gaschütz-Lubeseder-Baer-Typ, *Publ. Sec. Mat. Univ. Autònoma
     Barcelona* **29** (1985), no. 2–3, 39–76.
[6] W. Gaschütz, Zur Theorie der endlichen auflösbaren Gruppen, *Math. Z.* **80** (1963),
     300–305.
[7] W. Guo, *The theory of classes of groups*, Science Press–Kluwer Academic Publishers,
     Beijing–New York–Dordrecht–Boston–London, 2000.
[8] V. D. Mazurov and E. I. Khukhro (eds.), *Unsolved problems in group theory: The
     Kourovka notebook*, 12 ed., Institute of Mathematics, Sov. Akad., Nauk SSSR, Siberian
     Branch, Novosibirsk, SSSR, 1992.
[9] V. D. Mazurov and E. I. Khukhro (eds.), *Unsolved problems in group theory: The
     Kourovka notebook*, 14 ed., Institute of Mathematics, Sov. Akad., Nauk SSSR, Siberian
     Branch, Novosibirsk, SSSR, 1999.
[10] L. A. Shemetkov, The product of formations, *Dokl. Akad. Nauk BSSR* **28** (1984),
     no. 2, 101–103.
[11] L. A. Shemetkov and A. N. Skiba, *Formations of algebraic systems*, Nauka, Moscow,
     1989.
[12] A. N. Skiba and L. A. Shemetkov, Multiply $\mathfrak{L}$-composition formations of finite groups,
     Ukr. Math. J. **52** (2000), no. 6, 898–913.
[13] T. R. Vishnevskaya, On factorizations of one-generated p-local formations, *Izv. Gomel.
     Gos. Univ. Im. F. Skoriny Vopr. Algebry* **3** (2000), 88–92.

NEW RESULTS ON PRODUCTS OF FINITE GROUPS

A. BALLESTER-BOLINCHES*, JOHN COSSEY[†] and
M. C. PEDRAZA-AGUILERA[§] [1]

*Departament d'Àlgebra, Universitat de València, C/ Doctor Moliner 50, 46100 Burjassot
(València), Spain

[†]Mathematics Department, School of Mathematical Sciences, The Australian National University, Canberra, 0200, Australia

[§]ETS de Informática Aplicada, Departamento de Matemática Aplicada-IMPA, Universidad Politécnica de Valencia, Camino de Vera, s/n, 46022 Valencia, Spain

Email: mpedraza@mat.upv.es

The study of factorized groups has played an important role in the theory of groups. We can consider so relevant results as the Ito's Theorem about products of abelian groups or the celebrated Theorem of Kegel–Wielandt about the solubility of a product of two nilpotent groups. In the very much special case when the factors are normal and nilpotent, a well-known result due to Fitting shows that the product is nilpotent. Nevertheless it is not true in general that the product of two normal supersoluble subgroups of a group is a supersoluble group. To create intermediate situations it is usual to consider products of groups whose factors satisfy certain relations of permutability. Following Carocca [12] we say that a group $G = AB$ is the mutually permutable product of A and B if A permutes with every subgroup of B and vice versa. If, in addition, every subgroup of A permutes with every subgroup of B, we say that the group G is a totally permutable product of A and B.

In this context, we can consider as seminal the following results of Asaad and Shaalan.

Theorem A (Asaad and Shaalan [2])   (i) *Assume that a group $G = AB$ is the mutually permutable product of A and B. Suppose that A and B are supersoluble and that either A, B or G', the derived subgroup of G, is nilpotent. Then G is supersoluble.*

(ii) *If $G = AB$ is the totally permutable product of the supersoluble subgroups A and B, then G is supersoluble.*

These results are the beginning of a fruitful line of research in the area of factorised groups. Many contributions are made in the context of formation theory in the finite universe. More precisely Maier [17] prove that a totally permutable product of two $\mathcal{F}$-subgroups, where $\mathcal{F}$ is a saturated formation containing the class $\mathcal{U}$ of all supersoluble groups, belongs to the class $\mathcal{F}$ as well. Moreover, in [7] Ballester-Bolinches and Pérez-Ramos removed the hypothesis of saturation in the above theorem. On the other hand, Carocca (see [13]), generalizes Maier's result to an arbitrary number of factors and the same result for saturated formations

---

[1]This work is supported by Proyecto MTM2004-08219-C02-02 MEC (Spain) and FEDER (European Union).

or formations of soluble groups is proved by Ballester-Bolinches, Pedraza-Aguilera and Pérez-Ramos in [6]. In the same paper, it is studied the behaviour of totally permutable products respect to residuals, projectors and normalizers associated with saturated formations. Results in this line for mutually permutable products can be found in [5].

Extensions of some of the above results in the non-finite universe are proved by Beidleman, Galoppo, Heineken and Manfredino. (see [8], [10] and [11]). On the other hand, Hauck, Pérez-Ramos and Martínez-Pastor analyse totally permutable products through the context of Fitting classes [16].

Following Robinson [18], we say that a group is an SC-group if all of its chief factors are simple groups. Clearly a group is supersoluble if and only if it is a soluble SC-group. It is not difficult to see that the class of all SC-groups is a formation closed under taking normal subgroups. However it is neither an s-closed formation nor a saturated formation.

Example   Let G be a Frattini extension of A_5 for the prime number $p = 3$. Then $N = \Phi(G)$ is a minimal normal subgroup of G of order 3^4. Therefore G is not an SC-group. But $G/N \simeq A_5$.

Robinson [18] describes the structure of SC-groups. They can be characterized in the following way.

Proposition 1   A group G is an SC-group if and only if there is a perfect normal subgroup D such that G/D is supersoluble, $D/Z(D)$ is a direct product of G-invariant subgroups, and $Z(D)$ is supersolubly embedded in G (that is, there is a G-admissible series of $Z(D)$ with cyclic factors).

Note that in the above proposition $D = G^{\mathcal{U}} = G^{\mathcal{S}} = E(G)$, where $E(G)$ denotes the subgroup of G generated by all subnormal quasisimple subgroups of G.

Remark   If G is an SC-group and S denotes a component of G, that is, a subnormal quasisimple subgroup of G, then S is a normal subgroup of G.

The relation between the soluble residual and the soluble radical, i.e., the product of all soluble normal subgroups, of an SC-group is given in the following lemma.

Lemma 1 ([18])   If G is an SC-group and D is the soluble residual of G, then $C_G(D) = G_S$.

The above facts allows us to give some information about totally permutable products of SC-groups.

Theorem B ([3])   Let $G = AB$ the totally permutable product of the subgroups A and B. Then G is an SC-group if and only if A and B are SC-groups

This work has been continued by Beidleman, Hauck and Heineken in [9].

On the other hand, it is clear that a mutually permutable product whose factors intersect trivially is in fact a totally permutable one. Thus, in that case, if both factors are supersoluble, the group is also supersoluble. Taking this remark into account, Alejandre, Ballester-Bolinches and Cossey (see [1]) obtained that a mutually permutable product of supersoluble groups in which the intersection of the factors does not contain non-trivial normal subgroups is supersoluble.

Theorem C ([1]) *Let $G = AB$ be the mutually permutable product of the supersoluble subgroups A and B. If $\mathrm{Core}_G(A \cap B) = 1$, then G is supersoluble.*

It could be natural to ask for a SC-version of the above result. This follows after the analysis of the behaviour of the soluble residuals of the factors in mutually permutable products. That is the main result in [4].

Theorem 1 *Let the group $G = AB$ be the mutually permutable product of the subgroups A and B. Then $[A, B^S]$ and $[A^S, B]$ are contained in $\mathrm{Core}_G(A \cap B)$.*

Here A^S and B^S denote the soluble residuals of A and B, respectively.

Corollary 1 ([4]) *Let the group $G = AB$ be the mutually permutable of A and B. Then A^S and B^S are normal subgroups of G.*

As a consequence we have:

Theorem 2 ([4]) *Let the group $G = AB$ be the mutually permutable product of A and B. If A and B are SC-groups and $\mathrm{Core}_G(A \cap B) = 1$, then G is an SC-group.*

The converse holds with no restrictions.

Theorem 3 ([4]) *Let $G = AB$ be the mutually permutable product of the subgroups A and B. If G is an SC-group, then A and B are SC-groups.*

Theorem 1 also applies to get results about mutually permutable products of SC-groups in which one of the factors is quasinilpotent. These are also contained in [4].

Recall that a group G is said to be quasinilpotent if $\mathrm{Aut}_G(H/K) = \mathrm{Int}(H/K)$ for every chief factor of G. It is clear that a group is nilpotent if and only if it is a soluble quasinilpotent group and that quasinilpotent groups are particular SC-groups.

Theorem 4 *Let $G = AB$ be the mutually permutable product of A and B. Assume that A is an SC-group and B is quasinilpotent. Then G is an SC-group.*

Theorem 5 *Let $G = AB$ be the mutually permutable product of the subgroups A and B. If A and B are SC-groups and G', the derived subgroup of G, is quasinilpotent, then G is an SC-group.*

Notice that Theorem 4 and 5 are extension of the aforementioned results by Asaad and Shaalan.

References

[1] Manuel J. Alejandre, A. Ballester-Bolinches and John Cossey, Permutable products of supersoluble groups, *J. Algebra* **276** (2004), 453–461.

[2] M. Asaad and A. Shaalan, On the supersolvability of finite groups, *Arch. Math.* **53** (1989), 318–326.

[3] A. Ballester-Bolinches and John Cossey, Totally permutable products of finite groups satisfying SC or PST, *Monatshefte für Mathematik*, to appear.

[4] A. Ballester-Bolinches, John Cossey and M. C. Pedraza-Aguilera, On mutually permutable products of finite groups, *J. Algebra*, to appear.

[5] A. Ballester-Bolinches and M. C. Pedraza-Aguilera, Mutually permutable products of finite groups II, *J. Algebra* **218** (1999), 563–572.

[6] A. Ballester-Bolinches and M. C. Pedraza-Aguilera and M. D. Pérez-Ramos, Totally and mutually permutable products of finite groups, in *Groups St Andrews 1997 in Bath, Vol. I*, London Math. Soc. Lecture Notes **260** (1999), 65–68.

[7] A. Ballester-Bolinches and M. D. Pérez-Ramos, A Question of R. Maier concerning formations, *J. Algebra* **182** (1996), 738–747.

[8] J. C. Beidleman, A. Galoppo, H. Heineken and M. Manfredino, On certain products of soluble groups, *Forum Mathematicum*, to appear.

[9] J. C. Beidleman, P. Hauck and H. Heineken, Totally permutable products of certain classes of finite groups, *J. Algebra* **276** (2004), 826–835.

[10] J. C. Beidleman and H. Heineken, Survey of mutually and totally permutable products in infinite groups, *Topics in infinite groups*, 45–62, Quad. Mat., 8 Dept. Math., Seconda Univ. Napoli, Caserta, 2001.

[11] J. C. Beidleman and H. Heineken, Totally permutable torsion groups, *J. Group Theory* **2** (1999), 377–392.

[12] A. Carocca, p-supersolvability of factorized finite groups, *Hokkaido Math. J.* **21** (1992), 395–403.

[13] A. Carocca, A note on the product of $\mathcal{F}$-subgroups in a finite group, *Proc. Edinburgh Math. Soc.* **39** (1996), 37–42.

[14] A. Carocca and R. Maier, Theorems of Kegel–Wielandt type, in *Groups St Andrews 1997 in Bath, Vol. I*, London Math. Soc. Lecture Notes Series **260** (1999), 195–201.

[15] K. Doerk and T. Hawkes, *Finite Soluble Groups*, Walter De Gruyter, Berlin/New York, 1992.

[16] P. Hauck, A. Martínez-Pastor and M. D. Pérez-Ramos, Fitting classes and products of totally permutable groups, *J. Algebra* **252** (2002), 114–126.

[17] R. Maier, A completeness property of certain formations, *Bull. London Math. Soc.* **24** (1992), 540–544.

[18] D. J. S. Robinson, The structure of finite groups in which permutability is a transitive relation, *J. Austral. Math. Soc.* **70** (2001), 143–149.

RADICAL LOCALLY FINITE T-GROUPS

A. BALLESTER-BOLINCHES[*], H. HEINEKEN[‡] and TATIANA PEDRAZA[§1]

[*]Departament d'Àlgebra, Universitat de València, c/ Doctor Moliner 50, 46100 Burjassot (Valencia), Spain

[‡]Mathematisches Institut, Universität Würzburg, Am Hubland, 97074 Würzburg, Germany

[§]ETS de Informática Aplicada, Departamento de Matemática Aplicada-IMPA, Universidad Politécnica de Valencia, 46022 Valencia, Spain

Email: tapedraz@mat.upv.es

1 Introduction

A group G is said to be a T-*group* if every subnormal subgroup of G is normal in G, that is, if normality is a transitive relation in G. The study of this class of groups begins with the publication of a paper of Dedekind in 1896. He characterizes the finite groups in which every subgroup is normal. These groups, called *Dedekind groups*, are obvious examples of T-groups. The extension of Dedekind's result to infinite groups was proved by Baer in 1933.

Theorem (Dedekind, Baer) *All the subgroups of a group G are normal if and only if G is abelian or the direct product of a quaternion group of order 8, an elementary abelian 2-group and an abelian group with all its elements of odd order.*

In 1942, E. Best and O. Taussky [5] prove that every finite group with cyclic Sylow subgroups is a T-group. Later G. Zacher characterized soluble finite T-groups by means of Sylow towers properties (see [12]). However, the decisive result about the structure of T-groups in the finite soluble universe was obtained by Gaschütz in 1957 ([8]).

Theorem (Gaschütz) *Let G be a finite soluble group. Then G is a T-group if and only if it has an abelian normal Hall subgroup L of odd order such that G/L is a Dedekind group and the elements of G induce power automorphisms in L.*

As a consequence of Gaschütz's result, the class of all finite soluble T-groups is subgroup-closed. However, the class of all T-groups is not subgroup closed. For example, A_5 is a T-group but it has a subgroup isomorphic with A_4, which is not a T-group.

The study of the classes of groups in which the properties like normality, permutability or permutability with Sylow subgroups are transitive (T-, PT- and PST-groups) is an active area of research in group theory. Several characterizations of these classes have been obtained in the finite soluble universe. If we restrict our

---

[1]The first and third author are supported by Proyecto MTM2004-08219-C02-02 of MEC (Spain) and FEDER (European Union).

attention to local characterizations of T-groups, the results of Robinson [10] and Bryce and Cossey [6] are crucial.

Definition A group G satisfies $\mathcal{C}_p$ when every subgroup of a Sylow subgroup P of G is normal in $N_G(P)$.

Theorem (Robinson) *A finite group G is a soluble T-group if and only if it satisfies $\mathcal{C}_p$ for all primes p.*

Definition If p is a prime, $\mathcal{D}_p$ is the class of all groups satisfying
    (a) Sylow p-subgroups of G are Dedekind groups and
    (b) p-chief factors of G are cyclic and, as modules for G, form a single isomorphism class.

Theorem (Bryce and Cossey) *A finite soluble group G is a T-group if and only if G is in the class $\mathcal{D}_p$ for every prime p.*

In the general universe of all groups, the situation is more complicated. For instance, the class of soluble T-groups is not subgroup closed in general. Moreover, there exist different types of soluble infinite T-groups, some of which defy classification. In this sense, Robinson [9] has obtained important advances in the study and classification of these groups. He proves, in particular, that soluble T-groups are metabelian and that finitely generated soluble T-groups are either finite or abelian. On the other hand, it is known that finite soluble T-groups are supersoluble ([8]). This is not longer true in infinite groups. For example, the quasicyclic 2-group is a soluble T-group which is not supersoluble. However, Robinson [9] shows that every soluble T-group is locally supersoluble.

Our aim in this survey is to present some results about T-groups in a class of locally finite-soluble groups. This class of groups will be denoted by $c\bar{\mathfrak{L}}$ and it is composed by *radical locally finite groups with min-p for all primes p*. The class $c\bar{\mathfrak{L}}$ has nice properties of Sylow type. For instance, for every prime p, the Sylow p-subgroups of G are Chernikov groups and are all conjugate in G. These good properties have made possible a successful development of formation theory and nice results about generalized nilpotent groups and generalized supersoluble groups in this universe (see [1, 3, 4]). We refer the reader to the book of Dixon [7] for more information about this class of groups.

As we will see in the next section, T-groups in the universe $c\bar{\mathfrak{L}}$ are precisely groups in which every descendant subgroup is normal. We also present a characterization of T-groups in the universe $c\bar{\mathfrak{L}}$ similar to the Bryce and Cossey's characterization of finite soluble T-groups.

2 The results

All the results contained in this section are proved in [2].
    In the sequel, we tacitly assume that all groups belong to the class $c\bar{\mathfrak{L}}$.
    Let G be a group and consider H and K, two normal subgroups of G such that K is contained in H. Then H/K is called a δ-*chief factor of G* if H/K is either

a minimal normal subgroup of G/K or a divisibly irreducible $\mathbb{Z}G$-module, that is, H/K has no proper infinite G-invariant subgroups. In a $c\bar{\mathfrak{L}}$-group, every δ-chief factor is either an elementary abelian finite p-group ([7, (1.2.4)]) or a direct product of finitely many quasicyclic p-groups, for some prime p. In particular if $K = 1$, H is said to be a δ-*minimal normal subgroup of* G.

The *Wielandt subgroup* of a group G is defined to be the intersection of all normalizers of subnormal subgroups of G. This subgroup is denoted by $\omega(G)$ and it is, of course, a characteristic subgroup of G. Furthermore, $\omega(G)$ is a T-group and G is a T-group precisely when $G = \omega(G)$. Our first result shows that if G is a $c\bar{\mathfrak{L}}$-group, then $\omega(G)$ normalizes all descendant subgroups of G. As a consequence, G is a T-group precisely when every descendant subgroup of G is normal.

Theorem 2.1 *The Wielandt subgroup $\omega(G)$ of a group G coincides with the intersection of all normalizers of descendant subgroups of G.*

Corollary 2.2 *In the universe $c\bar{\mathfrak{L}}$, T-groups are precisely groups in which every descendant subgroup is normal.*

In particular, every descendant subgroup of a T-group is a T-group.

A result of Wielandt asserts that every minimal normal subgroup N of G normalizes every subnormal subgroup provided N satisfies the minimal condition on normal subgroups ([11, (13.3.7)]). If G is a $c\bar{\mathfrak{L}}$-group then G always contains minimal normal subgroups and every minimal normal subgroup of G satisfies the above property. Therefore the Wielandt subgroup $\omega(G)$ of a $c\bar{\mathfrak{L}}$-group G contains all the minimal normal subgroups of G (and therefore it is non-trivial). Our next result extends this property by showing that $\omega(G)$ also contains all normal subgroups of G which are divisibly irreducible as $\mathbb{Z}G$-modules.

Theorem 2.3 *Let N be a δ-minimal normal subgroup of a group G. Then N normalizes every descendant subgroup of G. Therefore $\omega(G)$ contains all the δ-minimal normal subgroups of G.*

As we have mentioned above, finite soluble T-groups are supersoluble. Now we obtain that, in our universe, T-groups are contained in a class of generalized supersoluble groups: the class $\mathcal{U}^*$ of all $c\bar{\mathfrak{L}}$-groups G in which every δ-chief factor of G is either a cyclic group of prime order or a quasicyclic group. This class of groups has been introduced and studied in [4] and it behaves as the class of finite supersoluble groups.

Lemma 2.4 *Let G be a T-group. Then the δ-chief factors of G are cyclic groups of prime order or quasicyclic groups. As a consequence, $G' \leq F(G)$.*

It is known that if G is a T-group then $F(G) = C_G(G')$ ([11, (13.4.1)]). Using this fact and Lemma 2.4 it follows easily that if G is a T-group in the class $c\bar{\mathfrak{L}}$ then G' is abelian. As a consequence, we obtain the following result on T-groups in the class $c\bar{\mathfrak{L}}$.

Corollary 2.5 *Let G be a T-group. Then G is metabelian.*

Our next result confirms the extension of Bryce and Cossey's result to our universe.

Theorem 2.6 *Let G be a group. Then G is a T-group if and only if*
  (i) *Sylow p-subgroups of G are T-groups for every prime p and*
  (ii) *δ-chief factors of G are cyclic groups of prime order or quasicyclic groups and two isomorphic chief factors are G-isomorphic.*

Finally, we obtain a characterization of T-groups similar to the well-known Gaschütz's characterization of finite soluble T-groups.

Theorem 2.7 *G is a T-group if and only if G has an abelian normal Sylow π-subgroup A with all elements of odd order such that $G = [A]D$, where D is a T-group and elements of G induce power automorphisms in A.*

Note that in the above theorem the condition "D is a T-group" cannot be replaced by "D is a Dedekind group" in order to get the same conclusion as the locally dihedral 2-group shows.

References

[1] A. Ballester-Bolinches and S. Camp-Mora, A Gaschütz–Lubeseder type theorem in a class of locally finite groups, *J. Algebra* **221** (1999), 562–569.

[2] A. Ballester-Bolinches, H. Heineken and T. Pedraza, T-groups in a class of locally finite groups, *Forum Math.* (to appear).

[3] A. Ballester-Bolinches and T. Pedraza, On a class of generalized nilpotent groups, *J. Algebra* **248** (2002), 219–229.

[4] A. Ballester-Bolinches and T. Pedraza, A class of generalized supersoluble groups, *Publ. Mat.* **49** (2005), 213–223.

[5] E. Best and O. Taussky, A class of groups, *Proc. Roy. Iris Acad. Sect. A* **47** (1942), 55–62.

[6] R. A. Bryce and J. Cossey, The Wielandt subgroup of a finite soluble group, *J. London Math. Soc.* **40** (1989), 244–256.

[7] M. R. Dixon, *Sylow Theory, Formations and Fitting classes in Locally Finite Groups.* Series in Algebra 2 (World Scientific, Singapore–New Jersey–London–Hong Kong 1994).

[8] W. Gaschütz, Gruppen in denen das Normalteilersein transitiv ist, *J. Reine Angew. Math.* **198** (1957), 87–92.

[9] D. J. S. Robinson, Groups in which normality is a transitive relation, *Proc. Cambridge Philos. Soc.* **60** (1964), 21–38.

[10] D. J. S. Robinson, A note on finite groups in which normality is transitive, *Proc. Amer. Math. Soc.* **19** (1968), 933–937.

[11] D. J. S. Robinson, *A course in the theory of groups* (Springer-Verlag, 1982).

[12] G. Zacher, Caratterizzazione dei t-gruppi risolubili, *Ricerche Mat.* **1** (1952), 287–294.

EXPLICIT TILTING COMPLEXES FOR THE BROUÉ CONJECTURE ON 3-BLOCKS

AYALA BAR-ILAN[2], TZVIYA BERREBI[2], GENADI CHERESHNYA[2],
RUTH LEABOVICH[2], MIKHAL COHEN[1] and MARY SCHAPS

Department of Mathematics, Bar-Ilan University, 52900 Ramat Gan, Israel
Email: mschaps@macs.biu.ac.il

Abstract

The Broué conjecture, that a block with abelian defect group is derived equivalent to its Brauer correspondent, has been proven for blocks of cyclic defect group and verified for many other blocks, mostly with defect group $C_3 \times C_3$ or $C_5 \times C_5$. In this paper, we exhibit explicit tilting complexes from the Brauer correspondent to the global block B for a number of Morita equivalence classes of blocks of defect group $C_3 \times C_3$. We also describe a database with data sheets for over a thousand blocks of abelian defect group in the ATLAS group and their subgroups.

1 Introduction

Let G be a finite group and let k be a field of characteristic p, where p divides $|G|$. Let $kG = \bigoplus B_i$ be a decomposition of the group algebra into blocks, and let D_i be the defect group of the block B_i, of order p^{d_i}. By Brauer's Main Theorems (see [1] for an accessible exposition) there is a one-to-one correspondence between blocks of kG with defect group D_i and blocks of $kN_G(D_i)$ with defect group D_i. Let b_i be the block corresponding to B_i, called its Brauer correspondent.

Broué [5] has conjectured that if D_i is abelian and B_i is a principal block, then B_i and b_i are derived equivalent, i.e., the bounded derived categories $D^b(B_i)$ and $D^b(b_i)$ are equivalent. In fact, it is generally believed by researchers in the field that the hypothesis that B_i be principal is unnecessary.

By a fundamental theorem of Rickard, if two algebras A and B are derived equivalent, then there are "two-sided" *tilting complexes*, complexes $_AX_B$ and $_BY_A$ of bimodules, projective as A- and B-modules, such that $_AX_B \hat{\otimes}_B {_BY_A} \overset{\sim}{\to} {_AA_A}$ in $D^b(A)$, and $_BY_A \hat{\otimes}_A {_AY_B} \overset{\sim}{\to} {_BB_B}$ in $D^b(B)$. Each two-sided tilting complex induces a "one-sided" tilting complex $_AT$ by ignoring one of the algebra actions. The one-sided complexes, after removing superfluous modules and maps, can be quite simple to describe. The $_AT$ are complexes of projective modules and in all the examples we will bring, with one exception, each indecomposable projective occurs in a unique degree. A one-sided complex $_AT$ satisfies

$$\operatorname{End}_A({_AT}) \overset{\sim}{\to} B.$$

[1]Partially supported by grants from the European Union and the Israel Science Foundation.
[2]Work done towards a master's degree at Bar-Ilan University.

In Sections 2–3, we will give explicit tilting complexes of a number for blocks. We are particularly interested in cases where one tilting complex can be obtained from another by an automorphism, as in the passage from the principal block of A_6 to that of S_6 and from the principal block of A_7 to that of S_7.

In some cases, the proof of the Broué conjecture for the block is due to Okuyama [24] and our contribution is to compute the tilting complex from his algorithm by taking mapping cones. In other cases, the experimental work to discover a conjectural tilting complex was done by G. Chereshnya and the proof by M. Schaps.

The blocks given here were chosen from a database of blocks of abelian defect group which will be described in Section 4. Using the sorting into Morita equivalences in the database, the examples in Sections 2–3 actually provide tilting complexes for numerous other blocks. The basic approach guiding our research is that outlined in [34], in which the column operation induced on the decomposition matrix by the tilting are taken as providing a first approximation to the tilting complex, up to a choice of folding. Thus whenever the Brauer character table and the decomposition matrix were available in GAP, we included the decomposition matrix on the data sheet for the block in the database. We also included the generalized decomposition matrices where we could generate them. These were useful in determining the permutations of the rows and columns of the decomposition matrix which occur when the rows and columns are reordered by degrees after tilting.

The verification of an explicit tilting complex permits machine calculation of the indecomposable projectives. We have written a program in MAGMA, based on Holloway's program "homotopy", to make these calculations for a few of the Brauer correspondents, as described in [12]. It is currently being tested and debugged.

2   Automizers C_4 and D_4

This section includes most of the blocks of the alternating and symmetric groups with defect group $C_3 \times C_3$. We assume a fixed ordering of the irreducibles of the character table of the group, as given in the ATLAS or the GAP Character Table library. The B_1 will always be the principal block, B_2 will contain the first irreducible which is not in the principal block, and so forth.

For the automizer C_4, the four projective modules $Q_1, \ldots, Q_4$ are of dimension 9, and each is diamond-shaped. We give the simples of Q_1, the structure of Q_2, Q_3 and Q_4 being obtained from that of Q_1 by cyclic permutation of $\{1,2,3,4\}$.

$$
\begin{array}{ccccc}
 & & 1 & & \\
 & 2 & & 4 & \\
3 & & 1 & & 3 \\
 & 2 & & 4 & \\
 & & 1 & &
\end{array}
$$

For the automizer D_4, there is one projective module P_1 of dimension 18 corresponding to the simple module of dimension 2, and there are four projective modules P_2, P_3, P_4, P_5 of dimension 9. We give the composition factors for P_1 and P_2.

The exact maps are given by the theorems in [12].

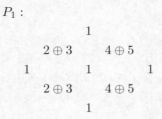

P_1 :

P_2 :

The principal blocks B_1 of A_7 and S_7

The tilting complex for B_1 of A_7 was given by Okuyama [24] as

$$
\begin{aligned}
Q_1' &: Q_3 \oplus Q_3 \to Q_1 \\
Q_2' &: \quad\quad Q_3 \to Q_2 \\
Q_3' &: \quad\quad Q_3 \\
Q_4' &: \quad\quad Q_3 \to Q_4
\end{aligned}
$$

The tilting complex B_1 for S_7 was also used by Okuyama [24], but not identified as belonging to S_7. It is

$$
\begin{aligned}
P_1' &: P_4 \oplus P_5 \to P_1 \\
P_2' &: P_4 \oplus P_5 \to P_2 \\
P_3' &: P_4 \oplus P_5 \to P_3 \\
P_4' &: P_4 \\
P_5' &: P_5
\end{aligned}
$$

Note that P_2' and P_3' are obtained by splitting Q_1', P_4' and P_5' are the splittings of P_3', and Q_2', Q_4' each give P_1'.

The principal blocks B_1 of A_6 and S_6

The algorithm for the principal blocks in these groups was given by Okuyama. In each case, one takes the elementary tilting complex giving B_1 of A_7 or S_7 and applies a further elementary tilting. The notation $P[n]$ indicates that the complex P is shifted n place to the left.

B_1 of A_6

$$Q_1'' = Q_2' \oplus Q_4' \to Q_1' : \qquad Q_2 \oplus Q_4 \to Q_1$$
$$Q_2'' = Q_2'[1] : \qquad Q_3 \to Q_2$$
$$Q_3'' = Q_2' \oplus Q_4' \to Q_3' : \quad Q_3 \to Q_2 \oplus Q_4$$
$$Q_4'' = Q_4'[1] \qquad Q_3 \to Q_4$$

B_1 of S_6

$$P_1'' = P_1'[1] : \quad P_4 \oplus P_5 \to P_1$$
$$P_2'' = P_1' \to P_2' : \qquad P_1 \to P_2$$
$$P_3'' = P_1' \to P_3' : \qquad P_1 \to P_3$$
$$P_4'' : P_1' \to P_4' : \qquad P_5 \to P_1$$
$$P_5'' : P_1' \to P_4' : \qquad P_4 \to P_1$$

If we analyze what has happened here, we see that Q_2'' or Q_4'' each give P_1'', that Q_1' splits into P_2'' and P_3'', while Q_3'' splits into P_4'' and P_5''.

$B2$ of $2.A_6$, with automizer C_4

This block is isomorphic to $2.SL(2,7)$, and thus the Broué conjecture was proven for it in [16]. The explicit tilting complex was given in [34]. The first step, with $I = \{3, 4\}$, produces the complex

$$Q_1' : \qquad Q_3 \oplus Q_4 \to Q_1$$
$$Q_2' : \qquad Q_3 \oplus Q_4 \to Q_2$$
$$Q_3' : \qquad Q_3$$
$$Q_4' : \qquad Q_4$$

The endomorphism ring of this complex in the derived category has the same decomposition matrix as $2.A_7$. A difficulty with the stable equivalence has prevented us from demonstrating that they are Morita equivalent, but we believe it to be true. Since, by [18], the blocks $2.A_7$ and $2.S_8$ are Morita equivalent, this would also establish an explicit tilting complex for $2.S_8$.

The second step is an elementary tilting with $I = \{1, 2\}$.

$$Q_1' : Q_3 \oplus Q_4 \to Q_1$$
$$Q_2' : Q_3 \oplus Q_4 \to Q_2$$
$$Q_3' : \qquad Q_4 \to Q_1$$
$$Q_4' : \qquad Q_3 \to Q_2$$

In this case the action of the automorphism giving $2.A_6.2$ does not act on the tilting complex by permutation.

Note that in all three cases involving A_6, the block is a result of two elementary tiltings of which the first gives the corresponding block for A_7.

3   Automizer Q_8

As with the automizer D_4, we have four projective modules of dimension 9 and one of dimension 18. The projective indecomposables look quite similar, but there is one major difference: all simples appear in each projective indecomposable. Thus for P_2, we have now

$$
\begin{array}{c}
2 \\
1 \\
3 \qquad 4 \oplus 5 \\
1 \\
2
\end{array}
$$

The composition factors for P_1 are the same as in the case of D_4, though the maps are slightly different. The exact maps can be obtained from the theorems in [12].

The principal block B_1 of M_{22}

The sequence of elementary tilting complexes was determined by Okuyama [24]: $I_0 = \{4, 5\}$. The tilting complexes are:

$$
\begin{aligned}
P_1^{(1)} &: P_4 \oplus P_5 \to P_1 \\
P_2^{(1)} &: P_4 \oplus P_5 \to P_2 \\
P_3^{(1)} &: P_4 \oplus P_5 \to P_3 \\
P_4^{(1)} &: P_4 \\
P_5^{(1)} &: P_5
\end{aligned}
$$

Then $I_1 = \{1\}$. Here we can determine that the multiplicity of $P_t^{(1)}$ into the other $P_j^{(1)}$ is always 1, because this gives the correct decomposition matrix when we apply column operations. Since the maps from $P_4 \oplus P_5$ into P_2 and P_3 factor through P_1, we get

$$
\begin{aligned}
P_1^{(2)} &: P_4 \oplus P_5 \to P_1 \\
P_2^{(2)} &: \qquad\qquad P_1 \to P_2 \\
P_3^{(2)} &: \qquad\qquad P_1 \to P_3 \\
P_4^{(2)} &: \qquad P_5 \to P_1 \\
P_5^{(2)} &: \qquad P_4 \to P_1
\end{aligned}
$$

Another block with the same invariants is B_5 of $2.L3(4).2_1$, treated in [12]. It has the same tilting complex.

The principal block B_1 of $PSL(3,4)$

The algorithm follows Okuyama. The first two steps are as for the principal block of M_{22} and give the above tilting complex. At this point, the decomposition matrix is

$$D = \begin{bmatrix} 1 & 1 & 1 & 1 & 1 \\ 0 & 1 & 0 & 0 & 0 \\ 0 & 0 & 1 & 0 & 0 \\ 1 & 0 & 0 & 0 & 1 \\ 1 & 0 & 0 & 1 & 0 \\ 0 & 1 & 1 & 1 & 1 \end{bmatrix}$$

In this case, the calculation of the elementary tilting complex was not straightforward. The next stage is the elementary tilting complex determined by the third column, $I_2 = 3$, followed by an elementary tilting complex given by the last two columns. There are two separate maps of $P_3^{(2)}$ into $P_2^{(2)}$, neither of which factors through the other. Using a variant of Holloway's MAGMA homotopy computation program developed at Bar-Ilan, we chose representatives; there are only zero maps from P_3 to P_2 in $P_2^{(3)}$, and both maps $P_1 \oplus P_1 \to P_1$ map into the radical squared. Thus when we take mapping cones, we get:

$$P_1^{(3)} = P_3^{(2)} \to P_1^{(2)}: \qquad P_4 \oplus P_5 \to P_3$$
$$P_2^{(3)} = P_3^{(2)} \oplus P_2^{(2)} \to P_1^{(2)}: \quad P_1 \oplus P_1 \to P_1 \oplus P_3 \oplus P_3 \to P_2$$
$$P_3^{(3)} = P_3^{(2)}[1]: \qquad P_1 \to P_3$$
$$P_4^{(3)} = P_3^{(2)} \to P_4^{(2)}: \qquad P_5 \to P_3$$
$$P_5^{(3)} = P_3^{(2)} \to P_5^{(2)}: \qquad P_4 \to P_3$$

The final stage in the algorithm is given by an elementary tilting with $I_3 = \{4,5\}$. The resulting mapping cones give the correct decomposition matrix and the final tilting complex

$$P_1^{(4)} = P_4^{(3)} \oplus P_5^{(3)} \to P_1^{(3)}: \qquad P_3$$
$$P_2^{(4)} = P_4^{(3)} \oplus P_5^{(3)} \to P_3^{(3)}: \quad P_4 \oplus P_5 \to P_1 \oplus P_1 \to P_1 \to P_2$$
$$P_3^{(4)} = P_4^{(3)} \oplus P_5^{(3)} \to P_3^{(3)}: \quad P_4 \oplus P_5 \to P_1 \oplus P_3$$
$$P_4^{(4)} = P_4^{(3)}[1]: \qquad P_5 \to \quad P_3$$
$$P_5^{(4)} = P_5^{(3)}[1]: \qquad P_4 \to \quad P_3$$

Note that in this example, unlike all the previous explicit tilting complexes, we have a projective, P_1, which appears in two different degrees.

The faithful blocks of $4.M_{22}$

This extremely interesting example is treated at length in [23]. The action of Q_8 on $C_3 \times C_3$ is a Frobenius action, and thus the results in [26] imply the existence of a stable equivalence given by restricting, tensoring with a 16-dimensional endopermutation module, and cutting to the correct block. The indecomposables of maximal vertex $C_3 \times C_3$ lead, using Okuyama's theorems, to the following explicit tilting complex.

$$P_2 \to P_1$$
$$P_1 \to P_3$$
$$P_1$$
$$P_4$$
$$P_5$$

This is the first non-cyclic defect group example of a split elementary tilting complex as defined in [34].

4   A database of blocks of abelian defect group

For experimental mathematics one must generate data either bit by bit as needed or *en masse*. For an investigation of blocks of abelian defect group, we chose the second option, creating a database of blocks of over a thousand blocks of abelian defect group for the 1129 groups in the GAP 4.2 Character Table Library [11]. The database was intended to help researchers working on the Broué conjecture. Since Rickard proved the conjecture for blocks of cyclic defect [29], the blocks of cyclic defect were treated separately. Blocks with non-cyclic abelian defect group existed only for primes 2, 3, 5, 7, 11, and 13. Since the case of prime 2 was well understood, we included only the odd primes in this range. For cyclic blocks of defect 1, the cases of primes 2 and 3 are trivial because there is only one Morita equivalence class of blocks, so we included only primes between 5 and 31.

The database also has some relevance for Donovan's conjecture and Puig's conjecture that for a given defect group there are only a finite number of equivalence classes under Morita or Puig equivalence [25].

The organization of the data

The first division of the data was by the prime p which was the characteristic of the field. For each prime p, the program IndexOfBlocks.gap, written in GAP 4.2, generated hypertext indices to a collection of data sheets for individual blocks. The hypertext indices for the various primes are as follows:

(1) Elementary abelian defect group, $d \geq 2$, for $p = 3, 5, 7, 11, 13$,
    BLOCKS/SortAt_p.html.

(2) Non-elementary abelian defect group, $d \geq 2$, for $p = 3, 5, 7$,
    BLOCKS/CyclicSortAt_p.html. Mostly defect group C_{p^2}.

(3) Cyclic defect group, $d = 1$, for $p = 5, 7, 11, 13, 17, 19, 23, 29, 31$,
    CYCLIC/SortAt_p.html.

(4) Nonabelian defect group, $d \geq 3$, for $p = 3, 5, 7$,
    BLOCKS/NonabSortAt_p.html.

For each block of a group in the Character Table Library, the program generated a text file data sheet containing various invariants of the groups and the block, including the decomposition matrix where known. Where the decomposition matrix was not known, the program at least calculated its dimensions $k(B)$ and $\ell(B)$. The invariant $k(B)$, the number of ordinary characters of the block, was available. The invariant $\ell(B)$ was calculated by using a method taken from [21] to get an integral matrix, restricting the character table to the given rows and the elements of order prime to p, and calculating the rank of the matrix. The first sorting of the data was by $(k(B), \ell(B))$ pairs.

The next problem was to sort the data into Morita equivalence classes. From the decomposition matrix one computes the diagonal of the Cartan matrix by taking the scalar product of each column with itself, a procedure for which the chosen ordering of the rows is irrelevant. If one sorts the Cartan diagonal in ascending order, the ordering of the columns is also irrelevant. The list of blocks was sorted by these sorted Cartan diagonals.

For blocks with noncyclic elementary abelian defect group, the sorted Cartan diagonals generally classed together blocks with the same decomposition matrix up to permutations of the rows and columns. A notable exception was the case $p = 3$, $k(B) = 6$, and $\ell(B) = 4$, which contained two sets of decomposition matrices, corresponding to the principal and non-principal blocks the $2.A_6$, the double cover of the alternating group on six numbers. Another interesting case was $p = 3$, $k(B) = 6$, and $\ell(B) = 2$. There are three possible Brauer correspondents, with differing shapes for their generalized decomposition matrices. For each sorted Cartan diagonal there is a unique decomposition matrix, but there may be up to three different Morita equivalence classes of blocks, which can be distinguished by the shapes of their generalized decomposition matrices.

For blocks with cyclic defect group, the Cartan diagonal is not very helpful in distinguishing Morita equivalence classes. In the cyclic case the Morita equivalence class are determined by the Brauer trees. For the sporadic groups the Brauer trees of the cyclic blocks are calculated in [14]. For the other blocks in our database the trees were calculated by Tal Kadaner [17] in her thesis, up to some uncertainty about the cyclic ordering of branches in complicated cases.

Finally, for a given Cartan diagonal, the blocks were sorted by the list of character degrees, after reduction by dividing out by the g.c.d. of the degrees. Ruth Leabovich undertook the study of blocks with the same reduced character degrees in the database which were neither Morita equivalent to the Brauer correspondent nor derived from blocks of cyclic defect. She was able to show in almost all cases that there was a Morita equivalence derived from Clifford theory.

Even for blocks for which the decomposition matrix was not available in GAP 4.2, we listed the blocks under the Cartan diagonal of a block with the same reduced degrees and indicated by "*" that the Cartan diagonal was only conjectural. For many of these blocks the Morita equivalence to a block in that group was established by other means.

Based on her work, the hypertext index was generated again with links to show the connections among the blocks leading back to a "root" block, usually in a group of minimal size. This was largely a matter of filtering out "noise", in order to reduce the totality of blocks to a nontrivial subset. The links were categorized into six possibilities:

(1) (IN) Inflation, where the block is linked to a block of a quotient group from which it is derived by inflation;

(2) (MU) The group is a direct product, and a block in one factor is "multiplied" by a defect zero block in the other factor, with the link given to the corresponding block of the first factor;

(3) (CL) More complicated Clifford theory, e.g., two blocks conjugate under an automorphism producing a single block in the extension by the automorphism;

(4) (MO) Application of Morita's theorem for grouping blocks by conjugacy classes of blocks of the maximal normal p' subgroup N, where the link is to the stabilzer of a representative, divided by N;

(5) (ISO) Isomorphism between blocks;

(6) (SC) Scopes reduction or some extension thereof [37].

In many cases the links were obvious from the group names, but in case (4) an analysis of the character table was required and (6) was more advanced.

We have not extended the system of links to nilpotent blocks, and certain other classes of blocks which were particularly problematical. Ruth Leabovich expects to make this extension as part of the work for her Ph.D. thesis.

Some of the root blocks were cases for which the Broué conjecture has been solved, and for these cases Mikhal Cohen supplied the references. Other cases which we believe to be open and interesting are marked with a "?".

Normalizers and Centralizers

The basic conception of the database, as a tool for working on the Broué conjecture, was to match up blocks with their Brauer correspondent, calculating the decomposition matrices for each, and comparing the centralizers of representatives of various conjugacy classes of nontrivial elements of the defect group.

One difficulty with working in the GAP Character Table Library is the lack of the group. This makes it difficult to calculate the Brauer correspondent. The calculation was done in two stages:

(1) *Via invariants.*

The normalizer $N = N_G(D)$ of the defect group D of a block B is a group with a normal subgroup. By a theorem of Külshammer, such a block is of the form $M_t(K^\alpha[D \rtimes H'])$, where H' is a p'-group and α is a cocycle in $H^2(H', K^*)$. Using a trick of Reynolds [27], we note that this is Morita equivalent to a block of a group $D \rtimes H$, where H is a central extension with cyclic kernel C_m and quotient H'

with $D \xrightarrow{\sim} C_p \times C_p$. For defect 2, we can find the candidate blocks by considering p'-subgroups of $GL(2,p)$ and their central extensions by C_2, a list of which are given at [35]. These groups produced $(k(b), \ell(b))$ pairs corresponding to all the global blocks B. For defect $d \geq 3$ the computation was more complicated and generally required determining the normalizer by one of the methods described below. Leabovitch [20] also showed that in many cases there were linear characters extending the central characters, and thus the various blocks of the normalizer were isomorphic.

The minimal group $D \rtimes H$ having a block Morita equivalent to the Brauer correspondent b was called the *Reynolds group* of B and tables of the Reynolds groups for the blocks with $p = 3, k(B) \leq 24$ and $p = 5, 7$ are given in [12].

Had there been a global blocks B for which we could not find a block with the same invariants in some group with normal subgroup D, this would have been a counter-example to the Broué conjecture, but in fact no such block was found. The next step, then, was to verify that the block of the Reynolds group was indeed Morita equivalent to a block of the normalizer of B, and choose among Reynolds groups when there was a choice.

(2) *Explicit calculation of normalizers and centralizers.*

For the root blocks, those not readily available from smaller groups by Clifford theory, many of the normalizers and centralizers were calculated explicitly by Ayala Bar-Ilan [2] and and Tzviya Berrebi [3]. This was done in one of four possible ways, depending on the group.

(a) *Symmetric and alternating groups and their covers.* In this case the groups were known from the combinatorial considerations. Let $c = (1 \ldots p)$ and $d = (p+1 \ldots 2p)$, and let D be the elementary abelian group $\langle c, d \rangle$.

(a.1) Assume $n \geq 2p$. Then

    (a) $N_{S_n}(D) = (C_p \rtimes C_{p-1}) \wr C_2 \times S_{n-2p}$,

    (b) $C_G(D) = D \times S_{n-2p}$,

    (c) $C_G(c) \xrightarrow{\sim} C_p \times S_{n-p}$, and

    (d) $C_G(cd) \xrightarrow{\sim} (D \rtimes C_2) \times S_{n-2p}$.

(a.2) For the alternating group A_n, we get the intersection of the above normalizers and centralizers with A_n. The cases $n = 2p, 2p + 1$ are special. Thereafter, every odd permutation in the normalizer of D in S_{2p} can be multiplied by a transposition in S_{n-2p}. The case of particular interest is the centralizer $C_{A_n}(cd) = (D \times A_{n-2p}) \rtimes C_2$, where the action of the nontrivial element of C_2 in D is to interchange c and d, while its projection onto S_{n-2p} is a transposition. This is actually isomorphic to $C_3 \times (C_3 \times A_{n-2p}) \rtimes C_2$,

(a.3) The covering groups $\tilde{A}_n$ and $\tilde{S}_n$. Here we get a central extension with kernel C_2 of the groups in (a.1) and (a.2). Let N denote the normalizer of the defect group D in S_{2p} and N_A the normalizer of D in A_{2p}. Let $\tilde{N}$ and $\tilde{N}_A$ be the preimages of N in $\tilde{S}_{2p}$ and $\tilde{A}_{2p}$, each a central extension by C_2 of the corresponding group N or N_A, using the rules given in [10]. Similarly, we can consider $\tilde{C}$, the preimage of the centralizer of cd in S_{2p}. If $p \equiv 1 \pmod 4$, we get $\tilde{C} \xrightarrow{\sim} D \rtimes C_2$, and if $p \equiv 3 \pmod 4$, then $\tilde{C} \xrightarrow{\sim} D \rtimes C_4$. In [15], Hoffman and Humphreys define an amalgamated product of covering groups, which is a direct product modulo the identified kernel, and which we will denote by "$\bar{Y}$". With this notation,

   (a) $N_{\tilde{S}_n}(D) = \tilde{N}\bar{Y}\tilde{S}_{n-2p}$.

   (b) $C_{\tilde{S}_n}(c) = C_p \times \tilde{S}_{n-p}$.

   (c) $C_{\tilde{S}_n}(cd) = \tilde{C}\bar{Y}\tilde{S}_{n-2p}$.

   (d) $N_{\tilde{A}_n}(D) = \tilde{N}_A\bar{Y}\tilde{A}_{n-2p}$.

   (e) $C_{\tilde{A}_n}(c) = C_p \rtimes \tilde{A}_{n-p}$.

   (f) $C_{\tilde{A}_n}(cd) = (D \times \tilde{A}_{n-p}).C_2$.

(b) *Special linear groups.* One common source of blocks with elementary abelian defect group are the special linear groups $SL(2, q)$, where $q = p^d$ is a non-trivial power of the odd characteristic p. The defect group is given by the upper triangular unipotent matrices and the normalizer is $N = (C_p)^d \rtimes C_{q-1}$. The library of groups contains various quotients and extensions, for which the group itself is not available in GAP. To determine the exact extension or quotient of N required an analysis of the character table and a study of the possible groups of the correct order with the proper invariants. Because the number of simples $\ell(B)$ in a block of normal defect group is the number of characters of the p' complement, there were usually few candidates. For this work, the advanced search capabilities in our database of groups of order up to 2000 [22] were useful, because we could filter out by number of characters and then compare the structure description of the groups with that of N. This database, like all the others, was created in GAP 4.2 and should be used subject to the GAP copyright.

(c) *Other groups for which suitable representations were known.* The first problem was to obtain the group; this was done whenever possible by downloading permutation representations from the ATLAS of finite group representations. (For calculating normalizers of small p-groups inside large groups, the matrix representations generally were too slow.)

Normalizers and centralizers are standard functions in GAP, so if the defect group was the p-Sylow subgroups it was straightforward to calculate them. When the defect groups was not a p-Sylow group, Bar-Ilan and Berrebi used a program of T. Breuer to calculate the defect classes, i.e., the conjugacy classes of p' order such that the p-Sylow subgroup of the centralizer of a representative is isomorphic to D. Then by comparing orders and centralizer sizes, the program located a corresponding conjugacy class representative in the group, and calculated the p-Sylow subgroup of its centralizer. The representations were downloaded by hand, but the program calculated the desired results for all the groups at once and stored them in a file to be read by the program which generated the data sheets. For these root blocks, the following subgroups were described:

   1) $H = N_G(D)$;

   2) $C_G(D)$;

   3) For each conjugacy class $[u]$ of p-elements in D, $C_G(u)$ and $C_H(u)$.

(d) *Other groups.* When the normalizer and the centralizer were too large, they could at times be identified from the list of maximal subgroups in the ATLAS and from the centralizer orders of elements.

Example 1 For the sporadic Held group (He) and prime $p = 3$, the normalizer of D can be identified as $((C_4 \times C_2) \rtimes C_4) \rtimes (S_3 \times C_2)$, but the centralizer of an element u of order 3 is 7560, too large for the Small Groups Library. However, the list of maximal subgroups contains a group of order 15120, identified as the normalizer of an element of order 3, with structure $C_3.S_7$, from which we deduce that the centralizer is $C_3.A_7$.

Example 2 Consider the principal 3-block of defect $d = 4$ in the fifth maximal subgroup G of the O'Nan group, a non-solvable group of order 25920. This is a block with invariants $k(B) = 24, l(B) = 14$. The sixth maximal subgroup N of the O'Nan group is a solvable group of the same order, which, according to the information in the ATLAS, is the normalizer of the 3-Sylow subgroup D, the defect group of the block. Thus $N_G(D) = G \cap N$. The structure of G as given in the ATLAS is $((C_3 : C_4) \times A_6).C_2$, so the structure of $N_G(D)$ must be $(C_3)^4 : (C_4 \times C_4).C_2$. In order to find the exact group, we downlowded a permutation representation of the O'Nan group, calculated the Sylow-3 group and then its normalizer N. The Sylow-2 subgroup of N had order 64. It was necessary to find a subgroup of order 32 with the proper structure, which, together with D, generated the desired normalizer, and this we did. It was of order 2592, too large for the GAP library of small groups.

For algebraic groups in non defining characteristic q, for which $p \mid q - 1$, it occasionally happened that the centralizer of an element was of the form $C_p \times G'$, where G' was a related algebraic group of smaller rank. For simple groups, centralizers of elements of low order play a role in the classification. So for some primes and some groups, we may be able to fill in the remaining gaps from the literature.

Generalized Decomposition Numbers

There is an extension of Broué's conjecture by Rickard [29], in which the two-sided tilting complex X is *splendid*, i.e., can be made up of modules which are direct summands of p-permutation modules. Let us suppose that X is a tilting complex from a symmetric algebra kGe to a symmetric algebra kHf, both of which are group blocks with a common abelian defect group D. Let Q be a non-trivial p-subgroup of D. The point to this is to ensure that the various restrictions X_Q of X as complexes of $C_G(Q)e$-$C_H(Q)f$-modules will also be tilting complexes. This means that the decomposition matrices of the corresponding blocks of the centralizers would also be also obtained, one from the other, by column operations, permutations and multiplication of rows by ± 1.

To understand the significance of this claim, let us consider, for each non-trivial u in D, the submatrix M_u of the character table consisting of rows of the block and columns which are conjugacy classes with representatives of the form uy for a p-regular element y, i.e., elements y of p' order. The part of the generalized decomposition matrix of the block corresponding to u describes how the rows of this matrix are generated by the Brauer characters of the corresponding blocks of $C_G(Q)$ for $Q = \langle u \rangle$ ([38],§43). The column operations taking the decomposition matrix of $C_G(Q)e$ to $C_H(Q)f$ determine a change of basis on the Brauer characters. Thus we expect corresponding parts of the generalized decomposition matrix to differ by column operations. Were we to find a case in which this did not hold, it would be a potential counterexample to the claim that there is a splendid tilting complex. So far we have not found any such example.

We have other reasons for wishing to know the complete generalized decomposition matrix. Our attempts to find column operations transforming the regular decomposition matrix is often hampered by difficulties in matching up the rows, which have undergone a permutation. Since the generalized decomposition matrix of a block is nonsingular, and often some of the sections of the generalized decomposition matrix have a single column, the information from the generalized decomposition matrix is often sufficient to determine the permutation.

Ayala Bar-Ilan wrote a program to construct the generalized decomposition matrix, but it requires character tables for the centralizers which are not currently available for each block. One of the most complicated parts of the program was an algorithm for matching up the conjugacy classes in the large group with the conjugacy classes in the centralizer of u.

Since we did not have the character tables of the centralizers for most of the groups in our database, this would have been quite complicated to implement as part of the program which constructs the database. The remaining problem would have been to find a basis for the set of Brauer characters of $C_G(u)$, and match them up with the columns of G.

From the examples calculated by Ayala Bar-Ilan, it became clear that these Brauer characters can almost always be found among the rows of the section of u, up to a possible multiple by -1 to make the value at u positive. (This is quite different from the decomposition matrix, where the values of the p-regular elements rarely contain a complete basis for a positive integral decomposition.) Therefore, we took as a working hypothsis that we could find the Brauer character among the rows we has, and tried to construct a conjectural generalized decomposition matrix for each block on this basis. Taking such a basic set of vectors for each section, we calculated the resulting decomposition of the rows of the section according to this basis. Occasionally there were failures involving arithmetic with algebraic integers. The function available in GAP was intended originally for rows of character tables. It attempts to convert the vector of algebraic integers into an integral vector using an integral basis as in [21]. Unfortunately, when u is not conjugate to all of its powers, one does not have all the necessary columns in M_u, and it was necessary to divide some of the rows by an algebraic integer before trying to decompose.

When the decomposition matrix was known, and the matrix obtained by adjoining the columns of this matrix to the columns of the decomposition matrix was square, we reproduced the total matrix at the bottom of the data scheet.

Example 3 Let G be S_{10} and let B be the first of the two blocks of defect 2.

The centralizer of the element (123) is isomorphic to $C_3 \times S_7$. The set of Brauer characters is the same as the set of Brauer characters for S_7. The submatrix M_u of the character table of S_{10} consisting of rows of the block and conjugacy classes with respresentatives of the form uy for y a p-regular element commuting with u is given by the rows of the following matrix:

$$
\begin{bmatrix}
6 & 2 & 0 & 1 & -1 & 4 & 0 & 2 & -1 \\
15 & -1 & -1 & 0 & 1 & -5 & 3 & -1 & 0 \\
6 & 2 & 0 & 1 & -1 & 4 & 0 & 2 & -1 \\
21 & 1 & -1 & 1 & 0 & -1 & 3 & 1 & -1 \\
15 & -1 & -1 & 0 & 1 & -5 & 3 & -1 & 0 \\
-21 & -1 & 1 & -1 & 0 & 1 & -3 & -1 & 1 \\
-6 & -2 & 0 & -1 & 1 & -4 & 0 & -2 & 1 \\
21 & 1 & -1 & 1 & 0 & -1 & 3 & 1 & -1 \\
-15 & 1 & 1 & 0 & -1 & 5 & -3 & 1 & 0
\end{bmatrix}
$$

The first two rows form a basis for the row space, and an examination of the character table of S_7 shows that they are also the Brauer characters of the centralizer.

When the inertial quotient acts freely on the defect group, e.g., a Frobenius action, the blocks of the centralizer are nilpotent and the sections of the generalized decomposition matrix for nontrivial u have dimension one [26]. Since the generalized decomposition matrix of a block is nonsingular [38], one of the rows must be a multiple by 1 or -1 of the unique Brauer character. Where, as in these cases of Frobenius action or other cases where we know the exact centralizers, we have succeeded in establishing that our matrix is identical with the generalized decomposition matrix, we label it as such. Mikhal Cohen has been working on increasing the number of blocks for which the generalized decomposition matrix is provided.

Applications

The database has so far been known only to our local research group. The applications to date have been as follows.

Explicit tilting complexes

In [34], we suggested an approach to the blocks of defect $C_p \times C_p$ using decomposition matrices to build up explicit tilting complexes using a sequence of elementary steps and simplifying the resulting mapping cones to get explicit one-step tilting complexes from the Brauer correspondent to the block B. The complexes which have been verified are given in the first part of this paper. The project has foundered on the difficulty with constructing stable equivalences for non-principal blocks. However, as part of the project, Genadi Chereshaya did find tilting complexes which, when applied to the Brauer correspondent b, produce blocks with the same decomposition matrix as B. This was done for various non-principal blocks with defect group $C_3 \times C_3$ and for blocks with defect group $C_5 \times C_5$ that have a maximal number of exceptional characters (i.e., characters producing multiple rows in the decomposition matrix). These have not yet been included in the database because they are still conjectural.

It is particularly easy to construct the conjectural tilting complexes for blocks with a maximal number of exceptional characters, i.e., characters whose multiplicity when restricted to the p' elements is greater than one. For example, when $p = 5$, all the blocks with $k(B) = 14$, $\ell(b) = 6$ have four pairs of exceptional characters. In these cases there is an actual algorithm to calculate the conjectural tilting complex [6]. When this tilting complex is stable under an outer automorphism, then it may also provide a conjectural tilting complex for a block of the group extended by the automorphism.

Morita equivalent families

In addition to the Clifford theory-type Morita equivalences, the categorization in the database has brought to light other structurally determined Morita equivalences, including a whole set of such equivalences between blocks of $\tilde{S}_n$ and blocks of $\tilde{A}_n$, where these are, respectively, the Schur covering groups of the symmetric and alternating groups [18].

The databases

The home page for the entire collection of databases, including a database of character tables, the p' subgroups of $GL(2, p)$ and some extensions, and the database of abelian non-cyclic blocks for 3, 5, 7, 11, and 13:

http://www.cs.biu.ac.il/~mschaps/math.html

In view of the results of Chuang–Rouquier on derived equivalent families of non-abelian defect group, there has been an extension for nonabelian defect groups, but it is not well developed as of this writing. Although the Broué conjecture has been solved for the cyclic blocks, other questions have arisen about the folding of the tilting complexes [36], [31], so we have added a database of cyclic blocks.

We would appreciate it if any further results using these databases would include a reference and that we be informed.

References

[1] J. Alperin, *Local Representation Theory*, Vol. 11, Cambridge Series.
[2] A. Bar-Ilan, *Generalized decomposition numbers in the case of defect group $C_p \times C_p$*, Master's thesis, Bar-Ilan University, 2005.
[3] T. Berrebi, *Normalizers and centralizers for elementary abelian defect groups in almost simple groups*, Master's thesis, Bar-Ilan University, 2005.
[4] M. Broué, Isométries parfaites, types de blocs, catégories dérivées, *Astérisque* **181–182** (1990), 61–92.
[5] M. Broué, Rickard equivalences and block theory, in *Groups '93, Galway/St Andrews, Vol. 1*, London Math. Soc. Series, Vol. 211, Cambridge University Press, 1995, 58–79.
[6] G. Chereshnya, *Conjectural titling complexes for defect group $C_p \times C_p$*, Master's thesis, Bar-Ilan University (2004).

[7] J. Chuang, Derived equivalence in $SL(2, p^2)$, preprint, Oxford, 1998.

[8] J. Chuang, The derived category of some blocks of symmetric groups and a conjecture of Broué, *J. Algebra* **217** (1999), 114–155.

[9] J. Chuang and R. Kessar, Symmetric groups, wreath products, Morita equivalence, and Broué's abelian conjecture, *Bull. London Math. Soc.*, to appear.

[10] J. H. Conway, R. T. Curtis, S. P. Norton, R. A. Parker, and R. A. Wilson, *ATLAS of finite groups*, Clarendon Press, Oxford, 1985.

[11] The GAP group, GAP, version 4.2, http://www-gap.dcs.st-and.ac.uk/~gap.

[12] M. Hassan Ali and M. Schaps, Lifting McKay graphs and relations to prime extension, *Rocky Mountain J. Math.*, to appear 2006.

[13] G. Hiss and R. Kessar, Scopes reduction and Morita equivalence classes of blocks in finite classical groups II, preprint.

[14] G. Hiss and K. Lux, *The Brauer Trees of Sporadic Groups*, Oxford University Press, New York, 1989.

[15] P. N. Hoffman and J. F. Humphreys, *Projective Representations of the Symmetric Groups*, Oxford Mathematical Momographs, Oxford Science Publications, The Clarendon Press, Oxford University Press, New York, 1992.

[16] M. Holloway, *Derived equivalences for group algebras*, thesis, Bristol (2001).

[17] T. Kadaner, *Natural Pointing for Brauer Trees*, Master's thesis, Bar-Ilan University, 2005.

[18] R. Kessar and M. Schaps, Crossover Morita equivalences for blocks of the covering groups of the symmetric and alternating groups, preprint.

[19] B. Külshammer, Crossed products and blocks with normal defect groups, *Comm. Algebra* **13** (1985), 147–168.

[20] R. Leabovich, *Morita equivalence classes for blocks of abelian defect group $C_p \times C_p$*, Master's thesis, Bar-Ilan University (2004).

[21] K. Lux and H. Pahlings, Computational aspects of representation theory of finite groups, in *Representation Theory of Finite Groups and Finite-Dimensional Algebras*, Progress in Mathematics, Birkhäuser, (1991), 37–64.

[22] A. Margolis, Y. Sokolitko, Z. Volgost, Advanced search for groups, http://www.cs.biu.ac.il/~mschaps/math.html

[23] J. Müller and M. Schaps, The Brauer conjecture for the faithful blocks of $4.M_{22}$, preprint.

[24] T. Okuyama, Some examples of derived equivalent blocks of finite groups, preprint, Hokkaido, 1998.

[25] L. Puig, Pointed groups and construction of characters, *Math. Z.* **176** (1981), 209–216.

[26] L. Puig, Une correspondence de modules pour les blocs à groupes de défaut abéliens, *Geom. Dedicata* **37** (1991), 9–43.

[27] W. F. Reynolds, Blocks and normal subgroups of finite groups, *Nagoya Math. J.* **22** (1963), 15–32.

[28] J. Rickard, Morita theory for derived equivalence, *J. London Math. Soc.* **39** (1989), 436–456.

[29] J. Rickard, Splendid equivalences: derived categories and permutation modules, *Proc. London Math. Soc. (3)* **72** (1996), 331–358.

[30] J. Rickard, Equivalences of derived categories for symmetric algebras, *J. Algebra* **257** (2002), no. 2, 460–481.

[31] J. Rickard and M. Schaps, Folded tilting complexes for Brauer tree algebras, *Adv. Math.* **171** (2002), 167–182.

[32] R. Rouquier, From stable equivalences to Rickard equivalences for blocks with cyclic defect, in *Groups '93, Galway/St Andrews, Vol. 2*, London Math. Soc. Series **212**, Cambridge University Press, 1995, 512–523.

[33] R. Rouquier, Block theory via stable and Rickard equivalences, preprint, Paris, 2000.

[34] M. Schaps, *Deformations, tiltings and decomposition matrices*, Fields Institute Publications (2005).

[35] M. Schaps, *Databases for the Broué conjecture*,
http://www.cs.biu.ac.il/~mschaps/math.html.

[36] M. Schaps and E. Zakay-Illouz, Pointed Brauer trees, *J. Algebra* **246** (2001), 647–672.

[37] J. Scopes, Cartan matrices and Morita equivalence for blocks of the symmetric group, *J. Algebra* **142** (1991), 441–455.

[38] J. Thévenaz, *G-algebras and modular representation theory*, Oxford University Press, 1995.

CONJUGACY CLASSES OF p-REGULAR ELEMENTS IN p-SOLVABLE GROUPS

ANTONIO BELTRÁN[†] and MARÍA JOSÉ FELIPE[§] [1]

[†]Departament de matemátiques, Universitat Jaume I de Castlló, Av. Sos Baynat s/n, 12071 Castellón, Spain
Email: `abeltran@mat.uji.es`

[§]Departamento de Matemática Aplicada-IMPA, Universidad Politécnica de Valencia, C/ Camino de Vera s/n, 46022 Valencia, Spain
Email: `mfelipe@mat.upv.es`

We shall assume that any group is finite. One of the classic problems in Group Theory is to study how the structure of a group G determines properties on its conjugacy class sizes and reciprocally how these class sizes influence the structure of G. During the nineties several authors studied this relation by defining and studying two graphs associated to the conjugacy class sizes. In 1990 (see [8]), E. Bertram, M. Herzog and A. Mann defined a graph $\Gamma(G)$ as follows: the vertices of $\Gamma(G)$ are represented by the non-central conjugacy classes of G and two vertices C and D are connected by an edge if $|C|$ and $|D|$ have a common prime divisor. Later, this graph was studied in [12] and also used in [9] to obtain properties on the structure of G when some arithmetical conditions are imposed on the conjugacy class sizes. On the other hand, in 1995, S. Dolfi [14] studied a dual graph, $\Gamma^*(G)$, defined in the following way: the set of vertices are the primes dividing some conjugacy class size of G and two primes r and s are joined by an edge if rs divides some conjugacy class size of G. Independently, G. Alfandary also obtained some properties of these graphs (see [1]).

We shall suppose that G is a p-solvable group for some prime p. We shall consider $\mathrm{Con}(G_{p'})$ the set of p-regular classes in G, that is, the conjugacy classes of p'-elements in G. Recently, several theorems have put forward that certain properties on the sizes of these classes are also reflected on the p-structure of G. In this survey, we analyze two new graphs, $\Gamma_p^*(G)$ and $\Gamma_p(G)$, defined by Z. Lu and J. Zhang in [18] and by A. Beltrán and M.J. Felipe in [2], respectively. Both graphs are associated to $\mathrm{Con}(G_{p'})$ and are defined in a similar way to the ordinary graphs. The vertices of $\Gamma_p(G)$ are the non-central elements in $\mathrm{Con}(G_{p'})$ and two vertices are connected by an edge if their sizes have a common prime divisor. On the other hand, the vertices of $\Gamma_p^*(G)$ are the prime numbers dividing the classes in $\mathrm{Con}(G_{p'})$ and two primes are joined by an edge if there exists some class in $\mathrm{Con}(G_{p'})$ whose size is divisible by both primes.

We would also like to mention the existence of some relation between the p-regular class sizes and the Brauer character degrees (for the prime p), in the same way that there are intimate connections between the set of irreducible character degrees of a group and the set of its conjugacy class sizes. There exist graphs for character degrees defined in a similar way to the ones for class sizes and in fact there

---

[1]Research partially supported by Ministerio de Educación y Ciencia and Fundació Caixa Castelló, Spain.

is a relation between them. For instance, it turns out that the set of primes dividing the character degrees is a subset of the set of primes dividing the class sizes in any group. In 1995, Dolfi proved in [13] that if two primes r and s divide the degree of some irreducible character of G, G being a solvable group, then rs divides some conjugacy class size of G. Consequently, the graph defined on the primes dividing the character degrees is a subgraph of $\Gamma^*(G)$ for any solvable group G. This is an unsolved problem when the solvability hypothesis is dropped. Recently, there has been some interest in studying this kind of relations between the graph $\Gamma^*_p(G)$ and the graph associated to the primes dividing the Brauer character degrees in a p-solvable group (see for instance [7] and [19]).

First we are going to develop the known properties of $\Gamma_p(G)$ and $\Gamma^*_p(G)$. We denote by $n(\Gamma_p(G))$ and $d(\Gamma_p(G))$ the number of connected components and the diameter of $\Gamma_p(G)$, respectively. Analogously, we shall use this notation for the dual graph $\Gamma^*_p(G)$. Lu and Zhang showed in [18] the following properties of $\Gamma^*_p(G)$.

Theorem 1 *Let G be a p-solvable group. Then*
  (a) $n(\Gamma^*_p(G)) \leq 2$.
  (b) *If $n(\Gamma^*_p(G)) = 1$, then $d(\Gamma^*_p(G)) \leq 6$.*
  (c) *If $n(\Gamma^*_p(G)) = 2$, then the diameter of each connected component is ≤ 3.*

Later, Beltrán and Felipe obtained in [2] the following bounds for the graph $\Gamma_p(G)$, which are also the best possible ones.

Theorem 2 *Let G be a p-solvable group. Then*
  (a) $n(\Gamma_p(G)) \leq 2$.
  (b) *If $n(\Gamma_p(G)) = 1$, then $d(\Gamma_p(G)) \leq 3$.*
  (c) *If $n(\Gamma_p(G)) = 2$, then each connected component is a complete graph.*

But in the same paper, the authors also improved the bounds initially given by Lu and Zhang for $\Gamma^*_p(G)$.

Theorem 3 *Let G be a p-solvable group.*
  (a) *If $n(\Gamma^*_p(G)) = 1$, then $d(\Gamma^*_p(G)) \leq 4$.*
  (b) *If $n(\Gamma^*_p(G)) = 2$, then each connected component is a complete graph.*

When $\Gamma^*_p(G)$ is disconnected, 1 is trivially the best bound possible. However, in the connected case it was unknown if the bound 4 for the diameter could indeed be shortened. In [3], it is proved that when $\Gamma^*_p(G)$ is connected, then $d(\Gamma^*_p(G)) \leq 3$, and in fact there exist examples showing that this is the best possible bound. We remark that in order to obtain this bound, some properties of the p-regular classes of maximal size were developed, as well as similar techniques to the ones used for bounding $d(\Gamma_p(G))$. Moreover, in [3] the authors also gave a new and simpler proof of a result of Dolfi (Theorem 17 of [14]) which asserts that $d(\Gamma^*(G)) \leq 3$ for every group G with $\Gamma^*(G)$ connected.

Groups with $\Gamma(G)$ disconnected (or equivalently with $\Gamma^*(G)$ disconnected) were completely characterized in [8] and previously in [16]. This happens if and only if G is a quasi-Frobenius group with abelian kernel and complement. We recall that a group G is said to be *quasi-Frobenius* if $G/\mathbf{Z}(G)$ is Frobenius and then the inverse image in G of the kernel and a complement are called the kernel and complement of G. This problem can be transferred to p-regular classes and the following question arises: What conditions are necessary or sufficient for a group G to have $\Gamma_p(G)$ (or equivalently $\Gamma_p^*(G)$) disconnected?

Following the ordinary case, one could think that if a p-solvable group G has a quasi-Frobenius p-complement with abelian kernel and complement, then $\Gamma_p(G)$ is disconnected. But this is not true. If we consider the standard wreath product $G = P \operatorname{Wr} S_3$, where P is a non-trivial p-group for some prime $p \neq 2, 3$, then the p-complements are Frobenius and $\operatorname{cs}(G_{p'}) = \{1, 2|P|^2, 3|P|\}$, so $\Gamma_p(G)$ is connected.

We fix the following notation. Whenever $\Gamma_p(G)$ is disconnected we shall denote by X_1 and X_2 the two connected components of $\Gamma_p(G)$ and assume that X_2 is the connected component which contains the classes of maximal size of $\operatorname{Con}(G_{p'})$. Let $\pi_i = \{q : q \text{ divides } |C| \text{ for some } C \in X_i \}$ for $i = 1, 2$. It trivially holds $\pi_1 \cap \pi_2 = \emptyset$. We denote by $Z_{p'} = \mathbf{Z}(G)_{p'}$ and by $\operatorname{cs}(G_{p'}) = \{ |C| : C \in \operatorname{Con}(G_{p'})\}$. In [5], Beltrán and Felipe obtained the following sufficient conditions.

Theorem 4 *Let G be a p-solvable group and assume that G has a quasi-Frobenius p-complement H with abelian kernel K and abelian complement T.*

(a) *If T is centralized by some Sylow p-subgroup of G, then $\Gamma_p(G)$ is disconnected and $p \notin \pi_2$.*

(b) *If every element of K is centralized by some Sylow p-subgroup, then $\Gamma_p(G)$ is disconnected and $p \notin \pi_1$.*

On the other hand, some necessary conditions are also obtained in [5].

Theorem 5 *Let G be a p-solvable group. Suppose that $\Gamma_p(G)$ is disconnected.*

(a) *If $p \notin \pi_1 \cup \pi_2$, then*

     1) *$G = P \times H$, with $P \in \operatorname{Syl}_p(G)$ and H a p-complement which is quasi-Frobenius with kernel K and complement T, both abelian.*

     2) *$\operatorname{cs}(G_{p'}) = \{1, |K/Z_{p'}|, |T/Z_{p'}|\}$.*

(b) *If $p \in \pi_1$, then*

     1) *G is p-nilpotent and its normal p-complement, H, is a quasi-Frobenius group, with abelian kernel K and abelian complement T, and such that $P \subseteq \mathbf{C}_G(T)$ for some $P \in \operatorname{Syl}_p(G)$. Moreover, $\mathbf{Z}(H) = Z_{p'}$.*

     2) *$\operatorname{cs}(G_{p'}) = \{1, |K/Z_{p'}|, |T/Z_{p'}|p^i\}$, where $i \in I \subseteq \mathbb{N}$, $I \neq \emptyset, \{0\}$.*

(c) *If $p \in \pi_2$ and G has an abelian Hall π-subgroup K where $\pi = \pi_2 - \{p\}$, then*

     1) *$K \trianglelefteq G$ and each $x \in K$ centralizes some Sylow p-subgroup of G.*

     2) *If H is a p-complement of G, then $\mathbf{Z}(H) = Z_{p'}$ and H is quasi-Frobenius with abelian kernel $KZ_{p'}$ and abelian complement T.*

     3) *$\operatorname{cs}(G_{p'}) = \{1, |T/Z_{p'}|, |KZ_{p'}/Z_{p'}|p^i\}$, where $i \in I \subseteq \mathbb{N}$, $I \neq \emptyset, \{0\}$.*

(d) *If $p \in \pi_2$ and $|\pi_2| \geq 3$, then*

1) *If H is a p-complement of G, then H is quasi-Frobenius with kernel K, normal in G, and complement T, both abelian. Moreover, $Z_{p'} = \mathbf{Z}(H)$.*

2) *G has a normal Sylow p-subgroup P, such that $P \subseteq \mathbf{C}_G(K)$.*

3) *$\mathrm{cs}(G_{p'}) = \{1, |T/Z_{p'}|, |K/Z_{p'}|p^i\}$, where $i \in I \subseteq \mathbb{N}$, $I \neq \emptyset, \{0\}$.*

It is proved that if $\Gamma_p^*(G)$ is disconnected, then the case $\pi_2 = \{p\}$ cannot happen. Thus, in view of the above results, it only remains one case to be studied: $\pi_2 = \{p, q\}$. The following questions arise in a natural way in this case:

i) Are the p-complements of G quasi-Frobenius groups with normal abelian kernel and abelian complement?

ii) Does G possess a normal Sylow p-subgroup?

iii) Is there any Sylow p-subgroup in G centralizing the kernel of a p-complement?

It is known that i) is true when G is p-nilpotent or when G has abelian Sylow q-subgroups, but in general for $\pi_2 = \{p, q\}$ this question is still open. On the other hand, ii) and iii) are false. The following group is a counterexample: Let G be the affine semilinear group $A\Gamma(q^{p^r})$, where p and q are distinct primes with $p \nmid (q-1)$ and $r \in \mathbb{Z}^+$. Then $\Gamma_p(G)$ is disconnected and G has normal p-complement which is a Frobenius group. But G has not a normal Sylow p-subgroup and there is no Sylow p-subgroup centralizing the kernel of the p-complement, as it occurs in case (d) of the above theorem.

Next, we are going to show some theorems on the structure of a p-solvable group, or of its p-complements, under certain arithmetical conditions on the p-regular class sizes. These results generalize known properties in the ordinary case, but we want to notice that this generalization may present some difficulties since for instance, if H is a p-complement of G then the class size of x in H may not divide the class size of x in G. We remark that in the proofs of some of these results for p-regular classes, new and more simplified proofs of the analogous theorems for ordinary classes are obtained. Furthermore, certain properties of the graphs $\Gamma_p(G)$ and $\Gamma_p^*(G)$ are used.

Let us denote by $\mathrm{cs}(G)$ the set of conjugacy class sizes of G. A classic result of N. Itô (see [17] or Theorem 33.6 of [15]) asserts that if G is a group with $\mathrm{cs}(G) = \{1, m\}$, then $m = q^a$ for some prime q and $G = Q \times A$ with $Q \in \mathrm{Syl}_q(G)$ and $A \subseteq \mathbf{Z}(G)$. In [4], the following extension appears.

Theorem 6 *Let G be a p-solvable group such that $\mathrm{cs}(G_{p'}) = \{1, m\}$. Then one of the following holds:*

(a) *$m = p^a$ and G has abelian p-complements.*

(b) *$m = q^b$, for some prime $q \neq p$, and G is nilpotent with all its Sylow subgroups abelian except at most for the prime q.*

(c) *$m = p^a q^b$, $q \neq p$ and $G = PQ \times A$, with $P \in \mathrm{Syl}_p(G)$, $Q \in \mathrm{Syl}_q(G)$ and $A \subseteq \mathbf{Z}(G)$.*

The following theorem (Theorem A of [6]) extends the main result of [10] for ordinary classes, by taking a prime p not dividing $|G|$.

Theorem 7 *Suppose that G is a p-solvable group. Let m and n be the two maximal sizes in $\text{Con}(G_{p'})$ with $m, n > 1$. Suppose that $(m, n) = 1$ and that p is not a prime divisor of m. Then G is solvable and*

(a) $\text{cs}(G_{p'}) = \{1, m, n\}$;

(b) *a p-complement of G is a quasi-Frobenius group with abelian kernel and complement. Furthermore, its conjugacy class sizes are $\{1, m, n_{p'}\}$.*

This result allows to obtain the structure of the p-complements of G when G has exactly two non-central p-regular class sizes which are coprime numbers.

Corollary 8 *Let G be a p-solvable group and suppose that $\text{cs}(G_{p'}) = \{1, m, n\}$ with $(m, n) = 1$, $m, n > 1$. Then G is solvable and the p-complements of G are quasi-Frobenius with abelian kernel and abelian complement.*

As an application of this corollary, among other results, in [6] it is also obtained the structure of p-solvable groups whose non-central p-regular class sizes are consecutive integers. The corresponding result for ordinary classes was already obtained by M. G. Bianchi et al. in 1992 (see [9]).

Theorem 9 *Let G be a p-solvable group and suppose that $\text{cs}(G_{p'}) = \{1, n, n + 1, \ldots, n + r\}$. Then one of the following possibilities holds:*

(a) $r = 0$, $n = p^a$, *for some $a \in \mathbb{Z}^+$ and G has abelian p-complements.*

(b) $r = 0$, $n = p^a q^b$, *for some prime $q \neq p$, $a \geq 0$ and $b \geq 1$, and $G = PQ \times A$, where $P \in \text{Syl}_p(G)$, $Q \in \text{Syl}_q(G)$, $A \leq \mathbf{Z}(G)$. Moreover, if $a = 0$, then $G = P \times Q \times A$.*

(c) $r = 1$ *and each p-complement of G is quasi-Frobenius with abelian kernel and complement. Furthermore, if p does not divide n, then G is p-nilpotent. If also $p \nmid n + 1$, then $G = P \times H$, where H is the p-complement of G and $P \in \text{Syl}_p(G)$.*

Finally, the next theorem (Theorem D of [6]) determines the structure of those groups having prime powers as p-regular class sizes, so it extends Theorem 2 and Corollary 2.2 of [11]. However, the proof is based on certain properties of the graph $\Gamma_p(G)$ and thus, it sufficiently differs from the proof of the mentioned results for ordinary conjugacy classes.

Theorem 10 *Let G be a p-solvable group. Then $|C|$ is a prime power for all $C \in \text{Con}(G_{p'})$ if and only if one of the following properties holds:*

(a) G *has abelian p-complements. This happens if and only if $|C|$ is a p-power for all $C \in \text{Con}(G_{p'})$.*

(b) G *is nilpotent with all its Sylow r-subgroups abelian for all primes $r \neq p, q$ and some prime $q \neq p$. This occurs if and only if $|C|$ is a q-power for all $C \in \text{Con}(G_{p'})$.*

(c) $G = P \times H$, *where $P \in \text{Syl}_p(G)$ and H is a p-complement of G, which is quasi-Frobenius with abelian kernel and complement. This happens if and only if $\text{cs}(G_{p'}) = \{1, q^n, r^m\}$, where $q \neq p \neq r$, and $n, m \in \mathbb{Z}^+$.*

References

[1] G. Alfandary, On graphs related to conjugacy classes, *Israel J. Math.* **86** (1994) 211–220.

[2] A. Beltrán and M. J. Felipe, On the diameter of a p-regular conjugacy class graph of finite groups, *Comm. Algebra* **30** (2002), no. 12, 5861–5873.

[3] A. Beltrán and M. J. Felipe, On the diameter of a p-regular conjugacy class graph of finite groups II, *Comm. Algebra* **31** (2003), no. 9, 4393–4403.

[4] A. Beltrán and M. J. Felipe, Finite groups with two p-regular conjugacy class lengths, *Bull. Austral. Math. Soc.* **67** (2003), 163–169.

[5] A. Beltrán and M. J. Felipe, Finite groups with a disconnected p-regular conjugacy class graph, *Comm. Algebra* **32** (2004), no. 9, 3503–3516.

[6] A. Beltrán and M. J. Felipe, Certain relations between p-regular class sizes and the p-structure of p-solvable groups, *J. Austral. Math. Soc.* **77** (2004), 387–400.

[7] A. Beltrán and M. J. Felipe, Prime factors of π-partial character degrees and conjugacy class lengths of π-elements, *Algebra Colloq.* **12** (2005), 699–707.

[8] E. A. Bertram, M. Herzog and A. Mann, On a graph related to conjugacy classes of groups, *Bull. London Math. Soc.* **22** (1990), 569–575.

[9] M. Bianchi, C. Chillag, A. G. B. Mauri, M. Herzog and C. M. Scoppola, Applications of a graph related to conjugacy classes in finite groups, *Arch. Math.* **58** (1992), 126–132.

[10] M. Bianchi, A. Gillio and C. Casolo, A note on conjugacy class sizes of finite groups, *Rend. Sem. Mat. Univ. Padova* **106** (2001), 255–260.

[11] D. Chillag and M. Herzog, On the lengths of the conjugacy classes of finite groups, *J. Algebra* **131** (1990), 110–125.

[12] D. Chillag, M. Herzog and A. Mann, On the diameter of a graph related to conjugacy classes of groups, *Bull. London Math. Soc.* **25** (1993), 255–262.

[13] S. Dolfi, Prime factors of conjugacy class lengths and irreducible character degrees, *J. Algebra* **174** (1995), 749–752.

[14] S. Dolfi, Arithmetical conditions on the length of the conjugacy classes of a finite group, *J. Algebra* **174** (1995), 753–771.

[15] B. Huppert, *Character theory of finite groups*, De Gruyter Expositions in Mathematics **25** (De Gruyter and Co., Berlin, New York, 1998).

[16] S. L. Kazarin, On groups with isolated conjugacy classes, *Izv. Vyssh. Uchebn. Zaved. Mat.* 1981, no. 7, 40–45.

[17] N. Itô, On finite groups with given conjugate types I, *Nagoya Math.* **6** (1953), 17–28.

[18] Z. Lu and J. Zhang, On the diameter of a graph related to p-regular conjugacy classes of finite groups, *J. Algebra* **231** (2000), 705–712.

[19] Z. Lu and J. Zhang, Irreducible modular characters and p-regular conjugacy classes, *Algebra Colloq.* **8**, (2001), no. 1, 55–61.

AN ALGORITHM FOR THE UNIT GROUP OF THE BURNSIDE RING OF A FINITE GROUP

ROBERT BOLTJE[*][1] and GÖTZ PFEIFFER[†]

[*]Department of Mathematics, University of California, Santa Cruz, CA 95064, U.S.A.
[†]National University of Ireland, Galway, Ireland
Email: `Goetz.Pfeiffer@nuigalway.ie`

Abstract

In this note we present an algorithm for the construction of the unit group of the Burnside ring $\Omega(G)$ of a finite group G from a list of representatives of the conjugacy classes of subgroups of G.

1 Introduction

Let G be a finite group. The Burnside ring $\Omega(G)$ of G is the Grothendieck ring of the isomorphism classes $[X]$ of the finite left G-sets X with respect to disjoint union and direct product. It has a $\mathbb{Z}$-basis consisting of the isomorphism classes of the transitive G-sets G/H, where H runs through a system of representatives of the conjugacy classes of subgroups of G.

The ghost ring of G is the set $\tilde{\Omega}(G)$ of functions f from the set of subgroups of G into $\mathbb{Z}$ which are constant on conjugacy classes of subgroups of G. For any finite G-set X, the function ϕ_X which maps a subgroup H of G to the number of its fixed points on X, i.e., $\phi_X(H) = \#\{x \in X : h.x = x \text{ for all } h \in H\}$, belongs to $\tilde{\Omega}(G)$. By a theorem of Burnside, the map $\phi : [X] \to \phi_X$ is an injective homomorphism of rings from $\Omega(G)$ to $\tilde{\Omega}(G)$. We identify $\Omega(G)$ with its image under ϕ in $\tilde{\Omega}(G)$, i.e., for $x \in \Omega(G)$, we write $x(H) = \phi(x)(H) = \phi_H(x)$.

The ghost ring has a natural $\mathbb{Z}$-basis consisting of the characteristic functions of the conjugacy classes of subgroups of G. The table of marks of G is defined as the square matrix $M(G)$ which records the coefficients when the transitive G-sets G/H are expressed as linear combinations of the characteristic functions. If G has r conjugacy classes of subgroups $M(G)$ is an $r \times r$ matrix over $\mathbb{Z}$ which is invertible over $\mathbb{Q}$.

Let $H_1, \ldots, H_r$ be representatives of the conjugacy classes of subgroups of G. Then we can further identify the ghost ring $\tilde{\Omega}(G)$ with $\mathbb{Z}^r$, where, for $x \in \tilde{\Omega}(G)$, we set $x_i = x(H_i)$, $i = 1, \ldots, r$. For $x \in \mathbb{Z}^r$, the product $xM(G)^{-1}$ yields the multiplicities of the transitive G-sets G/H_i in x. An element $x \in \tilde{\Omega}(G)$ thus lies in $\Omega(G)$ if and only if $xM(G)^{-1}$ consists of integers only.

The units of the ghost ring are $\{\pm 1\}^r \subseteq \mathbb{Z}^r$. We want to determine those ± 1-vectors which are contained in the image of $\Omega(G)$ in $\mathbb{Z}^r$. Of course every such vector can be tested with the table of marks. But this task grows exponentially with the number r of conjugacy classes.

---

[1]The first author is supported by the NSF, Grant 0200592 and 0128969.

A formula for the order of the unit group $\Omega^*(G)$ of $\Omega(G)$ in terms of normal subgroups of G has been given by Matsuda [2]. The following result of Yoshida [5] gives a necessary and sufficient condition, which will allow us to explicitly calculate a basis of $\Omega^*(G)$.

Theorem 1.1 *Let u be a unit in $\tilde{\Omega}(G)$. Then $u \in \Omega(G)$ if and only if, for every subgroup $H \leq G$, the function $\mu^H : N_G(H) \to \mathbb{C}$ defined by $\mu^H(n) = u(H \langle n \rangle)/u(H)$ is a linear character of $N_G(H)$.*

Here $u(H \langle n \rangle)$ is the value of u at the preimage in $N_G(H)$ of the cyclic subgroup of $N_G(H)/H$ generated by the coset Hn. The Theorem follows from a more general characterization of elements of the ghost ring which lie in the Burnside ring by certain congruences.

2 The algorithm

Let E be an elementary abelian 2-group of order 2^m, generated by $e_1, \ldots, e_m$. Every linear character λ of E is determined by its values on the e_i, which in turn can be chosen, independently, to be $+1$ or -1.

Given a subgroup $H \leq G$, to say that μ^H is a linear character of $N = N_G(H)$ amounts to the following. First, let $R \leq N$ be the minimal subgroup such that $H \leq R$ and N/R is an elementary abelian 2-group. Since μ^H has only values ± 1 it must have R in its kernel and can be regarded as a character of the elementary abelian 2-group $E := N/R$. Let $e_1, \ldots, e_m$ be a basis of E.

Let $n \in N$ and consider the coset $Hn \in N/H$. The element $Rn \in E$ can be expressed in a unique way as linear combination $Rn = e_1^{\alpha_1} \cdots e_m^{\alpha_m}$, with $\alpha_k \in \{0, 1\}$, $k = 1, \ldots, m$.

Let λ be a linear character of E. Then λ is determined by the values $\lambda(e_k)$, $k = 1, \ldots, m$ and $\lambda(Rn) = \lambda(e_1)^{\alpha_1} \cdots \lambda(e_m)^{\alpha_m}$.

Now μ^H is a linear character if and only if $\mu^H = \lambda$ for some choice of the values $\lambda(e_k)$, $k \in 1, \ldots, m$, i.e., $\mu^H(n) = \lambda(Rn) = \lambda(e_1)^{\alpha_1} \cdots \lambda(e_m)^{\alpha_m}$. Thus u must satisfy

$$u(H \langle n \rangle)/u(H) = \lambda(e_1)^{\alpha_1 (n)} \cdots \lambda(e_m)^{\alpha_m (n)}. \tag{1}$$

Let $p, q \in \{1, \ldots, r\}$ be such that H is a conjugate of H_p and $H \langle n \rangle$ is a conjugate of H_q. Then (1) can be written as a linear equation over $GF(2)$ in the unknowns $l_1, \ldots, l_m$ (such that $\lambda(e_k) = (-1)^{l_k}$, $k = 1, \ldots, m$), and $v_1, \ldots, v_r$ (such that $u(H_i) = (-1)^{v_i}$, $i = 1, \ldots, r$) as

$$\alpha_1 l_1 + \cdots + \alpha_m l_m + v_p + v_q = 0. \tag{*}$$

For a given subgroup $H \leq G$, each coset $Hn \in N/H$ contributes one such equation; conjugate elements of N/H of course yield the same equation. Since n can be chosen such that $Rn = e_k$, the system contains equations of the form

$$l_k + v_p + v_q = 0,$$

which allow us to express the l_k in terms of the v_i, for all $k = 1, \ldots, m$. What remains, for each subgroup H, is a (possibly trivial) system of homogeneous equations in the v_i only, which we denote by $\mathcal{E}(H)$. Of course, conjugate subgroups give rise to the same system of equations. The following theorem is now immediate.

Theorem 2.1 $u \in \Omega(G)$ *if and only if, for each subgroup $H \leq G$, it satisfies the conditions $\mathcal{E}(H)$.*

The algorithm is based on Theorem 2.1. Given a list $H_1, H_2, \ldots H_r$ of representatives of subgroups of G, the following steps are taken for each $H = H_i$, $i = 1, \ldots, r$.

1. Let $N = N_G(H)$ and $Q = N/H$. Let q_j, $j = 1, \ldots, l$, be representatives of the conjugacy classes of Q and let $C_j = H \langle q_j \rangle$ be the subgroup of G corresponding to the cyclic subgroup of Q generated by q_j. Then C_j is a conjugate of some H_k and $u(C_j) = u(H_k)$ for all $u \in \mathbb{Z}^r$.

2. Let $H \leq R \leq N$ be such that $E := N/R$ is the largest elementary abelian 2-quotient of N/H. Inside G, this subgroup R can be found as closure of H, the derived subgroup N' and the squares g^2 of all generators g of N.

3. Regard E as a $GF(2)$-vector space and find a basis $e_1, \ldots, e_m$. (This requires a search through the elements of E until a large enough linearly independent set has been found.) Now every element $e \in E$ can be described as a unique linear combination $e = \alpha_1 e_1 + \cdots + \alpha_m e_m$ of the basis elements with $\alpha_i \in \{0, 1\}$. In particular, for every representative q_j, we get such a decomposition of the coset $Rq_j \in E$.

4. For each q_j write down its equation $(*)$. Then eliminate the unknowns l_k to yield $\mathcal{E}(H)$.

Finally, it remains to solve the system $\bigcup_{i=1}^r \mathcal{E}(H_i)$: its nullspace corresponds to the group of units $\Omega^*(G)$.

3 Examples

Theorem 1.1 can be used to determine the units of the Burnside ring of an abelian group. The order of the unit group in the following theorem agrees with Matsuda's formula [2, Example 4.5].

Theorem 3.1 *If G is a finite abelian group whose largest elementary abelian 2-quotient has order 2^n, then $|\Omega^*(G)| = 2^{2^n}$. In particular, if G is an elementary abelian 2-group of order 2^n then $|\Omega^*(G)| = 2^{2^n}$.*

Proof Let $N_1, \ldots, N_{2^n-1} \leq G$ be the (maximal) subgroups of index 2 in G and define $\lambda_i \in \mathbb{Z}^r$ for $i = 1, \ldots, 2^n - 1$ as

$$\lambda_i(H) = \begin{cases} +1 & \text{if } H \leq N_i, \\ -1 & \text{otherwise.} \end{cases} \tag{2}$$

Furthermore set $\lambda_G := \prod_{i=1}^{2^n-1} \lambda_i$ if $n \geq 1$ and $\lambda_G := -1$ if $n = 0$. Then

$$\lambda_G(H) = \prod_{i=1}^{2^n-1} \lambda_i(H) = \begin{cases} -1 & \text{if } H = G, \\ +1 & \text{if } H \in \{N_1, \ldots, N_{2^n-1}\}. \end{cases} \tag{3}$$

We claim that the 2^n units $\mathcal{B} = \{-\lambda_i : i = 1, \ldots, 2^n - 1\} \cup \{\lambda_G\}$ form a basis of $\Omega^*(G)$.

First, we show that $\lambda_i \in \Omega(G)$. Fix $H \leq G$ and denote by μ^H the function $N_G(H) \to \mathbb{C}$ as defined in Theorem 1.1 for $u = \lambda_i$. Now, if $H \not\leq N_i$ then $U \not\leq N_i$ for all U with $H \leq U \leq G$. Hence $\lambda_i(U)/\lambda_i(H) = 1$ for all such U, i.e, μ^H is the trivial character of G/H. And if $H \leq N_i$ then μ^H is the linear character of G/H with kernel N_i/H. In any case, μ^H is a linear character of G/H, and from Theorem 1.1 then follows that $\lambda_i \in \Omega(G)$. Together with $-1 \in \Omega(G)$ this yields $\mathcal{B} \subseteq \Omega(G)$.

Next, note that $\mathcal{B}$ is linearly independent. For each such function, restricted to $\{N_i : i = 1, \ldots, 2^n - 1\} \cup \{G\}$ has exactly one value equal to -1.

Finally, every unit $u \in \Omega^*(G)$ is a linear combination of the $-\lambda_i$, $i = 1, \ldots, 2^n-1$, and λ_G. For the values of u at $\{N_i : i = 1, \ldots, 2^n - 1\} \cup \{G\}$ determine a unique linear combination v of $\mathcal{B}$ which coincides with u on $\{N_i : i = 1, \ldots, 2^n - 1\} \cup \{G\}$. Now it suffices to show that for every subgroup $H \leq G$ with $|G : H| > 2$ and for every unit w of $\Omega(G)$, the value $w(H)$ is already determined by the values $w(U)$ for subgroups U of G with $H < U$. To see this note that there must exist a subgroup U of G containing H such that U/H is either of odd prime order, or cyclic of order 4, or elementary abelian of order 4. From Theorem 1.1 we obtain a linear character μ^H on U/H with values ± 1. In the first case, this character is trivial which implies $w(H) = w(U)$. In the second case this character must be trivial on the subgroup V/H of U/H of order 2. This implies $w(H) = w(V)$. In the third case, observe that every linear character μ of U/H satisfies $\mu(U_1/H)\mu(U_2/H)\mu(U_3/H) = 1$, where $U_1/H, U_2/H, U_3/H$ are the subgroups of order 2 of U/H. This implies $w(H) = w(U_1)w(U_2)w(U_3)$. □

The argument which shows the linear independence of the set $\mathcal{B}$ is still valid in a general 2-group. Thus $\operatorname{rk} \Omega^*(G) \geq 2^n$ for any 2-group G with $|G/\Phi(G)| = 2^n$. It may however happen that $|N_G(H) : H| < 4$ and then the argument which shows that $\mathcal{B}$ spans the unit group breaks down. In fact, if G is the dihedral group of order 8 then $|G/\Phi(G)| = 4$ but $\operatorname{rk} \Omega^*(G) = 5$.

Let A be a finite abelian group of odd order, and let $i : A \to A$ be the automorphism of A which maps every element to its inverse, $i(a) = a^{-1}$, $a \in A$. Then let G be the semidirect product of A and $\langle i \rangle$. The conjugacy classes of subgroups of G are easy to describe in terms of the subgroups of A. For every subgroup N of A there are two conjugacy classes of subgroups of G. One consists of N only, since N is normal in G, and the other consists of $|A : N|$ conjugates of $\langle N, i \rangle$, which is a self-normalizing subgroup of G.

Let $u \in \Omega^*(G)$. It follows from Theorem 1.1 and the fact that the normalizer of every $N \leq A$ is G, that u is constant on $\{N : N \leq A\}$. Moreover, it is easy to see that for every $N \leq A$, the function $u_N \in \tilde{\Omega}(G)$ defined by

$$u_N(H) = \begin{cases} -1 & \text{if } H =_G \langle N, i \rangle, \\ 1 & \text{otherwise,} \end{cases}$$

is a unit in $\Omega(G)$. Thus $\mathrm{rk}_{\mathrm{GF}(2)} \Omega^*(G) = r + 1$, where r is the number of subgroups of A.

An implementation of the algorithm from section 2 in the GAP system for computational discrete algebra [4] allows us to calculate $\Omega^*(G)$ for particular groups G, given a list of representatives of the conjugacy classes of subgroups of G. GAP contains programs to calculate such a list for small groups. A procedure for the construction of a list of representatives of classes of subgroups (as well as the complete table of marks) of almost simple groups G has been described in [3].

The following table shows some of the results obtained.

G	$\mathrm{rk}\,\Omega^*(G)$	G	$\mathrm{rk}\,\Omega^*(G)$	G	$\mathrm{rk}\,\Omega^*(G)$	G	$\mathrm{rk}\,\Omega^*(G)$
A_3	1	S_3	3	M_{11}	18	J_1	15
A_4	2	S_4	6	M_{12}	49	J_2	38
A_5	5	S_5	10	M_{22}	59		
A_6	12	S_6	23				
A_7	20	S_7	34				
A_8	44	S_8	67				
A_9	66	S_9	110				

4 A conjecture

Let $\Omega_2(G)$ be the ring of monomial representations of G which are induced from linear representations of subgroups which have values ± 1 only. Then $\Omega_2(G)$ is a subring of the ring of all monomial representations of G containing the Burnside ring $\Omega(G)$. It has a basis labeled by the conjugacy classes of pairs (H, λ), where λ is a linear character of H with $\lambda(h) = \pm 1$ for all $h \in H$, or equivalently labeled by the conjugacy classes of pairs (H, K) where $K \leq H$ is such that $|H : K| \leq 2$ (corresponding to the kernel of λ).

Conjecture 4.1 *Let G be a finite group. Then*

$$\mathrm{rk}\,\Omega^*(G) - 1 \leq \mathrm{rk}\,\Omega_2(G) - \mathrm{rk}\,\Omega(G).$$

Using a result of Dress, the conjecture would imply immediately that any group G of odd order is solvable. For, if $|G|$ is odd no subgroup of G has a non-trivial linear character with values ± 1 or, equivalently, a subgroup of index 2. Hence $\mathrm{rk}\,\Omega_2(G) = \mathrm{rk}\,\Omega(G)$ and thus $\Omega^*(G) = \{\pm 1\}$. But if $\Omega(G)$ contains no non-trivial units, then it contains no non-trivial idempotents either (because a non-trivial idempotent e yields a non-trivial unit $2e - 1$). Solvability of G then follows by Dress's characterisation of solvable groups [1].

The formula clearly holds for 2-groups: if G is a 2-group then every non-trivial subgroup $H \leq G$ has a subgroup of index 2, whence $\Omega_2(G) - \mathrm{rk}\,\Omega(G) \geq \mathrm{rk}\,\Omega(G) - 1$. On the other hand, one always has $\mathrm{rk}\,\Omega^*(G) \leq \mathrm{rk}\,\Omega(G)$.

Of course most often a nontrivial subgroup H has many more than just one subgroup of index 2. In fact, for an elementary abelian group G of order 2^n one has

$$\mathrm{rk}\,\Omega_2(G) - \mathrm{rk}\,\Omega(G) = [n]_2 \sum_{k=0}^{n-1} \begin{bmatrix} n-1 \\ k \end{bmatrix}_2,$$

where $[k]_q = \frac{1-q^k}{1-q}$ and $[k]_q! = [1]_q[2]_q \cdots [k]_q$ and $\begin{bmatrix} n \\ k \end{bmatrix}_q = \frac{[n]_q!}{[k]_q![n-k]_q!}$. Thus, in this case, $\mathrm{rk}\,\Omega_2(G) - \mathrm{rk}\,\Omega(G)$ is a large multiple of $\mathrm{rk}\,\Omega^*(G) - 1 = [n]_2$. It follows from Theorem 3.1 that the conjecture is true for abelian groups. In fact, if G has odd order, this is clear; and if G has even order, let G/N be the largest elementary abelian 2-factor group and assume it has order 2^n. Then, using Theorem 3.1,

$$\mathrm{rk}\,\Omega^*(G) - 1 = |G/N| - 1 = [n]_2$$

$$\leq [n]_2 \sum_{k=0}^{n-1} \begin{bmatrix} n-1 \\ k \end{bmatrix}_2 = \mathrm{rk}\,\Omega_2(G/N) - \mathrm{rk}\,\Omega(G/N)$$

$$\leq \mathrm{rk}\,\Omega_2(G) - \mathrm{rk}\,\Omega(G),$$

where the last inequality follows from the fact that to each pair of subgroups $K/N \leq H/N$ of G/N such that K/N has index 2 in H/N corresponds at least one such pair (namely $K \leq H$) of subgroups of G.

The Feit–Thompson Theorem implies the conjecture for groups of odd order. Clearly there are no subgroups of index 2 in a group of odd order. Moreover, such a group admits only the trivial units in its Burnside ring, see Lemma 6.7 [5].

If G is the semidirect product of an abelian group A of odd order and the inversion i, we have seen in Section 3 that $\mathrm{rk}_{GF(2)}\,\Omega^*(G) = r + 1$, where r is the number of subgroups of A. Now each subgroup N of A occurs as a subgroup of index 2 in $\langle N, i \rangle$. It follows that $\mathrm{rk}\,\Omega_2(G) - \mathrm{rk}\,\Omega(G) = r$. So this class of groups provides infinitely many examples where the inequality in the conjecture becomes an equality. The only other known such example is the alternating group A_5.

In a slightly more general situation, let us suppose G has order $2m$ for an odd $m \in \mathbb{N}$. Then, using Feit–Thompson, G is solvable. Moreover, $\mathrm{rk}_{GF(2)}\,\Omega^*(G)$ equals the number of representatives H of conjugacy classes of subgroups of G which have no normal subgroup of index p for an odd prime p, see again Lemma 6.7 [5]. On the other hand $\mathrm{rk}\,\Omega_2(G) - \mathrm{rk}\,\Omega(G) = r$ equals the number of representatives H of conjugacy classes of subgroups of G which have a normal subgroup of index 2. Since, in a solvable group, every nontrivial subgroup has a normal subgroup of prime index, each representative which has no normal subgroup of index p for an odd prime p must have one of index 2. This shows the conjecture in that case.

If G is a solvable group, it is still true that $\mathrm{rk}\,\Omega^*(G)$ is less than or equal to the number of representatives H of conjugacy classes of subgroups of G which have no normal subgroup of index p for an odd prime p. Such a representative (except for the trivial subgroup) then has a normal subgroup of index 2. And on the other hand $\operatorname{rk}\Omega_2(G) - \operatorname{rk}\Omega(G) = r$ is greater or equal to the number of representatives H of conjugacy classes of subgroups of G which have a normal subgroup of index 2. This verifies the conjecture for all solvable groups G.

Does Feit–Thompson imply the conjecture for all finite groups G?

In general it seems that the larger the group the larger the difference between the two quantities. This is illustrated by the following table, if compared with the table in section 3.

G	$\operatorname{rk}\Omega_2(G)$ $-\operatorname{rk}\Omega(G)$	G	$\operatorname{rk}\Omega_2(G)$ $-\operatorname{rk}\Omega(G)$	G	$\operatorname{rk}\Omega_2(G)$ $-\operatorname{rk}\Omega(G)$	G	$\operatorname{rk}\Omega_2(G)$ $-\operatorname{rk}\Omega(G)$
A_3	0	S_3	2	M_{11}	36	J_1	29
A_4	2	S_4	11	M_{12}	221	J_2	178
A_5	4	S_5	19	M_{22}	217		
A_6	14	S_6	82	M_{23}	243		
A_7	27	S_7	153	M_{24}	5512		
A_8	199	S_8	699				
A_9	305	S_9	1328				

Moreover, the conjecture has been verified for all groups of order less than 960.

Acknowledgement   Most of the work leading to this paper was done when the authors were visiting the Centre Interfacultaire Bernoulli at the EPFL in Lausanne, Switzerland. Both authors would like to express their gratitude for the Institute's hospitality.

References

[1] Andreas Dress, A characterisation of solvable groups, *Math. Z.* **110** (1969), 213–217. MR 0248239 (40 #1491)
[2] Toshimitsu Matsuda, On the unit groups of Burnside rings, *Japan. J. Math. (N.S.)* **8** (1982), no. 1, 71–93. MR 722522 (85b:57047)
[3] Götz Pfeiffer, The subgroups of M_{24}, or how to compute the table of marks of a finite group, *Experiment. Math.* **6** (1997), no. 3, 247–270. MR 1481593 (98h:20032)
[4] Martin Schönert et al., *GAP – Groups, Algorithms, and Programming*, Lehrstuhl D für Mathematik, Rheinisch Westfälische Technische Hochschule, Aachen, Germany, fifth ed., 1995, Home page: http://www.gap-system.org.
[5] Tomoyuki Yoshida, On the unit groups of Burnside rings, *J. Math. Soc. Japan* **42** (1990), no. 1, 31–64. MR 1027539 (90j:20027)

INTEGRAL GROUP RING OF THE FIRST MATHIEU SIMPLE GROUP

VICTOR BOVDI* and ALEXANDER KONOVALOV†1

*Institute of Mathematics, University of Debrecen, P.O. Box 12, H-4010 Debrecen; Institute of Mathematics and Informatics, College of Nyíregyháza, Sóstói út 31/b, H-4410 Nyíregyháza, Hungary

†Department of Mathematics, Zaporozhye National University, 66 Zhukovskogo str., 69063, Zaporozhye, Ukraine; current address: Department of Mathematics, Vrije Universiteit Brussel, Pleinlaan 2, B-1050 Brussel, Belgium
Email: alexander.konovalov@gmail.com

Abstract

We investigate the classical Zassenhaus conjecture for the normalized unit group of the integral group ring of the simple Mathieu group M_{11}. As a consequence, for this group we confirm the conjecture by Kimmerle about prime graphs.

1 Introduction and main results

Let $V(\mathbb{Z}G)$ be the normalized unit group of the integral group ring $\mathbb{Z}G$ of a finite group G. The following famous conjecture was formulated by H. Zassenhaus in [15]:

Conjecture 1 (ZC) Every torsion unit $u \in V(\mathbb{Z}G)$ is conjugate within the rational group algebra $\mathbb{Q}G$ to an element of G.

This conjecture is already confirmed for several classes of groups but, in general, the problem remains open, and a counterexample is not known.

Various methods have been developed to deal with this conjecture. One of the original ones was suggested by I. S. Luthar and I. B. S. Passi [12, 13], and it was improved further by M. Hertweck [9]. Using this method, the conjecture was proved for several new classes of groups, in particular for S_5 and for some finite simple groups (see [4, 9, 10, 12, 13]).

The Zassenhaus conjecture appeared to be very hard, and several weakened variations of it were formulated (see, for example, [3]). One of the most interesting modifications was suggested by W. Kimmerle [11]. Let us briefly introduce it now.

Let G be a finite group. Denote by $\#(G)$ the set of all primes dividing the order of G. Then the *Gruenberg–Kegel graph* (or the *prime graph*) of G is a graph $\pi(G)$ with vertices labelled by primes from $\#(G)$, such that vertices p and q are adjacent if and only if there is an element of order pq in the group G. Then the conjecture by Kimmerle can be formulated in the following way:

Conjecture 2 (KC) If G is a finite group, then $\pi(G) = \pi(V(\mathbb{Z}G))$.

---

[1]The research was supported by OTKA T 037202, T 038059 and ADSI107(VUB).

For Frobenius groups and solvable groups this conjecture was confirmed in [11]. In the present paper we continue the investigation of **(KC)**, and confirm it for the first simple Mathieu group M_{11}, using the Luthar–Passi method. Moreover, this allows us to give a partial solution of **(ZC)** for M_{11}.

Our main results are the following:

Theorem 1 *Let $V(\mathbb{Z}G)$ be the normalized unit group of the integral group ring $\mathbb{Z}G$, where G is the simple Mathieu group M_{11}. Let u be a torsion unit of $V(\mathbb{Z}G)$ of order $|u|$. We have:*

  (i) *if $|u| \neq 12$, then $|u|$ coincides with the order of some element $g \in G$;*

  (ii) *if $|u| \in \{2, 3, 5, 11\}$, then u is rationally conjugate to some $g \in G$;*

  (iii) *if $|u| = 4$, then the tuple of partial augmentations of u belongs to the set*

$$\{\, (\nu_{2a}, \nu_{3a}, \nu_{4a}, \nu_{6a},\ \nu_{5a}, \nu_{8a}, \nu_{8b}, \nu_{11a}, \nu_{11b}) \in \mathbb{Z}^9 \mid \nu_{kx} = 0,$$
$$kx \notin \{2a, 4a\},\ (\nu_{2a}, \nu_{4a}) \in \{\, (0, 1),\, (2, -1) \,\} \,\};$$

  (iv) *if $|u| = 6$, then the tuple of partial augmentations of u belongs to the set*

$$\{\, (\nu_{2a}, \nu_{3a}, \nu_{4a}, \nu_{6a}, \nu_{5a}, \nu_{8a}, \nu_{8b}, \nu_{11a}, \nu_{11b}) \in \mathbb{Z}^9 \mid \nu_{kx} = 0,$$
$$kx \notin \{2a, 3a, 6a\},\ (\nu_{2a}, \nu_{3a}, \nu_{6a}) \in \{\, (-2, 3, 0), (0, 0, 1),$$
$$(0, 3, -2), (2, -3, 2), (2, 0, -1) \,\} \,\};$$

  (v) *if $|u| = 8$, then the tuple of partial augmentations of u belongs to the set*

$$\{\, (\nu_{2a}, \nu_{3a}, \nu_{4a}, \nu_{6a}, \nu_{5a}, \nu_{8a}, \nu_{8b}, \nu_{11a}, \nu_{11b}) \in \mathbb{Z}^9 \mid \nu_{kx} = 0,$$
$$kx \notin \{4a, 8a, 8b\},\ (\nu_{4a}, \nu_{8a}, \nu_{8b}) \in \{\, (0, 0, 1), (0, 1, 0),$$
$$(2, -1, 0), (2, 0, -1) \,\} \,\};$$

  (vi) *if $|u| = 12$, then the tuple of partial augmentations of u cannot belong to the set*

$$\mathbb{Z}^9 \setminus \{\, (\nu_{2a}, \nu_{3a}, \nu_{4a}, \nu_{6a}, \nu_{5a}, \nu_{8a}, \nu_{8b}, \nu_{11a}, \nu_{11b}) \in \mathbb{Z}^9 \mid \nu_{kx} = 0,$$
$$kx \notin \{2a, 4a, 6a\},\ (\nu_{2a}, \nu_{4a}, \nu_{6a}) \in \{\, (-1, 1, 1), (1, 1, -1) \,\} \,\}.$$

Corollary 1 *Let $V(\mathbb{Z}G)$ be the normalized unit group of the integral group ring $\mathbb{Z}G$, where G is the simple Mathieu group M_{11}. Then $\pi(G) = \pi(V(\mathbb{Z}G))$, where $\pi(G)$ and $\pi(V(\mathbb{Z}G))$ are prime graphs of G and $V(\mathbb{Z}G)$, respectively. Thus, for M_{11} the conjecture by Kimmerle is true.*

2   Notation and preliminaries

Let $u = \sum \alpha_g g$ be a normalized torsion unit of order k and let $\nu_i = \varepsilon_{C_i}(u)$ be a partial augmentation of u. By S. D. Berman's Theorem [2] we have that $\nu_1 = 0$ and

$$\nu_2 + \nu_3 + \cdots + \nu_m = 1. \tag{2.1}$$

For any character χ of G of degree n, we have that $\chi(u) = \sum_{i=2}^{m} \nu_i \chi(h_i)$, where h_i is a representative of the conjugacy class C_i.

We need the following results:

Proposition 1 (see [12]) *Suppose that u is an element of $\mathbb{Z}G$ of order k. Let z be a primitive k-th root of unity. Then for every integer l and any character χ of G, the number*

$$\mu_l(u, \chi) = \frac{1}{k} \sum_{d|k} \mathrm{Tr}_{\mathbb{Q}(z^d)/\mathbb{Q}} \{\chi(u^d) z^{-dl}\} \tag{2.2}$$

is a non-negative integer.

Proposition 2 (see [6]) *Let u be a torsion unit in $V(\mathbb{Z}G)$. Then the order of u divides the exponent of G.*

Proposition 3 (see [12] and Theorem 2.7 in [14]) *Let u be a torsion unit of $V(\mathbb{Z}G)$. Let C be a conjugacy class of G. If p is a prime dividing the order of a representative of C but not the order of u then the partial augmentation $\varepsilon_C(u) = 0$.*

M. Hertweck (see [10], Proposition 3.1; [9], Lemma 5.6) obtained the next result:

Proposition 4 *Let G be a finite group and let u be a torsion unit in $V(\mathbb{Z}G)$.*

(i) *If u has order p^n, then $\varepsilon_x(u) = 0$ for every x of G whose p-part is of order strictly greater than p^n.*

(ii) *If x is an element of G whose p-part, for some prime p, has order strictly greater than the order of the p-part of u, then $\varepsilon_x(u) = 0$.*

Note that the first part of Proposition 4 gives a partial answer to the conjecture by A. Bovdi (see [1]). Also M. Hertweck ([9], Lemma 5.5) gives a complete answer to the same conjecture in the case when $G = \mathrm{PSL}(2, \mathbb{F})$, where $\mathbb{F} = \mathrm{GF}(p^k)$.

In the rest of the paper, for the partial augmentation ν_i we shall also use the notation ν_{kx}, where k is the order of the representative of the i-th conjugacy class, and x is a distinguishing letter for this particular class with elements of order k.

Proposition 5 (see [12] and Theorem 2.5 in [14]) *Let u be a torsion unit of $V(\mathbb{Z}G)$ of order k. Then u is conjugate in $\mathbb{Q}G$ to an element $g \in G$ if and only if for each d dividing k there is precisely one conjugacy class C_{i_d} with partial augmentation $\varepsilon_{C_{i_d}}(u^d) \neq 0$.*

Proposition 6 (see [6]) *Let p be a prime, and let u be a torsion unit of $V(\mathbb{Z}G)$ of order p^n. Then for $m \neq n$ the sum of all partial augmentations of u with respect to conjugacy classes of elements of order p^m is divisible by p.*

The Brauer character table modulo p of the group M_{11} will be denoted by $\mathfrak{BCT}(p)$.

3   Proof of the Theorem

It is well known [7, 8] that $|G| = 2^4 \cdot 3^2 \cdot 5 \cdot 11$ and $\exp(G) = 1320$. The character table of G, as well as the Brauer character tables $\mathfrak{BCT}(p)$, where $p \in \{2, 3, 5, 11\}$, can be found using the computational algebra system GAP [7].

Since the group G possesses elements of orders 2, 3, 4, 5, 6, 8 and 11, first of all we shall investigate units with these orders. After this, by Proposition 2, the order of each torsion unit divides the exponent of G, so it will be enough to consider units of orders 10, 12, 15, 22, 24, 33 and 55, because if u will be a unit of another possible order, then there is $t \in \mathbb{N}$ such that u^t has an order from this list. We shall prove that units of all these orders except 12 do not appear in $V(\mathbb{Z}G)$. For units of order 12 we are not able to prove this, but we reduce this question to only two cases.

Let $u \in V(\mathbb{Z}G)$ have order k. By S. D. Berman's Theorem [2] and Proposition 3 we have $\nu_{1a} = 0$ and

$$
\begin{array}{lll}
\nu_{3a} = \nu_{5a} = \nu_{6a} = \nu_{11a} = \nu_{11b} = 0 & \text{when} & k = 2, 4, 8; \\
\nu_{2a} = \nu_{4a} = \nu_{5a} = \nu_{6a} = \nu_{8a} = \nu_{8b} = \nu_{11a} = \nu_{11b} = 0 & \text{when} & k = 3; \\
\nu_{2a} = \nu_{3a} = \nu_{4a} = \nu_{6a} = \nu_{8a} = \nu_{8b} = \nu_{11a} = \nu_{11b} = 0 & \text{when} & k = 5; \\
\nu_{5a} = \nu_{11a} = \nu_{11b} = 0 & \text{when} & k = 6; \\
\nu_{2a} = \nu_{3a} = \nu_{4a} = \nu_{5a} = \nu_{6a} = \nu_{8a} = \nu_{8b} = 0 & \text{when} & k = 11; \\
\nu_{3a} = \nu_{6a} = \nu_{11a} = \nu_{11b} = 0 & \text{when} & k = 10; \\
\nu_{5a} = \nu_{11a} = \nu_{11b} = 0 & \text{when} & k = 12; \\
\nu_{2a} = \nu_{4a} = \nu_{6a} = \nu_{8a} = \nu_{8b} = \nu_{11a} = \nu_{11b} = 0 & \text{when} & k = 15; \\
\nu_{2a} = \nu_{3a} = \nu_{4a} = \nu_{5a} = \nu_{6a} = 0 & \text{when} & k = 22; \\
\nu_{5a} = \nu_{11a} = \nu_{11b} = 0 & \text{when} & k = 24; \\
\nu_{2a} = \nu_{4a} = \nu_{5a} = \nu_{6a} = \nu_{8a} = \nu_{8b} = 0 & \text{when} & k = 33; \\
\nu_{2a} = \nu_{3a} = \nu_{4a} = \nu_{6a} = \nu_{8a} = \nu_{8b} = 0 & \text{when} & k = 55.
\end{array}
\tag{3.1}
$$

It follows immediately by Proposition 5 that the units of orders 3 and 5 are rationally conjugate to some element of G.

Now we consider each case separately:

- Let u be an involution. Then using (3.1) and Proposition 4 we obtain that $\nu_{4a} = \nu_{8a} = \nu_{8b} = 0$, so $\nu_{2a} = 1$.

- Let u be a unit of order 4. Then by (3.1) and Proposition 4 we have $\nu_{2a} + \nu_{4a} = 1$. By (2.2), $\mu_0(u, \chi_3) = \frac{1}{4}(-4\nu_{2a} + 8) \geq 0$ and $\mu_2(u, \chi_3) = \frac{1}{4}(4\nu_{2a} + 8) \geq 0$, so $\nu_{2a} \in \{-2, -1, 0, 1, 2\}$. Now using the inequalities

$$
\mu_0(u, \chi_5) = \tfrac{1}{4}(6\nu_{2a} - 2\nu_{4a} + 14) \geq 0;
$$
$$
\mu_2(u, \chi_5) = \tfrac{1}{4}(-6\nu_{2a} + 2\nu_{4a} + 14) \geq 0,
$$

we get that there are only two integral solutions $(\nu_{2a}, \nu_{4a}) \in \{(0, 1), (2, -1)\}$ satisfying (2.1) and Proposition 6, such that all $\mu_i(u, \chi_j)$ are non-negative integers.

- Let u be a unit of order 6. Then by (2.1), (2.2) and Proposition 4 we obtain

$$
\nu_{2a} + \nu_{3a} + \nu_{6a} = 1.
$$

Now, using $\mathfrak{BCT}(11)$ from the system of inequalities $\mu_0(u, \chi_6) = \frac{1}{6}(-4\nu_{3a} + 12) \geq 0$ and $\mu_3(u, \chi_6) = \frac{1}{6}(4\nu_{3a} + 12) \geq 0$, we have that $\nu_{3a} \in \{-3, 0, 3\}$. Furthermore, from the system of inequalities

$$\mu_3(u, \chi_2) = \tfrac{1}{6}(-2\nu_{2a} + 4\nu_{6a} + 8) \geq 0;$$
$$\mu_0(u, \chi_2) = \tfrac{1}{6}(2\nu_{2a} - 4\nu_{6a} + 10) \geq 0;$$
$$\mu_1(u, \chi_2) = \tfrac{1}{6}(\nu_{2a} - 2\nu_{6a} + 8) \geq 0,$$

we get that $\nu_{2a} - 2\nu_{6a} \in \{-2, 4\}$, so $\nu_{6a} \in \{-2, -1, 0, 1, 2\}$. Using the inequalities

$$\mu_0(u, \chi_3) = \tfrac{1}{6}(-4\nu_{2a} + 2\nu_{3a} + 2\nu_{6a} + 10) \geq 0;$$
$$\mu_2(u, \chi_3) = \tfrac{1}{6}(2\nu_{2a} - \nu_{3a} - \nu_{6a} + 7) \geq 0,$$

we obtain only the following integral solutions $(\nu_{2a}, \nu_{3a}, \nu_{6a})$:

$$\{\, (-2, 3, 0),\ (0, 0, 1),\ (0, 3, -2),\ (2, -3, 2),\ (2, 0, -1) \,\}, \qquad (3.2)$$

such that all $\mu_i(u, \chi_j)$ are non-negative integers.

Using the GAP package LAGUNA [5], we tested all possible $\mu_i(u, \chi_j)$ for all tuples $(\nu_{2a}, \nu_{3a}, \nu_{6a})$ from (3.2), and were not able to produce a contradiction. Thus, in this case, as well as in the case of elements of order 4, the Luthar–Passi method is not enough to prove the rational conjugacy.

• Let u be a unit of order 8. By (3.1) and Proposition 4 we have

$$\nu_{2a} + \nu_{4a} + \nu_{8a} + \nu_{8b} = 1.$$

Since $|u^2| = 4$, by (3.2) it yields that $\chi_j(u^2) = \bar{\nu}_{2a}\chi_j(2a) + \bar{\nu}_{4a}\chi_j(4a)$. Now using $\mathfrak{BCT}(3)$, by (2.2) in the case when $(\bar{\nu}_{2a}, \bar{\nu}_{4a}) = (0, 1)$ we obtain

$$\mu_0(u, \chi_5) = \tfrac{1}{8}(-8\nu_{2a} + 8) \geq 0; \qquad \mu_4(u, \chi_5) = \tfrac{1}{8}(8\nu_{2a} + 8) \geq 0;$$
$$\mu_0(u, \chi_4) = \tfrac{1}{8}(8\nu_{2a} + 8\nu_{4a} + 16) \geq 0; \quad \mu_4(u, \chi_4) = \tfrac{1}{8}(-8\nu_{2a} - 8\nu_{4a} + 16) \geq 0;$$
$$\mu_1(u, \chi_2) = \tfrac{1}{8}(4\nu_{8a} - 4\nu_{8b} + 4) \geq 0; \quad \mu_4(u, \chi_7) = \tfrac{1}{8}(-8\nu_{8a} - 8\nu_{8b} + 24) \geq 0;$$
$$\mu_5(u, \chi_2) = \tfrac{1}{8}(-4\nu_{8a} + 4\nu_{8b} + 4) \geq 0; \quad \mu_0(u, \chi_7) = \tfrac{1}{8}(8\nu_{8a} + 8\nu_{8b} + 24) \geq 0.$$

It follows that $-1 \leq \nu_{2a} \leq 1$, $-3 \leq \nu_{4a} \leq 3$, $-2 \leq \nu_{8a}, \nu_{8b} \leq 2$. Considering the additional inequality

$$\mu_0(u, \chi_2) = \tfrac{1}{8}(4\nu_{2a} - 4\nu_{4a} - 4\nu_{8a} - 4\nu_{8b} + 4) \geq 0,$$

and using Proposition 6, it is easy to check that this system has the following integral solutions $(\nu_{2a}, \nu_{4a}, \nu_{8a}, \nu_{8b})$:

$$\begin{array}{cccc} \{ & (0, 2, 0, -1), & (0, 2, -1, 0), & (0, -2, 1, 2), \\ & (0, -2, 2, 1), & (0, 0, 1, 0), & (0, 0, 0, 1) & \}. \end{array} \qquad (3.3)$$

In the case when $(\overline{\nu}_{2a}, \overline{\nu}_{4a}) = (2, -1)$ using $\mathfrak{BCT}(3)$, by (2.2) we obtained that $\mu_0(u, \chi_5) = -\mu_4(u, \chi_5) = -\nu_{2a} = 0$ and

$$\mu_0(u, \chi_4) = \tfrac{1}{8}(8\nu_{4a} + 16) \geq 0; \qquad \mu_4(u, \chi_4) = \tfrac{1}{8}(-8\nu_{4a} + 16) \geq 0;$$
$$\mu_1(u, \chi_2) = \tfrac{1}{8}(4\nu_{8a} - 4\nu_{8b} + 4) \geq 0; \quad \mu_5(u, \chi_2) = \tfrac{1}{8}(-4\nu_{8a} + 4\nu_{8b} + 4) \geq 0.$$

It is easy to check that $-2 \leq \nu_{4a}, \nu_{8a}, \nu_{8b} \leq 2$, and this system has the following integral solutions:

$$\{ \quad (0, 2, -1, 0), \quad (0, 2, 0, -1), \quad (0, 0, 0, 1), \qquad\qquad (3.4)$$
$$(0, -2, 2, 1), \quad (0, 0, 1, 0), \quad (0, -2, 1, 2) \quad \}.$$

Now using $\mathfrak{BCT}(11)$ in the case when $(\overline{\nu}_{2a}, \overline{\nu}_{4a}) = (0, 1)$, by (2.2) we get

$$\mu_0(u, \chi_3) = \tfrac{1}{8}(-8\nu_{2a} + 8) \geq 0; \qquad \mu_4(u, \chi_3) = \tfrac{1}{8}(8\nu_{2a} + 8) \geq 0;$$
$$\mu_1(u, \chi_3) = \tfrac{1}{8}(4t + 12) \geq 0; \qquad \mu_5(u, \chi_3) = \tfrac{1}{8}(-4t + 12) \geq 0;$$
$$\mu_0(u, \chi_2) = \tfrac{1}{8}(4v - 4w + 12) \geq 0; \quad \mu_4(u, \chi_2) = \tfrac{1}{8}(-4v + 4w + 12) \geq 0;$$
$$\mu_0(u, \chi_5) = \tfrac{1}{8}(4z - 4w + 12) \geq 0; \quad \mu_4(u, \chi_5) = \tfrac{1}{8}(-4z + 4w + 12) \geq 0,$$

where $t = \nu_{8a} - \nu_{8b}$, $z = 3\nu_{2a} - \nu_{4a}$, $v = \nu_{2a} + \nu_{4a}$ and $w = \nu_{8a} + \nu_{8b}$. From this it follows that $-1 \leq \nu_{2a} \leq 1$, $-8 \leq \nu_{2a} \leq 10$, $-4 \leq \nu_{8a}, \nu_{8b} \leq 4$, and, using Proposition 6, it is easy to check that this system has the following integral solutions $(\nu_{2a}, \nu_{4a}, \nu_{8a}, \nu_{8b})$:

$$\{ \quad (0, 2, 1, -2), \quad (0, 2, -2, 1), \quad (0, 2, 0, -1), \quad (0, 2, -1, 0), \qquad (3.5)$$
$$(0, 0, 1, 0), \quad (0, 0, 2, -1), \quad (0, 0, 0, 1), \quad (0, 0, -1, 2) \quad \}.$$

In the case when $(\overline{\nu}_{2a}, \overline{\nu}_{4a}) = (2, -1)$, first using $\mathfrak{BCT}(11)$, by (2.2) we obtain $\mu_0(u, \chi_3) = -\mu_4(u, \chi_3) = -\nu_{2a} = 0$ and

$$\mu_1(u, \chi_3) = \tfrac{1}{8}(4\nu_{8a} - 4\nu_{8b} + 12) \geq 0;$$
$$\mu_5(u, \chi_3) = \tfrac{1}{8}(-4\nu_{8a} + 4\nu_{8b} + 12) \geq 0;$$
$$\mu_0(u, \chi_2) = \tfrac{1}{8}(4\nu_{4a} - 4\nu_{8a} - 4\nu_{8b} + 12) \geq 0;$$
$$\mu_4(u, \chi_2) = \tfrac{1}{8}(-4\nu_{4a} + 4\nu_{8a} + 4\nu_{8a} + 12) \geq 0;$$
$$\mu_0(u, \chi_5) = \tfrac{1}{8}(-4\nu_{4a} - 4\nu_{8a} - 4\nu_{8a} + 28) \geq 0;$$
$$\mu_4(u, \chi_5) = \tfrac{1}{8}(4\nu_{4a} + 4\nu_{8a} + 4\nu_{8a} + 28) \geq 0.$$

It is easy to check that $-7 \leq \nu_{4a} \leq 9$, $-4 \leq \nu_{8a}, \nu_{8b} \leq 4$, and the system has the following integral solutions $(\nu_{2a}, \nu_{4a}, \nu_{8a}, \nu_{8b})$:

$$\{ \quad (0, 2, -1, 0), \quad (0, 0, -1, 2), \quad (0, 2, 0, -1), \quad (0, 0, 0, 1), \qquad (3.6)$$
$$(0, 0, 2, -1), \quad (0, 2, -2, 1), \quad (0, 0, 1, 0), \quad (0, 2, 1, -2) \quad \}.$$

It follows from (3.3)–(3.6) that the only four solutions which appear in both cases when $p = 3$ and $p = 11$ are the following ones: $\nu_{2a} = 0$ and

$$(\nu_{4a}, \nu_{8a}, \nu_{8b}) \in \{ (0, 0, 1), (0, 1, 0), (2, -1, 0), (2, 0, -1) \}.$$

Again, we were not able to produce a contradiction computing all possible $\mu_i(u, \chi_j)$ for all above listed tuples $(\nu_{4a}, \nu_{8a}, \nu_{8b})$ for the ordinary character table of G as well as for $\mathfrak{BCT}(p)$, where $p \in \{3, 5, 11\}$.

- Let u be a unit of order 11. Then using $\mathfrak{BCT}(3)$, by (2.2) we have

$$\mu_1(u, \chi_2) = \tfrac{1}{11}(6\nu_{11a} - 5\nu_{11b} + 5) \geq 0;$$
$$\mu_2(u, \chi_2) = \tfrac{1}{11}(-5\nu_{11a} + 6\nu_{11b} + 5) \geq 0,$$

which has only the following trivial solutions $(\nu_{11a}, \nu_{11b}) = \{ (1, 0), (0, 1) \}$.

For all the above mentioned cases except elements of orders 4, 6 and 8 we see that there is precisely one conjugacy class with non-zero partial augmentation. Thus, by Proposition 5, part (ii) of the Theorem is proved.

It remains to prove parts (i) and (vi), considering units of $V(\mathbb{Z}G)$ of orders 10, 12, 15, 22, 24, 33 and 55. Now we treat each of these cases separately:

- Let u be a unit of order 10. Then by (2.1), (3.1) and Proposition 4 we get $\nu_{2a} + \nu_{5a} = 1$. Using $\mathfrak{BCT}(3)$, by (2.2) we have the system of inequalities

$$\mu_5(u, \chi_4) = \tfrac{1}{10}(-8\nu_{2a} + 8) \geq 0;$$
$$\mu_0(u, \chi_4) = \tfrac{1}{10}(8\nu_{2a} + 12) \geq 0;$$
$$\mu_2(u, \chi_2) = \tfrac{1}{10}(-\nu_{2a} + 6) \geq 0,$$

which has no integral solutions such that $\mu_5(u, \chi_4)$, $\mu_0(u, \chi_4)$, $\mu_2(u, \chi_2) \in \mathbb{Z}$.

- Let u be a unit of order 12. By (2.1), (3.1) and Proposition 4, we obtain that

$$\nu_{2a} + \nu_{3a} + \nu_{4a} + \nu_{6a} = 1.$$

Since $|u^2| = 6$ and $|u^3| = 4$, by (3.2) it yields that

$$\chi_j(u^2) = \overline{\nu}_{2a}\chi_j(2a) + \overline{\nu}_{3a}\chi_j(3a) + \overline{\nu}_{6a}\chi_j(6a)$$

and $\chi_j(u^3) = \widetilde{\nu}_{2a}\chi_j(2a) + \widetilde{\nu}_{4a}\chi_j(4a)$.

Consider the following four cases from parts (ii) and (iii) of the Theorem:

1. Let $(\overline{\nu}_{2a}, \overline{\nu}_{3a}, \overline{\nu}_{6a}) \in \{ (0, 0, 1), (2, -3, 2), (-2, 3, 0) \}$ and suppose that

$$(\widetilde{\nu}_{2a}, \widetilde{\nu}_{4a}) \in \{ (0, 1), (2, -1) \}.$$

Then by (2.2) we have $\mu_0(u, \chi_2) = \tfrac{1}{2} \notin \mathbb{Z}$, a contradiction.

2. Let $(\overline{\nu}_{2a}, \overline{\nu}_{3a}, \overline{\nu}_{6a}) = (2, 0, -1)$ and $(\widetilde{\nu}_{2a}, \widetilde{\nu}_{4a}) \in \{ (0, 1), (2, -1) \}$. Then by (2.2) we have $\mu_1(u, \chi_3) = \tfrac{1}{2} \notin \mathbb{Z}$, a contradiction.

3. Let $(\overline{\nu}_{2a}, \overline{\nu}_{3a}, \overline{\nu}_{6a}) = (0, 3, -2)$ and $(\widetilde{\nu}_{2a}, \widetilde{\nu}_{4a}) = (0, 1)$. According to (2.2), $\mu_6(u, \chi_6) = -\mu_0(u, \chi_6) = \tfrac{2}{3}\nu_{3a} = 0$, so $\nu_{3a} = 0$, and we have the system

$$\mu_2(u, \chi_5) = \tfrac{1}{12}(6\nu_{2a} - 2\nu_{4a} + 8) \geq 0;$$
$$\mu_4(u, \chi_5) = \tfrac{1}{12}(-6\nu_{2a} + 2\nu_{4a} + 4) \geq 0;$$
$$\mu_2(u, \chi_3) = \tfrac{1}{12}(-4\nu_{2a} + 2\nu_{6a} + 6) \geq 0;$$
$$\mu_4(u, \chi_3) = \tfrac{1}{12}(4\nu_{2a} - 2\nu_{6a} + 6) \geq 0;$$
$$\mu_4(u, \chi_2) = \tfrac{1}{12}(-4\nu_{2a} - 4\nu_{4a} + 2\nu_{6a} + 10) \geq 0;$$
$$\mu_2(u, \chi_2) = \tfrac{1}{12}(4\nu_{2a} + 4\nu_{4a} - 2\nu_{6a} + 2) \geq 0,$$

that has only two solutions $\{ (-1, 0, 1, 1), (1, 0, 1, -1) \}$ with $\mu_i(u, \chi_j) \in \mathbb{Z}$.

4. Let $(\overline{\nu}_{2a}, \overline{\nu}_{3a}, \overline{\nu}_{6a}) = (0, 3, -2)$ and $(\tilde{\nu}_{2a}, \tilde{\nu}_{4a}) = (2, -1)$. Using (2.2), we get $\mu_6(u, \chi_6) = -\mu_0(u, \chi_6) = \frac{2}{3}\nu_{3a} = 0$, so $\nu_{3a} = 0$. Put $t = 2\nu_{2a} - \nu_{6a}$. Then by (2.2)

$$\mu_0(u, \chi_3) = \tfrac{1}{12}(-4t + 4) \geq 0; \qquad \mu_4(u, \chi_3) = \tfrac{1}{12}(2t - 2) \geq 0,$$

so $2\nu_{2a} - \nu_{6a} = 1$. Now by (2.2) we have

$$\mu_2(u, \chi_5) = \tfrac{1}{12}(2(3\nu_{2a} - \nu_{4a}) - 8) \geq 0,$$
$$\mu_0(u, \chi_9) = \tfrac{1}{12}(-4(3\nu_{2a} - \nu_{4a}) + 28) \geq 0,$$

and $3\nu_{2a} - \nu_{4a} = 4$. Using (2.1) we obtain that $\nu_{2a} = \nu_{4a} = -\nu_{6a} = 1$. Finally, $\mu_4(u, \chi_9) = \tfrac{1}{12}(6\nu_{2a} - 2\nu_{4a} + 28) = \frac{8}{3} \notin \mathbb{Z}$, a contradiction. Thus, part (vi) of the Theorem is proved.

- Let u be a unit of order 15. Then by (2.1) and (3.1) we have $\nu_{3a} + \nu_{5a} = 1$. Now using the character table of G, by (2.2) we get the system of inequalities

$$\mu_0(u, \chi_2) = \tfrac{1}{15}(8\nu_{3a} + 12) \geq 0; \quad \mu_5(u, \chi_2) = \tfrac{1}{15}(-4\nu_{3a} + 9) \geq 0,$$

that has no integral solutions such that $\mu_0(u, \chi_2), \mu_5(u, \chi_2) \in \mathbb{Z}$.

- Let u be a unit of order 22. Then by (2.1), (3.1) and Proposition 4 we obtain that

$$\nu_{2a} + \nu_{11a} + \nu_{11b} = 1.$$

In (2.2) we need to consider two cases: $\chi(u^2) = \chi(11a)$ and $\chi(u^2) = \chi(11b)$, but in both cases by (2.2) we have

$$\mu_0(u, \chi_2) = -\mu_{11}(u, \chi_2) = \tfrac{1}{22}(20\nu_{2a} - 10\nu_{11a} - 10\nu_{11b} + 2) = 0.$$

It yields $10\nu_{2a} - 5\nu_{11a} - 5\nu_{11b} = -1$, that has no integral solutions.

- Let u be a unit of order 24. Then by (2.1) and (3.1) we have

$$\nu_{2a} + \nu_{3a} + \nu_{4a} + \nu_{6a} + \nu_{8a} + \nu_{8b} = 1.$$

Since $|u^2| = 12$, $|u^4| = 6$, $|u^3| = 8$, $|u^6| = 4$, and G has two conjugacy classes of elements of order 8, we need to consider 40 various cases defined by parts (iii)–(vi) of the Theorem.

Let $(\overline{\nu}_{2a}, \overline{\nu}_{3a}, \overline{\nu}_{6a}) \in \{ (0, 3, -2), (2, 0, -1) \}$, where

$$\chi_j(u^4) = \overline{\nu}_{2a}\chi_j(2a) + \overline{\nu}_{3a}\chi_j(3a) + \overline{\nu}_{6a}\chi_j(6a).$$

According to (2.2) we have that $\mu_1(u, \chi_2) = \frac{1}{2} \notin \mathbb{Z}$, a contradiction. In the remaining cases similarly we obtain that $\mu_1(u, \chi_2) = \frac{1}{4} \notin \mathbb{Z}$, which is also a contradiction.

- Let u be a unit of order 33. Then by (2.1) and (3.1) we have $\nu_{3a} + \nu_{11a} + \nu_{11b} = 1$. Again, in (2.2) we need to consider two cases: $\chi(u^3) = \chi(11a)$ and $\chi(u^3) = \chi(11b)$, but both cases lead us to the same system of inequalities

$$\mu_1(u, \chi_5) = \tfrac{1}{33}(2\nu_{3a} + 9) \geq 0; \quad \mu_{11}(u, \chi_5) = \tfrac{1}{33}(-20\nu_{3a} + 9) \geq 0,$$

which has no integral solutions with $\mu_1(u, \chi_5), \mu_{11}(u, \chi_5) \in \mathbb{Z}$.

- Let u be a unit of order 55. Then by (2.1) and (3.1) we have $\nu_{5a}+\nu_{11a}+\nu_{11b} = 1$. Considering two cases when $\chi(u^5) = \chi(11a)$ and $\chi(u^5) = \chi(11b)$, we get the same systems of inequalities

$$\mu_5(u, \chi_8) = \tfrac{1}{55}(4\nu_{5a} + 40) \geq 0; \quad \mu_5(u, \chi_5) = \tfrac{1}{55}(-4\nu_{5a} + 15) \geq 0,$$

which also has no integral solutions such that $\mu_5(u, \chi_5), \mu_5(u, \chi_8) \in \mathbb{Z}$.

Thus, the theorem is proved, and now the corollary follows immediately.

References

[1] V. A. Artamonov and A. A. Bovdi, Integral group rings: groups of invertible elements and classical K-theory, in *Algebra. Topology. Geometry, Vol. 27 (Russian)*, 3–43, 232, Akad. Nauk SSSR, Vsesoyuz. Inst. Nauchn. i Tekhn. Inform., Moscow. MR1039822 (91e:16028). Translated in *J. Soviet Math.* **57** (1991), no. 2, 2931–2958.

[2] S. D. Berman, On the equation $x^m = 1$ in an integral group ring, *Ukrain. Mat. Ž.* **7** (1955), 253–261. MR0077521 (17,1048g)

[3] F. M. Bleher and W. Kimmerle, On the structure of integral group rings of sporadic groups, *LMS J. Comput. Math.* **3** (2000), 274–306 (electronic). MR1783414 (2001i:20006)

[4] V. Bovdi, C. Höfert and W. Kimmerle, On the first Zassenhaus conjecture for integral group rings, *Publ. Math. Debrecen* **65** (2004), no. 3–4, 291–303. MR2107948 (2006f:20009)

[5] V. Bovdi, A. Konovalov, R. Rossmanith and C. Schneider, *LAGUNA – Lie AlGebras and UNits of group Algebras*, v.3.3.3; 2006 (http://ukrgap.exponenta.ru/laguna.htm).

[6] J. A. Cohn and D. Livingstone, On the structure of group algebras, I, *Canad. J. Math.* **17** (1965), 583–593. MR0179266 (31 #3514)

[7] The GAP Group, *GAP – Groups, Algorithms, and Programming, Version 4.4*; 2006 (http://www.gap-system.org).

[8] D. Gorenstein, *The classification of finite simple groups, Vol. 1*, Plenum, New York, 1983. MR0746470 (86i:20024)

[9] M. Hertweck, Partial augmentations and Brauer character values of torsion units in group rings, to appear (2005), 26 pages.

[10] M. Hertweck, On the torsion units of some integral group rings, *Algebra Colloq.* **13** (2006), no. 2, 329–348. MR2208368

[11] W. Kimmerle, On the prime graph of the unit group of integral group rings of finite groups, in *Groups, rings and algebras*, Contemporary Mathematics, AMS, to appear.

[12] I. S. Luthar and I. B. S. Passi, Zassenhaus conjecture for A_5, *Proc. Indian Acad. Sci. Math. Sci.* **99** (1989), no. 1, 1–5. MR1004634 (90g:20007)

[13] I. S. Luthar and P. Trama, Zassenhaus conjecture for S_5, *Comm. Algebra* **19** (1991), no. 8, 2353–2362. MR1123128 (92g:20003)

[14] Z. Marciniak, J. Ritter, S. K. Sehgal and A. Weiss, Torsion units in integral group rings of some metabelian groups, II, *J. Number Theory* **25** (1987), no. 3, 340–352. MR0880467 (88k:20019)

[15] H. Zassenhaus, On the torsion units of finite group rings, in *Studies in mathematics (in honor of A. Almeida Costa) (Portuguese)*, 119–126, Inst. Alta Cultura, Lisbon. MR0376747 (51 #12922)

EMBEDDING PROPERTIES IN DIRECT PRODUCTS

B. BREWSTER[*], A. MARTÍNEZ-PASTOR[†] and M. D. PÉREZ-RAMOS[◦] [1]

[*]Department of Mathematical Sciences, Binghamton University-SUNY, Binghamton, NY 13902-6000, U.S.A.

[†]Escuela Técnica Superior de Informática Aplicada, Departamento de Matemática Aplicada, Universidad Politécnica de Valencia, Camino de Vera, s/n, 46022 Valencia, Spain
Email: anamarti@mat.upv.es

[◦]Departament d'Àlgebra, Universitat de València, C/ Doctor Moliner 50, 46100 Burjassot (València), Spain

1 Introduction

This paper is a survey article containing an up-to-date account of recent achievements regarding embedding properties in direct products of groups. In the last years, several authors are carrying out a systematic study with the aim of understanding how subgroups with various embedding properties can be detected and characterized in the subgroup lattice of a direct product of two groups in terms of the subgroup lattices of the two groups.

Unless otherwise stated all groups considered in this paper are finite.

Direct products are maybe the easiest way to construct new groups from given ones and in spite of the simplicity of this construction, their structures are sometimes surprising.

The subgroup structure of direct products is well-known by a classical result due to Goursat. In this paper $G_1 \times G_2 = \{(g_1, g_2) \mid g_i \in G_i, i = 1, 2\}$ will always denote the direct product of the groups G_1 and G_2 and π_i will denote the canonical projection $\pi_i : G_1 \times G_2 \to G_i$, for $i = 1, 2$. For a subgroup U of $G_1 \times G_2$:

- $\pi_i(U) = UG_j \cap G_i$, $\{i, j\} = \{1, 2\}$,
- $U \cap G_i \trianglelefteq \pi_i(U)$, for $i = 1, 2$.

Goursat's theorem states that, apart from the direct product of subgroups of the direct factors, only 'diagonal' subgroups appear in a direct product.

Theorem 1.1 (Goursat)   1) *For a subgroup U of $G_1 \times G_2$, there exists an isomorphism $\sigma : \pi_1(U)/U \cap G_1 \to \pi_2(U)/U \cap G_2$.*

2) *If $U_1 \leq G_1$, $W \trianglelefteq U_2 \leq G_2$, and $\alpha : U_1 \to U_2/W$ is an epimorphism, then $U := D(U_1, \alpha) = \{(x, y) \mid x \in U_1, y \in x^\alpha\}$ is a subgroup of $G_1 \times G_2$, with $\pi_i(U) = U_i$, $i = 1, 2$, $W = U \cap G_2$ and $\operatorname{Ker} \alpha = U \cap G_1$.*

(We will denote simply $U = D(U_1)$, when the role of α does not need consideration.)

Also normal subgroups are well characterized in direct products, as we will detail later. It seems then natural to investigate another subgroups with embedding

[1]This research has been supported by Proyecto MTM2004-06067-C02-02, Ministerio de Educación y Ciencia and FEDER, Spain.

properties related to, but weaker than normality. We will report in this paper about a current activity in this framework. The main reference for the basic terminology and results about embedding properties will be [4] and we refer to [14] for the results on permutability.

2 Normal and subnormal subgroups of direct products

The description of the subgroup lattice of a direct product of groups is enriched with the characterization of normal subgroups.

Lemma 2.1 *A subgroup U of $G_1 \times G_2$ is normal in $G_1 \times G_2$ if and only if $[U, G_i] = [\pi_i(U), G_i] \leq U \cap G_i$, for $i = 1, 2$, or equivalently, $U \cap G_i \trianglelefteq G_i$ and $\pi_i(U)/U \cap G_i \leq Z(G_i/U \cap G_i)$, for $i = 1, 2$.*

On the other hand, subnormal subgroups of direct products were studied by P. Hauck [9] in 1987 motivated by a conjecture stated by T. O. Hawkes in 1977 about normally detectable groups. A group G is called *normally detectable* if the only normal subgroups in any direct product $G_1 \times \ldots \times G_n$, where $G_i \cong G$, for $i = 1, \ldots, n$, are just the direct factors $G_1, \ldots, G_n$. The motivation for this study arises in relation with certain Fitting classes constructions. For details the reader is referred to [4]. Hawkes conjectured that a group G is normally detectable if and only if G is directly indecomposable and $|G/G'|$ and $|Z(G)|$ are coprime. Hauck verified this conjecture in a number of special cases. Moreover he introduced the concept of subnormally detectable groups in the natural way and proved that a group G is subnormally detectable if and only if $|G/G'|$ and $|F(G)|$ are coprime.

Going on with our purposes, Hauck pointed out the following extension for subnormality of the stated characterization of normal subgroups of direct products.

Lemma 2.2 *A subgroup U of $G_1 \times G_2$ is subnormal in $G_1 \times G_2$ (of defect r) if and only if $[G_i, \pi_i(U), \overset{r}{\cdots}, \pi_i(U)] \leq U \cap G_i$ for $i = 1, 2$.*

Following J. Evan [7], we call a subgroup U of $G_1 \times G_2$ a *diagonal-type* subgroup if $U \cap G_i = 1$, for $i = 1, 2$. For this particular subgroups, the aforementioned characterization of normal subgroups of a direct product and a corollary to the work of Hauck [9] lead to the following result.

Lemma 2.3 *Let U be a diagonal-type subgroup of $G_1 \times G_2$. Then:*
  (i) *U is normal in $G_1 \times G_2$ if and only if $\pi_i(U) \leq Z(G_i)$, for $i = 1, 2$, or equivalently, $U \leq Z(G_1 \times G_2)$.*
  (ii) *U is subnormal in $G_1 \times G_2$ if and only if $\pi_i(U) \leq F(G_i)$, for $i = 1, 2$, or equivalently, $U \leq F(G_1 \times G_2)$.*

3 Permutable subgroups of direct products

J. Evan started the study of permutable subgroups in direct products in his Ph.D. Dissertation [5] under supervision of B. Brewster. Evan took further this research in [6], [7] and [8], which is reported next in this section.

Definition 3.1 A subgroup M of a group G is called permutable if $MX = XM$ (denoted by $M \perp X$) for every subgroup X of G.

Permutable subgroups were initially studied by O. Ore [12], who called them quasinormal, in 1939. Permutability is an embedding property which lies between normality and subnormality. Then, in a natural way, the first aim of Evan was to find characterizations of permutable subgroups of direct products like those for normality and subnormality.

3.1   Permutability. Diagonal-type subgroups

Evan's aim was nicely achieved for diagonal-type subgroups in [7]. He obtained a characterization for permutable diagonal-type subgroups similar to those pointed out above for normality and subnormality, by considering the norm to play the role of the center and the Fitting subgroup.

First studied by R. Baer [1] in 1934, the *norm* of a group G, denoted by $N(G)$, is the intersection of the normalizers of all subgroups:

$$N(G) = \{g \in G \mid \text{ for all } X \leq G, \, g \in N_G(X)\}.$$

Then, if a subgroup is contained in the norm, it is permutable.

Since permutable subgroups behave well under epimorphic images, Evan's result is in fact extended to subgroups whose intersections with the direct factors are normal.

Theorem 3.2 *Let G_1 and G_2 be (not necessarily finite) groups and let U be a subgroup of $G_1 \times G_2$ such that $U \cap G_i \trianglelefteq G_i$ for $i = 1, 2$. Then U is permutable in $G_1 \times G_2$ if and only if $U/((U \cap G_1) \times (U \cap G_2)) \leq N(G_1 \times G_2/((U \cap G_1) \times (U \cap G_2)))$.*
Further, if $U/((U \cap G_1) \times (U \cap G_2))$ contains an element of infinite order, then U is permutable in $G_1 \times G_2$ if and only if $U \trianglelefteq G_1 \times G_2$.

The norm of a group is close to its center. In fact, E. Scheckman [13] proved in 1960 that the norm is contained in the second term of the ascending central series, i.e., for any group G, $N(G) \leq Z_2(G)$, and so $Z(G) \leq N(G) \leq F(G)$. However unlike the center and the Fitting subgroup, the norm of a direct product is not in general the direct product of the norms of the direct factors.

Example   Let $E = \langle a, b \mid a^9 = 1 = b^3, a^b = a^4 \rangle$ be an extraspecial group of order 27 and exponent 9. Then $N(E) = \langle a^3 \rangle \langle b \rangle$ but $N(E \times E) = (\langle a^3 \rangle \times \langle a^3 \rangle) \langle (b, b) \rangle$.

We gather here the results for diagonal-type subgroups.

Theorem 3.3 *If U is a diagonal-type subgroup of $G_1 \times G_2$ then:*
  (i) *U is normal in $G_1 \times G_2$ if and only if $U \leq Z(G_1 \times G_2)$.*
  (ii) *U is subnormal in $G_1 \times G_2$ if and only if $U \leq F(G_1 \times G_2)$.*
  (iii) *U is permutable in $G_1 \times G_2$ if and only if $U \leq N(G_1 \times G_2)$.*

3.2   Permutability. Direct products of subgroups

Permutability is however not easy to handle and its behaviour is not always as one could expect. For instance if U is a permutable subgroup of a group G, it is not true in general that U is permutable in $G \times H$, for a group H.

Example   Let $E = \langle a, b \mid a^9 = 1 = b^3, a^b = a^4 \rangle$ be an extraspecial group of order 27 and exponent 9. Then $\langle b \rangle$ is permutable in E but $\langle b \rangle$ is not permutable in $E \times E$.

In fact, as a consequence of Corollary 3.6 below, if U is permutable in G, then U is permutable in $G \times G$ if and only if U is normal in G.

Then in a first step Evan studied in [6] the permutability of subgroups which are direct product of subgroups of the direct factors. From basic properties of permutable subgroups, it can be deduced that if U is permutable in $G_1 \times G_2$, then $\pi_i(U)$ and $U \cap G_i$ are permutable subgroups of G_i, for $i = 1, 2$. In particular, if $U_i \leq G_i$, for $i = 1, 2$, in order for $U_1 \times U_2$ to be permutable in $G_1 \times G_2$ it is necessary that U_i is a permutable subgroup of G_i, for $i = 1, 2$. However, as the above example shows, this is not a sufficient condition. Indeed, Evan obtained the following result.

Theorem 3.4 *If U_i is a subgroup of G_i for $i = 1, 2$, then the subgroup $U_1 \times U_2$ is permutable in $G_1 \times G_2$ if and only if U_i is a permutable subgroup of $G_i \times (G_j / \operatorname{Core}_{G_j}(U_j))$, for $\{i, j\} = \{1, 2\}$.*

This result focuses the attention on the permutability of a subgroup of a single direct factor in the whole direct product. In order to study this, useful conditions for a subgroup of $G_1 \times G_2$ to permute with a subgroup of one of the direct factors are first provided.

Theorem 3.5 *If $U_1 \leq G_1$ and $S \leq G_1 \times G_2$, then $S \perp U_1$ if and only if $\pi_1(S) \subseteq N_{G_1}((S \cap G_1)U_1)$.*

Here it is not needed to require that $(S \cap G_1)U_1$ is a subgroup but this is in fact a consequence when S permutes with U_1, by the Dedekind's modular law. This result has been also useful in the study of normal embedding in direct products.

We notice that $\langle (g, h) \rangle \cap G_1 = \langle g^{o(h)} \rangle$, for every $(g, h) \in G_1 \times G_2$. Then the previous theorem for cyclic subgroups particularizes as follows.

Corollary 3.6 *If $U_1 \leq G_1$ and $(g, h) \in G_1 \times G_2$, $\langle (g, h) \rangle$ permutes with U_1 if and only if $g \in N_G(\langle g^{o(h)} \rangle U_1)$.*

We recall that the permutability of a subgroup in a finite group is equivalent to the permutability with all the cyclic subgroups of prime power order. As a consequence of all these facts Evan obtained the desired characterization of subgroups of one direct factor which are permutable in the direct product.

Theorem 3.7 *Let U_1 be a permutable subgroup of G_1. Then U_1 is permutable in $G_1 \times G_2$ if and only if for each prime p dividing $|G_2|$ and $g \in G_1$ of p-power order, $g \in N_G(\langle g^{\exp P} \rangle U_1)$, where P a Sylow p-subgroup of G_2.*

3.3 Permutability. The general case

After considering direct products of subgroups and diagonal-type subgroups. Evan obtained a necessary and suficient condition for an arbitrary subgroup of a direct product to be permutable in [8].

At this point we recall first the well-known result of R. Maier and P. Schmid [10] which states that if H is a permutable subgroup of a group G, then $H/\operatorname{Core}_G(H) \le Z_\infty(G/\operatorname{Core}_G(H))$. This fact allows to restrict to p-groups, for a prime p, when studying the permutability of a subgroup. Then the main result of [8] provides a characterization for an arbitrary subgroup of a direct product of p-groups. for a prime p, to be permutable.

Theorem 3.8 *Let U be a subgroup of the p-group $G_1 \times G_2$, for a prime p. Assume $\exp(G_1/\operatorname{Core}_{G_1}(U \cap G_1)) \ge \exp(G_2/\operatorname{Core}_{G_2}(U \cap G_2))$. Then U is permutable in $G_1 \times G_2$ if and only if for all $(g, h) \in G_1 \times G_2$, one of the following two statements is true:*

1) $U \le N_{G_1 \times G_2}(((U \cap G_1) \times (U \cap G_2))\langle(g, h)\rangle)$, *or*

2) $|\langle g \rangle/(\langle g \rangle \cap U)| > \exp(G_2/\operatorname{Core}_{G_2}(U \cap G_2))$, *and there exists $i \ge 0$ such that $h^{p^i} \in \pi_{G_2}((\langle g \rangle \times \langle h \rangle) \cap U)$ and*
$U \le N_{G_1 \times G_2}(((U \cap G_1) \times (U \cap G_2))\langle(1, h^{p^i})\rangle\langle(g, h)\rangle)$.

Though the conditions in this theorem may look a bit technical, different attempts by Evan [8] to simplify them were not successful. For instance, we mention the following conjecture:

Conjecture *U is a permutable subgroup of $G_1 \times G_2$ if and only if $(U \cap G_1) \times (U \cap G_2)$ is permutable in $G_1 \times G_2$ and for all $S \le G_1 \times G_2$ such that $(U \cap G_1) \times (U \cap G_2) \le S$, we have $U \le N_{G_1 \times G_2}(S)$.*

This would mean that the first condition in the Theorem 3.8 for all elements in $G_1 \times G_2$ would be enough to characterize the permutability of the subgroup. On the other hand this result would fit with the characterization of normal subgroups in the following sense. A subgroup N is normal in $G_1 \times G_2$ if and only if N normalizes the set $((N \cap G_1) \times (N \cap G_2))(g, h)$, for all $(g, h) \in G_1 \times G_2$. (In this case $N(g, h) = (g, h)N$ holds.) In the case of permutability we would have that U would normalize $((U \cap G_1) \times (U \cap G_2))\langle(g, h)\rangle$. (Certainly in this case $U\langle(g, h)\rangle = \langle(g, h)\rangle U$.)

We refer to Evan [8] for an example showing that this conjecture is not true but its truthfulness is verified in some special cases.

Theorem 3.9 *Let U be a subgroup of $G_1 \times G_2$. Assume that one of the following conditions hold:*

i) $\operatorname{Core}_{G_i}(\pi_{G_i}(U)) = 1$, *for $i = 1, 2$, or*

ii) $\pi_{G_i}(U)$ *is cyclic, for $i = 1, 2$.*

Then Conjecture holds.

Still with respect to the failure of the conjecture, the example constructed by Evan exhibits a group $G_1 \times G_2$ with a subgroup U such that U is permutable in $G_1 \times G_2$, but $(U \cap G_1) \times (U \cap G_2)$ is not permutable in $G_1 \times G_2$. Again a positive result in this direction can be obtained in some particular cases.

Theorem 3.10 *Let G_1 and G_2 be groups of odd order and let U be a permutable subgroup of $G_1 \times G_2$. Assume that one of the following conditions hold:*

    i) $U \cap G_i$ *is cyclic, for* $i = 1, 2$, *or*

    ii) G_1 *and* G_2 *have modular subgroup lattices.*

Then $(U \cap G_1) \times (U \cap G_2)$ *is permutable in* $G_1 \times G_2$.

4 Cover-avoidance property in direct products

The cover-avoidance property was considered in relation to the subgroup lattice of direct products by J. Petrillo in his Ph.D. Dissertation [11] under direction of B. Brewster.

Definition 4.1 A subgroup U of a group G is said to have the cover-avoidance property (U is a CAP-subgroup of G), if U covers or avoids every chief factor H/K of G. Equivalently, $K(U \cap H) \trianglelefteq G$, for every chief factor H/K of G.

The cover-avoidance property is again an extension of normality and plays an important role mainly in the study of solvable groups. In this framework maximal subgroups, Sylow subgroups, system normalizers and injectors associated to Fitting classes are example of subgroups with the cover-avoidance property.

The initial aim of Petrillo was to characterize a CAP-subgroup U in a direct product $G_1 \times G_2$ in terms of conditions on the subgroups $\pi_i(U)$ and $U \cap G_i$, $i = 1, 2$. It is an easy remark that if U is a CAP-subgroup of $G_1 \times G_2$, then $\pi_i(U)$ and $U \cap G_i$ are CAP-subgroups of G_i, for $i = 1, 2$. But these are not sufficient conditions to assure that U is a CAP-subgroup of $G_1 \times G_2$.

Example Let $G_1 = \langle x, y \rangle \cong Z_2 \times Z_2$, $G_2 = [\langle a, b \rangle]\langle c \rangle \cong \mathrm{Alt}(4)$, the alternating group on 4 letters, and $U = D(G_1)$. Then $\pi_1(U) = G_1$, $\pi_2(U) = \langle a, b \rangle$ and $U \cap G_1 = U \cap G_2 = 1$, so $\pi_i(U)$ and $U \cap G_i$ are CAP-subgroups of G_i, $i = 1, 2$. But U is not a CAP-subgroup of $G_1 \times G_2$, since U neither covers nor avoids the chief factor $\langle x \rangle\langle a, b \rangle / \langle x \rangle$.

After a detailed study of the different kinds of chief factors in a direct product, Petrillo found out the following characterization of CAP-subgroups in direct products in terms of the subgroup lattice in the sections $\pi_i(U)/U \cap G_i$, for $i = 1, 2$.

Theorem 4.2 *If U is a subgroup of $G_1 \times G_2$, U is a CAP-subgroup of $G_1 \times G_2$ if and only if $UM \cap G_1$ is a CAP-subgroup of G_1 for every $M \trianglelefteq G_2$, and $UN \cap G_2$ is a CAP-subgroup of G_2 for every $N \trianglelefteq G_1$.*

Notice that, under the isomorphism between $\pi_1(U)/U \cap G_1$ and $\pi_2(U)/U \cap G_2$, the section $UX \cap G_i/U \cap G_i$ is in correspondence with the section $(U \cap G_j)(X \cap \pi_j(U))/(U \cap G_j)$ for every $X \trianglelefteq G_j$ and $\{i,j\} = \{1,2\}$.

Particular consequences of this theorem are the following.

Corollary 4.3 (1) *If $U_i \leq G_i$, $i = 1, 2$, then $U_1 \times U_2$ is a CAP-subgroup of $G_1 \times G_2$ if and only if U_i is a CAP-subgroup of G_i, for $i = 1, 2$.*

(2) *If U is a subdirect subgroup of $G_1 \times G_2$ (i.e., $\pi_i(U) = G_i$, for $i = 1, 2$), then U is a CAP-subgroup of $G_1 \times G_2$.*

5 Normally embedded subgroups in direct products

We report in this section about normally embedded subgroups in direct products, considered by the authors in [2].

Definition 5.1 A subgroup U is normally embedded in a group G if, for every prime p, a Sylow p-subgroup U_p of U is a Sylow p-subgroup of its normal closure $\langle U_p^G \rangle$.

Again the fact that $\pi_i(U)$ and $U \cap G_i$ are normally embedded in G_i for $i = 1, 2$, does not characterize the normal embedding of the subgroup U in the group $G_1 \times G_2$.

Example Let $G_i = \mathrm{Sym}(3)$, the symmetric group on 3 letters, for $i = 1, 2$, $P_1 = O_3(G_1)$ and $U = D(P_1)$. Then $\pi_i(U) \trianglelefteq G_i$, $U \cap G_i = 1$, for $i = 1, 2$, but U is not a normally embedded subgroup of $G_1 \times G_2$.

We recall that normal embedding implies the cover-avoidance property. But an analogous characterization to the one for CAP-subgroups in Theorem 4.2 is not further true for normally embedded subgroups.

Example Let $G_1 = Z_3$, $G_2 = \mathrm{Sym}(3)$ and $U = D(G_1)$. Then it holds that $UM \cap G_1$ is normally embedded in G_1 for every $M \trianglelefteq G_2$, and $UN \cap G_2$ is normally embedded in G_2 for every $N \trianglelefteq G_1$, but U is not normally embedded in $G_1 \times G_2$.

A key structural property of normally embedded subgroups in direct products is the following.

Proposition 5.2 *If U is a normally embedded subgroup of $G = G_1 \times G_2$, then:*

(a) $U \trianglelefteq \pi_1(U) \times \pi_2(U)$.

(b) $\pi_i(U)/U \cap G_i$ *is an abelian group, for $i = 1, 2$.*

We remark that Evan proved in [5] that property (b) is also true for permutable subgroups of direct products. In particular, in contrast to Corollary 4.3.(2), the only subdirect subgroups of a direct product which are either normally embedded or permutable are the normal subgroups.

A characterization of normally embedded subgroups in direct products is given next.

Theorem 5.3 *Let U be a subgroup of $G = G_1 \times G_2$. Then the following statements are pairwise equivalent:*

(1) *U is normally embedded in G.*

(2) *For $i = 1, 2$, the following holds:*

    (2.a) *$\pi_i(U)$ is normally embedded in G_i, and*

    (2.b) *$\pi_i(U_p) \cap [\pi_i(U_p), G_i] \leq U$, for every $U_p \in \mathrm{Syl}_p(U)$ and for every prime number p.*

   In fact, (2.b) is equivalent to (2.b'):

    (2.b') *$\pi_i(U_p) \cap [\pi_i(U_p), G_i] \leq U_p$, for every $U_p \in \mathrm{Syl}_p(U)$ and for every prime number p.*

In view of this result one could think of a global condition characterizing normal embedding. In this sense it is clear that if $\pi_i(U) \cap [\pi_i(U), G_i] \leq U$, then property (2.b) in the above theorem holds. However, this is not a necessary condition for a subgroup U to be normally embedded, unless U is a p-group for some prime p.

Example Consider $G_1 = G_2 = SL(2,3) = QC$ where Q is a quaternion group of order 8 and C is a cyclic group of order 3. Now let $G = G_1 \times G_2$ and $U = D(Z(Q)C)$. Then U is normally embedded in G but $\pi_i(U) \cap [\pi_i(U), G_i]$ is not contained in U, for $i = 1, 2$.

5.1 Normally embedded subgroups in solvable direct products

Now the discussion narrows to solvable groups. We introduce first some notation.

Notation Whenever G is a solvable group and given a Hall system Σ of G, we denote

$\mathrm{Hall}_\pi(G) \cap \Sigma = \{G_\pi\}$, for every $\pi \subseteq \sigma(G)$,

$\mathrm{Syl}_p(G) \cap \Sigma = \{G_p\}$, for every prime $p \in \sigma(G)$,

$N_U(\Sigma) = U \cap N_G(\Sigma)$, for every $U \leq G$.

With this notation, we present a characterization of normally embedded subgroups in direct products of solvable groups.

Theorem 5.4 *If U is a subgroup of the solvable group $G = G_1 \times G_2$, the following statements are pairwise equivalent:*

(i) *U is normally embedded in G.*

(ii) *There exists a Hall system Σ of G such that:*

$$U = (U \cap G_1)(U \cap G_2)N_U(\Sigma) \tag{1}$$

$$[U \cap G_p, G_p] \leq (U \cap G_1)(U \cap G_2) \text{ for all } p \in \sigma(U) \tag{2}$$

(iii) *There exists a Hall system Σ of G reducing into U such that (1),(2) hold for Σ.*

(iv) *If Σ is a Hall system of G reducing into U, then* (1),(2) *hold for Σ.*

In particular, for diagonal-type subgroups we get the following.

Corollary 5.5 *If U is a diagonal-type subgroup of a solvable group $G = G_1 \times G_2$, then U is normally embedded in G if and only if for any Hall system Σ of G reducing into U, $U \leq Z(N_G(\Sigma))$ and $U \cap G_p \leq Z(G_p)$, for every prime p.*

The above theorem rests on the fact that U is a normally embedded subgroup of a solvable group G if and only if there exists a Hall system Σ of G such that $U \perp \Sigma$ and $U_p = U \cap G_p \trianglelefteq G_p$, for every prime $p \in \sigma(U)$. This last condition on normality is equivalent to condition (2) in Theorem 5.4. On the other hand, condition (1) characterizes system permutability in direct products of solvable groups, as stated next.

Proposition 5.6 *If $G = G_1 \times G_2$ is solvable, $U \leq G$ and Σ is a Hall system of G, the following statements are pairwise equivalent:*
  (i) *Σ permutes with U.*
  (ii) *$U \leq \bigcap_{G_\pi \in \Sigma} N_G((U \cap G_1)(U \cap G_2)G_\pi)$.*
  (iii) *$U = (U \cap G_1)(U \cap G_2)N_U(\Sigma)$.*

This result is achieved after studying conditions for the permutability of a subgroup with a single Hall subgroup in a direct product of groups.

The next remark lays out explicitily how normally normally embedded subgroups in solvable direct products can be constructed.

Remark Let $G = G_1 \times G_2$ be a solvable group. Take H_i a normally embedded subgroup of G_i, for $i = 1, 2$, and let Σ be a Hall system of G reducing into $H_1 H_2$. Let $U \leq G$ such that

$$H_1 H_2 \leq U \trianglelefteq H_1 H_2 N_G(\Sigma) \quad \text{and} \quad U \cap G_p \trianglelefteq G_p, \text{ for all } p \in \sigma(G).$$

Then U is normally embedded in G and every normally embedded subgroup in G can be constructed in this way.

It is known that normally embedded subgroups play a significant role in the theory of Fitting sets and injectors. In fact normally embedded subgroups are exactly injectors for subgroup-closed Fitting sets (see [4]). The study carried out in [2] is applied to give descriptions of Fitting sets with normally embedded injectors in a direct product of two solvable groups.

6   Further results

Finally, it is to be mentioned that a research concerning pronormality and local pronormality is being currently carried out by the authors ([3]). We announce in advance a characterization of pronormality in direct products of groups.

Theorem 6.1 *For a subgroup U of a solvable group $G_1 \times G_2$ the following statements are equivalent:*

1) U *is pronormal in* $G_1 \times G_2$;

2) U *satisfies the following properties:*
   (i) $\pi_i(U)$ *is pronormal in* G_i, *for* $i = 1, 2$,
   (ii) $N_G(U) = N_{G_1}(\pi_1(U)) \times N_{G_2}(\pi_2(U))$.

As a previous step it is also proved that if U is pronormal in the solvable group $G_1 \times G_2$, then $\pi_i(U)/U \cap G_i$ is an abelian group, for $i = 1, 2$.

Acknowledgements   We thank J. Evan and J. Petrillo for their kind consent to include their work in this survey.

References

[1] R. Baer, Der Kern eine charakteristiche Untergruppe, *Compositio Math.* **1** (1934), 254–283.

[2] B. Brewster, A. Martínez-Pastor and M. D. Pérez-Ramos, Normally embedded subgroups in direct products of groups, *J. Group Theory*, to appear.

[3] B. Brewster, A. Martínez-Pastor and M. D. Pérez-Ramos, Pronormal subgroups in direct products of groups (preprint).

[4] K. Doerk and T. Hawkes, *Finite soluble groups* (Walter De Gruyter, 1992).

[5] J. Evan, Permutability in direct products of finite groups, Dissertation, Binghamton University, 2000.

[6] J. Evan, Permutability of subgroups of $G \times H$ that are direct products of subgroups of the direct factors, *Arch. Math.* **77** (2001), 449–455.

[7] J. Evan, Permutable diagonal-type subgroups of $G \times H$, *Glasgow Math. J.* **45** (2003), 73–77.

[8] J. Evan, Permutable subgroups of a direct product, *J. Algebra* **265** (2003), 734–743.

[9] P. Hauck, Subnormal subgroups in direct products, *J. Austral. Math. Soc. (Ser. A)* **42** (1987), 147–172.

[10] R. Maier and P. Schmid, The embedding of quasinormal subgroups in finite groups, *Math. Z.* **131** (1973), 269–272.

[11] J. Petrillo, The Cover-Avoidance Property in Finite Groups, Dissertation, Binghamton University, 2003.

[12] O. Ore, Contributions to the theory of groups, *Duke Math. J.* **5** (1939), 431–460.

[13] E. Schenkman, On the norm of a group, *Illinois J. Math.* **4** (1960), 150–152.

[14] R. Schmidt, *Subgroup Lattices of Groups* (Walter De Gruyter, 1994).

MALCEV PRESENTATIONS FOR SUBSEMIGROUPS OF GROUPS — A SURVEY

ALAN J. CAIN[1]

School of Mathematics and Statistics, University of St Andrews, North Haugh, St Andrews, Fife KY16 9SS, United Kingdom
Email: alanc@mcs.st-andrews.ac.uk

Abstract

This paper introduces and surveys the theory of Malcev presentations. [Malcev presentations are a species of presentation that can be used to define any semigroup that can be embedded into a group.] In particular, various classes of groups and monoids all of whose finitely generated subsemigroups admit finite Malcev presentations are described; closure and containment results are stated; links with the theory of automatic semigroups are mentioned; and various questions asked. Many of the results stated herein are summarized in tabular form.

1   Introduction

A Malcev presentation is a presentation of a special type for a semigroup that embeds in a group. Informally, a Malcev presentation defines a semigroup by means of generators, defining relations, and the unwritten rule that the semigroup so defined must be embeddable in a group. This rule of group-embeddability is worth an infinite number of defining relations, in the sense that a semigroup may admit a finite Malcev presentation but no finite 'ordinary' presentation. Spehner [31] introduced Malcev presentations, though they are based on Malcev's necessary and sufficient condition for a semigroup to be embeddable into a group [22]. Spehner exhibited an example of a finitely generated submonoid of a free monoid that admitted a finite Malcev presentation but which was not finitely presented. He later showed that all finitely generated submonoids of free monoids have finite Malcev presentations [32]. Until the recent work of Cain, Robertson & Ruškuc [12, 11, 10, 9, 8], Spehner's two articles represented the whole of the literature on Malcev presentations.

The aim of this survey is to introduce the theory of Malcev presentations and summarize both Spehner's work and the recent progress in the field, and to state and speculate about currently open questions. Section 2 defines Malcev presentations and establishes their basic properties. Section 3 discusses Malcev presentations for finitely generated subsemigroups of virtually free and virtually nilpotent groups. Sections 4, 5, and 6 do the same for free products, direct products, and Baumslag–Solitar semigroups and groups. Section 7 describes a link between the theory of Malcev presentations and the active area of automatic semigroups. The interaction of Malcev presentations with subsemigroups and extensions of finite

[1]The author acknowledges the support of the Carnegie Trust for the Universities of Scotland.

Rees index is considered in Section 8. Section 9 contains further open problems. The various results and questions are tabulated for the purposes of comparison and reference in Section 10.

2   Fundamentals

A loose definition of a Malcev presentation was given in Section 1: the semigroup defined by a Malcev presentation has the given generators, satisfies the given defining relations, and is embeddable into a group. The first task is to formalize this notion. The second is to state the syntactic rules that determine when two words represent the same element of the semigroup defined by the presentation. These rules are more complex than the corresponding ones for 'ordinary' semigroup presentations.

This section assumes that the reader is familiar with the theory of semigroup presentations; for the necessary background, see [28].

First of all, some information is required about universal groups: if S is a semigroup that embeds in a group, then the *universal group* U of S is the largest group into which S embeds and which it generates, in the sense that all other such groups are homomorphic images of U. Alternatively, the universal group of S is the group defined by treating any semigroup presentation for S as a group presentation. (Actually, universal groups are defined for all semigroups, not just those embeddable into groups. For the formal definition of universal groups of semigroups, and for further information on the subject, see [15, Chapter 12].)

Let S be a subsemigroup of a group G, and let H be the subgroup of G generated by S. In general, H is *not* isomorphic to the universal group U of S; see [9, Example 0.9.2]. The subgroup H will, however, be a homomorphic image of U.

The free semigroup with basis A is denoted by A^+; the free monoid with the same basis by A^*. Identify A^+ with the set of non-empty words over the alphabet A and A^* with the set of all words over A.

Let A be an alphabet representing a set of generators for a semigroup S. For any word $w \in A^+$, denote by $\overline{w}$ the element of S represented by w. For any set of words W, $\overline{W}$ is the set of all elements represented by at least one word in W.

Definition 2.1 Let S be any semigroup. A congruence σ on S is a *Malcev congruence* if the corresponding factor semigroup S/σ is embeddable in a group.

If $\{\sigma_i : i \in I\}$ is a set of Malcev congruences on S, then $\sigma = \bigcap_{i \in I} \sigma_i$ is also a Malcev congruence on S. This is true because S/σ_i embeds in a group G_i for each $i \in I$, so S/σ embeds in $\prod_{i \in I} S/\sigma_i$, which in turn embeds in $\prod_{i \in I} G_i$. The following definition therefore makes sense.

Definition 2.2 Let A^+ be a free semigroup; let $\rho \subseteq A^+ \times A^+$ be any binary relation on A^+. Denote by ρ^{M} the smallest Malcev congruence containing ρ — namely,

$$\rho^{\mathrm{M}} = \bigcap \left\{ \sigma : \sigma \supseteq \rho, \ \sigma \text{ is a Malcev congruence on } A^+ \right\}.$$

Then $\mathrm{SgM}\langle A | \rho\rangle$ is a *Malcev presentation* for [any semigroup isomorphic to] A^+/ρ^{M}. If both A and ρ are finite, then the Malcev presentation $\mathrm{SgM}\langle A | \rho\rangle$ is said to be finite.

The notation $\mathrm{SgM}\langle A | \rho\rangle$ distinguishes the Malcev presentation with generators A and defining relations ρ from the ordinary semigroup presentation $\mathrm{Sg}\langle A | \rho\rangle$, which defines $A^+/\rho^{\#}$. (The smallest congruence containing ρ is denoted by $\rho^{\#}$.) Similarly, $\mathrm{Gp}\langle A | \rho\rangle$ denotes the group presentation with the same set of generators and defining relations. Let X be a subset of a group G. Denote by $\mathrm{Sg}\langle X\rangle$ the subsemigroup generated by X and by $\mathrm{Gp}\langle X\rangle$ the subgroup generated by X.

Fix A^+ and ρ as in the Definition 2.2 and let $S = A^+/\rho^{\mathrm{M}}$. Let $A^{\mathrm{L}}, A^{\mathrm{R}}$ be two sets in bijection with A under the mappings $a \mapsto a^{\mathrm{L}}$, $a \mapsto a^{\mathrm{R}}$, respectively, with $A, A^{\mathrm{L}}, A^{\mathrm{R}}$ being pairwise disjoint. Extend the mappings $a \mapsto a^{\mathrm{L}}$, $a \mapsto a^{\mathrm{R}}$ to anti-isomorphisms from A^* to $(A^{\mathrm{L}})^*$ and $(A^{\mathrm{R}})^*$, respectively. Let

$$\tau = \rho \cup \left\{ (bb^{\mathrm{R}}a, a), (abb^{\mathrm{R}}, a), (b^{\mathrm{L}}ba, a), (ab^{\mathrm{L}}b, a) : a \in A \cup A^{\mathrm{L}} \cup A^{\mathrm{R}}, b \in A \right\}.$$

Let $G = \mathrm{Sg}\langle A \cup A^{\mathrm{L}} \cup A^{\mathrm{R}} | \tau\rangle$. The semigroup G is actually the universal group of S, with $G \simeq \mathrm{Gp}\langle A | \rho\rangle$. It can be shown that

$$\rho^{\mathrm{M}} = \tau^{\#} \cap (A^+ \times A^+),$$

or equivalently that

$$\rho^{\mathrm{M}} = \{(u, v) : u, v \in A^+ \text{ represent the same element of } \mathrm{Gp}\langle A | \rho\rangle\}. \tag{1}$$

Therefore, two words $u, v \in A^+$ represent the same element of S if and only if there is a sequence

$$u = u_0 \to u_1 \to \ldots \to u_n = v,$$

with $n \geq 0$, where, for each $i \in \{0, \ldots, n-1\}$, there exist $p_i, q_i, q_i', r_i \in (A \cup A^{\mathrm{L}} \cup A^{\mathrm{R}})^*$ such that $u_i = p_i q_i r_i$, $u_{i+1} = p_i q_i' r_i$, and $(q_i, q_i') \in \tau$ or $(q_i', q_i) \in \tau$. In fact, it can be shown that $u, v \in A^+$ represent the same element of S if any only if there exists such a sequence with $p_i \in (A \cup A^{\mathrm{L}})^*$ and $r_i \in (A \cup A^{\mathrm{R}})^*$. This restriction on the letters that can appear in p_i and r_i simply means that no changes can occur to the left of an a^{L} or to the right of an a^{R}. Such a sequence is called a *Malcev ρ-chain* (or simply a *Malcev chain*) from u to v. If $(u, v) \in \rho^{\mathrm{M}}$, then (u, v) is said to be a *Malcev consequence* of ρ.

[The reader should be aware of disagreement between various authors [3, 15, 31] on the correct terminology relating to 'Malcev chains'. The problem is discussed fully in [9, Section 1.2].]

Example 2.3 Let F be the free semigroup $\{x, y, z, t\}^+$. Suppose the alphabet $A = \{a, b, c, d, e, f\}$ represents elements of F as follows:

$$\bar{a} = x^2yz, \qquad\qquad \bar{d} = x^2y,$$
$$\bar{b} = yz, \qquad\qquad \bar{e} = zy,$$
$$\bar{c} = yt^2, \qquad\qquad \bar{f} = zyt^2,$$

and let S be the subsemigroup of F generated by $\overline{A}$. Elementary reasoning shows that S has presentation $\mathrm{Sg}\langle A \mid \mathcal{R} \rangle$, where

$$\mathcal{R} = \{(ab^\alpha c, de^\alpha f) : \alpha \in \mathbb{N} \cup \{0\}\}.$$

The elements $\overline{ab^\alpha} = x^2 yz(yz)^\alpha$ and $\overline{b^\alpha c} = (yz)^\alpha yx^2$ have unique representatives over the alphabet A. Therefore no valid relations hold in S that can be applied to a proper subword of $ab^\alpha c$. Each of the words $ab^\alpha c$ must therefore appear as one side of a defining relation in a presentation for S on the generating set $\overline{A}$. The semigroup S is thus not finitely presented.

However, S has a finite Malcev presentation $\mathrm{SgM}\langle A \mid \mathcal{Q} \rangle$, where

$$\mathcal{Q} = \{(ac, df), (abc, def)\}.$$

Each defining relation in $\mathcal{R}$ is a Malcev consequence of the two defining relations in $\mathcal{Q}$ — that is, $\mathcal{R} \subseteq \mathcal{Q}^{\mathrm{M}}$. One can easily prove this by induction on α: assume that, for $\beta < \alpha$, the relations $(ab^\beta c, de^\beta f)$ are in $\mathcal{Q}^{\mathrm{M}}$. Then

$$
\begin{aligned}
ab^\alpha c &\to ab^{\alpha-1} cc^{\mathrm{R}} bc \\
&\to de^{\alpha-1} fc^{\mathrm{R}} bc && \text{(by the induction hypothesis)} \\
&\to de^{\alpha-1} d^{\mathrm{L}} df c^{\mathrm{R}} bc \\
&\to de^{\alpha-1} d^{\mathrm{L}} acc^{\mathrm{R}} bc \\
&\to de^{\alpha-1} d^{\mathrm{L}} abc \\
&\to de^{\alpha-1} d^{\mathrm{L}} def && \text{(by the induction hypothesis)} \\
&\to de^\alpha f,
\end{aligned}
$$

and so $(ab^\alpha c, de^\alpha f) \in \mathcal{Q}^{\mathrm{M}}$.

3   Malcev coherent groups & monoids

Recall that a group is coherent if all of its finitely generated subgroups are finitely presented [29].

Definition 3.1 A group — or more generally a group-embeddable semigroup — is *Malcev coherent* if all of its finitely generated subsemigroups admit finite Malcev presentations.

As the universal group of a group is simply that group itself, it follows from the properties of Malcev presentations described in Section 2 that every Malcev coherent group is coherent.

Example 2.3 illustrates a finitely generated subsemigroup of a free semigroup that admits a finite Malcev presentation. Spehner proved that the same is true of all such subsemigroups:

Theorem 3.2 ([32]) *Every free semigroup is Malcev coherent.*

This result, together with examples in Spehner's earlier paper [31], formed the whole of the theory of Malcev presentations in early 2003. Since then, a great deal of progress, surveyed in the remainder of the present article, has been made.

As part of a wider investigation into subsemigroups of free and virtually free groups [12, 11], Cain, Robertson & Ruškuc proved the following generalization of Theorem 3.2:

Theorem 3.3 ([12, Theorem 3]) *Every virtually free group is Malcev coherent.*

[Recall that a group is *virtually* $\mathfrak{P}$ if it contains a finite-index subgroup with property $\mathfrak{P}$.] The proof of Theorem 3.3 in [12] relies on the fact that virtually free groups have context-free word problem [24]. A second proof uses Theorem 7.1.

The Malcev coherence of abelian groups is an immediate consequence of Rédei's Theorem [25], which asserts that all finitely generated commutative semigroups are finitely presented. A more general result holds:

Theorem 3.4 ([11, Theorem 1]) *Every nilpotent-by-finite group is Malcev coherent.*

[Rédei's Theorem does not extend to nilpotent groups: finitely generated subsemigroups that do not admit finite presentations can be found inside nilpotent groups. For example, let G be the Heisenberg group

$$\mathrm{Gp}\langle x, y, z \mid (yx, xyz), (yz, zy), (zx, xz)\rangle,$$

which is nilpotent of class 2. The subsemigroup $S = \mathrm{Sg}\langle \overline{x}, \overline{y}\rangle$ of G is finitely generated but not finitely presented [11, Example 4.5].]

Nilpotent-by-finite groups have a special property: the subgroup generated by a subsemigroup of such a group coincides with the universal group of that subsemigroup. [Recall that this is not generally true: the subgroup envelope may be a proper homomorphic image of the universal group.] Thus the coherence of nilpotent-by-finite groups implies their Malcev coherence.

4   Free products

Free groups and abelian groups are coherent; the free product of a free group and an abelian group is therefore coherent by the Kurosh Subgroup Theorem. Free groups are Malcev coherent by Theorem 3.3; abelian groups by Theorem 3.4. One therefore naturally asks whether the free product of a free group and an abelian group is Malcev coherent, and more generally whether the class of Malcev coherent groups is closed under free products. The following example answers this question negatively:

Example 4.1 ([11, Section 7]) Let F be the free product $\mathrm{FG}(x, y, z, s, t) * (\mathbb{Z} \times \mathbb{Z} \times \mathbb{Z})$. Identify elements of F with alternating products of elements of the free group $\mathrm{FG}(x, y, z, s, t)$ (viewed as reduced words) and of $\mathbb{Z} \times \mathbb{Z} \times \mathbb{Z}$ (viewed as triples of integers).

Let $A = \{a, b, c, d, e, f, g, h, i, j\}$ be an alphabet, and let this alphabet represent elements of F in the following way:

$$\overline{a} = x^2 y, \qquad\qquad\qquad \overline{f} = x^2 s,$$
$$\overline{b} = y^{-1}(1, 0, 1)y, \qquad\qquad \overline{g} = s^{-1}(1, 0, 0)s,$$
$$\overline{c} = y^{-1}z, \qquad\qquad\qquad \overline{h} = s^{-1}t,$$
$$\overline{d} = z^{-1}(0, 1, 0)z, \qquad\qquad \overline{i} = t^{-1}(0, 1, 1)t,$$
$$\overline{e} = z^{-1}x^2, \qquad\qquad\qquad \overline{j} = t^{-1}x^2.$$

Let S be the subsemigroup of F generated by $\overline{A}$.

Elementary but tedious reasoning shows that the semigroup S is presented by $\mathrm{Sg}\langle A \mid \mathcal{R}\rangle$, where

$$\mathcal{R} = \{(ab^\alpha cd^\alpha e, fg^\alpha hi^\alpha j) : \alpha \in \mathbb{N} \cup \{0\}\}.$$

The universal group of the semigroup S is $U = \mathrm{Gp}\langle A \mid \mathcal{R}\rangle$, which is simply the amalgamated free product

$$\mathrm{FG}(a, b, c, d, e) *_K \mathrm{FG}(f, g, h, i, j),$$

where $K \simeq \mathrm{Gp}\langle ab^\alpha cd^\alpha e, \alpha \in \mathbb{N} \cup \{0\}\rangle \simeq \mathrm{Gp}\langle fg^\alpha hi^\alpha j, \alpha \in \mathbb{N} \cup \{0\}\rangle$. Using the elementary theory of free groups (see, for example, [21, Section I.2]), one can show that the amalgamated subgroup K is not finitely generated. A theorem of Baumslag [5] then shows that U is not finitely presented. Since a Malcev presentation for S is simply a presentation for U, the semigroup S cannot admit a finite Malcev presentation.

This example yields several important results. Observing that free groups and abelian groups are Malcev coherent by Theorems 3.3 and 3.4 produces the first result:

Theorem 4.2 *The class of Malcev coherent groups is not closed under forming free products.*

The Kurosh Subgroup Theorem implies that the class of coherent groups *is* closed under forming free products. Consequently, the free product of a free group and an abelian group is coherent. Therefore F is an example of a coherent group that is not Malcev coherent.

Theorem 4.3 *The class of Malcev coherent groups is properly contained in the class of coherent groups.*

Although the free product of a free group and an abelian group is not in general Malcev coherent, if one replaces the free group with a free monoid, one recovers Malcev coherence:

Theorem 4.4 ([11, Theorem 6]) *Every [monoid] free product of a free monoid and an abelian group is Malcev coherent.*

Problem 4.5 Is the free product of a free monoid and a virtually abelian group, or a nilpotent or virtually nilpotent group Malcev coherent? More generally, is the class of Malcev coherent monoids closed under forming free products with a free monoid?

5 Direct products

The direct product of two free non-abelian groups is not coherent; see [18]. However, this does not preclude the possibility that the direct product of two free semigroups is Malcev coherent. However, it is possible to construct an example that proves otherwise.

Theorem 5.1 ([10, Theorem 1]) *The direct product of two free semigroups of rank at least 2 is not Malcev coherent.*

Since any polycyclic group that is not virtually nilpotent admits a free subsemigroup of rank 2 [27, Theorem 4.12], and since the class of polycyclic groups is closed under forming direct products, the following important corollary is established:

Corollary 5.2 ([10, Theorem 2]) *Polycyclic groups are not in general Malcev coherent.*

[Polycyclic groups are known to be coherent; see, for example, [30, Section 9.3].]
The direct product of a [virtually] free group and a polycyclic group is coherent by a variation on a theorem of Miller [23]. Corollary 5.2 implies that this result does not generalize to Malcev coherence. The following partial generalization has been established:

Theorem 5.3 ([10, Theorem 3]) *Direct products of virtually free groups and abelian groups are Malcev coherent.*

Problem 5.4 Is every direct product of a [virtually] free group and a virtually abelian group Malcev coherent? Is every direct product of a [virtually] free group and a [virtually] nilpotent group Malcev coherent?

6 Baumslag–Solitar semigroups and groups

Baumslag & Solitar [7] introduced groups with presentations of the form

$$\mathrm{Gp}\langle x, y \mid (yx^m, x^n y)\rangle, \tag{2}$$

where m and n are natural numbers. Denote by $\mathrm{BSG}(m, n)$ the Baumslag–Solitar group presented by (2). Analogously, the Baumslag–Solitar semigroups are those semigroups

$$\mathrm{BSS}(m, n) = \mathrm{Sg}\langle x, y \mid (yx^m, x^n y)\rangle,$$

where $m, n \in \mathbb{N}$.

The Baumslag–Solitar groups are known to be coherent [20]. The question of their Malcev coherence has not yet been settled (see Problem 6.2). However, the following, more specialized, result has been established:

Theorem 6.1 ([8]) *All Baumslag–Solitar semigroups* BSS(m, n), *where* $m \neq n$, *are Malcev coherent.*

Problem 6.2 Are the Baumslag–Solitar semigroups BSS(m, m) Malcev coherent? Are the Baumslag–Solitar groups BSG(m, n) Malcev coherent?

Baumslag [6, Section B] asks whether all one-relator groups are coherent. This problem remains unanswered, but it is natural, although perhaps precipitate, to ask the following questions:

Problem 6.3 Are all-one relator groups Malcev coherent? Are all one-relator cancellative semigroups Malcev coherent? [One-relator cancellative semigroups are always group-embeddable [1, 3].]

7   Links with automatic semigroup theory

The concept of an automatic structure has been generalized from groups [17] to semigroups [14]. Automatic groups are always finitely presented [17, Theorem 2.3.12]; automatic semigroups, in general, are not [14, Examples 3.9, 4.4, and 4.5].

Theorem 7.1 ([11, Theorem 2]) *Every [synchronously or asynchronously] automatic group-embeddable semigroup admits a finite Malcev presentation.*

This theorem has proven to be a very useful tool: it is sometimes easier to prove that all finitely generated subsemigroups of a particular group or semigroup are automatic or asynchronously automatic than to attempt to prove Malcev coherence directly. Theorems 4.4 and 6.1 were originally proven in this way, and Theorem 3.3 can also be proven thus [8, Section 6].

8   Large subsemigroups and small extensions

Definition 8.1 Let S and T be semigroups with T being contained in S. The *Rees index* of T in S is $|S - T| + 1$. If this Rees index is finite, then the semigroup S is a *small extension* of T, and the semigroup T is a *large subsemigroup* of S.

Many properties of semigroups are known to be preserved under constructing small extensions or passing to large subsemigroups. Finite generation is such a property [13], as is automatism [16]. Ruškuc [19, Theorem 1.1] showed that finite presentability is also thus preserved. Anyone familiar with Malcev presentations will therefore ask whether the property of admitting a finite Malcev presentation is preserved under taking large subsemigroups or small extensions. Such preservation does indeed occur:

Theorem 8.2 *Let S be a semigroup that embeds in a group. Let T be a subsemigroup of S of finite Rees index. Then S has a finite Malcev presentation if and only if T has a finite Malcev presentation.*

Theorem 8.2 is actually an immediate consequence of the following result:

Theorem 8.3 *Let S be a semigroup that embeds in a group. Let T be a subsemigroup of S. Suppose that $|T| > |S - T|$. [This includes the possibility that T is infinite and $S - T$ finite.] Then the universal groups of S and T are isomorphic.*

[The results in this section have appeared in the author's Ph.D. thesis [9, Section 9.2], but are otherwise unpublished.]

9   A final open problem

In the author's opinion, the most important unanswered question in the theory of Malcev presentations is the following:

Problem 9.1 Is the class of Malcev coherent groups closed under forming finite extensions?

[The class of coherent groups is closed under forming finite extensions as a consequence of the Reidermeister–Schreier Theorem (see, for example, [21, Section III.3]).] Two of the three classes of Malcev coherent groups known — namely, the virtually free and virtually nilpotent groups — are by definition closed under forming finite extensions. Thus, with the current state of knowledge of Malcev coherence, only the class of direct products of virtually free and abelian groups might hypothetically yield an example to negatively answer Problem 9.1. On the other hand, an attempt to positively answer this question by means of a proof paralleling the equivalent result for coherent groups will certainly fail; see [9, Section 9.5] for details.

10   Précis

The purpose of the present section is to gather and summarize the examples and results for the purposes of reference and comparison. Three tables contain the various data:

- **Table 1** summarizes the closure of the classes of coherent and Malcev coherent groups under free product, finite extension, and direct product.
- **Table 2** compares the Malcev coherence of groups satisfying various properties $\mathfrak{P}$, virtually $\mathfrak{P}$ groups, and direct products of these groups with free groups.
- **Table 3** describes the Malcev coherence of groups and semigroups of various classes.

| | CLASS OF GROUPS | |
CLOSURE UNDER	COHERENT	MALCEV COHERENT
Free product	Y (K)	N (Th. 4.2)
Finite extension	Y (R–S)	? (Pr. 9.1)
Direct product	N [18]	N [18]

Table 1. Closure of various classes of groups under certain constructions. [Y = Yes, N = No, ? = Undecided; K = Kurosh Subgroup Theorem, R–S = Reidermeister–Schreier Theorem]

| | MALCEV COHERENT | | | |
$\mathfrak{P}$	$\mathfrak{P}$	VIRT. $\mathfrak{P}$	FREE $\times$ $\mathfrak{P}$ [1]	FREE $\times$ VIRT. $\mathfrak{P}$ [1]
Free	Y (Th. 3.3)	Y (Th. 3.3)	N (Incoherent)	N (Incoherent)
Abelian	Y (Th. 3.4)	Y (Th. 3.4)	Y (Th. 5.3)	? (Pr. 5.4)
Nilpotent	Y (Th. 3.4)	Y (Th. 3.4)	? (Pr. 5.4)	? (Pr. 5.4)
Polycyclic	N (Co. 5.2)	N (Co. 5.2)	N (Co. 5.2)	N (Co. 5.2)
One-relator	? (Pr. 6.3)	? (Pr. 6.3)	N (Incoherent)	N (Incoherent)

Table 2. Malcev coherence of groups of certain classes. [Y = Yes, N = No, ? = Undecided.] It is unknown whether all [virtually] one-relator groups are coherent. Except for the classes so marked, groups of the other classes in this table are known to be coherent. An entry 'N' in this table does not preclude a particular semigroup of the class in question from being Malcev coherent. It merely asserts that not all such semigroups have that property.

---

[1]The term 'free' could be replaced by 'virtually free' in the headings of the rightmost two columns of Table 2 without altering its validity. However, it is conceivable that — for example — direct products of free groups and nilpotent groups are Malcev coherent but that there exists a direct product of a *virtually* free group and a nilpotent group that is not Malcev coherent. This situation, however, cannot arise if Open Problem 9.1 has a positive answer.

SEMIGROUP	MALCEV COHER.
Virtually free group	Y (Th. 3.3)
Virt. abelian/nilpotent/polycyclic group:	
Abelian group	Y (Th. 3.4)
Virtually abelian group	Y (Th. 3.4)
Nilpotent/virt. nilpotent group	Y (Th. 3.4)
Polycyclic group	N (Co. 5.2)
Direct product:	
Free semigroup $\times$ Free semigroup	N (Th. 5.1)
Free semigroup $\times$ Natural numbers	Y (Th. 5.3)
Virt. free group $\times$ Abelian group	Y (Th. 5.3)
Virt. free group $\times$ Virt. abel. group	? (Pr. 5.4)
Virt. free group $\times$ Nil./virt. nil. group	? (Pr. 5.4)
Free product:	
Free group $*$ Abelian group	N (Ex. 4.1)
Free monoid $*$ Abelian group	Y (Th. 4.4)
Free monoid $*$ Virt. abelian. group	? (Pr. 4.5)
Free monoid $*$ Nil./virt. nil. group	? (Pr. 4.5)
Free monoid $*$ Malcev coher. monoid	? (Pr. 4.5)
Baumslag–Solitar semigroup/group:	
$\mathrm{BSS}(m,n)$, $m > n$	Y (Co. 6.1)
$\mathrm{BSS}(m,n)$, $m < n$	Y (Co. 6.1)
$\mathrm{BSS}(m,m)$, $m > 1$	? (Pr. 6.2)
$\mathrm{BSG}(m,n)$, $m \neq n$	? (Pr. 6.2)
$\mathrm{BSG}(m,m)$, $m > 1$	? (Pr. 6.2)
One-relator cancellative semigroup	? (Pr. 6.3)
One-relator group	? (Pr. 6.3)

Table 3. Malcev coherence of various semigroups. [Y = Yes, N = No, ? = Undecided.] An entry 'N' in this table does not preclude a particular semigroup of the class in question from being Malcev coherent. It merely asserts that not all such semigroups have that property.

References

[1] S. I. Adyan, On the embeddability of semigroups in groups, *Doklady Akademii Nauk SSSR* **1** (1960), 819–821. [In Russian. See [2] for a translation.].

[2] S. I. Adyan, On the embeddability of semigroups in groups, *Soviet Math. Dokl.* **1** (1960), 819–821, 1960. [Translated from the Russian.].

[3] S. I. Adjan, Defining relations and algorithmic problems for groups and semigroups, *Trudy Mat. Inst. Steklov.* **85** (1966). [In Russian. See [4] for a translation.].

[4] S. I. Adjan, Defining relations and algorithmic problems for groups and semigroups, *Proceedings of the Steklov Institute of Mathematics* **85** (1966). [Translated from the Russian by M. Greendlinger.].

[5] G. Baumslag, A remark on generalized free products, *Proc. Amer. Math. Soc.* **13** (1962), 53–54.

[6] G. Baumslag, Some problems on one-relator groups, In *Proceedings of the Second International Conference on the Theory of Groups (Australian National University, Canberra, 1973)*, volume 372 of *Lecture Notes in Mathematics*, pages 75–81, Berlin, 1974. Springer.

[7] G. Baumslag and D. Solitar, Some two-generator one-relator non-Hopfian groups, *Bull. Amer. Math. Soc.* **68** (1962), 199–201.

[8] A. J. Cain, Automatism of subsemigroups of Baumslag–Solitar semigroups, 2005. In preparation.

[9] A. J. Cain, *Presentations for Subsemigroups of Groups*, Ph.D. Thesis, University of St Andrews, 2005.

[10] A. J. Cain, Subsemigroups of direct products of coherent groups, 2005. In preparation.

[11] A. J. Cain, E. F. Robertson, and N. Ruškuc, Subsemigroups of groups: presentations, Malcev presentations, and automatic structures, *J. Group Theory* **9** (2006), 397–426.

[12] A. J. Cain, E. F. Robertson, and N. Ruškuc, Subsemigroups of virtually free groups: finite Malcev presentations and testing for freeness, *Math. Proc. Cambridge Philos. Soc.* **141** (2006), 57–66.

[13] C. M. Campbell, E. F. Robertson, N. Ruškuc, and R. M. Thomas, Reidemeister–Schreier type rewriting for semigroups, *Semigroup Forum* **51** (1995), no. 1, 47–62.

[14] C. M. Campbell, E. F. Robertson, N. Ruškuc, and R. M. Thomas, Automatic semigroups, *Theoret. Comput. Sci.* **250** (2001), no. 1–2, 365–391.

[15] A. H. Clifford and G. B. Preston, *The Algebraic Theory of Semigroups* (Vol. II), Number 7 in Mathematical Surveys, American Mathematical Society, Providence, R.I., 1967.

[16] A. J. Duncan, E. F. Robertson, and N. Ruškuc, Automatic monoids and change of generators, *Math. Proc. Cambridge Philos. Soc.* **127** (1999), no. 3, 403–409.

[17] D. B. A. Epstein, J. W. Cannon, D. F. Holt, S. V. F. Levy, M. S. Paterson, and W. P. Thurston, *Word Processing in Groups*, Jones & Bartlett, Boston, Mass., 1992.

[18] F. J. Grunewald, On some groups which cannot be finitely presented, *J. London Math. Soc. (2)* **17** (1978), no. 3, 427–436.

[19] M. Hoffmann, R. M. Thomas, and N. Ruškuc, Automatic semigroups with subsemigroups of finite Rees index, *Internat. J. Algebra Comput.* **12** (2002), no. 3, 463–476.

[20] P. H. Kropholler, Baumslag–Solitar groups and some other groups of cohomological dimension two, *Comment. Math. Helv.* **65** (1990), no. 4, 547–558.

[21] R. C. Lyndon and P. E. Schupp, *Combinatorial Group Theory*, volume 89 of *Ergebnisse der Mathematik und ihrer Grenzgebiete*, Springer-Verlag, Berlin, 1977.

[22] A. I. Malcev, On the immersion of associative systems in groups, *Mat. Sbornik* **6 (48)** (1939), 331–336. [In Russian].

[23] C. F. Miller, III, Subgroups of direct products with a free group, *Q. J. Math.* **53** (2002), no. 4, 503–506.

[24] D. E. Muller and P. E. Schupp, Groups, the theory of ends, and context-free languages, *J. Comput. System Sci.* **26** (1983), no. 3, 295–310.

[25] L. Rédei, *Theorie der Endlich Erzeugbaren Kommutativen Halbgruppen*, volume 41 of *Hamburger Mathematische Einzelschriften*, Physica-Verlag, Würzburg, 1963. [In German. See [26] for a translation.].

[26] L. Rédei, *The Theory of Finitely Generated Commutative Semigroups*, Pergamon Press, Oxford, 1965. [Translated from the German. Edited by N. Reilly.].

[27] J. M. Rosenblatt, Invariant measures and growth conditions, *Trans. Amer. Math. Soc.* **193** (1974), 33–53.

[28] N. Ruškuc, *Semigroup Presentations*, Ph.D. Thesis, University of St Andrews, 1995.

[29] J.-P. Serre, Problem section, In J. Cossey, editor, *Proceedings of the Second International Conference on the Theory of Groups (Australian National University, Can-

berra, 1973), volume 372 of *Lecture Notes in Mathematics*, pages 733–740, Berlin, 1974. Springer.

[30] C. C. Sims, *Computation with Finitely Presented Groups*, volume 48 of *Encyclopedia of Mathematics and its Applications*, Cambridge University Press, Cambridge. 1994.

[31] J.-C. Spehner, Présentations et présentations simplifiables d'un monoïde simplifiable, *Semigroup Forum* **14** (1977), no. 4, 295–329. [In French].

[32] J.-C. Spehner, Every finitely generated submonoid of a free monoid has a finite Malcev's presentation, *J. Pure Appl. Algebra* **58** (1989), no. 3, 279–287.

FINITE GROUPS WITH EXTREMAL CONDITIONS ON SIZES OF CONJUGACY CLASSES AND ON DEGREES OF IRREDUCIBLE CHARACTERS

DAVID CHILLAG* and MARCEL HERZOG†

*Department of Mathematics, Technion, Israel Institute of Technology, Haifa 32000, Israel

†School of Mathematical Sciences, Raymond and Beverly Sackler Faculty of Exact Sciences, Tel-Aviv University, Tel-Aviv, Israel

Email: `herzogm@post.tau.ac.il`

I Introduction

In this survey G denotes a finite group of order g, with k conjugacy classes and center $Z(G)$ of order z. Denote the order of G', the commutator subgroup of G, by g' and assume that $g > 1$. Denote by $Cls(G) = \{c_1 = 1, c_2, \ldots, c_k\}$ the multiset composed of the sizes of the conjugacy classes of G (with $c_1 = |\{1\}|$) and by $Chd(G) = \{x_1 = 1, x_2, \ldots, x_k\}$ the multiset composed of the degrees of the irreducible characters of G (with $x_1 = 1_G(1)$). The influence of the arithmetical structure of the c_i's and the x_i's on the group-theoretical structure of G has been investigated in many papers. For example, the following results concerning the class sizes in G were proved in [4]. Here, and in the sequel, by "a class" we mean "a conjugacy class" and by "a prime" we mean "a non-necessarily fixed prime". For additional information, see [4], [5], [6] and [8].

Theorem 1 *The following statements hold:*
1. *If c_i equals 1 or a prime for each $c_i \in Cls(G)$, then either G is nilpotent of class ≤ 2 or $G/Z(G)$ is a Frobenius group of order pq, where p and q are distinct primes.*

2. *If c_i equals 1 or a prime power for each $c_i \in Cls(G)$, then either G is nilpotent or $G/Z(G)$ is a solvable Frobenius group.*

3. *If c_i is a squarefree number for each $c_i \in Cls(G)$, then G is supersolvable and $dl(G) \leq 3$.*

4. *If p is a fixed prime and $p \nmid c_i$ for each $c_i \in Cls(G)$, then the Sylow p-subgroup of G is central in G.*

5. *If p is a fixed prime and $p \mid c_i$ for each $c_i \in Cls(G)$ satisfying $c_i \neq 1$, then $C_G(P) \leq Z(G)$ for each Sylow p-subgroup P of G.*

6. *If $4 \nmid c_i$ for each $c_i \in Cls(G)$, then G is solvable.*

Similar results were obtained by many authors concerning the irreducible character degrees of G. We summarize the best known ones in the following omnibus theorem. The names of the authors and the references can be found in [4], [8] and [9].

Theorem 2 *The following statements hold:*

1. If x_i equals 1 or a prime for each $x_i \in Chd(G)$, then G is solvable and $dl(G) \le 3$.

2. If x_i equals 1 or a prime power for each $x_i \in Chd(G)$, then G is not necessarily solvable. If G is assumed to be solvable, then $dl(G) \le 5$.

3. If x_i is a squarefree number for each $x_i \in Chd(G)$, then again G is not necessarily solvable. If G is assumed to be solvable, then $dl(G) \le 4$.

4. If p is a fixed prime and $p \nmid x_i$ for each $x_i \in Chd(G)$, then G has a normal abelian Sylow p-subgroup.

5. If p is a fixed prime and $p \mid x_i$ for each $x_i \in Chd(G)$ satisfying $x_i \ne 1$, then G has a normal p-complement.

6. If $4 \nmid x_i$ for each $x_i \in Chd(G)$, then either G is solvable or $G/N \cong A_7$ for some solvable $N \lhd G$.

II   Extremal conditions

In this note we wish to consider groups satisfying some extremal conditions with respect to the sizes of $Cls(G)$ and $Chd(G)$ viewed **as sets**. It is clear that either of the conditions $|Cls(G)| = 1$ (i.e., all classes of G are of size 1) and $|Chd(G)| = 1$ (i.e., all irreducible characters of G are linear) is equivalent to G being abelian. Therefore we shall deal with nonabelian groups G satisfying one of the following conditions:

  (1) $|Cls(G)| = 2$ (i.e., all noncentral classes are of the same size);

  (2) $|Chd(G)| = 2$ (i.e., all nonlinear irreducible characters are of the same degree);

  (3) $|Cls(G)| = k$ (i.e., all classes are of distinct sizes);

  (4) $|Chd(G)| = k$ (i.e., all irreducible characters are of distinct degrees).

We shall consider two additional conditions with a similar flavor. We first define $Cls^*(G)$ as the set of sizes of the noncentral classes of G and $Chd^*(G)$ as the set of degrees of the nonlinear irreducible characters of G. The remaining conditions are:

  (5) $|Cls^*(G)| = k - z$ (i.e., all noncentral classes are of distinct sizes); and

  (6) $|Chd^*(G)| = k - g/g'$ (i.e., all nonlinear irreducible characters are of distinct degrees).

Some additional conditions will be considered in the next section, which deals with groups of odd order.

We shall consider now each case separately.

  (1) $|Cls(G)| = 2$.

In this case, it was shown by N. Ito [11] that G is nilpotent with a unique nonabelian Sylow p-subgroup, which implies that the class sizes are 1 and p^r for some integer $r > 0$. Thus the problem reduces to a p-group problem and is **still open**, as far as a complete classification is considered. L. Verardi [14] showed that $G/Z(G)$ is of exponent p and recently K. Ishikawa [10] proved that $cl(G) \le 3$. For Isaacs' simpler proof of Ishikawa's result see [1] and for Mann's generalization see [13].

(2) $|Chd(G)| = 2$.

This problem is also **still open**. Let the character degrees of G be 1 and m. Isaacs and Passman proved that G' is abelian and either G is nilpotent, with $m = p^a$ for some prime p, or there exists an abelian normal subgroup A of G with $[G : A] = m$ (see [9, Chapter 12]). Recently, the nonnilpotent groups were completely classified by Bianchi, Gillio, Herzog, Qian and Shi in [3]. They proved:

Theorem 3 *Let G be a nonnilpotent group. Then the irreducible character degrees of G are 1 and $m > 1$ if and only if G' is abelian and one of the following holds.*

1. *$m = p$, a prime, and there exists an abelian normal subgroup A of G with $[G : A] = p$.*

2. *$G' \cap Z(G) = 1$ and $G/Z(G)$ is a Frobenius group with the kernel $G' \times Z(G)/Z(G)$ and a cyclic complement of order $[G : G' \times Z(G)] = m$.*

(3) $|Cls(G)| = k$.

This problem is **still open** in the nonsolvable case. In the solvable case it was proved by Zhang [15] and independently by Knörr, Lempken and Thielcke [12] that $G \cong S_3$. It is generally believed that no nonsolvable group satisfies this condition.

(4) $|Chd(G)| = k$.

No such group $G \neq 1$ exists, since it was proved in [2] that groups with nonlinear irreducible characters of distinct degrees are solvable (see Theorem 5 below) and hence contain more than one linear character.

(5) $|Cls^*(G)| = k - z$.

This condition implies that $z = 1$ by the following result of Herzog and Schönheim in [7]:

Theorem 4 *Let G be a nonabelian finite group and suppose that G contains at most two noncentral classes of each size. Moreover, suppose that if $x, y \in G - Z(G)$ and the classes of x and y are of equal sizes, then x and y are of the same order. Then $Z(G) = 1$.*

Hence conditions (3) and (5) are identical.

(6) $|Chd^*(G)| = k - g/g'$.

In this case, Berkovich, Chillag and Herzog proved in [2] the following classification theorem.

Theorem 5 *Suppose that G is a nonabelian group with nonlinear irreducible characters of distinct degrees. Then G is one of the following groups:*

1. *An extraspecial 2-group of order 2^{2m+1}, with a unique nonlinear irreducible character of degree 2^m.*

2. *A Frobenius group of order $p^n(p^n - 1)$ for some prime p, with an elementary abelian kernel G' of order p^n, a cyclic complement and a unique nonlinear irreducible character of degree $p^n - 1$.*

3. *The Frobenius group of order 72, with a complement isomorphic to the quaternion group of order 8 and two nonlinear irreducible characters of degrees 2 and 8.*

III   Nonabelian groups of odd order

If G is a nonabelian group of odd order g, then it has at least two nonidentity classes of each size and at least two nonprincipal irreducible characters of each degree. Therefore the extremal conditions (3)–(6) correspond to the following conditions:

(7) $|Cls(G)| = (k+1)/2$ (i.e., there are exactly two nonidentity classes of each size);

(8) $|Chd(G)| = (k+1)/2$ (i.e., there are exactly two nonprincipal irreducible characters of each degree);

(9) $|Cls^*(G)| = (k-z)/2$ (i.e., there are exactly two noncentral classes of each size); and

(10) $|Chd^*(G)| = (kg'-g)/2g'$ (i.e., there are exactly two nonlinear irreducible characters of each degree).

We shall consider now each case separately. First we notice that cases (7) and (9) are identical. Indeed, if G satisfies (7), then it clearly satisfies (9) and if G satisfies (9), then by Theorem 4, $z = 1$ and G satisfies (7). Thus it suffices to consider G satisfying

(9) $|Cls^*(G)| = (k-z)/2$.

This problem was solved in [7]. They proved:

Theorem 6 *Let G be a nonabelian group of odd order. Then G has exactly two noncentral classes of each size if and only if G is the nonabelian group of order 21.*

It remains to deal with conditions (8) and (10). We begin with

(10) $|Chd^*(G)| = (kg'-g)/2g'$.

Also this problem was solved recently. In [4], Chillag and Herzog proved:

Theorem 7 *Let G be a nonabelian group of odd order. Then G has exactly two nonlinear irreducible characters of each degree if and only if G is one of the following groups:*

1. *An extraspecial 3-group, with exactly two nonlinear irreducible characters of degree $\sqrt{|G|/3}$.*

2. *A Frobenius group of odd order $\frac{p^n-1}{2}p^n$ for some odd prime p, with an abelian kernel G' of order p^n and exactly two nonlinear irreducible characters of degree $\frac{p^n-1}{2}$.*

Finally, consider the case

(8) $|Chd(G)| = (k+1)/2$.

By the Feit–Thompson theorem, groups of odd order are solvable and consequently G must satisfy condition (10) and $g/g' = 3$. Hence Theorem 7 implies that $g = \frac{p^n-1}{2}p^n$ for some odd prime p, $g' = p^n$ and $(p^n-1)/2 = 3$. Thus we get

Theorem 8 *Let G be a nonabelian group of odd order. Then G has exactly two nonprincipal irreducible characters of each degree if and only if G is the nonabelian group of order* 21.

The irreducible character degrees of such G are $\{1, 1, 1, 3, 3\}$.

References

[1] Y. Barnea and I. M. Isaacs, Lie algebras with few centralizer dimensions, *J. Algebra* **259** (2003), 284–299.

[2] Y. Berkovich, D. Chillag and M. Herzog, Finite groups in which the degrees of the nonlinear irreducible characters are distinct, *Proc. Amer. Math. Soc.* **115** (1992), 955–959.

[3] M. Bianchi, A. Gillio, M. Herzog, G. Qian and W. Shi, Characterization of non-nilpotent groups with two irreducible character degrees, *J. Algebra* (to appear).

[4] D. Chillag and M. Herzog, Finite groups with almost distinct character degrees, *J. Algebra* (to appear).

[5] S. Dolfi, Prime factors of conjugacy class lengths and irreducible character-degrees in finite solvable groups, *J. Algebra* **174** (1995), 749–752.

[6] S. Dolfi, Arithmetical conditions on the lengths of the conjugacy classes of a finite group, *J. Algebra* **174** (1995), 753–771.

[7] M. Herzog and J. Schönheim, Groups of odd order with exactly two non-central conjugacy classes of the same size, *Arch. Math.* (to appear).

[8] B. Huppert, *Character Theory of Finite Groups*, Walter de Gruyter, Berlin (1998).

[9] I. M. Isaacs, *Character Theory of Finite Groups*, Dover Publications, Inc., New York (1994).

[10] K. Ishikawa, On finite p-groups which have only two conjugacy lengths, *Israel J. Math.* **129** (2002), 119–123.

[11] N. Ito, On finite groups with given conjugate types I, *Nagoya Math. J.* **6** (1953), 17–28.

[12] R. Knörr, W. Lempken and B. Thielcke, The S_3 conjecture for solvable groups, *Israel J. Math.* **91** (1995), 61–76.

[13] A. Mann, Minimal elements in finite p-groups, Personal communication.

[14] L. Verardi, On groups whose noncentral elements have the same finite number of conjugates, *Bull. Unione Mat. Italia (7)* **2A** (1988), 391–400.

[15] J. P. Zhang, Finite groups with many conjugate elements, *J. Algebra* **170** (1994), 608–624.

CONJUGACY CLASS STRUCTURE IN SIMPLE ALGEBRAIC GROUPS

MARTIN COOK[1]

Department of Mathematics and Statistics, Lancaster University, Lancaster LA1 4YF, United Kingdom
Email: m.cook@lancaster.ac.uk

Abstract

Let $G(k)$ be a simple algebraic group over an algebraically closed field k with prime characteristic. We look into aspects of the conjugacy class structure of $G(k)$ that are independent of the characteristic of k. This gives rise to some interesting combinatorics.

1 Introduction

The present paper gives some results from the Ph.D. thesis of the author. This thesis grew out of recent work by Lawther [6]; we begin with the main result of that work.

Let G be a simple algebraic group over an algebraically closed field k of characteristic p; let Φ be the root system of G, and take $t \in \mathbb{N}$. Define

$$G_{[t]} = \{g \in G : g^t = 1\} \quad \text{and} \quad G_{(t)} = \{g \in G : o(g) = t\}.$$

Then the main result of [6] is the following.

Theorem 1 *Given G, Φ and t as above, there is a number $d_{\Phi,t}$, depending only on Φ and t and satisfying $d_{\Phi,t} \geq |\Phi|/t$, with the property that $\operatorname{codim} G_{[t]} \geq d_{\Phi,t}$; if G is of adjoint type we in fact have $\operatorname{codim} G_{[t]} = d_{\Phi,t}$, and if in addition $G_{(t)} \neq \emptyset$, then $\operatorname{codim} G_{(t)} = d_{\Phi,t}$.*

Notice in particular that the above statement is independent of the characteristic p. This was observed for t small and prime by Liebeck. We also see that $G_{(t)}$ is a disjoint union of conjugacy classes and we want to see if a similar result is true if $G_{(t)}$ is replaced by one of these classes. This was our initial problem.

Throughout we consider t to be a prime power. We may then observe that an element of order p^i in a group over a field of characteristic $q \neq p$ is semisimple (diagonalizable), however if the characteristic is p then the element is unipotent (having its eigenvalues all equal to 1). Now using work by Hartley and Kuzucuoğlu [3] (following Deriziotis [2]) we know the possible isomorphism types of connected centralizers of a semisimple element s and the possible orders of semisimple elements with a given type of connected centralizer (we will only be interested in

---

[1]The author gratefully acknowledges support from the Engineering and Physical Sciences Research Council.

$C_G(s)^\circ$, so from now on when we say centralizer we mean connected centralizer). We also have the Jordan block structure of the unipotent conjugacy classes in all characteristics for the natural modules of the classical groups (Hesselink [4]) and the adjoint modules of the exceptionals (Lawther [5]). Section 2 will recall this background and state our main conjecture. Section 3 will then deal with the conjecture for type A and introduce some combinatorics. Section 4 will deal with type C and review progress to date.

2   Background

We begin with the work of Hartley and Kuzucuoğlu. We wish to find all possible centralizers of semisimple elements s of order t in an adjoint simple algebraic group G (throughout this article all groups are assumed to be adjoint). It is a well-known theorem of Steinberg ([1], section 3.5) that $C_G(s)^\circ$ is a reductive group of rank equal to Rank $G = n$. To determine the possible semisimple parts of these centralizers we begin by taking the extended Dynkin diagram of G with the ith node weighted by c_i, where c_i is the coefficient of the root α_i in the highest root α_0. The weight on the extending node (denoted c_0) is 1. Below we give these weighted diagrams for types A and C.

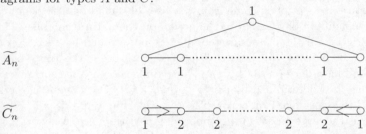

We then put labels $b_0, \ldots, b_n \geq 0$ on the nodes such that at most one of the b_i is greater than 1, $b_j = 1$ if $b_i = 0$ for all $i \neq j$ and $b_0 c_0 + \cdots + b_n c_n = t$. The graph spanned by those nodes with $b_i = 0$ is the Dynkin diagram of the semisimple part of $C_G(s)^\circ$. The possible choices of the b_is correspond to the possible structures of centralizers. This algorithm is slightly different from that seen in [3], but it is not too difficult to see they are equivalent.

Using this algorithm we can see that every centralizer of a semisimple element in $G = A_n$ is of the form $X = A_{n_1} \cdots A_{n_r} T_l$, where $n_1 \geq \cdots \geq n_r \geq 1$, $r \leq l + 1$ and $n_1 + \cdots + n_r + l = n$. We also see that if G is over a field of characteristic p, then for any t coprime to p and at least $l + 1$, a semisimple element s may be constructed with $o(s) = t$ and $C_G(s)^\circ \cong X$. Similarly, every centralizer of a semisimple element in $G = C_n$ has the form $X = A_{n_1} \cdots A_{n_r} C_{m_1} C_{m_2} T_l$, where $n_1 \geq \cdots \geq n_r \geq 1$, $m_1 \geq m_2 \geq 0$, $r \leq l$ and $n_1 + \cdots + n_r + m_1 + m_2 + l = n$. Here if $m_1 = m_2 = 0$ then for any t coprime to p and at least $2l$, a semisimple element may be constructed with $o(s) = t$ and $C_G(s)^\circ \cong X$. If $m_1 > 0 = m_2$ then a semisimple element may be constructed with $C_G(s)^\circ \cong X$ and $o(s) = t$, for any t coprime to p and at least $2l + 1$. Finally, if $m_1 \geq m_2 \geq 1$ then a semisimple element may be constructed with $C_G(s)^\circ \cong X$ and $o(s) = t$, for any *even* t coprime to p and at least $2l + 2$.

We now turn our attention to unipotent conjugacy classes. The material on type C is due to Hesselink [4]. Now in groups of type A_n, i.e., $PGL_{n+1}(k)$ for some n, unipotent classes are parameterized by partitions of $n + 1$. Furthermore the dimension of the centralizer of any element in the class corresponding to the partition $\mathbf{p}$ is given by $\sum_{i=1}^{z} x_i^2 - 1$, where $(x_1, \ldots, x_z)$ is the dual partition of $\mathbf{p}$ (see [1], section 13.1). The order of any element in this class is given by $\min\{q^i : q^i \geq p_1\}$, where $\operatorname{char} k = q$ and p_1 is the largest part in $\mathbf{p}$.

Unipotent classes in groups of type C_n are parameterized by symbols $\mathbf{p} = p_1{}_{\chi_1}^{m_1} \cdots p_z{}_{\chi_z}^{m_z}$. Here $p_1{}^{m_1} \cdots p_z{}^{m_z}$ is a partition of $2n$ where each odd part has even multiplicity and $\chi : p_j \mapsto \chi_j$ is a function from the parts of $p_1{}^{m_1} \cdots p_z{}^{m_z}$ to the natural numbers. The function χ satisfies the following rule:

$$\chi(p) = \begin{cases} \frac{1}{2}(p-1) & \text{if } p \text{ is odd;} \\ \frac{1}{2}p & \text{if } p \text{ is even and has odd multiplicity;} \\ \frac{1}{2}(p-2) \text{ or } \frac{1}{2}p & \text{if } p \text{ is even and has even multiplicity.} \end{cases}$$

Unipotent classes in groups over fields of odd characteristic are parameterized by those symbols where $\chi(p) = \frac{1}{2}p$ for all even p. We briefly note that in both types A and C, the partitions mentioned come from the sizes of the Jordan blocks in the normal form of elements in our classes in the action on the natural module. Thus if the characteristic is odd the classes are completely determined by their Jordan structure. Now if, as will always be our convention, we order the parts of the partition in descending order of size $(p_1, \ldots, p_z)$ the dimension of the centralizer of any element in the class determined by $\mathbf{p}$ is $\sum_{i=1}^{z}(ip_i - \chi_i)$. Once again the order of any element in the class determined by $\mathbf{p}$ is given by $\min\{q^i : q^i \geq p_1\}$, where $\operatorname{char} k = q$ and p_1 is the largest part in the underlying partition of $\mathbf{p}$. We now make some general definitions. So let Φ be a root system and let $\Pi(X, \Phi, p)$ be the set of semisimple elements in $G(\Phi)_p$ with a centralizer isomorphic to X. Next let $s_{\min}$ be an element of minimum order in $\Pi(X, \Phi, p)$ and set $q^{(X,p)} = \min\{q^e : q^e \geq o(s_{\min})\}$.

We can now state our main conjecture.

Conjecture 2 *Let $G(\Phi)_p$ and $G(\Phi)_q$ be simple algebraic groups with the same root system Φ and over algebraically closed fields of characteristic p and q respectively. Let $X \leq G(\Phi)_p$ be the centralizer of a semisimple element. Then there exists a unipotent conjugacy class $\mathcal{C} \subset G(\Phi)_q$ such that for any $u \in \mathcal{C}$*

$$\dim C_{G(\Phi)_q}(u) = \dim X \qquad \text{and} \qquad o(u) \text{ divides } q^{(X,p)}.$$

3 Type A

The following theorem proves our conjecture for type A.

Theorem 3.1 *Define a function f_A from the isomorphism classes of centralizers of semisimple elements in $G(A_n)_p$ to the unipotent conjugacy classes in $G(A_n)_q$ as follows. Let $X = A_{n_1} \ldots A_{n_r} T_l$ represent the isomorphism class of the centralizer of a* *semisimple element in $G(A_n)_p$ and let $\mathbf{n}^{\perp}$ be the dual partition of $\mathbf{n} = (n_1, \ldots, n_r)$, then set $f_A(X)$ to be the partition $(l+1, \mathbf{n}^{\perp})$ and take u in the class determined by $f_A(X)$. Then f_A is a bijection such that $\dim C_{G(A_n)_p}(s_{\min}) = \dim C_{G(A_n)_q}(u)$, and moreover $q^{(X,p)} = o(u)$ unless $p = q$ and $l + 1 = q^i$ for some $i \in \mathbb{N}$, in which case $q^{(X,p)} = q.o(u)$. Furthermore if $p \neq q$ then there exists a semisimple element $s \in G(A_n)_p$ such that $o(s) = o(u)$ and $\dim C_{G(A_n)_p}(s) = \dim C_{G(A_n)_q}(u)$.*

We omit the proof since it is not very difficult or particularly interesting. However our function f_A does have a distinguished property. Let $X = A_{n_1} \ldots A_{n_r} T_l$, $\mathbf{m} = f_A(X)$ and let $\mathbf{m}'$ be the image of a centralizer X' obtained from X by increasing the rank of one of the factors by 1. Intuitively we think that X and X' are close and that a 'sensible' function would ensure that $\mathbf{m}$ and $\mathbf{m}'$ are also close. We will now formalize this. Recall that the set of partitions has a natural ordering: if $\mathbf{n} = (n_1, \ldots, n_r)$ and $\mathbf{m} = (m_1, \ldots, m_z)$ then $\mathbf{n} \preceq \mathbf{m}$ if and only if $r \leq z$ and $n_i \leq m_i$ for all $1 \leq i \leq r$. So let $\mathrm{Cent}(A)$ be the set of possible centralizer types of semisimple elements in groups of type A. We can identify this set with the collection of ordered pairs $(\mathbf{n}, l)$ where $\mathbf{n} = (n_1, \ldots, n_r)$ and $r \leq l + 1$. This set can then be given a partial order by declaring $(\mathbf{n}, l) \preceq (\mathbf{n}', l')$ if $\mathbf{n} \preceq \mathbf{n}'$ and $l \leq l'$. Next let $\mathrm{uccl}(A)$ be the set of all partitions of size at least two. This set clearly parameterizes the unipotent classes in groups of type A and it is also partially ordered by the standard partition ordering. Now let $\Lambda = A$ and $r(\mathbf{m}) = |\mathbf{m}| - 1$ (where $|\mathbf{m}|$ is the sum of all the parts of $\mathbf{m}$) and consider functions

$$f : \mathrm{Cent}(\Lambda) \to \mathrm{uccl}(\Lambda)$$

satisfying the following conditions:
(S1) If $X \preceq X'$ then $f(X) \preceq f(X')$;
(S2) If $\mathrm{Rank}\, X = n$ then $r(f(X)) = n$;
(S3) For any $u \in f(X)$

$$\dim C_{G(\Lambda_n)}(u) = \dim X.$$

(S1) and (S2) formalize the notion of a sensible function; the third axiom is there so that the function might be used to prove our conjecture. By convention when we write '$u \in f(X)$' we mean u is in the unipotent class determined by $f(X)$. This motivates us to make the following definition.

Definition 3.2 Let $\Lambda \in \{A, B, C, D, E\}$, write $\mathrm{Cent}(\Lambda)$ for the set of possible centralizer types of semisimple elements in groups of type Λ and let $\mathrm{uccl}(\Lambda)$ be a set parameterizing pairs consisting of a unipotent conjugacy class and the rank (denoted $r(m)$ for $m \in \mathrm{uccl}(\Lambda)$) of the group of type Λ containing it. If Ω_1 and Ω_2 are partially ordered subsets of $\mathrm{Cent}(\Lambda)$ and $\mathrm{uccl}(\Lambda)$ respectively, then a function $f : \Omega_1 \to \Omega_2$ is called **sensible with respect to the partial orders**, or simply **sensible**, if the axioms (S1)–(S3) are satisfied.

We now get to the distinguished property mentioned earlier.

Theorem 3.3 *There is a unique sensible function $f : \mathrm{Cent}(A) \to \mathrm{uccl}(A)$ and it is equal to f_A.*

This also has quite an easy proof, so we omit it to save space.

4 Type C

In this section we will sketch the uniqueness proof for the sensible function in type C and then see how this is used to prove our conjecture in that type. So let Cent(C) be the set of all isomorphism classes of centralizers of semisimple elements in groups of type C. Recalling section 2 we can identify this set with ordered quadruples of the form $(\mathbf{n}, m_1, m_2, l)$ where $\mathbf{n} = (n_1, \ldots, n_r)$, $m_1 \geq m_2$ are nonnegative integers and $r \leq l$. The set can then be given a partial order by declaring $(\mathbf{n}, m_1, m_2, l) \preceq (\mathbf{n}', m_1', m_2', l')$ if $\mathbf{n} \preceq \mathbf{n}'$, $m_1 \leq m_1'$, $m_2 \leq m_2'$ and $l \leq l'$. The set Cent$_{\text{odd}}(C)$ of classes of centralizers of semisimple elements of odd order may be identified with those quadruples where $m_2 = 0$.

Turning to unipotent classes, we let uccl(C) be the set of all Hesselink symbols, as defined in section 2. The set uccl(C) is partially ordered by declaring $\mathbf{p} \preceq \mathbf{p}'$ if and only if the number of parts in $\mathbf{p}$ (denoted z) is less than or equal to the number of parts in $\mathbf{p}'$ and both $p_i \leq p_i'$ and $\chi_i \leq \chi_i'$ for all $1 \leq i \leq z$. We define uccl$_{\text{odd}}(C)$ to be the subset of uccl(C) consisting of those symbols where $\chi(p) = \frac{1}{2}p$ for all even p. The partial order can obviously be restricted to this subset.

The elements and the partial order of uccl(C) can be seen diagrammatically as follows.

Example

$$2_0^2 \, 1_0^2 \qquad\qquad \preceq \qquad\qquad 2_1^4$$

The columns in the diagrams are in one-to-one correspondence with the parts in the partitions: the ith column contains p_i squares if and only if the ith part has size p_i. The value in the bottom square of the ith column is $\chi(p_i)$. One symbol precedes another if the diagram of the first symbol is contained in the diagram of the second, and the values in the bottom squares of the first diagram are less than or equal to the corresponding values in the second. In the above diagram the black squares and bold values show the parts and χ-values of the second symbol which are greater than those of the first symbol.

We may now define a function $f_C : \text{Cent}(C) \to \text{uccl}(C)$. If $\mathbf{n} = (n_1, \ldots, n_r)$, where $n_1 \geq \cdots \geq n_r \geq 1$, and $2(l, \mathbf{n}^\perp) = (2p_1, \ldots, 2p_z)$, where $(l, \mathbf{n}^\perp) = (p_1, \ldots, p_z)$, and if we set the χ-values of $2(l, \mathbf{n}^\perp)$ to take their largest possible values, then

$$f_C(A_{n_1} \ldots A_{n_r} T_l) = 2(l, \mathbf{n}^\perp)$$
$$= 2l_l \, 2r_r^{n_r} \, 2(r-1)_{r-1}^{n_{r-1}-n_r} \cdots 2_1^{n_1-n_2}.$$

Next we define $f_C(A_{n_1} \ldots A_{n_r} C_{m_1} C_{m_2} T_l)$, where m_1 and m_2 are arbitrary. Unlike the above we will not give an explicit formula for $f_C(X)$ here, instead we will give an algorithm for producing it. Now we know what $f_C(X)$ is when $m_1 = m_2 = 0$. So begin with $X_0 = A_{n_1} \ldots A_{n_r} C_{m_1} C_{m_2} T_l$ where $m_1 = m_2 = 0$;

also let $m^{(v)}$ be the rank of the vth C-factor in X. We will increase the m_v until we reach $(m^{(1)}, m^{(2)})$. We will also give the corresponding alterations to make to $f_C(X_0)$. So proceed as follows.

Step one If $m^{(2)} = m_2$ then go to Step two. Otherwise increase m_1 and m_2 by 1, and replace the parts $2s_s\, 2t_t$ of $f_C(X_0)$ in positions $2m_2 - 1$ and $2m_2$ by $2(s+1)_s^2$ if $s = t$, by $(2s+1)_s^2$ if $s = t+1$ and by $2s_s\, 2(t+2)_{t+2}$ if $s \geq t+2$. Repeat. (Note that s and t are allowed to be zero here.)

Step two If $m^{(1)} = m_1$ then we are done. Otherwise increase m_1 by 1, and replace the parts $2s_s\, 2t_t$ in positions $2m_1 - 1$ and $2m_1$ by $(2s+1)_s^2$ if $s = t$ and by $2s_s\, 2(t+1)_{t+1}$ if $s \geq t+1$. Repeat. (Again s and t are allowed to be zero.)

Below we illustrate each of the possibilities for Step one and Step two diagrammatically. In each case the first column represents the part in position $2m_2 - 1$ or $2m_1 - 1$ as appropriate.

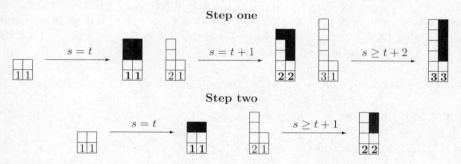

Here is an example of the algorithm at work.

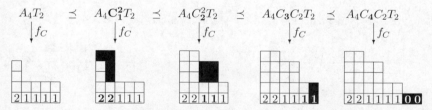

It can be shown, though it is time consuming, that the above function is a sensible function. In fact it is the unique sensible function from $\mathrm{Cent}(C)$ to $\mathrm{uccl}(C)$. We will now sketch the proof of this. So from now on we let f denote an arbitrary sensible function from $\mathrm{Cent}(C)$ to $\mathrm{uccl}(C)$. Our first observation is that by listing all unipotent classes in $G(C_2)_2$ it can be seen that only one has centralizers of dimension 2. Thus by (S3) we have $f(T_2) = 4_2$. We can now use the axioms to determine what values f must produce for all other centralizers in $\mathrm{Cent}(C)$. First we record the effects on the dimension of increasing l, n_i or m_j for some i or j.

$$\dim T_{l+1} - \dim T_l = (l+1) - l = 1;$$
$$\dim A_{n_i+1} - \dim A_{n_i} = (n_i + 1)(n_i + 3) - n_i(n_i + 2) = 2n_i + 3;$$
$$\dim C_{m_j} - \dim C_{m_j-1} = m_j(2m_j + 1) - (m_j - 1)(2m_j - 1) = 4m_j - 1.$$

Now to match the above information we must look at the possible images $\mathbf{p} =$

$f(X') \succ f(X)$ of size $2n + 2$, where $X \leq G(C_n)$ precedes $X' \leq G(C_{n+1})$, and the corresponding dimension increases.

There are two options for obtaining $\mathbf{p}$ from $f(X)$: increase one part by 2 or increase two parts by 1. In the first case the increased part p must be even, since the parity of the multiplicity of p and of $p + 2$ changes. In the second case we can either have the two parts of equal size, in which case they must be adjacent to preserve the descending order of the parts, or their sizes must differ by 1 and the larger part must be odd. This follows since, as above, for each increased part p_i the parity of the multiplicity of p_i and of $p_i + 1$ changes; thus to avoid creating an odd part of odd multiplicity the increased parts must differ by 1, with the larger part being odd.

We must also consider the χ-values of $\mathbf{p}$ and $f(X)$: any even part p of $f(X)$ with even multiplicity and $\chi(p) = \frac{1}{2}(p - 2)$ may have its χ-value increased by 1 in $\mathbf{p}$. Performing this increase will be called option (A). Now if x_j denotes the part in position j then the following options (which we call *basic*), used in conjunction with option (A) give all possible ways of constructing $\mathbf{p}$ from $f(X)$.

(1) $x_j = 2m_m$ is replaced by $2(m + 1)_{m+1}$.

(2U) $x_j = 2m_m$ and $x_j + 1 = (2m+1)_m$ are replaced by $(2m+1)_m$ and $2(m+1)_{m+1}$.

(2E$^+$) $2m_m^2$ in positions j and $j + 1$ is replaced by $(2m + 1)_m^2$.

(2E$^-$) $2m_{m-1}^2$ in positions j and $j + 1$ is replaced by $(2m + 1)_m^2$.

(2O$^+$) $(2m + 1)_m^2$ in positions j and $j + 1$ is replaced by $2(m + 1)_m^2$.

(2O$^-$) $(2m + 1)_m^2$ in positions j and $j + 1$ is replaced by $2(m + 1)_{m+1}^2$.

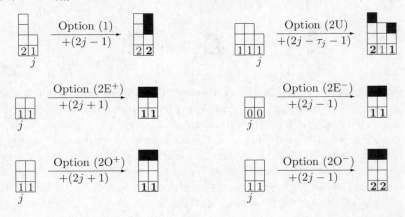

Above we show each of the basic options diagrammatically; we have labelled the part in position j according to the above and given the centralizer dimension increase associated to each option. This is easily calculated from the formula for centralizer dimension given in section 2, i.e., $\sum_{i=1}^{z}(ip_i - \chi_i)$. For option (2U) the multiplicity of $x_j + 1$ is denoted τ_j. The values in bold are those which the option has acted upon.

We may now return to determining f. So we know that $f(T_2) = 4_2$ and to find $f(T_3)$ we can use option (1) with $j = 1$ or 2, or option $(2E^+)$ with $j = 2$. These give centralizer dimension increases of 1, 3 and 5 respectively, so by (S3) we have $f(T_3) = 6_3$. Using this technique we can determine f on all elements in $\mathrm{Cent}_{\mathrm{odd}}(C)$, i.e., those centralizers with at most one factor of type C. In this way we can prove the following proposition.

Proposition 4.1 *If* $f : \mathrm{Cent}(C) \rightarrow \mathrm{uccl}(C)$ *is a sensible function then* $f(X) = f_C(X)$ *for all* $X \in \mathrm{Cent}_{\mathrm{odd}}(C)$.

Determining f for centralizers in $\mathrm{Cent}(C) \setminus \mathrm{Cent}_{\mathrm{odd}}(C)$ is rather more difficult however and we will now look at this case.

Let $X = A_{n_1} \ldots A_{n_r} C_{m_1} C_{m_2} T_l$. In order to determine $f(X)$ we shall use induction together with Proposition 4.1. We begin by defining some centralizers associated to X: for $m \geq 1$ write $n_i' = \min\{n_i, 2m - 1\}$ and set

$$X_m = A_{n_1'} \ldots A_{n_r'} C_m C_{m-1} T_l;$$
$$X_m^+ = A_{n_1'} \ldots A_{n_r'} C_m^2 T_l;$$
$$X_\alpha = A_{n_1} \ldots A_{n_r} T_l.$$

Notice that $f(X_\alpha)$ has already been shown to agree with the image under the known sensible function f_C by Proposition 4.1. Now the final induction will be completed later, but before that we must lay some groundwork. In particular we will prove the following property $\mathbb{P}$:

$$\text{if } f(X_m) = f_C(X_m) \text{ then } f(X_m^+) = f_C(X_m^+).$$

So assume that $f(X_m) = f(A_{n_1'} \ldots A_{n_r'} C_m C_{m-1} T_l)$ has been determined and follows our algorithm. Our task is to determine what $f(X_m^+) = f(A_{n_1'} \ldots A_{n_r'} C_m^2 T_l)$ must be. Now we note that replacing C_{m-1} by C_m will increase the centralizer dimension by $4m - 1$, so we must match this on the corresponding symbol. Here we observe that for each of the basic options we can see what j must be to achieve such an increase. The use of option (A) decreases the centralizer dimension and hence when used with the basic options can permit a larger j than is normally allowable. However the values of j calculated without the influence of option (A) still provide a lower bound: the dimension increase (which we want to be $4m - 1$) is always less than or equal to $2j + 1$; thus $j \geq 2m - 1$. We can now go ahead with our case analysis.

Case i: *The part in position* $2m - 1$ *of* $f(X_m)$ *is odd and of size* ≥ 3.

Now we are assuming that $f(X_m)$ has been determined according to the algorithm given above. This means that the the symbol has been constructed by taking the symbol for $f((X_m)_\alpha)$ and applying Step one of our algorithm $m - 1$ times, followed by Step two once. Thus since the part in position $2m - 1$ is of the form $2s + 1$, where $s \geq 1$, we must have had the parts in positions $2m - 1$ and $2m$ equal before we applied the algorithm. So if Step one had acted upon these parts they would be of the form $2(s+1)_s^2$ and hence we want to show that option $(2O^+)$ with $j = 2m - 1$ is the only way to proceed. This is illustrated below (the first and second horizontal arrows representing 'undoing' Step two and performing Step one respectively).

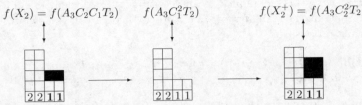

$$f(X_2) = f(A_3C_2C_1T_2) \qquad f(A_3C_1^2T_2) \qquad\qquad f(X_2^+) = f(A_3C_2^2T_2)$$

Now to prove this we introduce some more notation. So let s be as above (then the parts in positions $2m - 1$ and $2m$ of $f(X_m)$ have size $2s + 1$), let $z \geq m$ and recall that $n_i' = \min\{n_i, 2m - 1\}$ $(\leq 2m - 1)$. Next define

$$
\begin{aligned}
X_{(m)}^{s\times z} &= A_{2z-1}^s A_{n_{s+1}'} \ldots A_{n_r'} C_z T_l; \\
X_m^{s\times z} &= A_{2z-1}^s A_{n_{s+1}'} \ldots A_{n_r'} C_z C_{m-1} T_l; \\
X_m^{s\times z+} &= A_{2z-1}^s A_{n_{s+1}'} \ldots A_{n_r'} C_z C_m T_l.
\end{aligned}
$$

As with the above we note that $f(X_{(m)}^{s\times z})$ has already been shown to agree with the image under f_C by Proposition 4.1. We also define l.u.b$\{\mathbf{a}, \mathbf{b}\}$ to be the symbol whose jth part (and χ-value) is the larger of the jth parts (and χ-values) of $\mathbf{a}$ and $\mathbf{b}$; note that this may not always be a valid symbol. For example l.u.b$\{4_2^2, 4_2\,3_1^2\} = 4_2^2\,3_1$, which has an odd part of odd multiplicity. However if l.u.b$\{\mathbf{a}, \mathbf{b}\}$ is in the poset uccl(C) then it will clearly be the unique element satisfying $\mathbf{p} \succeq$ l.u.b$\{\mathbf{a}, \mathbf{b}\} \succeq \mathbf{a}, \mathbf{b}$ for all $\mathbf{p} \succeq \mathbf{a}, \mathbf{b}$.

We claim that $f(X_m^{s\times z}) =$ l.u.b$\{f(X_m), f(X_{(m)}^{s\times z})\}$. To prove this we use the following lemma.

Lemma 4.2 *If $x \succeq y, z$ and $f_C(x) =$ l.u.b$\{f(y), f(z)\}$ then $f(x) = f_C(x)$ and hence also equals* l.u.b$\{f(y), f(z)\}$.

This follows since f_C is known to be a sensible function and hence we have that l.u.b$\{f(y), f(z)\} = f_C(x) \in$ uccl(C); the centralizer dimension of l.u.b$\{f(y), f(z)\}$ $= f_C(x)$ is equal to $\dim x$ and $2\,\text{Rank}\,x = |f_C(x)| = |\text{l.u.b}\{f(y), f(z)\}|$. So (S1) forces us to have $f(x) \succeq$ l.u.b$\{f(y), f(z)\}$ since $x \succeq y, z$ implies $f(x) \succeq f(y), f(z)$. Also by (S2) we have $|f(x)| = 2\,\text{Rank}\,x = |\text{l.u.b}\{f(y), f(z)\}|$. Therefore the underlying partitions of $f(x)$ and l.u.b$\{f(y), f(z)\}$ are equal, however the χ-values of some parts of $f(x)$ may still be greater than those of l.u.b$\{f(y), f(z)\}$. This does not occur since we know that the centralizer dimension of l.u.b$\{f(y), f(z)\}$, henceforth denoted c.dim l.u.b$\{f(y), f(z)\}$, equals $\dim x$ and increasing any χ-value would decrease the centralizer dimension. So since (S3) forces c.dim $f(x) = \dim x$ we must have $f(x) =$ l.u.b$\{f(y), f(z)\}$. This completes the proof of the lemma.

Now we can easily see that $X_m^{s\times z} \succeq X_m$, $X_{(m)}^{s\times z}$ and so we must show that $f_C(X_m^{s\times z}) =$ l.u.b$\{f(X_m), f(X_{(m)}^{s\times z})\}$. This can be done since we know how f_C acts and hence we know what $f_C(X_m^{s \times z})$ looks like. Also since we are assuming that $f(X_m) = f_C(X_m)$ and Proposition 4.1 tells us that $f(X_{(m)}^{s \times z}) = f_C(X_{(m)}^{s \times z})$ we know what l.u.b$\{f(X_m), f(X_{(m)}^{s \times z})\}$ looks like. See the example below.

$$f(X_2^{1 \times 4}) = f(A_7 C_4 C_1 T_2)$$
$$\|$$
$$f(X_2) = f(A_3 C_2 C_1 T_2) \preceq \text{l.u.b}\{f(X_2), f(X_{(2)}^{1 \times 4})\} \succeq f(X_{(2)}^{1 \times 4}) = f(A_7 C_4 T_2)$$

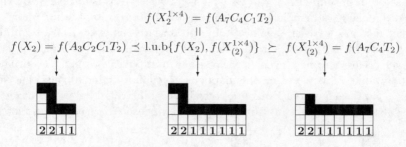

We are now in a position to determine the value of $f(X_m^+)$. From the above we know that we must use options with $j \geq 2m - 1$ to provide a centralizer dimension increase of $4m - 1$. Thus we can only use options $(2O^+)$ with $j = 2m - 1$, or options (1) or $(2E^+)$ with $j = 2m + 1$. Option (2U) with $j = 2m + 1$ is not possible, since we are assuming that the parts in positions $2m - 1$ and $2m$ have size greater than 1. Also option $(2O^-)$ with $j = 2m - 1$ does not give a large enough dimension increase. Options with $j = 2m + 1$ give increases which are too large, but this may be compensated for by increasing a χ-value (if an appropriate value is available). Thus we can either use $(2O^+)$ with $j = 2m - 1$ (which gives the correct increase of $4m - 1$) or one of the options with $j = 2m + 1$ and option (A) to compensate. We claim that the latter leads to a contradiction. To prove this our strategy will be to show that using option (A) here forces us to use it again when determining $f(X_m^{s \times z +})$ for $z = m + 1$ and this forces us to use it again when determining $f(X_m^{s \times z +})$ for $z = m + 2$, etc. We then observe that each step of this process leaves us with fewer χ-values suitable for compensation. Thus we must eventually find a z such that there is no value of $f(X_m^{s \times z +})$ that is consistent with the axioms. This will give us our contradiction.

So we observe that the 'excess' centralizer dimension increase when using option (1) with $j = 2m - 1 + 2a$ is $2j - 1 - (4m - 1) = 4a - 2$. When using option $(2E^+)$ with $j = 2m - 1 + 2a$ the excess is $2j + 1 - (4m - 1) = 4a$. At this point we also observe that $X_m = X_m^{s \times m}$ and $X_m^+ = X_m^{s \times m+}$. So assume that $f(X_m^+) = f(A_{2m-1}^s A_{n'_{s+1}} \cdots A_{n'_r} C_m^2 T_l)$ has been produced by using an option with $j = 2m + 1$ and option (A) to compensate, then set $z = m + a$ with $a = 1$ and consider the value of $f(X_m^{s \times z +}) = f(A_{2z-1}^s A_{n'_{s+1}} \cdots A_{n'_r} C_z C_m T_l)$. Now given that we know what the value of $f(X_m^{s \times z}) = f(A_{2z-1}^s A_{n'_{s+1}} \cdots A_{n'_r} C_z C_{m-1} T_l)$ is by the above, we can see that the parts in position j, where $2m - 1 \leq j \leq 2z$, all have size $2s + 1$. Thus as above, to match replacing C_{m-1} with C_m we have a choice between using option $(2O^+)$ with $j = 2m - 1$, or options (1) or $(2E^+)$ with $j = 2z + 1$. However by (S1) we see that $f(X_m^{s \times z +})$ must succeed $f(X_m^+)$ in the partial order, since $X_m^{s \times z +} \succeq X_m^+$. Thus the χ-value increased in $f(X_m^+)$ must also be increased in $f(X_m^{s \times z +})$, meaning that we require a centralizer dimension increase of $4m - 1 + b$ for $b \in \{4a - 2, 4a\}$. Therefore option $(2O^+)$ with $j = 2m - 1$ no longer gives a large enough centralizer dimension increase and we must use an option with $2z + 1$. However the minimum centralizer dimension increase of an option with $j = 2z + 1 = 2m + 1 + 2a$ is $4m + 1 + 4a$ and the maximum centralizer dimension increase of an option with $j = 2(z - 1) + 1 = 2m - 1 + 2a$ is $4m - 1 + 4a$; thus the increased χ-value is insufficient to compensate for an option with $j = 2z + 1$. So we must increase another χ-value (if an appropriate one exists). We then increase a by 1 and repeat this paragraph. Since by increasing the value of z in $f(X_m^{s \times z})$ we never create any new even parts of even multiplicity we must eventually run out of suitable χ-values; when this happens we cannot define a consistent value of $f(X_m^{s \times z+})$ and we have our contradiction. This is illustrated below, the italic values represent those that have been increased to compensate for the excess increase.

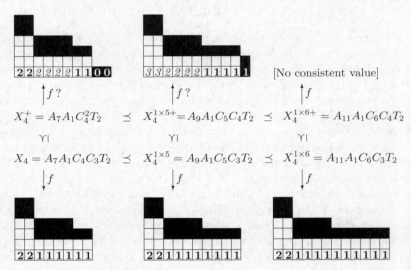

[No consistent value]

$$X_4^+ = A_7 A_1 C_4^2 T_2 \ \preceq\ X_4^{1 \times 5+} = A_9 A_1 C_5 C_4 T_2 \ \preceq\ X_4^{1 \times 6+} = A_{11} A_1 C_6 C_4 T_2$$

$$X_4 = A_7 A_1 C_4 C_3 T_2 \ \preceq\ X_4^{1 \times 5} = A_9 A_1 C_5 C_3 T_2 \ \preceq\ X_4^{1 \times 6} = A_{11} A_1 C_6 C_3 T_2$$

Case ii: *The part in position $2m - 1$ of $f(X_m)$ has size 1.*

As in case i we are assuming that the the symbol has been constructed by taking the symbol for $f((X_m)_\alpha)$ and applying Step one of our algorithm $m - 1$ times, followed by Step two once. Thus since the part in position $2m - 1$ has odd size, we must have had the parts in positions $2m - 1$ and $2m$ equal before we applied the algorithm. So if Step one had acted upon these parts they would be equal to 2_1^2; hence we want to show that option $(2O^+)$ with $j = 2m - 1$ is the only way to proceed. This is illustrated below, as in case i.

$$f(X_2) = f(C_2 C_1 T_2) \qquad f(C_1^2 T_2) \qquad f(X_2^+) = f(C_2^2 T_2)$$

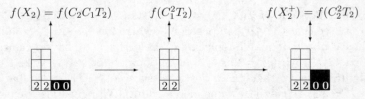

Here we construct a new centralizer X_m^{ii} which is equal to X_m, but with an extra A-factor of rank $2m - 1$ and the value of l increased by 1 (to ensure we still have

l greater than or equal to the number of A-factors and therefore still have a valid centralizer). This new symbol is constructed so the parts in positions $2m - 1$ and $2m$ have size equal to 3; thus we may use case i above to show that to match replacing C_{m-1} with C_m in X_m^{ii} we use option $(2O^+)$ with $j = 2m - 1$. So let $X_m^{\text{ii}+}$ denote X_m^{ii} with C_{m-1} replaced by C_m. Then since $f(X_m^{\text{ii}+})$ (which has $2m$ parts) succeeds the symbol for $f(X_m^+)$ we see that we cannot get from $f(X_m)$ to $f(X_m^+)$ by increasing any parts with $j > 2m$. The result now follows. This is illustrated below.

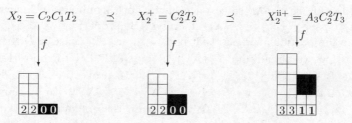

So summarizing cases i and ii: *if the part in position $2m - 1$ is odd then we must use option $(2O^+)$ with $j = 2m - 1$.* We may also reduce to case i to prove the following (the diagrams provide illustrative examples, as in cases i and ii).

Case iii: *If the parts in positions $2m - 1$ and $2m$ are equal and even then we must use option $(2E^+)$ with $j = 2m - 1$.*

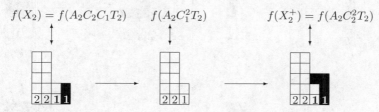

Case iv: *If the parts in positions $2m - 1$ and $2m$ are even, but unequal then we must use option (1) with $j = 2m$.*

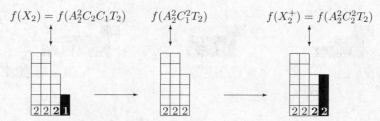

This completes our proof of property $\mathbb{P}$. We now come to the final induction. We begin by defining two more centralizers:

$$X_{(m)} = A_{n_1'} \ldots A_{n_r'} C_m T_l \quad \text{and} \quad X^{(m)} = A_{n_1} \ldots A_{n_r} C_m T_l,$$

where we recall that $n_i' = \min\{n_i, 2m - 1\}$. We again note that by Proposition

4.1 both $f(X_{(m)})$ and $f(X^{(m)})$ agree with the images under our known sensible function f_C. To complete our uniqueness proof we will use induction on m to prove that $f(X_m) = f_C(X_m)$ and then use that equality to prove that $f(X) = f_C(X)$.

So first observe that $X_1 = X_{(1)}$ and $f(X_{(1)}) = f_C(X_{(1)})$ by the above. Next assume that $f(X_m) = f_C(X_m)$ for some m. Now by property $\mathbb{P}$ we see that $f(X_m^+) = f_C(X_m^+)$. We claim that $f(X_{m+1}) = \text{l.u.b}\{f(X_m^+), f(X_{(m+1)})\}$. To prove this we shall use Lemma 4.2 again: as before it is straightforward to see that $X_{m+1} \succeq X_m^+, X_{(m+1)}$ so we must show that $f_C(X_{m+1}) = \text{l.u.b}\{f(X_m^+), f(X_{(m+1)})\}$. We may do this as before since once again both symbols are known. Thus by induction we have $f(X_m) = f_C(X_m)$ for all m. By property $\mathbb{P}$ we also have $f(X_m^+) = f_C(X_m^+)$ and in particular we see that if $X = A_{n_1} \ldots A_{n_r} C_{m_1} C_{m_2} T_l$ then $f(X_{m_2}^+) = f_C(X_{m_2}^+)$. Now we already have $f(X^{(m_1)}) = f_C(X^{(m_1)})$ by Proposition 4.1, so we claim that $f(X) = \text{l.u.b}\{f(X_{m_2}^+), f(X^{(m_1)})\}$. To prove this we shall once again use Lemma 4.2: as before we clearly have $X \succeq X_{m_2}^+, X^{(m_1)}$ so we must show that $f_C(X) = \text{l.u.b}\{f(X_{m_2}^+), f(X^{(m_1)})\}$. Once again these symbols are known and so our proof is complete. The figure below illustrates, in the case of the centralizer $X = A_8 C_6 C_3 T_2$, the steps we go through to prove that $f(X) = f_C(X)$.

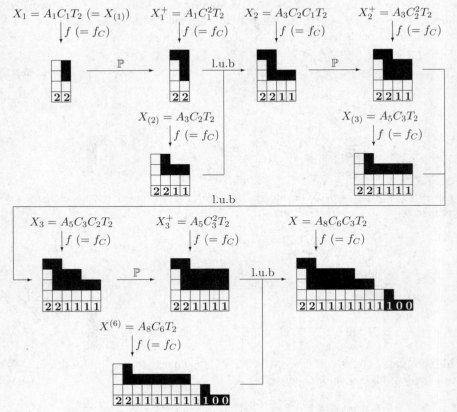

So $f = f_C$ is the unique sensible function from $\text{Cent}(C)$ to $\text{uccl}(C)$ and we can now see how it is used to prove Conjecture 2. Recall that the definition of a sensible function made no restriction on element orders in our conjugacy classes. Nevertheless f_C does preserve order up to a point. So given a centralizer X, we call a prime q *admissible* if there exists a semisimple element s such that $C_{G(C_n)}(s) = X$ and $o(s) = q^i$ for some i. Note that if X has two C-factors then the only admissible prime is 2, otherwise q can be any prime. We may now complete the proof of our conjecture for type C.

Theorem 4.3 *Let $f_C : \mathrm{Cent}(C) \to \mathrm{uccl}(C)$ be the unique sensible function and for each $X \in \mathrm{Cent}(C)$ take a unipotent element $u \in f_C(X)$. Also let q be admissible for X and take $f_C(X)$ to be a class in a group with defining characteristic q. Then $q^{(X,p)} = o(u)$ unless we have one of the following:*

1. *X has no C-factor, $p = q = 2$ and $l = 2^i$ for some i;*

2. *X has one C-factor, $p = q$ and $2l + 1 = q^i$ for some i;*

3. *X has one C-factor, $l > r$ and $(q, l) = (2, 2^i)$ for some i;*

4. *X has two C-factors, $l = r + 1$, $p = q$ and $(q, r + 2) = (2, 2^i)$ for some i;*

5. *X has two C-factors, $l > r + 1$ and $(q, l) = (2, 2^i)$ for some i.*

In these cases we have $q^{(X,p)} = q.o(u)$. Furthermore if $p \neq q$ then there exists a semisimple element s such that $C_{G(C_n)}(s) = X$ and $o(s) = q^{(X,p)}$.

This is proven by case analysis.

Finally we review the progress made so far. We are currently in the final stages of completing similar uniqueness results in type D and it is expected that similar results can also be obtained in type B. Conjecture 2 has been proven for all exceptional groups and a classification of sensible functions in type E is planned (we do not believe uniqueness holds in this case). Types A and C are completely finished and the results of this paper and those just mentioned will be made available in full detail at a later date.

References

[1] R. Carter, *Finite Groups of Lie Type: Conjugacy Classes and Complex Characters*, Wiley-Interscience, London, 1985.

[2] D. Deriziotis, Centralizers of semisimple elements in a Chevalley group, *Comm. Algebra* **9** (1981), 1997–2014.

[3] B. Hartley and M. Kuzucuoğlu, Centralizers of elements in locally finite simple groups, *Proc. London Math. Soc.* **62** (1991), no. 3, 301–324.

[4] W. Hesselink, Nilpotency in classical groups over a field of characteristic 2, *Math. Z.* **166** (1979), 165–181.

[5] R. Lawther, Jordan block sizes of unipotent elements in exceptional algebraic groups, *Comm. Algebra* **23** (1995), no. 11, 4125–4156.

[6] R. Lawther, Elements of specified order in simple algebraic groups, *Trans. Amer. Math. Soc.* **357** (2004), no. 1, 221–245.

ON AUTOMORPHISMS OF PRODUCTS OF GROUPS

JILL DIETZ

Department of Mathematics, Statistics, and Computer Science, St. Olaf College, Northfield, MN 55057, U.S.A.
Email: dietz@stolaf.edu

Abstract

When the finite group G can be written as a product $G = PQ$ (a direct, semi-direct, or central product), with $Q \triangleleft G$, we investigate the extent to which $\mathrm{Aut}(P)$ and $\mathrm{Aut}(Q)$ figure in the structure of $\mathrm{Aut}(G)$. In particular, we study the image of the map $\rho : \mathrm{Aut}(G; Q) \to \mathrm{Aut}(G/Q) \times \mathrm{Aut}(Q)$, where $\mathrm{Aut}(G; Q)$ is the subgroup of automorphisms of G that restrict to automorphisms of Q.

1 Introduction

In this study, we are interested in the relationships among $\mathrm{Aut}(G)$ and the automorphisms of subgroups and subquotients of G. In particular, if $G = PQ$ we would like to understand the extent to which $\mathrm{Aut}(P)$ and $\mathrm{Aut}(Q)$ figure in the structure of $\mathrm{Aut}(G)$. There is, *a priori*, no reason to believe that $\mathrm{Aut}(P)$ and $\mathrm{Aut}(Q)$ have any influence on $\mathrm{Aut}(G)$, but, indeed, there are conditions under which the influence can be both felt and described.

In the case that $G = P \times Q$, the relationship among $\mathrm{Aut}(P)$, $\mathrm{Aut}(Q)$, and $\mathrm{Aut}(G)$ is easy to discern when one of P or Q is characteristic. More generally, there are conditions under which information on $\mathrm{Aut}(G/\Phi(G))$ can be obtained from understanding the automorphisms of P, Q, or their Frattini quotients. For some algebraic topologists, this latter kind of information suffices for gaining knowledge about the stable homotopy decomposition of the classifying spaces of direct products of groups (see [3] and [6]).

In the case that $G = P \ltimes Q$, the influence of $\mathrm{Aut}(P)$ and $\mathrm{Aut}(Q)$ on the structure of $\mathrm{Aut}(G)$ depends on how automorphisms of P (respectively Q) interact with elements of Q (respectively P).

Finally, in the case that $G = P \circ_N Q$ is a central product with $P \cap Q = N$, direct information connecting $\mathrm{Aut}(P)$ and $\mathrm{Aut}(Q)$ to $\mathrm{Aut}(G)$ is difficult to obtain because of the "inseparability" of P from Q. Factoring out the influence of N by considering $\mathrm{Aut}_N G$ — the automorphisms of G that fix N elementwise — we can again find conditions under which automorphisms of P, Q, and P/N help determine $\mathrm{Aut}_N G$. Indeed, a special case (when G is the central product of an extra-special p-group with certain types of abelian p-groups) that appears in [2] is the motivation for this project.

In the next section we provide some background information and notation. Sections 3, 4, and 5 consider G as a direct product, semi-direct product, and central product respectively.

2   Background Information

The material referenced below is best found in [5]. The work in that paper is a modern interpretation of the original research first published by Wells [9] and extended by Buckley [1].

Let

$$E : 1 \to A \xrightarrow{i} G \xrightarrow{\pi} B \to 1$$

be an extension of the group A by the group B. The *twisting* χ of E is the homomorphism $\chi : B \to \mathrm{Out}\, A$ where $\chi(b)$ is the equivalence class of the homomorphism defined by $\chi(b)(a) = i^{-1}(g^{-1}i(a)g)$, where $g \in \pi^{-1}(b)$ and $a \in A$. Let $\mathcal{E}(B, A)$ be the set of equivalence classes of such extensions, where E_1 is equivalent to E_2 if there exists $\alpha : G_1 \to G_2$ which restricts to the identity on A and induces the identity on B. Equivalent extensions have the same twisting.

A left action of $\mathrm{Aut}(B) \times \mathrm{Aut}(A)$ on $\mathcal{E}(B, A)$ is given by $(\sigma, \tau)[E] = [\sigma E \tau^{-1}]$, where $\sigma \in \mathrm{Aut}(B)$, $\tau \in \mathrm{Aut}(A)$, and $\sigma E \tau^{-1}$ is the extension

$$1 \to A \xrightarrow{i \circ \tau^{-1}} G \xrightarrow{\sigma \circ \pi} B \to 1.$$

The twisting of $\sigma E \tau^{-1}$ is $c_{\bar{\tau}^{-1}} \chi \sigma^{-1}$, where $c_{\bar{\tau}^{-1}}$ is conjugation by $\bar{\tau}^{-1}$, the image of τ^{-1} in $\mathrm{Out}\, A$. Let C_χ be the subgroup of $\mathrm{Aut}(B) \times \mathrm{Aut}(A)$ consisting of pairs (σ, τ) which preserve the twisting on E. The subgroup C_χ acts on $\{[E] \mid E \text{ has twisting } \chi\}$.

Consider the homomorphism

$$\rho : \mathrm{Aut}(G; A) \xrightarrow{\rho_B \times \rho_A} \mathrm{Aut}(B) \times \mathrm{Aut}(A)$$

where $\mathrm{Aut}(G; A)$ is the subgroup of $\mathrm{Aut}(G)$ consisting of those automorphisms which map A into A. If $\alpha \in \mathrm{Aut}(G; A)$, then $\rho_A(\alpha)$ is simply restriction of α to A, while $\rho_B(\alpha) = \bar{\alpha}$ is the induced map on $B \cong G/A$. It is easy to see that $\mathrm{Im}\, \rho \le C_\chi$.

B acts on $Z(A)$ by

$$\bar{\chi} : B \xrightarrow{\chi} \mathrm{Out}\, A \xrightarrow{\overline{res}} \mathrm{Aut}(Z(A)),$$

where $\overline{res}$ is induced by the restriction map $\mathrm{Aut}(A) \to \mathrm{Aut}(Z(A))$. Let $Z^1_{\bar{\chi}}(B, Z(A))$ denote the group of derivations $\mathrm{Der}(B, Z(A))$ (functions $d : B \to Z(A)$ satisfying $d(b_1 b_2) = d(b_1) \bar{\chi}(b_1)(d(b_2))$). There is a homomorphism $\mu : Z^1_{\bar{\chi}}(B, Z(A)) \to \mathrm{Aut}(G; A)$ given by $\mu(d)(g) = i(d(\pi(g))) \cdot g$, where d is a derivation and $g \in G$.

There is a set map $\epsilon : C_\chi \to H^2_{\bar{\chi}}(B, Z(A))$ defined by $\epsilon((\sigma, \tau)) = (\sigma, \tau)[E]$.

The following sequence was obtained by Wells, with Buckley recognizing $\mathrm{Im}\, \rho$ as a stabilizer.

Theorem 2.1 (Wells [9], Buckley [1]) *There is an exact sequence*

$$1 \to Z^1_{\bar{\chi}}(B, Z(A)) \xrightarrow{\mu} \mathrm{Aut}(G; A) \xrightarrow{\rho} C_\chi \xrightarrow{\epsilon} H^2_{\bar{\chi}}(B, A)$$

where ϵ is only a set map, and $\mathrm{Im}\, \rho = (C_\chi)_{[E]}$, the stabilizer subgroup of C_χ fixing $[E]$.

In the case that $G = PQ$ with $Q \triangleleft G$, we consider the extension

$$1 \to Q \to G \to G/Q \to 1.$$

Let $N = P \cap Q$ so that $G/Q \cong P/N$. We get the map

$$\rho : \mathrm{Aut}(G; Q) \xrightarrow{\rho_P \times \rho_Q} \mathrm{Aut}(P/N) \times \mathrm{Aut}(Q).$$

Note that we use the notation ρ_P for the induced map $\mathrm{Aut}(G; Q) \to \mathrm{Aut}(P/N)$.

We can use the Wells exact sequence to obtain information about $\mathrm{Aut}(G; Q)$ from knowledge of $H_{\chi}^2(P/N, Z(Q))$. However, while $\mathrm{Im}\, \rho \le \mathrm{Aut}(P/N) \times \mathrm{Aut}(Q)$, none of the published research gives a sense of how $\mathrm{Im}\, \rho$ may be obtained as a product involving subgroups of $\mathrm{Aut}(P/N)$ and $\mathrm{Aut}(Q)$. In this paper, we aim to address the product question precisely.

3   Direct Products

Suppose $G = P \times Q$, then N is trivial and the map

$$\rho : \mathrm{Aut}(G; Q) \xrightarrow{\rho_P \times \rho_Q} \mathrm{Aut}(P) \times \mathrm{Aut}(Q)$$

is surjective. Moreover, ρ is split by the inclusion $\mathrm{Aut}(P) \times \mathrm{Aut}(Q) \hookrightarrow \mathrm{Aut}(G; Q)$. In this case we get

$$\mathrm{Aut}(G; Q) \cong \mathrm{Ker}\, \rho \rtimes (\mathrm{Aut}(P) \times \mathrm{Aut}(Q)),$$

where $\mathrm{Ker}\, \rho \cong \mathrm{Hom}(P, Z(Q))$ since P acts trivially on $Z(Q)$.

Example 3.1 Let G have the following presentation:

$$\langle a, b, c, d \mid a^4 = b^2 = c^2 = d^2 = 1,\ [a, b] = a^2,\ \text{and}\ a^2, c, d \in Z(G) \rangle.$$

This is group 32/8 in Thomas and Wood [8], and is SmallGroup(32,46) in the small group library of the program GAP [4]. We can write $G = P \times Q$ where $P = \langle a, b \rangle \cong D_4$ and $Q = \langle c, d \rangle \cong \mathbb{Z}_2 \times \mathbb{Z}_2$. The discussion above shows that

$$\mathrm{Aut}(G; Q) \cong \mathrm{Hom}(P, Q) \rtimes (\mathrm{Aut}(P) \times \mathrm{Aut}(Q)) \cong (\mathbb{Z}_2)^4 \rtimes (D_4 \times S_3).$$

We see that the order of $\mathrm{Aut}(G; Q)$ is 768, In Example 5.14, we will see that the order of the full automorphism group is 3072.

Further information on $\mathrm{Aut}(P \times Q)$ and on the image of this group in $\mathrm{Aut}(G/\Phi(G))$ can be found in [6] and [3].

4   Semi-direct Products

In this and subsequent sections, we will use the following notation:
- c_x is a conjugation map: $c_x(y) = x^{-1}yx$
- $[a, b] = a^{-1}b^{-1}ab$

- $x^y = y^{-1}xy$
- In $G = P \ltimes Q$ the product is given by $(p_1 q_1)(p_2 q_2) = (p_1 p_2)(q_1^{p_2} q_2)$

Suppose $G = P \ltimes Q$, where $P \leq G$ and $Q \triangleleft G$. Consider the extension

$$E : 1 \to Q \xrightarrow{i} G \xrightarrow{\pi} P \to 1,$$

where i is inclusion and π is projection. The twisting of E is $\chi : P \to \operatorname{Out} Q$, where $\chi(x)$ is the equivalence class of c_x.

As in the direct product case, we have a homomorphism $\rho : \operatorname{Aut}(G; Q) \xrightarrow{\rho_P \times \rho_Q} \operatorname{Aut}(P) \times \operatorname{Aut}(Q)$, and an extension $F : 1 \to \operatorname{Ker} \rho \to \operatorname{Aut}(G; Q) \to \operatorname{Im} \rho \to 1$.

We would like to understand $\operatorname{Im} \rho$ and determine when the extension F splits (which we will also call a splitting of ρ). The map ρ is not generally surjective. A pair $(\sigma, \tau) \in \operatorname{Aut}(P) \times \operatorname{Aut}(Q)$ is called *inducible* if $\rho(\alpha) = (\sigma, \tau)$ for some $\alpha \in \operatorname{Aut}(G; Q)$, and we call such an α a *lifting*.

Lifting (σ, τ) would ideally lead to a splitting $s : \operatorname{Im} \rho \to \operatorname{Aut}(G; Q)$. A naive lifting would be $s(\sigma, \tau) : G \to G$ defined by $s(\sigma, \tau)(xy) = \sigma(x)\tau(y)$, where $x \in P$ and $y \in Q$.

Let $g_1, g_2 \in G$, where $g_1 = x_1 y_1$ and $g_2 = x_2 y_2$ for some unique $x_1, x_2 \in P$ and $y_1, y_2 \in Q$. If s is defined as above, then $s(\sigma, \tau)(g_1 g_2) = \sigma(x_1 x_2)\tau(y_1^{x_2} y_2)$, while $s(\sigma, \tau)(g_1)s(\sigma, \tau)(g_2) = \sigma(x_1 x_2)\tau(y_1)^{\sigma(x_2)}\tau(y_2)$. We see that

$$s(\sigma, \tau) \in \operatorname{Aut}(G; Q) \text{ if and only if } \tau \circ c_x = c_{\sigma(x)} \circ \tau, \forall x \in P. \tag{4.1}$$

In this case, it is easy to see that s is a homomorphism and $\rho \circ s(\sigma, \tau) = (\sigma, \tau)$.

We will find conditions under which $s(\sigma, \tau) \in \operatorname{Aut}(G; Q)$ by considering componentwise lifts. That is $(\sigma, 1_Q)$ has a naive lift (that we will denote $\hat{\sigma}$) if and only if $c_x = c_{\sigma(x)}$ for all $x \in P$. Thus we must have $\sigma(x) \equiv x \mod C_G(Q)$. Similarly, $(1_P, \tau)$ has a naive lift (that we will denote $\hat{\tau}$) if and only if $\tau \circ c_x = c_x \circ \tau$ for all $x \in P$.

Recall that an automorphism α of a group G is called *central* if α commutes with all inner automorphisms of G. Equivalently, for all $g \in G$, $\alpha(g) = gz$, for some $z \in Z(G)$. The subgroup of all central automorphisms of G is denoted $\operatorname{Aut}_c G$.

Definition 4.1 Say $\sigma \in \operatorname{Aut}(P)$ is *Q-central* if for all $p \in P$, $\sigma(p) = pc$ for some $c \in C_G(Q) \cap P$. Let S be the subgroup of $\operatorname{Aut}(P)$ consisting of all automorphisms of P which are Q-central. If $S = \operatorname{Aut}(P)$ then we say $\operatorname{Aut}(P)$ is Q-central.

Definition 4.2 Say $\tau \in \operatorname{Aut}(Q)$ is *P-central* if τ commutes with all $c_x \in \operatorname{Aut}(Q)$, where $x \in P$. Let T be the subgroup of $\operatorname{Aut}(Q)$ consisting of all automorphisms of Q which are P-central. If $T = \operatorname{Aut}(Q)$ then we say $\operatorname{Aut}(Q)$ is P-central.

The centrality conditions defined above are exactly what we need in order to obtain (naive) lifts of elements of $\operatorname{Aut}(P)$ and $\operatorname{Aut}(Q)$.

The following conditions that imply P- and Q-centrality are fairly obvious.

Proposition 4.3 $\operatorname{Aut}(P)$ *is Q-central if* $C_G(Q) \cap P \leq Z(P)$ *and* $\operatorname{Aut}(P) = \operatorname{Aut}_c P$.

Proposition 4.4 $\mathrm{Aut}(Q)$ *is* P*-central if any of the following conditions holds*

1. $[P,Q] = 1$ *(in which case* $G = P \times Q$*)*
2. $\mathrm{Aut}(Q)$ *is abelian*
3. $\mathrm{Aut}(Q) = \mathrm{Aut}_c Q$ *and* $[Z(Q), P] = 1$
4. $\mathrm{Aut}(Q) = \mathrm{Inn}\, Q$ *and* $[P,Q] \le Z(Q)$

Theorem 4.5 (1) *If either* $\mathrm{Aut}(P)$ *is* Q*-central or* $\mathrm{Aut}(Q)$ *is* P*-central, then* $\mathrm{Im}\,\rho = \mathrm{Im}\,\rho_P \times \mathrm{Im}\,\rho_Q$. (2) *If both* $\mathrm{Aut}(P)$ *is* Q*-central and* $\mathrm{Aut}(Q)$ *is* P*-central, then* $\mathrm{Im}\,\rho = \mathrm{Aut}(P) \times \mathrm{Aut}(Q)$ *and* $\mathrm{Aut}(G;Q) \cong \mathrm{Ker}\,\rho \rtimes \mathrm{Im}\,\rho$.

Proof First assume $\mathrm{Aut}(P)$ is Q-central, then clearly $\mathrm{Im}\,\rho_P = \mathrm{Aut}(P)$. Let $(\sigma, \tau) \in \mathrm{Aut}(P) \times \mathrm{Im}\,\rho_Q$. There is a lift of σ to $\hat{\sigma} \in \mathrm{Aut}(G;Q)$ so that $\rho(\hat{\sigma}) = (\sigma, 1_Q)$. Since $\tau \in \mathrm{Im}\,\rho_Q$, there exists $\alpha \in \mathrm{Aut}(G;Q)$ such that $\rho_Q(\alpha) = \tau$. Say $\rho(\alpha) = (\sigma_1, \tau)$ for some $\sigma_1 \in \mathrm{Aut}(P)$. There is a lift of σ_1 to $\hat{\sigma}_1 \in \mathrm{Aut}(G;Q)$ such that $\rho(\hat{\sigma}_1) = (\sigma_1, 1_Q)$.

Now $\rho(\hat{\sigma} \circ \hat{\sigma}_1^{-1} \circ \alpha) = (\sigma, 1_Q)(\sigma_1^{-1}, 1_Q)(\sigma_1, \tau) = (\sigma, \tau)$, so we get $\mathrm{Im}\,\rho = \mathrm{Aut}(P) \times \mathrm{Im}\,\rho_Q$.

Next assume $\mathrm{Aut}(Q)$ is P-central, then clearly $\mathrm{Im}\,\rho_Q = \mathrm{Aut}(Q)$. Let $(\sigma, \tau) \in \mathrm{Im}\,\rho_P \times \mathrm{Aut}(Q)$. There exists $\alpha_1 \in \mathrm{Aut}(G;Q)$ such that $\rho_P(\alpha_1) = \sigma$. Say $\rho(\alpha_1) = (\sigma, \tau_1)$ for some $\tau_1 \in \mathrm{Aut}(Q)$.

Since τ is P-central, it lifts to $\hat{\tau} \in \mathrm{Aut}(G;Q)$ such that $\rho(\hat{\tau}) = (1_P, \tau)$. Similarly, τ_1 lifts to $\hat{\tau}_1 \in \mathrm{Aut}(G;Q)$ such that $\rho(\hat{\tau}_1) = (1_P, \tau_1)$.

Now $\rho(\alpha_1 \circ \hat{\tau}_1^{-1} \circ \hat{\tau}) = (\sigma, \tau_1)(1_P, \tau_1^{-1})(1_P, \tau) = (\sigma, \tau)$, so we get $\mathrm{Im}\,\rho = \mathrm{Im}\,\rho_P \times \mathrm{Aut}(Q)$.

Finally, when both $\mathrm{Aut}(P)$ is Q-central and $\mathrm{Aut}(Q)$ is P-central, $s : \mathrm{Im}\,\rho \to \mathrm{Aut}(G;Q)$ (as defined earlier) is a homomorphism and hence a splitting. $\square$

In some sense, the second part of the theorem above is ideal in that $\mathrm{Aut}(G;Q)$ is completely determined from maps involving P and Q. Unfortunately, it will hold only in very special circumstances. When Q is abelian, though, we can get a similar result with relaxed conditions.

4.1 When Q is abelian

First of all, the extension F will always split.

Theorem 4.6 *When* Q *is abelian,* $\mathrm{Aut}(G;Q) \cong \mathrm{Ker}\,\rho \rtimes \mathrm{Im}\,\rho$.

Proof Let $(\sigma, \tau) \in \mathrm{Im}\,\rho$, so $\rho(\alpha) = (\sigma, \tau)$ for some $\alpha \in \mathrm{Aut}(G;Q)$. By Theorem 2.1, $(\sigma^{-1}, \tau^{-1}) \in (C_\chi)_{[E]}$, thus the twisting on $\sigma^{-1} E \tau$ equals the twisting on E. We see that $c_\tau \circ \chi \circ \sigma(x) \equiv \chi(x) \mod \mathrm{Inn}\, Q$ for all $x \in P$. Since $\chi(x) = c_x$ we get $\tau^{-1} \circ c_{\sigma(x)} \circ \tau \equiv c_x \mod \mathrm{Inn}\, Q$.

Since Q is abelian, $\mathrm{Inn}\, Q = \{1\}$, so $\tau \circ c_x = c_{\sigma(x)} \circ \tau$ for all $x \in P$. By 4.1, s is a splitting of F and we have a semi-direct product. $\square$

When Q is abelian, we can identify $\operatorname{Ker}\rho$ and $\operatorname{Im}\rho$ in convenient ways. First, in terms of automorphisms of G, we see that $\operatorname{Ker}\rho = \{\alpha \in \operatorname{Aut}(G) \,|\, \alpha|_Q = 1_Q$ and $\bar{\alpha} = 1_{G/Q}\}$. While $\operatorname{Im}\rho \cong \operatorname{Im}s = \{\gamma \in \operatorname{Aut}(G) \,|\, \gamma(P) = P$ and $\gamma(Q) = Q\}$.

Second, in terms of P and Q, we see that $\operatorname{Ker}\rho = \operatorname{Der}(P, Z(Q))$. While we will see $\operatorname{Im}\rho$ is sometimes equal to $S \times T$.

Theorem 4.7 *If Q is abelian and either $\operatorname{Aut}(P)$ is Q-central or $\operatorname{Aut}(Q)$ is P-central, then $\operatorname{Im}\rho = S \times T$.*

Proof First assume $\operatorname{Aut}(P)$ is Q-central, then $S = \operatorname{Aut}(P)$. By Theorem 4.5, we need only show $\operatorname{Im}\rho_Q = T$.

Clearly $T \leq \operatorname{Im}\rho_Q$. Now suppose $\tau \in \operatorname{Im}\rho_Q$ where $\rho_Q(\alpha) = \tau$, for some $\alpha \in \operatorname{Aut}(G; Q)$. Say $\rho(\alpha) = (\sigma, \tau)$. There is a lift of σ to $\hat{\sigma}$. Consider $\rho(\hat{\sigma}^{-1} \circ \alpha) = (1, \tau)$, then $(1, \tau) \in (C_\chi)_{[E]}$. Comparing twistings, we see that $\tau \circ c_x = c_x \circ \tau$ mod $\operatorname{Inn}Q$. Since Q is abelian $\tau \circ c_x = c_x \circ \tau$ and we have $\tau \in T$.

Next assume $\operatorname{Aut}(Q)$ is P-central, then $T = \operatorname{Aut}(Q)$. We need only show $\operatorname{Im}\rho_P = S$.

Suppose $\sigma \in \operatorname{Im}\rho_Q$ where $\rho_P(\alpha) = \sigma$, for some $\alpha \in \operatorname{Aut}(G; Q)$. Say $\rho(\alpha) = (\sigma, \tau)$. There is a lift of τ to $\hat{\tau}$. Consider $\rho(\alpha \circ \hat{\tau}^{-1}) = (\sigma, 1)$, then $(\sigma, 1) \in (C_\chi)_{[E]}$. Comparing twistings, we see that $c_x = c_{\sigma(x)}$ and we have $\sigma \in S$. $\square$

Remark 4.8 It is not generally true that $\operatorname{Im}\rho = S \times T$. Consider the extraspecial group G of order 27 and exponent 3 with presentation

$$\langle a, b \,|\, a^3 = b^3 = 1, [b, a] = c, c \in Z(G) \rangle.$$

We have $G = P \ltimes Q$ where $P = \mathbb{Z}_3 = \langle a \rangle$, $Q = \mathbb{Z}_3 \times \mathbb{Z}_3 = \langle b, c \rangle$.

Define $\gamma : G \to G$ by $\gamma(a) = a^2$, $\gamma(b) = b$, and $\gamma(c) = c^2$. It is easy to check that $\gamma \in \operatorname{Aut}(G)$, and $\rho(\gamma) = (\sigma, \tau)$ where neither σ nor τ is trivial. Since $C_G(Q) \cap P = \langle c \rangle \cap \langle a \rangle = \{1\}$, the only element of S is the identity element. Thus we see that $(\sigma, \tau) \in \operatorname{Im}\rho$ but $(\sigma, \tau) \notin S \times T$.

More generally, $\operatorname{Im}\rho \neq M \times N$ for any $M \leq \operatorname{Im}\rho_Q$ and $N \leq \operatorname{Im}\rho_Q$. Let γ be as above, then $(\sigma, 1_Q) \in \operatorname{Im}\rho_P \times \operatorname{Im}\rho_Q$, but it is not an inducible pair. Any $\beta \in \operatorname{Aut}(G; Q)$ satisfying $\rho(\beta) = (\sigma, 1_Q)$ must be of the form $\beta(a) = a^2 b^j c^k$, $\beta(b) = b$ and $\beta(c) = c$. But then $\beta([b, a]) = [b, a^2 b^j c^k] = [b, a^2] = c^2$, so β is not a homomorphism.

4.2 Examples

Example 4.9 Consider the p-group G with presentation

$$\langle x, y \,|\, x^{p^m} = y^p = 1, \; yxy^{-1} = x^{p^{m-1}+1} \rangle,$$

where $m \geq 2$.

G is a metacyclic group, and if $m = 2$ then it is the unique extraspecial p-group of exponent p^2. Letting $Q = \langle x \rangle$ and $P = \langle y \rangle$, we see that $G = P \ltimes Q$.

Since Q is abelian Theorem 4.6 says $\operatorname{Aut}(G; Q) \cong \operatorname{Ker} \rho \rtimes \operatorname{Im} \rho$. The group $\operatorname{Aut}(Q)$ is abelian, hence P-central, thus Theorem 4.7 says $\operatorname{Im} \rho = S \times \operatorname{Aut}(Q)$.

Now $\operatorname{Aut}(Q) = U(p^m)$, the multiplicative group of integers mod p^m. Since $C_G(Q) \cap P = \langle x \rangle \cap \langle y \rangle = \{1\}$, the only element of $\operatorname{Aut}(P)$ which is Q-central is the identity homomorphism. Lastly, suppose $d \in \operatorname{Der}(P, Q)$ is defined by $d(y) = x^r$, for some $r \in \mathbb{Z}_{p^m}$. Using the fact that $d(ab) = d(a)a^{-1}d(b)a$, we see that

$$d(y^i) = x^{r(1+\beta+\beta^2+\cdots+\beta^{i-1})},$$

where $x^\beta = (x^{p^{m-1}+1})^{-1}$. Now

$$1 = d(1) = d(1 \cdot y^p) = d(1)d(y^p) = x^{r(1+\beta+\beta^2+\cdots+\beta^{p-1})}.$$

Lemma 7 of [7] (applied to $\Lambda(-1, p)$) shows that

$$(1 + \beta + \beta^2 + \cdots + \beta^{p-1}) \equiv p \bmod p^{m-1}.$$

Thus we must have $r \equiv 0 \bmod p^{m-1}$, and $\operatorname{Ker} \rho \cong \mathbb{Z}_p$. We conclude

$$\operatorname{Aut}(G; Q) \cong \mathbb{Z}_p \rtimes U(p^m).$$

We can actually say more about the exact nature of the automorphisms in $\operatorname{Aut}(G; Q)$. A typical element $\tau : x \mapsto x^j$ of $\operatorname{Aut}(Q)$ lifts to an automorphism of G which is τ on Q and the identity on P. A typical element $d : y \mapsto x^r$ of $\operatorname{Der}(P, Q)$ maps via μ to $\hat{d} \in \operatorname{Aut}(G; Q)$ defined by $\hat{d}(y^i x^j) = d(y^i)y^i x^j$. Thus any $\phi \in \operatorname{Aut}(G; Q)$, is defined by

$$\begin{aligned} \phi(x) &= x^j, \text{ where } j \in U(p^m) \\ \phi(y) &= yx^r, \text{ where } r \in \mathbb{Z}_{p^m} \text{ and } r \equiv 0 \bmod p^{m-1}. \end{aligned}$$

Note that in [7], all of $\operatorname{Aut}(G)$ is computed. It turns out that $\phi \in \operatorname{Aut}(G)$ is defined by

$$\begin{aligned} \phi(x) &= y^i x^j, \text{ where } i \in \mathbb{Z}_p \text{ and } j \in U(p^m) \\ \phi(y) &= yx^r, \text{ where } r \in \mathbb{Z}_{p^m} \text{ and } r \equiv 0 \bmod p^{m-1}. \end{aligned}$$

Example 4.10 Let $G = \langle a, b, c, d \rangle$ be the same group as in Example 3.1.

In this example we consider the group as a semi-direct product: $G = P' \ltimes Q'$, where $P' = \langle b, c, d \rangle \cong (\mathbb{Z}_2)^3$ and $Q' = \langle a \rangle \cong \mathbb{Z}_4$.

As in the example above, Q' and $\operatorname{Aut} Q'$ are both abelian, so

$$\operatorname{Aut}(G; Q) \cong \operatorname{Ker} \rho \rtimes (S \times \operatorname{Aut}(Q')).$$

In this case $\operatorname{Ker} \rho = \operatorname{Der}(P', Q') = \operatorname{Hom}(P', Q') = (\mathbb{Z}_2)^3$ and $\operatorname{Aut}(Q') = U(4) \cong \mathbb{Z}_2$. Now $\operatorname{Aut}(P') = GL_3(\mathbb{F}_2)$, but the subgroup S consists of those linear transformations σ which for all $p \in P'$ satisfy $\sigma(p) = px$, for some $x \in C_G(Q') \cap P' = \langle c, d \rangle$. Thinking of b as the standard basis element $(1, 0, 0) \in \mathbb{F}_2 \oplus \mathbb{F}_2 \oplus \mathbb{F}_2$, we get that S is the set of all 3×3 invertible matrices over $\mathbb{F}_2$ of the form

$$\begin{bmatrix} 1 & 0 & 0 \\ a_{21} & a_{22} & a_{23} \\ a_{31} & a_{32} & a_{33} \end{bmatrix}.$$

The order of S is 24, and the order of $\operatorname{Aut}(G; Q)$ is 384.

5 Central Products

In this section we assume N is non-trivial and that P and Q commute with one another. In this case, G is a central product and we write $G = P \circ_N Q$ or just $G = P \circ Q$.

One difficulty in dealing with central products is that an element $g \in G$ does not have a unique representation as $g = pq$ for some $p \in P$ and $q \in Q$. Projecting automorphisms of G to the subgroups P and Q, as well as lifting automorphisms of P and Q to G are nearly impossible tasks unless we impose some restrictions on the structure of G. Until subsection 5.2, we assume G is a p-group and $N = \Phi(P) = \Phi(Q)$. One way of obtaining a group with the proper structure is by amalgamating two groups along their common Frattini subgroup.

Both P/N and Q/N are elementary abelian, with p-rank r and s respectively. Let $\{p_i\}_{i=1}^r$ be a set of generators of P which induce a basis of P/N, and let $\{q_j\}_{j=1}^s$ be a similar set of generators for Q.

In general, if H is a p-group with $\{h_1, h_2, \ldots, h_k\}$ a set of generators which induce a basis of $H/\Phi(H)$, then every element $x \in H$ can be written as $x = h_1^{a_1} h_2^{a_2} \cdots h_k^{a_k} z$ where $0 \le a_i < p$, $z \in \Phi(H)$, and the a_i can be chosen uniquely.

The condition $N = \Phi(P) = \Phi(Q)$ implies $\Phi(G) = N$ and $G/\Phi(G) = P/\Phi(P) \times Q/\Phi(Q)$. The set $\{p_1, \ldots, p_r, q_1, \ldots, q_s\}$ is a set of generators of G which induces a basis of $G/\Phi(G)$, hence every element $g \in G$ can be written in the form

$$g = \left(\prod_{i=1}^r p_i^{k_i}\right)\left(\prod_{j=1}^s q_j^{l_j}\right) n$$

where $n \in N$, $0 \le k_i, l_j < p$, and the k_i and l_i can be chosen uniquely.

Description 5.1 Let

$$g_1 = \left(\prod_{i=1}^r p_i^{k_i}\right)\left(\prod_{j=1}^s q_j^{l_j}\right) n_1 \quad \text{and} \quad g_2 = \left(\prod_{i=1}^r p_i^{k_i'}\right)\left(\prod_{j=1}^s q_j^{l_j'}\right) n_2$$

where $0 \le k_i, k_i', l_j, l_j' < p$ and $n_i \in N$. Then

$$
\begin{aligned}
g_1 g_2 &= \left[\left(\prod_{i=1}^r p_i^{k_i}\right)\left(\prod_{j=1}^s q_j^{l_j}\right) n_1\right]\left[\left(\prod_{i=1}^r p_i^{k_i'}\right)\left(\prod_{j=1}^s q_j^{l_j'}\right) n_2\right] \\
&= \left[\left(\prod_{i=1}^r p_i^{k_i}\right)\left(\prod_{i=1}^r p_i^{k_i'}\right)\right]\left[\left(\prod_{j=1}^s q_j^{l_j}\right)\left(\prod_{j=1}^s q_j^{l_j'}\right)\right] n_1 n_2 \\
&= \left(\prod_{i=1}^r p_i^{k_i+k_i'} c_1\right)\left(\prod_{j=1}^s q_j^{l_j+l_j'} c_2\right) n_1 n_2 \qquad \text{(where } c_1, c_2 \text{ are described below)} \\
&= \left(\prod_{i=1}^r p_i^{\overline{k_i+k_i'}} c_1 d_1\right)\left(\prod_{j=1}^s q_j^{\overline{l_j+l_j'}} c_2 d_2\right) n_1 n_2 \qquad \text{(where } d_1, d_2 \text{ are described below)}
\end{aligned}
$$

$$= \left(\prod_{i=1}^{r} p_i^{\overline{k_i + k_i'}} \right) \left(\prod_{j=1}^{s} q_j^{\overline{l_j + l_j'}} \right) n_1 n_2 c_1 d_1 c_2 d_2$$

where

$$c_1 = \prod_{u=1}^{r-1} \prod_{v=u+1}^{r} [p_v^{k_v}, p_u^{k_u'}],$$

$$c_2 = \prod_{u=1}^{s-1} \prod_{v=u+1}^{s} [q_v^{l_v}, q_u^{l_u'}],$$

$$d_1 = \prod_{i=1}^{r} p_i^{\delta_i^P p} \quad \text{where } k_i + k_i' = \overline{k_i + k_i'} + \delta_i^P p \text{ with } 0 \le \overline{k_i + k_i'} < p$$

$$d_2 = \prod_{j=1}^{s} q_j^{\delta_j^Q p} \quad \text{where } l_j + l_j' = \overline{l_j + l_j'} + \delta_j^Q p \text{ with } 0 \le \overline{l_j + l_j'} < p.$$

Note that $\delta_i^P, \delta_j^Q = 0$ or 1.

5.1 An analogue of ρ

In general, it is nearly impossible to construct a lift of an automorphism of P or of Q to the full group G as is automatic with direct products, and is achieved for semi-direct products under the centrality conditions. If $\tau \in \text{Aut}(Q)$, for example, moves elements of the intersection, then constructing a lift of τ which behaves appropriately on the intersection requires detailed knowledge of the groups involved. (In [10] and [2], such lifts were constructed for very specific group examples.) But if we restrict to looking at automorphisms which fix the intersection pointwise, then we can get good information about the influence $\text{Aut}(P)$ and $\text{Aut}(Q)$ have on the structure of $\text{Aut}(G)$.

For any group H with subgroups K and L, we have the following notation:
- $\text{Aut}_L H$ is the subgroup of automorphisms of H which fix L pointwise
- $\text{Aut}_L(H; K)$ is the subgroup of automorphisms of H which map K to itself and fix L pointwise.

Restricting the homomorphism ρ to $\text{Aut}_N(G; Q)$, we get

$$\eta : \text{Aut}_N(G; Q) \xrightarrow{(\eta_P \times \eta_Q)} \text{Aut}(P/N) \times \text{Aut}_N Q.$$

It is easy to see that $\text{Ker}\,\eta = \text{Ker}\,\rho \cong \text{Der}(P/N, Z(Q))$. Since P acts trivially on Q when G is a central product, we have $\text{Der}(P/N, Z(Q)) = \text{Hom}(P/N, Z(Q))$.

The rest of this section concerns the image of η.

Proposition 5.2 $\text{Im}\,\eta_Q = \text{Aut}_N Q.$

Proof Let $\tau \in \text{Aut}_N Q$ and define a map $\hat{\tau} : G \to G$ to be the identity on P and τ on Q. More explicitly, let $g_1, g_2 \in G$ be as in Description 5.1. Define

$$\hat{\tau}(g_1) = \prod_{i=1}^{r} p_i^{k_i} \prod_{j=1}^{s} \tau(q_j)^{l_j} n_1.$$

It is easy to see that $\hat{\tau}$ is one-to-one and onto, so it remains to show that $\hat{\tau}$ is a homomorphism.

By definition,

$$\hat{\tau}(g_1 g_2) = (\prod_{i=1}^{r} p_i^{\overline{k_i + k_i'}})(\prod_{j=1}^{s} \tau(q_j)^{\overline{l_j + l_j'}}) n_1 n_2 c_1 d_1 c_2 d_2.$$

While

$$\hat{\tau}(g_1)\hat{\tau}(g_2) = \left[\prod_{i=1}^{r} p_i^{k_i} \prod_{j=1}^{s} \tau(q_j)^{l_j} n_1\right]\left[\prod_{i=1}^{r} p_i^{k_i'} \prod_{j=1}^{s} \tau(q_j)^{l_j'} n_2\right]$$

$$= \left[\prod_{i=1}^{r} p_i^{k_i} \prod_{i=1}^{r} p_i^{k_i'}\right]\left[\prod_{j=1}^{s} \tau(q_j)^{l_j} \prod_{j=1}^{s} \tau(q_j)^{l_j'}\right] n_1 n_2$$

$$= \left[\prod_{i=1}^{r} p_i^{\overline{k_i + k_i'}} c_1 d_1\right]\left[\tau\left(\prod_{j=1}^{s} q_j^{l_j}\right)\tau\left(\prod_{j=1}^{s} q_j^{l_j'}\right)\right] n_1 n_2$$

$$= \left[\prod_{i=1}^{r} p_i^{\overline{k_i + k_i'}} c_1 d_1\right]\left[\tau\left(\left(\prod_{j=1}^{s} q_j^{l_j}\right)\left(\prod_{j=1}^{s} q_j^{l_j'}\right)\right)\right] n_1 n_2$$

$$= \left[\prod_{i=1}^{r} p_i^{\overline{k_i + k_i'}} c_1 d_1\right]\left[\tau\left(\prod_{j=1}^{s} q_j^{\overline{l_j + l_j'}} c_2 d_2\right)\right] n_1 n_2$$

$$= \left[\prod_{i=1}^{r} p_i^{\overline{k_i + k_i'}} c_1 d_1\right]\left[\tau\left(\prod_{j=1}^{s} q_j^{\overline{l_j + l_j'}}\right)\right] n_1 n_2 c_2 d_2$$

$$= \left[\prod_{i=1}^{r} p_i^{\overline{k_i + k_i'}}\right]\left[\prod_{j=1}^{s} \tau(q_j)^{\overline{l_j + l_j'}}\right] n_1 n_2 c_1 d_1 c_2 d_2.$$

$\square$

Remark 5.3 A similar argument shows that any element σ of $\text{Aut}_N P$ also lifts to $\hat{\sigma} \in \text{Aut}_N(G; Q)$, which is σ on P and the identity on Q.

Theorem 5.4 $\text{Im}\,\eta = \text{Im}\,\eta_P \times \text{Im}\,\eta_Q$.

Proof We need only show $\text{Im}\,\eta_P \times \text{Im}\,\eta_Q \leq \text{Im}\,\eta$. Let $(\sigma, \tau) \in \text{Im}\,\eta_P \times \text{Im}\,\eta_Q$, with $\alpha_1, \alpha_2 \in \text{Aut}_N(G; Q)$ satisfying $\eta_P(\alpha_1) = \sigma$ and $\eta_Q(\alpha_2) = \tau$.

Suppose $\eta_Q(\alpha_1) = \tau_1$. Let $\tau_2 = (\tau_1)^{-1} \circ \tau$. By the proposition above, τ_2 lifts to an element $\hat{\tau}_2 \in \text{Aut}_N(G; Q)$. Letting $\gamma = \alpha_1 \circ \hat{\tau}_2$, we see that $\eta(\gamma) = (\sigma, \tau)$.

$\square$

This theorem seems a bit innocuous, but Remark 4.8 shows that in the semi-direct product case it is far from assured that one can obtain $\text{Im}\,\rho$ as a product.

Corollary 5.5 *The homomorphism η splits if and only if η_P splits.*

Proof This is easy to see since the proof of Proposition 5.2 implies that η_Q splits. $\square$

Since P/N is elementary abelian of p-rank r, we know that $\mathrm{Aut}(P/N) = GL_r(\mathbb{F}_p)$. We will usually denote this general linear group by GL_r.

Description 5.6 Let $M = (m_{ij}) \in GL_r$. Define a map $\mu : P \to P$ as follows. A typical element $x \in P$ can be written in the form

$$x = \prod_{i=1}^{r} p_i^{k_i} n,$$

where $n \in N$ and $0 \le k_i < p$. Set

$$\mu(x) = \prod_{i=1}^{r} \left(\prod_{t=1}^{r} p_t^{m_{ti}} \right)^{k_i} n.$$

So μ is given by the left action of the matrix M on the generators of P.

Lemma 5.7 *The map μ in Description 5.6 is an element of $\mathrm{Aut}_N P$ if and only if* (1) $\mu(p_i)^p = p_i^p$ *for all* $i = 1, \ldots, r$ *and* (2) $[\mu(p_i), \mu(p_j)] = [p_i, p_j]$ *for all* $i, j = 1, \ldots, r$.

Proof If $\mu \in \mathrm{Aut}_N P$ then μ is a homomorphism fixing N, so (1) and (2) hold.

Let $x_1 = \prod_{i=1}^{r} p_i^{k_i} n_1$ and $x_2 = \prod_{i=1}^{r} p_i^{k_i'} n_2$ be two elements of P. Modifying the computation of $g_1 g_2$ in Description 5.1, we see that

$$x_1 x_2 = \prod_{i=1}^{r} p_i^{\overline{k_i + k_i'}} n_1 n_2 c_1 d_1.$$

By definition,

$$\mu(x_1 x_2) = \prod_{i=1}^{r} \left(\prod_{t=1}^{r} p_t^{m_{ti}} \right)^{\overline{k_i + k_i'}} n_1 n_2 c_1 d_1.$$

While,

$$\mu(x_1)\mu(x_2) = \left[\prod_{i=1}^{r} \underbrace{\left(\prod_{t=1}^{r} p_t^{m_{ti}} \right)^{k_i}}_{:=a_i} n_1 \right] \left[\prod_{i=1}^{r} \left(\prod_{t=1}^{r} p_t^{m_{ti}} \right)^{k_i'} n_2 \right]$$

$$= \prod_{i=1}^{r} a_i^{k_i} \prod_{i=1}^{r} a_i^{k_i'} n_1 n_2$$

$$= \prod_{i=1}^{r} a_i^{\overline{k_i + k_i'}} n_1 n_2 n_3 n_4,$$

where $n_3 = \prod_{u=1}^{r-1} \prod_{v=u+1}^{r} [a_v^{k_v}, a_u^{k_u'}]$ and $n_4 = \prod_{i=1}^{r} a_i^{\delta_i^P p}$.

Now $a_i = \mu(p_i)$, so

$$
\begin{aligned}
[a_v^{k_v}, a_u^{k'_u}] &= [\mu(p_v)^{k_v}, \mu(p_u)^{k'_u}] \\
&= [\mu(p_v), \mu(p_u)]^{k_v k'_u} \quad \text{since } N \le Z(G) \\
&= [p_v, p_u]^{k_v k'_u} \quad\quad\;\; \text{by property (2)} \\
&= [p_v^{k_v}, p_u^{k'_u}].
\end{aligned}
$$

Thus $n_3 = c_1$. Also,

$$
a_i^{\delta_i^P p} = \mu(p_i)^{\delta_i^P p} = p_i^{\delta_i^P p},
$$

and we see that $n_4 = d_1$.

To show μ is onto, we let ν be the homomorphism obtained via Description 5.6 from the matrix M^{-1}. In P/N, $\bar{p}_i = M(M^{-1}(\bar{p}_i))$, so $p_i = \mu(\nu(p_i)) \bmod N$. Say $p_i = \mu(\nu(p_i))n$ for some $n \in N$. Then $p_i = \mu(\nu(p_i n))$ and μ is surjective. $\qquad\square$

Definition 5.8 Let Γ be the subgroup of GL_r consisting of all matrices M that have lifts to $\mathrm{Aut}_N P$ of the form μ, as in Description 5.6.

Let $\Phi_P : \mathrm{Aut}(P) \to \mathrm{Aut}(P/N)$ be the usual map that takes an automorphism of P to the induced map on the Frattini quotient. Denote by Φ_{NP} the reduction of Φ_P to the subgroup $\mathrm{Aut}_N P$.

Corollary 5.9 $\Gamma \le \mathrm{Im}\,\Phi_{NP} \le \mathrm{Im}\,\eta_P$.

Proof Clearly $\Gamma \le \mathrm{Im}\,\Phi_{NP}$. Now let $M \in \mathrm{Im}\,\Phi_{NP}$ with $\Phi_{NP}(\beta) = M$ for some $\beta \in \mathrm{Aut}_N P$. From Remark 5.3, we see that β lifts to $\hat{\beta} \in \mathrm{Aut}_N(G; Q)$. $\qquad\square$

Under certain circumstances, we can get more explicit relationships among Γ, $\mathrm{Im}\,\Phi_{NP}$, and $\mathrm{Im}\,\eta_P$.

If P is a characteristic subgroup of G, then $\sigma = \eta_P(\alpha) \in \mathrm{Im}\,\eta_P$ satisfies $\sigma = \Phi_{NP}(\alpha|_P)$ where $\alpha|_P$ is the restriction of α to P. Thus we have the following proposition.

Proposition 5.10 *If P is a characteristic subgroup of G, then $\mathrm{Im}\,\Phi_{NP} = \mathrm{Im}\,\eta_P$.*

A much better result is obtained when we insist that the exponent of N is equal to p.

Theorem 5.11 *If $exp(N) = p$ then $\Gamma = \mathrm{Im}\,\Phi_{NP} = \mathrm{Im}\,\eta_P$. Moreover,*
$\mathrm{Aut}_N(G; Q) \cong \mathrm{Hom}(P/N, Z(Q)) \rtimes (\Gamma \times \mathrm{Aut}_N Q)$.

Proof We will show $\mathrm{Im}\,\eta_P \le \Gamma$. Let $\sigma \in \mathrm{Im}\,\eta_P$ with $\eta_P(\alpha) = \sigma$ for some $\alpha \in \mathrm{Aut}_N(G; Q)$. If $M = (m_{ij})$ is the matrix in GL_r corresponding to σ, construct $\mu \in \mathrm{Aut}_N P$ from M as in Description 5.6. We must show μ satisfies the two properties from Lemma 5.7.

Denote the generators of $G/Q = P/N$ by $\{\overline{p_i}\}_{t=1}^r$. Then $\sigma(\overline{p_i}) = \prod_{t=1}^r \overline{p_t}^{m_{ti}}$. On the other hand, $\sigma(\overline{p_i}) = \alpha(p_i)N$. We see that α must satisfy $\alpha(p_i) = \prod_{t=1}^r p_t^{m_{ti}} y_i$ for some $y_i \in N$.

Consider

$$
\begin{aligned}
p_i^p = \alpha(p_i^p) &= \left(\prod_{t=1}^r p_t^{m_{ti}} y_i \right)^p \\
&= \left(\prod_{t=1}^r p_t^{m_{ti}} \right)^p (y_i)^p \\
&= \left(\prod_{t=1}^r p_t^{m_{ti}} \right)^p \qquad \text{since } exp(N) = p \\
&= \mu(p_i)^p.
\end{aligned}
$$

Furthermore,

$$
\begin{aligned}
[p_i, p_j] = [\alpha(p_i), \alpha(p_j)] &= \left[\prod_{t=1}^r p_t^{m_{ti}} y_i, \prod_{t=1}^r p_t^{m_{tj}} y_j \right] \\
&= \left[\prod_{t=1}^r p_t^{m_{ti}}, \prod_{t=1}^r p_t^{m_{tj}} \right] [y_i, y_j] \\
&= \left[\prod_{t=1}^r p_t^{m_{ti}}, \prod_{t=1}^r p_t^{m_{tj}} \right] \qquad \text{since } N \text{ abelian} \\
&= [\mu(p_i), \mu(p_j)].
\end{aligned}
$$

Finally, η_P is split by $s : \mathrm{Aut}(P/N) \to \mathrm{Aut}_N(G; Q)$, defined by $s(M) = \hat{\mu}$. □

Remark 5.12 The condition $exp(N) = p$ is not necessary to get the splitting in Theorem 5.11; we need only have $\mathrm{Im}\,\eta_P = \Gamma$.

Example 5.13 Let P have the following presentation:

$$\langle a_1, b_1 \mid a_1^8 = b_1^2 = 1, b_1 a_1 = a_1^5 b_1 \rangle.$$

P has order 16 and is the group $\Gamma_2 d$ in the Thomas & Wood notation [8]. Let Q have the same presentation, but with generators a_2 and b_2. Then $\Phi(P) = \langle a_1^2 \rangle \cong \langle a_2^2 \rangle = \Phi(Q)$. Amalgamate P and Q along their common Frattini subgroup N to get the group $G = P \circ_N Q$ with presentation

$$\langle a_1, b_1, a_2, b_2 \mid a_1^8 = a_2^8 = b_1^2 = b_2^2 = 1, a_1^2 = a_2^2, b_1 a_1 = a_1^5 b_1, b_2 a_2 = a_2^5 b_2,$$
$$[a_1, a_2] = [a_1, b_2] = [b_1, a_2] = [b_1, b_2] = 1 \rangle.$$

G has order 64 and is SmallGroup(64,249) in the small group library of GAP [4].

The full automorphism group of Q is isomorphic to $D_4 \times \mathbb{Z}_2$ with generators f_1, f_2, and g defined by $f_1(a_2) = a_2^3 b_2$, $f_1(b_2) = b_2$, $f_2(a_2) = a_2$, $f_2(b_2) = a_2^4 b_2$,

$g(a_2) = a_2^3$, and $g(b_2) = a_2^4 b_2$. It is easy to check that both f_1 and f_2 fix $N = \langle a_2^2 \rangle$, while g does not. Hence $\mathrm{Aut}_N Q \cong D_4 = \langle f_1, f_2 \rangle$.

The computation above also shows that $\mathrm{Aut}_N P \cong D_4$. It is also easy to see that $\mathrm{Im}\Phi_{NP}$ is isomorphic to $\mathbb{Z}_2$ with generator $\Phi(f_1)$. By Corollary 5.9 we know $\mathrm{Im}\,\Phi_{NP} \le \mathrm{Im}\,\eta_P$.

We can actually show $\mathrm{Im}\,\Phi_{NP} = \mathrm{Im}\,\eta_P$. Since $\mathrm{Aut}(P/N) = GL_2(\mathbb{F}_2) \cong S_3$, it suffices to show that the element γ of order 3 in S_3 given by $\gamma(\overline{a_1}) = \overline{a_1 b_1}$ and $\gamma(\overline{b_1}) = \overline{a_1}$ is not in the image of η_P.

Suppose $\alpha \in \mathrm{Aut}_N(G; Q)$ satisfies $\eta(\alpha) = (\gamma, \tau)$ where γ is defined above and τ is some element of $\mathrm{Aut}_N Q$. Now $\eta(\alpha \circ \hat{\tau}^{-1}) = (\gamma, 1)$, so without loss of generality we can assume α is the identity on Q. Thus, α must have the following form:

$$
\begin{aligned}
\alpha(a_1) &= a_1^v b_1 a_2^i b_2^j \\
\alpha(b_1) &= a_1^w a_2^r b_2^s \\
\alpha(a_2) &= a_2 \\
\alpha(b_2) &= b_2
\end{aligned}
$$

where v and w are odd.

The relation $[b_1, a_2] = 1$ implies $s = 0$. The relation $[b_1, b_2] = 1$ implies $r \equiv 0$ mod 2. The relation $b_2^2 = 1$ implies $2w + r + r5^s \equiv 0$ mod 8. Hence $2w + 2r \equiv 0$ mod 8, and we see that w cannot be odd. Thus, such an α cannot exist and $\mathrm{Im}\,\eta_P = \mathrm{Im}\,\Phi_{NP} \cong \mathbb{Z}_2$. Γ is non-trivial, so $\Gamma = \mathrm{Im}\,\eta_P$ too.

Finally, $\mathrm{Ker}\,\eta \cong \mathrm{Hom}(P/N, Z(Q)) \cong \mathrm{Hom}(\mathbb{Z}_2 \times \mathbb{Z}_2, \mathbb{Z}_4) = \mathbb{Z}_2 \times \mathbb{Z}_2$.

Since $\Gamma = \mathrm{Im}\,\eta_P$, Remark 5.12 implies

$$\mathrm{Aut}_N(G; Q) \cong \mathrm{Hom}(P/N, Z(Q)) \rtimes (\Gamma \times \mathrm{Aut}_N Q) \cong (\mathbb{Z}_2 \times \mathbb{Z}_2) \rtimes (\mathbb{Z}_2 \times D_4),$$

and $|\mathrm{Aut}_N(G; Q)| = 64$.

5.2 When Q is Elementary Abelian

If we continue to assume $N = \Phi(P) = \Phi(Q)$ when Q is elementary abelian, then $\Phi(Q) = \{1\}$ and $P \circ_N Q = P \times Q$ (a direct product, that was studied in section 3). However, we can relax the condition on N to simply assuming $\Phi(G) = \Phi(P) \le N$ and still recover the efforts of the previous subsections for a non-direct central product involving an elementary abelian Q. For the rest of this subsection, assume Q is elementary abelian and $\Phi(G) = \Phi(P) \le N$.

Let $\{q_1, q_2, \ldots, q_{s-t}, n_1, n_2, \ldots, n_t\}$ generate Q, where $\{n_1, n_2, \ldots, n_t\}$ is a minimal generating set for N. Any element $g \in G = P \circ_N Q$ can be written uniquely as

$$g = \prod_{i=1}^{r} p_i^{k_i} \prod_{j=1}^{s-t} q_i^{l_i} \prod_{u=1}^{t} n_u^{m_u}$$

with $0 \le k_i, l_j, m_u < p$. This is essentially the same situation as described in Section 5.1. It is not hard to see that all of the work in that section holds in this alternative case. In particular, we will always get the conclusion of Theorem 5.11.

Example 5.14 Again, let $G = \langle a, b, c, d \rangle$ as in Example 3.1.

This time we consider the group as a central product $G = P'' \circ_N Q''$, where $P'' = \langle a, b \rangle \cong D_4$ and $Q'' = \langle a^2, b, d \rangle$. The center of G is Q'', so it is a characteristic subgroup of G. Furthermore, $N = \langle a^2 \rangle = \Phi(G)$ is characteristic in G. Since $\mathrm{Aut}(N) \cong \mathbb{Z}_2$, every automorphism of G actually fixes N pointwise. Thus $\mathrm{Aut}_N(G; Q) = \mathrm{Aut}(G)$.

Since Q'' is elementary abelian, N has exponent p, and we see that $\mathrm{Im}\,\eta = \Gamma \times \mathrm{Aut}_N Q$. It is routine to compute $\Gamma \cong \mathbb{Z}_2$. The full automorphism group of Q is $GL_3(\mathbb{F}_2)$. If we think of a^2 as corresponding to the standard basis vector $(1, 0, 0) \in \mathbb{F}_2 \oplus \mathbb{F}_2 \oplus \mathbb{F}_2$, we get that $\mathrm{Aut}_N Q$ is the set of all 3×3 invertible matrices over $\mathbb{F}_2$ of the form

$$\begin{bmatrix} 1 & a_{12} & a_{13} \\ 0 & a_{22} & a_{23} \\ 0 & a_{32} & a_{33} \end{bmatrix}.$$

The order of $\mathrm{Aut}_N Q$ is 24.

The kernel of η is $\mathrm{Hom}(P, Q) \cong (\mathbb{Z}_2 \times \mathbb{Z}_2)^3$.

Finally, we use Theorem 5.11 to conclude that

$$\mathrm{Aut}(G) \cong (\mathbb{Z}_2 \times \mathbb{Z}_2)^3 \rtimes (\mathbb{Z}_2 \times \mathrm{Aut}_N Q)$$

and $|\mathrm{Aut}(G)| = 4^3 \cdot 2 \cdot 24 = 3072$.

Example 5.15 Let $G = P \circ Q$, where P is an extraspecial p-group, $Q = Z(G)$ is elementary abelian, and $N = \Phi(P) = \Phi(G) = \mathbb{Z}_p$.

By Theorem 5.11, we see

$$\mathrm{Aut}_N(G; Q) = \mathrm{Aut}_N G \cong \mathrm{Hom}(P, A) \rtimes (\Gamma \times \mathrm{Aut}_N Q).$$

Let $|P| = p^{2n+1}$ and $|Q| = p^{m+1}$. Then $|\mathrm{Hom}(P, A)| = (p^{2n})^{m+1}$ because each of the $2n$ generators of P can be mapped to p^{m+1} different places. Since one of the generators of Q gets fixed by elements of $\mathrm{Aut}_N Q$, we see that $|\mathrm{Aut}_N Q| = p^m * |GL_m(\mathbb{F}_p)|$. Thus, the order of $\mathrm{Aut}_N(G; Q)$ is $|\Gamma| * |GL_m(\mathbb{F}_p)| * p^{2nm+2n+m}$.

On the other hand, work in [2] (with $G = E_1 \circ A_1$) shows that

$$\mathrm{Aut}_N G \cong K \rtimes (\Gamma \times I)$$

where I is a subgroup of $GL_{m+1}(\mathbb{F}_p)$ isomorphic to $GL_m(\mathbb{F}_p)$, and K is a normal subgroup of order $p^{2nm+2n+m}$.

While the structure of $\mathrm{Aut}_N G$ obtained via Theorem 5.11 is slightly different than the structure obtained via [2], this paper may be seen as a generalization of [2].

Example 5.16 Here is an unorthodox use of the theory developed. Let G be the group with presentation

$$\langle a, b, c \mid a^{p^2} = b^{p^2} = c^p = 1, \ [b, a] = c, \ [a, c] = [b, c] = 1 \rangle.$$

Let $P = \langle a, b, c \rangle$ and $Q = \langle a^p, b^p, c \rangle$. Note that $N = Q = \Phi(P)$ and $P/N = \mathbb{Z}_p \times \mathbb{Z}_p$.

Since N has exponent p, Theorem 5.11 shows that

$$\operatorname{Aut}_N(G; Q) = \operatorname{Aut}_N G \cong \operatorname{Hom}(P/N, Q) \rtimes (\Gamma \times \operatorname{Aut}_N Q).$$

Now $N = Q$ implies $\operatorname{Aut}_N Q = \{1\}$. We also know that

$$\operatorname{Hom}(P/N, Q) = \operatorname{Hom}((\mathbb{Z}_p)^2, (\mathbb{Z}_p)^3) = \operatorname{Mat}_{3,2}(\mathbb{F}_p).$$

If $M = \begin{bmatrix} i & r \\ j & s \end{bmatrix} \in \Gamma$ then the condition $\mu(a^p) = a^p$ implies $\hat{\mu} \in \operatorname{Aut}_N G$ must satisfy $\hat{\mu}(a)^p = a^p$ and $\hat{\mu}(b)^p = b^p$. A quick calculation shows these equations hold if and only if $M = I_2$, the identity matrix.

We conclude that $\operatorname{Aut}_N G \cong \operatorname{Hom}(P/N, Q) = \operatorname{Mat}_{3,2}(\mathbb{F}_p)$, and has order p^6. In [5] the full automorphism group is computed to have order $p^6 |GL_2(\mathbf{F}_p)|$.

Let (m_{ij}) be a matrix in $\operatorname{Hom}(P/N, Q) = \operatorname{Mat}_{3,2}(\mathbb{F}_p)$, and let M be its corresponding linear transformation. Recall that the homomorphism $\mu : \operatorname{Ker} \eta \to \operatorname{Aut}_N(G; Q)$ is defined by $\mu(M(g)) = M(\pi(g))g$. Thus, the matrix (m_{ij}) induces an element $\alpha \in \operatorname{Aut}_N G$ defined by

$$\begin{aligned} \alpha(a) &= (a^p)^{m_{11}} (b^p)^{m_{21}} c^{m_{31}} a &= a^{pm_{11}+1} b^{pm_{21}} c^{pm_{21}+m_{31}} \\ \alpha(b) &= (a^p)^{m_{12}} (b^p)^{m_{22}} c^{m_{32}} b &= a^{pm_{12}} b^{pm_{22}+1} c^{m_{32}} \\ \alpha(c) &= c. \end{aligned}$$

These are the p^6 elements of $\operatorname{Aut}_N G$.

Let $f \in \operatorname{Aut}(G)$ be defined by $f(a) = a$, $f(b) = ab$, $f(c) = c$, and let $H = \langle f \rangle$. Note that $f^i(b) = a^i b$, so the order of f is p^2 and $\operatorname{Aut}_N G \cap H = \langle f^p \rangle$. Since $N = Z(G)$ is characteristic in G, $\operatorname{Aut}_N G \triangleleft \operatorname{Aut}(G)$. Thus, $H \operatorname{Aut}_N G$ is a subgroup of $\operatorname{Aut}(G)$. The order of $H \operatorname{Aut}_N G$ is p^7, so it is a p-Sylow subgroup of $\operatorname{Aut}(G)$.

5.3 Analog of the Wells Exact Sequence

We get an analog of Theorem 2.1 using the map η.

Restrict the action of $\operatorname{Aut}(P/N) \times \operatorname{Aut}(Q)$ on $\mathcal{E}(P/N, Q)$ to an action of $\operatorname{Aut}(P/N) \times \operatorname{Aut}_N Q$ on $\mathcal{E}(P/N, Q)$. Since the twisting homomorphism χ is trivial in the case $G = P \circ Q$, any element of $\operatorname{Aut}(P/N) \times \operatorname{Aut}_N Q$ acting on the extension $E : 1 \to Q \to G \to P/N \to 1$ will preserve the twisting of E. Let $A_{[E]}$ be the stabilizer subgroup of $[E]$, so

$$A_{[E]} = \{(\sigma, \tau) \in \operatorname{Aut}(P/N) \times \operatorname{Aut}_N Q \mid [\sigma E \tau^{-1}] = [E]\}.$$

Theorem 5.17 *There is an exact sequence*

$$1 \to \operatorname{Hom}(P, Z(Q)) \xrightarrow{\mu'} \operatorname{Aut}_N(G; Q) \xrightarrow{\eta} \operatorname{Aut}(P/N) \times \operatorname{Aut}_N Q \xrightarrow{\epsilon'} H^2(P/N, Q)$$

where ϵ' is only a set map, and $\operatorname{Im} \eta = A_{[E]}$.

Proof Since χ is trivial, $\mathrm{Der}(P/N, Z(Q)) = \mathrm{Hom}(P/N, Z(Q))$. One can see from the definition of μ in Section 2 that if $f \in \mathrm{Hom}(P/N, Z(Q))$, then $\mu(f)$ is the identity on Q. Thus $\mathrm{Im}\,\mu \leq \mathrm{Aut}_N(G; Q)$, and we set $\mu' = \mu$.

The following diagram is clearly commutative:

$$
\begin{array}{ccc}
\mathrm{Aut}(G; Q) & \overset{\rho}{\longrightarrow} & \mathrm{Aut}(P/N) \times \mathrm{Aut}(Q) \\
\uparrow i_1 & & \uparrow i_2 \\
\mathrm{Aut}_N(G; Q) & \overset{\eta}{\longrightarrow} & \mathrm{Aut}(P/N) \times \mathrm{Aut}_N Q
\end{array}
$$

where i_1 and i_2 are inclusion maps.

Now

$$\mathrm{Ker}\,\eta = \mathrm{Ker}\,\rho \cap \mathrm{Aut}_N(G; Q) = \mathrm{Im}\,\mu \cap \mathrm{Aut}_N(G; Q) = \mathrm{Im}\,\mu',$$

so the sequence is exact at $\mathrm{Aut}_N(G; Q)$.

The map ϵ' is just ϵ restricted to $\mathrm{Aut}(P/N) \times \mathrm{Aut}_N Q \leq C_\chi = \mathrm{Aut}(P/N) \times \mathrm{Aut}(Q)$. Now

$$
\begin{aligned}
\mathrm{Ker}\,\epsilon' &= \mathrm{Ker}\,\epsilon \cap (\mathrm{Aut}(P/N) \times \mathrm{Aut}_N Q) \\
&= \mathrm{Im}\,\rho \cap (\mathrm{Aut}(P/N) \times \mathrm{Aut}_N Q) \\
&\geq \mathrm{Im}(\rho \circ i_1) \cap (\mathrm{Aut}(P/N) \times \mathrm{Aut}_N Q) \\
&= \mathrm{Im}(i_2 \circ \eta) \cap (\mathrm{Aut}(P/N) \times \mathrm{Aut}_N Q) \\
&= \mathrm{Im}\,\eta.
\end{aligned}
$$

Finally, to show $\mathrm{Ker}\,\epsilon' \leq \mathrm{Im}\,\eta$, let $(\sigma, \tau) \in \mathrm{Aut}(P/N) \times \mathrm{Aut}_N Q$ such that $\epsilon'((\sigma, \tau)) = 0$. Then $\epsilon((\sigma, \tau)) = 0$ and we see that $(\sigma, \tau) \in (C_\chi)_{[E]}$. But $\tau \in \mathrm{Aut}_N Q$ implies $(\sigma, \tau) \in A_{[E]}$, and so $(\sigma, \tau) \in \mathrm{Im}\,\eta$.

$\square$

Acknowledgements

The bulk of this research was completed while on sabbatical at the University of Otago in Dunedin, New Zealand. I would like to thank everyone in the Department of Mathematics and Statistics there for their hospitality and collegiality. Thanks also to the referee for a careful reading of the paper, and helpful comments.

References

[1] J. Buckley, Automorphism groups of isoclinic p-groups, *J. London Math. Soc.* **12** (1975), 37–44.

[2] J. Dietz, Automorphisms of p-groups given as cyclic-by-elementary abelian central extensions, *J. Algebra* **242** (2001), 417–432.

[3] J. Douma, *Automorphisms of products of finite p-groups with applications to algebraic topology*, Ph.D. dissertation, Northwestern University, 1998.

[4] The GAP Group, *GAP – Groups, Algorithms, and Programming, Version 4.3*; 2002, (http://www.gap-system.org).

[5] J. Martino and S. Priddy, Group extensions and automorphism group rings, to appear.

[6] J. Martino, S. Priddy, and J. Douma, On stably decomposing products of classifying spaces, *Math. Zeit.* **235** (2000), 435–453.

[7] M. Schulte, Automorphisms of metacyclic p-groups with cyclic maximal subgroups, *Rose-Hulman Undergraduate Research Journal* **2** (2001), no. 2.

[8] A. D. Thomas and G. V. Wood, *Group Tables*, Shiva Math. Series **2**, Shiva Publishing, Orpington, Kent, 1980.

[9] C. Wells, Automorphisms of group extensions, *Trans. Amer. Math. Soc.* **155** (1971), 189–194.

[10] D. Winter, The automorphism group of an extraspecial p-group, *Rocky Mountain J. Math.* **2** (1972), 159–168.

LINEAR GROUPS WITH INFINITE CENTRAL DIMENSION

MARTYN R. DIXON* and LEONID A. KURDACHENKO[†]

*Department of Mathematics, University of Alabama Tuscaloosa, AL 35487-0350, U.S.A.
Email: Mdixon@gp.as.ua.edu

[†] Department of Algebra, University of Dnepropetrovsk, Vulycya Naukova 13,
Dnepropetrovsk 50, 49050, Ukraine

Abstract

The authors report recent work concerning infinite dimensional linear groups with various finiteness conditions on subgroups of infinite central dimension.

1 Introduction

In this paper we give a survey of recent work of the authors and others. We recall that a group G that is isomorphic to a group of automorphisms of a vector space A over a field F is called a linear group and we denote the group of all such automorphisms by $GL(F, A)$. If $\dim_F A$, the dimension of A over F, is finite, n say, then we say that G is a *finite dimensional* linear group and it is then well-known that $GL(F, A)$ can be identified with the group of $n \times n$ matrices with entries in F. From the outset, finite dimensional linear groups have played an important role in mathematics. This is partly due to the correspondence mentioned above, but also because of the rich interplay between geometrical and algebraic ideas associated with such groups.

The study of the subgroups of $GL(F, A)$ in the case when A is infinite dimensional over F has been much more limited and normally requires some additional restrictions. The circumstances here are similar to those present in the early development of Infinite Group Theory. One approach there consisted in the application of finiteness conditions to the study of infinite groups. One such restriction that has enjoyed considerable attention in linear groups is the notion of a finitary linear group, where a group G is called *finitary* if, for each element $g \in G$, the subspace $C_A(g)$ has finite codimension in A, and the reader is referred to the paper [10] and the survey [11] to see the type of results that have been obtained. This is a good example of the effectiveness of finiteness conditions in the study of infinite dimensional linear groups.

From our point of view a finitary linear group can be viewed as the linear analogue of an FC-group (that is, a group with finite conjugacy classes); this association suggests that it would be reasonable to start a systematic investigation of these "infinite dimensional linear groups" analogous to the fruitful study of finiteness conditions in infinite group theory. Historically, FC-groups were first studied by R. Baer [1]. But the first important problems in the theory of groups with finiteness conditions were concerned with the maximal and minimal conditions, which were first studied in ring and module theory. Soluble groups with the maximal condition were considered by K. A. Hirsch, and S. N. Černikov began the investigation of groups with the minimal condition. Connected with these problems was the celebrated problem of O. Yu. Schmidt concerning groups all of whose proper subgroups are finite. The investigations which resulted from these problems determined, in many respects, the further development of the theory of groups with finiteness conditions. (Incidentally, although Schmidt was German, he studied at Kiev and together with P. S. Aleksandrov helped create close ties between German and Soviet mathematicians for many years.)

In this survey we discuss recent research investigating similar problems for linear groups. If H is a subgroup of $GL(F, A)$ then H really acts on the quotient space $A/C_A(H)$ in a natural way. We denote $\dim_F(A/C_A(H))$ by centdim$_F$ H. If $\dim_F(A/C_A(H))$ is finite we say that H has *finite central dimension*; otherwise we shall say that H has *infinite central dimension*. A group G is therefore finitary linear precisely when each of its cyclic subgroups has finite central dimension. We remark that the central dimension of a subgroup depends on the particular vector space on which it is acting.

If $H \leq GL(F, A)$ has finite central dimension then it is easy to see that if $K \leq H$ then K also has finite central dimension. Let $G \leq GL(F, A)$ and let

$$\mathcal{L}_{id}(G) = \{H \leq G \mid H \text{ has infinite central dimension}\}.$$

As a first step we consider linear groups that are close to finite dimensional; that is, we consider linear groups G in which the set $\mathcal{L}_{id}(G)$ is "very small" in some sense. According to the above analogy with groups with finiteness conditions the following problems arise naturally:

- The study of linear groups in which every proper subgroup has finite central dimension (a linear analogue of Schmidt's problem).

- The study of linear groups in which the set of all subgroups having infinite central dimension satisfies the minimal condition (a linear analogue of Černikov's problem).

- The study of linear groups, in which the set of all subgroups having infinite central dimension satisfies the maximal condition (a linear analogue of Baer's problem).

If $\mathcal{L}_{id}(G)$ (partially ordered by set inclusion) satisfies the minimal condition (respectively, maximal condition) we say that G satisfies the minimal condition (respectively, maximal condition) on subgroups with infinite central dimension or, more briefly, that G satisfies min-id (respectively, max-id). We here outline recent work on these finiteness conditions and others.

We remark also that if $G \leq GL(F, A)$ then G also acts naturally on the space $A/A(\omega FG)$, where here ωFG is the augmentation ideal of the group ring FG. The reader is referred to the recent paper [6] where groups with the minimal condition on subgroups of infinite augmentation dimension are considered.

2   Linear groups satisfying min-id

In this section we consider linear groups satisfying min-id. Naturally if $G \leq GL(F, A)$ is a Černikov group then G satisfies min-id. We are concerned with almost locally soluble groups and also locally finite groups. These groups have been considered in the paper [3] and here we exhibit the main results of that paper.

Theorem 2.1 *Let G be a subgroup of $GL(F, A)$ and suppose that G satisfies min-id. Then either G satisfies the minimal condition or G is a finitary linear group.*

We note that locally soluble groups with the minimal condition on subgroups are soluble Černikov groups. The structure of soluble groups with min-id follows a general pattern that is similar to the structure of finite dimensional soluble linear groups. The results fall into two cases depending on the characteristic of the field F. In the case when F has positive characteristic the groups are close to being Černikov, in some sense. We recall that a group G is almost locally soluble if G has a normal locally soluble subgroup of finite index.

Theorem 2.2 *Let F be a field of characteristic $p > 0$ and let $G \leq GL(F, A)$ be an almost locally soluble group of infinite central dimension. Suppose also that G satisfies min-id and is not Černikov. Then G has a series of normal subgroups $P \leq D \leq G$ satisfying the following conditions:*

(i) *P is a nilpotent bounded p-group,*

(ii) *$D = P \rtimes Q$ for some non-trivial divisible Černikov p'-subgroup Q, and G/D is finite,*

(iii) *P has finite central dimension and satisfies min-Q (the minimal condition on Q-invariant subgroups), and Q has infinite central dimension.*

In particular G is nilpotent-by-abelian-by-finite and satisfies the minimal condition on normal subgroups.

An example is given in [3] illustrating this result. When the field F has characteristic 0 the situation is simpler.

Theorem 2.3 *Let G be an almost locally soluble subgroup of $GL(F, A)$ where F has characteristic 0. Suppose that G has infinite central dimension and satisfies min-id. Then G is a Černikov group.*

We deduce the following corollary concerning groups with infinite central dimension in which all proper subgroups have finite central dimension; thus we obtain an exact analogue of Schmidt's results.

Corollary 2.4 *Let F be a field of characteristic $p \geq 0$ and let $G \leq GL(F, A)$ be an almost locally soluble group with infinite central dimension such that every proper subgroup of G has finite central dimension. Then $G \cong C_{q^\infty}$ for some prime $q \neq p$.*

And finally we consider locally finite groups satisfying min-id.

Theorem 2.5 *Let G be a locally finite subgroup of $GL(F, A)$. Suppose that G has infinite central dimension and satisfies min-id. Then G is almost soluble.*

Thus if F has characteristic 0 a locally finite subgroup of $GL(F, A)$ satisfying min-id is Černikov. We remark that a locally finite group with the minimal condition is Černikov (see [5, 5.8 Theorem]).

3 Linear groups satisfying max-id

In this section we consider linear groups satisfying max-id. Such groups have been considered in the paper [7]. The main results of that paper show that the situation with max-id is somewhat more complicated. As in Section 2, if $G \leq GL(F, A)$, an important role is played by the normal subgroup

$$FD(G) = \{x \in G \mid \langle x \rangle \text{ has finite central dimension}\}.$$

We call $FD(G)$ the *finitary radical* of the linear group G. First we note that the relationship between finitary linear groups and linear groups satisfying max-id is analogous to that given in Theorem 2.1.

Theorem 3.1 *Let G be a subgroup of $GL(F, A)$ and suppose that G satisfies max-id. Then either G is finitely generated or G is a finitary linear group.*

In order to determine the structure of soluble linear groups satisfying max-id we require a number of cases, each of which is quite complicated. We first consider the case when G/G' is not finitely generated.

Theorem 3.2 *Let $G \leq GL(F, A)$ and suppose that G is a soluble group satisfying max-id. If G has infinite central dimensional and G/G' is not finitely generated, then G contains normal subgroups $H \leq R \leq V$ such that G/V is finite and $V/R \cong C_{q^\infty}$, where $q \neq \operatorname{char} F$. Also $\operatorname{centdim}_F R$ is finite, R/H is finitely generated and V/H is abelian. Moreover if $\operatorname{char} F = 0$ then H is torsion-free; if $\operatorname{char} F = p > 0$ then H is a bounded p-group. Also A has a series $0 \leq B \leq C$ of FG-submodules such that $B = \bigoplus_{n \in \mathbb{N}} B_n$ where B_n is a simple FG-submodule having finite F-dimension, for $n \in \mathbb{N}$, $C(\omega FG) \leq B$ and C has finite codimension.*

Next we consider the case when G is not finitely generated.

Theorem 3.3 *Let $G \leq GL(F, A)$ and suppose that G is a soluble group satisfying max-id. If G has infinite central dimension and G is not finitely generated, then there is a normal subgroup S such that G/S is a finitely generated, abelian-by-finite group and S/S' is not finitely generated.*

Of course by Theorem 3.1 G is finitary, whence so is S and moreover S has infinite central dimension also. Thus S satisfies the conditions of Theorem 3.2. Next we consider the case when G is finitely generated. Here we also have two cases.

Theorem 3.4 *Let $G \leq GL(F, A)$, and suppose that G is a finitely generated soluble group satisfying max-id. If G has infinite central dimension but $\mathrm{centdim}_F(FD(G))$ is finite, then the following conditions hold:*

(i) *G contains a normal subgroup U such that G/U is polycyclic. If char $F = 0$, then U is torsion-free and if char $F = p > 0$, then U is a bounded p-subgroup;*

(ii) *there is an integer $m \in \mathbb{N}$ such that $A(x - 1)^m = 0$ for each $x \in U$. Hence U is nilpotent;*

(iii) *if $1 = Z_0 \leq Z_1 \leq \cdots \leq Z_m = U$ is the upper central series of U, then Z_{j+1}/Z_j is a Noetherian $\mathbb{Z}\langle g \rangle$-module for each element $g \in G \setminus FD(G)$, for $0 \leq j \leq m - 1$.*

Finally in this section we have

Theorem 3.5 *Let $G \leq GL(F, A)$, and suppose that G is a finitely generated soluble group satisfying max-id. If $\mathrm{centdim}_F(G)$ and $\mathrm{centdim}_F(FD(G))$ are infinite, then G contains a normal subgroup L satisfying the following conditions:*

(i) *G/L is abelian-by-finite;*

(ii) *$L \leq FD(G)$ and L has infinite central dimension;*

(iii) *L/L' is not finitely generated;*

(iv) *L satisfies max-$\langle g \rangle$ for each element $g \in G \setminus FD(G)$.*

4   Linear groups satisfying certain rank restrictions

Finally we consider linear groups satisfying certain restrictions on the ranks of their subgroups.

A group G is said to have *finite 0-rank*, $r_0(G) = r$, if G has a finite subnormal series with exactly r infinite cyclic factors, all other factors being periodic. We note that every refinement of one of these series has only r factors that are infinite cyclic. By the Schreier Refinement Theorem the 0-rank is independent of the chosen series. This numerical invariant is also known as the torsion-free rank of G. In the case of polycyclic-by-finite groups the 0-rank is simply the Hirsch number. It is well-known that if G is a group of finite 0-rank, $H \leq G$ and $L \lhd G$ then H and G/L also have finite 0-rank. Furthermore $r_0(H) \leq r_0(G)$ and $r_0(G) = r_0(L) + r_0(G/L)$.

Let p be a prime. A group G has finite p-rank, $r_p(G) = r$, if every elementary abelian p-section of G is finite of order at most p^r and there is an elementary abelian p-section U/V such that $|U/V| = p^r$. It is well-known that if G is a group of finite p-rank, $H \leq G$ and $L \lhd G$ then H and G/L also have finite p-rank. Furthermore $r_p(H) \leq r_p(G)$ and $r_p(G) \leq r_p(L) + r_p(G/L)$.

In this section when speaking of the p-rank we shall assume that $p = 0$ or that p is a prime, being more specific as needed. Rather a lot is known concerning soluble groups of finite torsion-free rank (see [8, Section 5.2] for example). This theory has been taken somewhat further in the paper [4] where even locally (soluble-by-finite) groups of finite torsion-free rank are considered. We are here interested in those groups G that are of infinite central dimension and of infinite p-rank in which the proper subgroups of infinite p-rank themselves have finite central dimension.

Thus the proper subgroups of infinite p-rank, although large, are not too large. Naturally our results depend upon p. The proofs of the results mentioned here are given in [2].

Theorem 4.1 *Let $G \leq GL(F, A)$ be soluble. Suppose that every proper subgroup of infinite 0-rank has finite central dimension. Then either G has finite central dimension or $r_0(G)$ is finite.*

Using [4, Lemma 2.12] this result extends to locally soluble groups as follows.

Theorem 4.2 *Let $G \leq GL(F, A)$ be of infinite central dimension and suppose that $r_0(G)$ is infinite. Suppose that the maximal normal torsion subgroup of G is trivial and that if H is a proper subgroup of G such that $r_0(H)$ is infinite then H has finite central dimension. If G is locally soluble then G is soluble.*

Now let $p > 0$. In [2] we give an example to show that in this case there are groups G of infinite central dimension such that $r_p(G)$ is infinite having the property that every proper subgroup of infinite p-rank has finite central dimension. However the structure of such groups is somewhat restricted as the following result shows.

Theorem 4.3 *Let $G \leq GL(F, A)$ be soluble and let $p > 0$. Suppose that $r_p(G)$ and $\mathrm{centdim}_F(G)$ are infinite. If every proper subgroup of infinite p-rank has finite central dimension then G satisfies the following conditions:*

(i) $G = H \rtimes Q$ *where $Q \cong C_{q^\infty}$ for some prime q;*

(ii) *H is a p-group of finite central dimension where $p = \mathrm{char}\, F \neq q$;*

(iii) *$K = H \cap Z(G)$ is finite;*

(iv) *H/K is an infinite elementary abelian p-group;*

(v) *H/K is a minimal normal subgroup of G/K.*

There is an appropriate version of this result for locally soluble groups, but we shall omit this here.

A group G is said to have *finite section rank* if $r_0(G)$ is finite and $r_p(G)$ is finite for each prime p. Thus if G is soluble and has infinite section rank then there is a prime p such that $r_p(G)$ is infinite. If H is a subgroup of infinite p-rank then H has infinite section rank. A group G is said to have *finite special rank*, $r(G) = r$, if every finitely generated subgroup of G can be generated by r elements and r is the least positive integer with this property. This notion is due to A.I. Mal'cev [9]. The special rank of a group is called also the Prüfer–Mal'cev rank (or just the rank when the context is clear). By applying Theorem 4.3 we may deduce the following result.

Theorem 4.4 *Let $G \leq GL(F, A)$ be soluble and suppose that G has infinite section (respectively special) rank and $\mathrm{centdim}_F(G)$ is infinite. If every proper subgroup of infinite section (respectively special) rank has finite central dimension then G satisfies the following conditions:*

(i) $G = H \rtimes Q$ *where $Q \cong C_{q^\infty}$ for some prime q;*

    (ii) H *is a p-group of finite central dimension where* $p = \operatorname{char} F \neq q;$

  (iii) $K = H \cap Z(G)$ *is finite;*

  (iv) H/K *is an infinite elementary abelian p-group;*

   (v) H/K *is a minimal normal subgroup of* G/K.

    Theorem 4.4 has an appropriate version for locally soluble groups. Note that linear groups of finite central dimension having finite rank are soluble-by-finite by a result of V. Platonov (see [13, Theorem 10.9]). This latter result has been generalized to finitary linear groups in [12].

References

[1] R. Baer, Finiteness properties of groups, *Duke Math. J.* **15** (1948), 1021–1032.

[2] O. Yu. Dashkova, M. R. Dixon, and L. A. Kurdachenko, Infinite dimensional linear groups satisfying certain rank restrictions, preprint.

[3] M. R. Dixon, M. J. Evans, and L. A. Kurdachenko, Linear groups with the minimal condition on subgroups of infinite central dimension, *J. Algebra* **277** (2004), 172–186.

[4] S. Franciosi, F. De Giovanni, and L. A. Kurdachenko, The Schur property and groups with uniform conjugacy classes, *J. Algebra* **174** (1995), 823–847.

[5] O. H. Kegel and B. A. F. Wehrfritz, *Locally Finite Groups*, North-Holland Mathematical Library, North-Holland, Amsterdam, London, 1973, Volume 3.

[6] L. A. Kurdachenko, Linear groups with the minimal condition on subgroups of infinite augmentation dimension, *Ukrainian Math. J.*, to appear.

[7] L. A. Kurdachenko and I. Ya. Subbotin, Linear groups with the maximal condition on subgroups of infinite central dimension, preprint.

[8] J. C. Lennox and D. J. S. Robinson, *The Theory of Infinite Soluble Groups*, Oxford Mathematical Monographs, Oxford University Press, Oxford, 2004.

[9] A. I. Mal'cev, On groups of finite rank, *Mat. Sbornik* **22** (1948), 351–352.

[10] R. E. Phillips, The structure of groups of finitary transformations, *J. Algebra* **119** (1988), 400–448.

[11] R. E. Phillips, Finitary linear groups: a survey, in *Finite and Locally Finite Groups (Istanbul 1994)* (B. Hartley, G. M. Seitz, A. V. Borovik, and R. M. Bryant, eds.), NATO Adv. Sci. Inst. Ser. C Math. Phys. Sci., vol. 471, 111–146, Kluwer Acad. Publ., Dordrecht, 1995.

[12] C. J. E. Pinnock, Lawlessness and rank restrictions in certain finitary groups, *Proc. Amer. Math. Soc.* **130** (2002), 2815–2819.

[13] B. A. F. Wehrfritz, *Infinite Linear Groups*, Ergebnisse der Mathematik und ihrer Grenzgebiete, Springer-Verlag, New York, Heidelberg, Berlin, 1973, Band 76.

G-AUTOMATA, COUNTER LANGUAGES AND THE CHOMSKY HIERARCHY

MURRAY ELDER[1]

School of Mathematics and Statistics, University of St Andrews, Scotland
Email: murrayelder@gmail.com

Abstract

We consider how the languages of G-automata compare with other formal language classes. We prove that if the word problem of G is accepted by a machine in the class $\mathcal{M}$ then the language of any G-automaton is in the class $\mathcal{M}$. It follows that the so called *counter languages* (languages of $\mathbb{Z}^n$-automata) are context-sensitive, and further that counter languages are indexed if and only if the word problem for $\mathbb{Z}^n$ is indexed.

AMS Classification: 20F65, 20F10, 68Q45
Keywords: G-automaton; counter language; word problem for groups; Chomsky hierarchy

1 Introduction

In this article we compare the languages of G-automata, which include the set of *counter languages*, with the formal language classes of context-sensitive, indexed, context-free and regular. We prove in Theorem 6 that if the word problem of G is accepted by a machine in the class $\mathcal{M}$ then the language of any G-automaton is in the class $\mathcal{M}$. It follows that the counter languages are context-sensitive. Moreover it follows that counter languages are indexed if and only if the word problem for $\mathbb{Z}^n$ is indexed.

The article is organized as follows. In Section 2 we define G-automata, linearly bounded automata, nested stack, stack, and pushdown automata, and the word problem for a finitely generated group. In Section 3 we prove the main theorem, and give the corollary that counter languages are indexed if and only if the word problem for $\mathbb{Z}^n$ is indexed for all n.

2 Definitions

If Σ is a set define the set Σ^* to be the set of all words consisting of letters in Σ, including the *empty word* ϵ which contains no letters. If G is a group with generating set $\mathcal{G}$, we say two words u, v are equal in the group, or $u =_G v$, if they represent the same group element. We say u and v are identical if they are equal in the free monoid, that is, they are equal in $\mathcal{G}^*$.

---

[1]Supported by EPSRC grant GR/S53503/01

Definition 1 (Word problem) The *word problem* of a group G with respect to a finite generating set $\mathcal{G}$ is the set of all elements w of $\mathcal{G}^*$ which are equal to the identity in the group. Note that the word problem for a finitely generated group is a language over a finite alphabet.

Definition 2 (G-automaton) Let G be a group and Σ a finite set. A (non-deterministic) G-*automaton* A_G over Σ is a finite directed graph with a distinguished *start vertex* q_0, some distinguished *accept vertices*, and with edges labeled by elements of $(\Sigma \cup \{\epsilon\}) \times G$. If p is a path in A_G, the element of $(\Sigma^{\pm 1})$ which is the first component of the label of p is denoted by $w(p)$, and the element of G which is the second component of the label of p is denoted $g(p)$. If p is the empty path, then $g(p)$ is the identity element of G and $w(p)$ is the empty word. The G-automaton A_G is said to *accept* a word $w \in \Sigma^*$ if there is a path p from the start vertex to some accept vertex such that $w(p) = w$ and $g(p) =_G 1$.

Definition 3 (Counter language) A language is k-*counter* if it is accepted by some $\mathbb{Z}^k$-automaton. We call the (standard) generators of $\mathbb{Z}^k$ *counters*. A language is *counter* if it is k-counter for some $k \geq 1$.

These definitions are due to Mitrana and Striebe [9]. Note that in these counter automata, the values of the counters are not accessible until the final accept/fail state. For this reason they are sometimes called *blind*. Elston and Ostheimer [2] proved that the word problem of G is deterministic counter with an extra "inverse" property if and only if G is virtually abelian. Recently Kambites [8] has shown that the inverse property restriction can be removed from this theorem.

It is easy to see that the word problem for G is accepted by a G-automaton.

Lemma 4 *The word problem for a finitely generated group G is accepted by a deterministic G-automaton.*

Proof Construct a G-automaton with one state and a directed loop labeled by (g, g) for each generator g. The state is both start and accept. A word in the generators is accepted by this automaton if and only if it represents the identity, by definition. $\square$

Recall the definitions of the formal language classes of recursively enumerable, decidable, context-sensitive, indexed, stack, context-free, and regular. Each of these can be defined as the languages of some type of restricted Turing machine as follows.

Consider a machine consisting of finite *alphabet* Σ, a finite *tape alphabet* Γ, a *finite state control* and an infinite *work tape*, which operates as follows. The finite state control is a finite graph with a specified *start node*, some specified *accept nodes*, and edges labeled by an alphabet letter and one or more instructions for the work tape. The instructions in general are of the form **read**, **write**, **move left**, **move right** and one reads/writes letters from Γ on the tape. The tape starts out blank.

One inputs a finite string in Σ^* one letter at a time, read from left to right. For each letter $x \in \Sigma$, the finite state control performs some instructions on the work tape corresponding to an edge whose label is $(x, \texttt{instructions})$, and moves to the target node of the edge. One starts at the start node, and *accepts* the string if there is some path from the start node to an accept node labeled by the letters of the string. The *language* of a machine is the set of strings in Σ^* which the machine accepts.

If the finite state control is such that more than one path is possible on a given input string, then such machines (and their languages) are called *non-deterministic*. A string is accepted if it labels *some* path from the start state to an accept state. For example, we may have two edges with the same letter in the first coordinate of the edge label emanating from the same state, or edges with ϵ in the first coordinate of the edge label leading to a choice of states.

If we allow no further restrictions on how this machine operates, then we have a (deterministic/non-deterministic) *Turing machine*. By placing increasingly strict restrictions on the machine, we obtain a hierarchy of languages corresponding to the machines, sometimes called the *Chomsky hierarchy*. If the Turing machine always reaches an accept state in a finite number of steps, or *halts*, on strings that are to be accepted, but may or may not run indefinitely otherwise, then the language is *recursively enumerable* or *r.e.*, and if it halts on both accepted and rejected strings the language is *decidable* or *recursive*. If we restrict the number of squares of tape that can be used to be a constant multiple of the length of the input string, then we obtain a *linearly bounded automaton*, and the languages of these are called *context-sensitive*. If we make the tape act as a *nested stack* (see [5]) then the language of such a machine is called *indexed*.

If the tape is a stack (first in last out) where the pointer may read but not write on any square, then the machine and its languages are called *stack*. If the tape is a stack such that the pointer can only read the top square, we have a *pushdown automaton*, the languages of which are *context-free*. Finally, if we remove the tape altogether we are left with just the finite state control, which we call a *finite state automaton*, languages of which are *regular*.

For more precise definitions see [5] for nested stack automata, [6] for stack, [12] for regular and context-free and [7] for these plus linearly bounded automata. Two good survey articles are [11] and [4].

We end this section with the following fact.

Lemma 5 *If the word problem for a group G with respect to a finite generating set $\mathcal{G}$ is in the language class $\mathcal{M}$, then so is the word problem for G for any other finite generating set.*

Proof Assume the word problem for G with respect to $\mathcal{G}$ is accepted by a machine M, and let $\mathcal{Y}$ be another finite inverse-closed generating set. Modify M as follows. Choose a representative in $\mathcal{G}^*$ for each letter of $\mathcal{Y}$. Then for each input letter, run the machine M on the input for the representative word. Accept when M accepts. $\qquad\square$

3   Main theorem

Theorem 6 *Let $\mathcal{M}$ be a formal language class: (regular, context-free, stack, indexed, context-sensitive, decidable, r.e.) and let G be a finitely generated group. The word problem for G is in $\mathcal{M}$ if and only if the language of every G-automaton is in $\mathcal{M}$.*

Proof By Lemma 4 the word problem of G is accepted by a G-automaton, so one direction is done.

Let L be a language over an alphabet Σ accepted by a G-automaton P. Fix a finite generating set $\mathcal{G}$ for G which includes all elements of G that are the second coordinate of an edge label in P. Let N be an $\mathcal{M}$-automaton which accepts the word problem for G with respect to this generating set.

Construct a machine M from the class $\mathcal{M}$ to accept L as follows. The states of M are of the form (p_i, q_j) where p_i is a state of P and q_i is a state of N. The start state is (p_S, q_S), and accept states are (p_A, q_A) where p_S, p_A, q_S, q_A are start and accept states in P and N.

The transitions are defined as follows. For each edge (x, g) in P from p to p', and for each edge $(g, \texttt{instruction})$ in N from q to q', add an edge from (p, q) to (p', q') in M labeled $(x, \texttt{instruction})$.

Then M accepts a string in Σ^* if there is a path in M corresponding to paths in P from the start state to an accept state such that the labels of the second coordinate of the edges give a word w in $\mathcal{G}^*$, and a path in N from a start state to an accept state labeled by w which evaluates to the identity of G and respects the tape instructions. Then by Lemma 5 we can modify M to obtain a G-automaton to accept the word problem of G with respect to any finite generating set.

Note that M is deterministic if both N and P are deterministic and no state in P has two outgoing edges with the same group element in the second coordinate of the edge labels.   □

It is easy to see that if G is a finite group, then the construction describes a finite state automaton (we don't need any tape). It is known that the word problem for a group is regular if and only if the group is finite [1], and that G is virtually free if and only if G has a context-free word problem [10], so the language of every G-automaton for G virtually free is context-free.

Since the word problem for G virtually abelian is context-sensitive, we get the following corollary.

Corollary 7 *Counter languages are context-sensitive.*

More generally, Gersten, Holt and Riley [3, Corollary B.2] have shown that finitely generated nilpotent groups of class c have context-sensitive word problems, so the language of every G-automaton for G finitely generated nilpotent is context-sensitive.

Figure 1 shows how counter languages fit into the hierarchy.

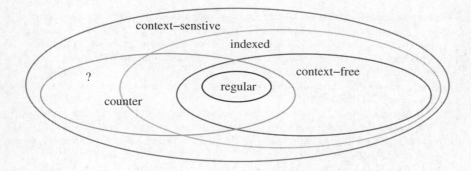

Figure 1. How counter languages fit into the hierarchy

Corollary 8 *The word problem for $\mathbb{Z}^n$ is indexed for all $n \geq 1$ if and only if every counter language is indexed.*

Proof By Theorem 6 if the word problem for $\mathbb{Z}^n$ is indexed then the language of every $\mathbb{Z}^n$-automaton is indexed, proving one direction.

Since the word problem of $\mathbb{Z}^n$ is an n-counter language then if it is not indexed then not every counter language is indexed.   □

By Gilman and Shapiro, if the word problem of G is accepted by a nested stack automaton with some extra restrictions, then it is virtually free. We still do not know whether word problem of $\mathbb{Z}^n$ is accepted by a nested stack automaton without extra restrictions. The author guesses that this is not the case.

Conjecture 9 The word problem for $\mathbb{Z}^2$ is not indexed.

4   Acknowledgments

Thanks to Ray Cho, Andrew Fish, Bob Gilman, Claas Röver, Sarah Rees and Gretchen Ostheimer for many fruitful discussions about counter, context-sensitive and indexed languages and word problems. Special thanks to the reviewer for helpful corrections and suggestions.

References

[1] V. A. Anisimov, Group Languages, *Cybernetics* **7** (1971), 594–601 (translation to *Kybernetica* **4** (1971) 18–24).

[2] Gillian Elston and Gretchen Ostheimer, On groups whose word problem is solved by a counter automaton, *Theoret. Comput. Sci.* **320** (2004) 175–185.

[3] Steve Gersten, Derek Holt and Tim Riley, Isoperimetric inequalities for nilpotent groups, *Geom. Funct. Anal.* **13** (2003), 795–814.

[4] Robert Gilman, Formal languages and infinite groups, in *Geometric and Computational Perspectives on Infinite Groups*, DIMACS Ser. Discrete Math. Theoret. Comput. Sci. **25** (1996), 27–51.

[5] Robert Gilman and Michael Shapiro, On groups whose word problem is solved by a nested stack automaton, arXiv: math.GR/9812028.

[6] Derek Holt and Claas Röver, Groups with indexed co-word problem, submitted.

[7] John Hopcroft and Jeffery Ullman, *Introduction to Automata Theory, Languages and Computation*, Addison-Wesley, 1979.

[8] Mark Kambites, Word problems recognizable by deterministic blind monoid automata, arXiv: math.GR/0506137.

[9] Victor Mitrana and Ralf Stiebe, The accepting power of finite automata over groups, in *New trends in formal languages*, Lecture Notes in Comput. Sci. **1218**, 39–48, 1997.

[10] David Muller and Paul Schupp, Groups, the theory of ends and context-free languages, *J. Comput. System Sci.* **26** (1983), 295–310.

[11] Sarah Rees, Hairdressing in groups: a survey of combings and formal languages, *Geom. Topol. Monogr.* **1** (1998), 493–509.

[12] Michael Sipser, *Introduction to the Theory of Computation*, PWS Publishing Co., 1997.

AN EMBEDDING THEOREM FOR GROUPS UNIVERSALLY EQUIVALENT TO FREE NILPOTENT GROUPS

BENJAMIN FINE*, ANTHONY M. GAGLIONE† and DENNIS SPELLMAN*

*Department of Mathematics, Fairfield University, Fairfield, CT 06430, U.S.A.

†Department of Mathematics, U.S. Naval Academy, Annapolis, MD 21402, U.S.A.
Email: amg@usna.edu

Abstract

Let F be a finitely generated nonabelian free group. Kharlampovich and Myasnikov proved that any finitely generated group H containing a distinguished copy of F is universally equivalent to F in the language of F if and only if there is an embedding of H into Lyndon's free exponential group $F^{\mathbf{Z}[t]}$ which is the identity on F. Myasnikov posed the question to the third author as to whether or not a similar result holds for finitely generated free nilpotent groups with Lyndon's group replaced by Philip Hall's completion with respect to a suitable binomial ring. We answer his question in the affirmative.

0 Introduction

Let G be a group. A *G-group* H shall be a pair (H, φ) where H is a group and $\varphi : G \to H$ is a monomorphism. *G-subgroup, G-homomorphism, G-normal subgroup* and *G-quotient group* are defined in the obvious way. Henceforth we shall suppress the monomorphism φ and consider the distinguished copy $\varphi(G)$ of G in H understood and identify it with G. The *language of G*, $L[G]$, shall be the first order language with equality containing a binary operation symbol $\cdot$, a unary operation symbol $^{-1}$, and a set of constant symbols

$$C_G = \{c_g : g \in G\}$$

in bijective correspondence with G, $g \to c_g$. Assume a fixed bijection. Following standard terminology in model theory (see, e.g., [2]) the *diagram* of G (with respect to $L[G]$) shall be the set of all atomic sentences and negated atomic sentences of $L[G]$ true in G whenever each c_g is interpreted in G as the corresponding element g. A model H of the group theory axioms

$$\forall x, y, z((x \cdot y) \cdot z = x \cdot (y \cdot z))$$
$$\forall x(x \cdot c_1 = x)$$
$$\forall x(x \cdot x^{-1} = c_1)$$

together with the diagram of G is a G-group and conversely. The distinguished copy of G is precisely the set of values in H of the c_g as g varies over G. Henceforth we shall commit the abuse of identifying c_g with g. By a *ring* we shall always mean an associative ring with 1. Subrings and homomorphisms are required to respect the multiplicative identity 1. Clearly there is an appropriate first order language with equality for rings, which we choose not to make explicit here. A commutative integral domain B of characteristic 0 is a *binomial ring* provided, for each element $b \in B$ and each positive integer m, B contains the "binomial coefficient" $_bC_m = b(b-1)\ldots(b-m+1)/m!$. If A is a commutative integral domain of characteristic 0, then the *binomial closure* of A shall be the least binomial subring B of the field of fractions of A such that B contains A as a subdomain. Clearly binomial rings are generalizations of the ring $\mathbb{Z}$ of integers.

Suppose G is a finitely generated free nilpotent group. Suppose $a_1,\ldots,a_r$ is a fixed (ordered) set of free generators for G with respect to the variety N_c. Let $a_1,\ldots,a_r,c_{r+1},\ldots,c_N$ be a fixed set of Hall basic commutators for G beginning with the ordered basis $a_1 < \cdots < a_r$. Then every element of G is uniquely expressible in the form $a_1^{e_1}\ldots a_r^{e_r}c_{r+1}^{e_{r+1}}\ldots c_N^{e_N}$ where $(e_1,\ldots,e_N) \in \mathbb{Z}^N$. Multiplying $a_1^{e_1}\ldots a_r^{e_r}c_{r+1}^{e_{r+1}}\ldots c_N^{e_N}$ by $a_1^{f_1}\ldots a_r^{f_r}c_{r+1}^{f_{r+1}}\ldots c_N^{f_N}$ yields $a_1^{g_1}\ldots a_r^{g_r}c_{r+1}^{g_{r+1}}\ldots c_N^{g_N}$ where the g's are fixed integral polynomials in the e's and f's; moreover, if n is an integer, then $(a_1^{e_1}\ldots a_r^{e_r}c_{r+1}^{e_{r+1}}\ldots c_N^{e_N})^n = a_1^{h_1}\ldots a_r^{h_r}c_{r+1}^{h_{r+1}}\ldots c_N^{h_N}$ where the h's are fixed integral polynomials in the e's and f's and binomial coefficients $_nC_m$. (Here $_nC_m$ is *defined* as $n(n-1)\ldots(n-m+1)/m!$ for all integers n and all positive integers m.) Clearly these operations can be carried out with $\mathbb{Z}$ replaced by any binomial ring B. When this is done we get a group G^B containing G and free in the variety N_c^B of nil-c groups admitting exponents from B. This is due to Philip Hall [4] and G^B is the *Hall completion* of G with respect to the binomial ring B. Making the notion of *exponential group* more precise, we give Lyndon's definition. Let R be a commutative ring with $1 \neq 0$. An *R-group* shall be a group H admitting a mapping $R \times H \to H$, $(r,h) \to h^r$ satisfying the following conditions:

$$h^r h^s = h^{r+s} \quad \forall h \in H; (r,s) \in R^2$$
$$(h^r)^s = h^{rs} \quad \forall h \in H; (r,s) \in R^2$$
$$h^1 = h \quad \forall h \in H$$
$$(k^{-1}hk)^r = k^{-1}h^r k \quad \forall (h,k) \in H^2 \ ; \ r \in R.$$

Thus, G^B is a B-group and N_c^B is the variety of all nil-c B-groups.

A ring R of characteristic 0 is *ω-residually* $\mathbb{Z}$ provided to any finite nonempty set $S \subseteq R\backslash\{0\}$ of nonzero elements there is a retraction $\rho : R \to \mathbb{Z}$ such that $\rho(s) \neq 0$ for all $s \in S$. It is not difficult to prove that R is ω-residually $\mathbb{Z}$ if and only if R is simultaneously residually $\mathbb{Z}$ and a commutative integral domain. Of course, $\mathbb{Z}$ itself is ω-residually $\mathbb{Z}$. Less trivial examples are the integral polynomial rings $\mathbb{Z}[t_1,\ldots,t_k]$ where $t_1,\ldots,t_k$ are algebraically independent commuting indeterminates. Moreover, it is not difficult to show that the binomial closure of an ω-residually $\mathbb{Z}$ ring is also ω-residually $\mathbb{Z}$. A *Peano ring* shall be a ring $^*\mathbb{Z}$ elementarily equivalent to the ring $\mathbb{Z}$ of integers. In symbols: $^*\mathbb{Z} \equiv \mathbb{Z}$. Of course, $\mathbb{Z}$ itself is a Peano ring. A nontrivial example may be found by taking an ultrapower $^*\mathbb{Z}$ of $\mathbb{Z}$ with respect to a nonprincipal ultrafilter. (The reader is refered to Bell and

Slomson [1] for all facts we shall need about ultrapowers and ultralimits.) Since being a commutative integral domain of characteristic 0 is first order expressible every Peano ring is such. It follows that every subring of a Peano ring is a commutative integral domain of characteristic 0. More, of course, is true. From the fact that, for each positive integer m, $\forall x \exists y (m!y = x(x-1)\ldots(x-m+1))$ is true in $\mathbb{Z}$ the above sentence must hold in every Peano ring. That is, every Peano ring is a binomial ring.

In order to have a context general enough to simultaneously talk about G-groups and rings we now introduce some standard notions of universal algebra. An *operator domain* is a triple $\Omega = (F, C, d)$ where F and C are sets and d is a function from F into the positive integers. An Ω-*algebra* is a triple $\mathbf{A} = (A, (f_{\mathbf{A}})_{f \in F}, (c_{\mathbf{A}})_{c \in C})$ where A is a nonempty set, $c_{\mathbf{A}} \in A$ for all $c \in C$ and for all $f \in F$, if $d(f) = n$, then $f_{\mathbf{A}} : A^n \to A$ is an n-ary operation on A. Subalgebra, direct product and homomorphism are defined in the obvious ways. Generally, the operations and distinguished elements are understood from the context and we henceforth write A to signify the algebra $\mathbf{A}$. (We have already tacitly done that in the cases of groups, rings and G-groups.)

Remark 0.1 If R is a nonzero commutative ring, then R-groups may be put into this context by viewing each $r \in R$ as determining a unary operation

$$x \to x^r.$$

Clearly, for each operator domain Ω, there is a first order language with equality L_Ω appropriate for Ω-algebras.

Definition 0.2 ([1]) Let A and B be Ω-algebras. *A universally covers B* provided every universal sentence of L_Ω true in A is also true in B. We write $A \forall B$ in this case.

Definition 0.3 Let A and B be Ω-algebras. *A and B are universally equivalent* or *have the same universal theory* provided they satisfy the same universal sentences of L_Ω. We write $A \equiv_\forall B$ in this event.

Remark 0.4 Observe that $A \equiv_\forall B$ if and only if $A \forall B$ and $B \forall A$ hold simultaneously.

Remark 0.5 Since the negation of a universal sentence is logically equivalent to an existential sentence and vice-versa, $A \equiv_\forall B$ if and only if A and B satisfy the same existential sentences of L_Ω.

Lemma 0.6 ([1]) *Let A and B be Ω-algebras. $A \forall B$ if and only if there is an ultrapower $^*A$ of A admitting an Ω-monomorphism $B \to {}^*A$.*

Lemma 0.7 *$A \equiv_\forall B$ if and only if, for every finite system Σ of equations and inequations, it is the case that Σ has a solution in A if and only if Σ has a solution in B.*

Proof $A \equiv_\forall B$ if and only if A and B satisfy the same existential sentences of L_Ω. Every existential sentence ψ of L_Ω is equivalent to one whose *matrix* is in *disjunctive normal form*. That is, ψ may be viewed as $\exists \bar{x}(\bigvee_k \psi_k)$ where each ψ_k is a conjunction of atomic formulas and negated atomic formulas. That is, ψ_k is $\bigwedge_i (P_{k,i}(\bar{x}) = p_{k,i}(\bar{x})) \wedge \bigwedge_j (Q_{k,j}(\bar{x}) \neq q_{k,j}(\bar{x}))$ where the $P_{k,i}$, $p_{k,i}$, $Q_{k,j}$ and $q_{k,j}$ are terms of L_Ω. Now a disjunction is true if and only if at least one of the disjuncts is true. Thus, $A \equiv_\forall B$ if and only if A and B satisfy the same *primitive* sentences of L_Ω: $\exists \bar{x}(\bigwedge_i (P_i(\bar{x}) = p_i(\bar{x})) \wedge \bigwedge_j (Q_j(\bar{x}) \neq q_j(\bar{x}))$. That is, $A \equiv_\forall B$ if and only if, for every finite system Σ:

$$P_i(\bar{x}) = p_i(\bar{x}), \ 1 \leq i \leq I$$
$$Q_j(\bar{x}) \neq q_j(\bar{x}), \ 1s < j \leq J$$

it is the case that Σ has a solution in A if and only if it has a solution in B. $\square$

Remark 0.8 It easily follows from Lemma 0.7 that, if A is a subalgebra of the Ω-algebra B, then a sufficient condition for $A \equiv_\forall B$ is that there be a *discriminating family* Ψ of Ω-homomorphisms $B \to A$. Here Ψ is *discriminating* provided to every finite nonempty set $\{(b_{1,1}, b_{1,2}), \ldots, (b_{n,1}, b_{n,2})\}$ of pairs of unequal elements $b_{i,1} \neq b_{i,2}$ of B there is $\psi \in \Psi$ such that $\psi(b_{i,1}) \neq \psi(b_{i,2})$ for all $i = 1, \ldots, n$. In particular, if R is any ω-residually $\mathbb{Z}$ ring, then $R \equiv_\forall \mathbb{Z}$.

Theorem 0.9 (Remeslennikov) *Every finitely generated subring of a Peano ring is ω-residually $\mathbb{Z}$.*

A proof may be found in [3]. It is now not difficult to deduce using Lemmas 0.6 and 0.7 and Theorem 0.9 the following.

Corollary 0.10 *The following three statements are pairwise equivalent.*

1. $R \equiv_\forall \mathbb{Z}$
2. R *is embeddable in a Peano ring* $^*\mathbb{Z}$.
3. R *is locally ω-residually $\mathbb{Z}$.*

Remark 0.11 If $R \equiv_\forall \mathbb{Z}$ and $R \neq \mathbb{Z}$, then R must contain a copy of the integral polynomial ring $\mathbb{Z}[t]$.

To see this suppose $R \equiv_\forall \mathbb{Z}$ and $\theta \in R \backslash \mathbb{Z}$. We claim that θ is transcendental over $\mathbb{Z}$. Suppose not. Then there is a nonzero integral polynomial $a_n x^n + \cdots + a_0$ having θ as a root in R. Thus $\exists x(a_n x^n + \cdots + a_0 = 0)$ is true in R. So it must also be true in $\mathbb{Z}$. Suppose that $m_1, \ldots, m_k$ are the distinct roots of $a_n x^n + \cdots + a_0$ in $\mathbb{Z}$. Then $\forall x((a_n x^n + \cdots + a_0 = 0) \to \bigvee_{1 \leq i \leq k}(x = m_i))$ is true in $\mathbb{Z}$. Hence it must be true in R. Hence, θ must be one of the m_i. But this is a contradiction since θ was assumed not in $\mathbb{Z}$.

If t is an indeterminate over $\mathbb{Z}$, then the class of $\mathbb{Z}[t]$-groups is a variety of algebras of a suitable type. For each integer $r \geq 2$ a free algebra of rank r exists in this variety. If $F = \langle a_1, \ldots, a_r ; \rangle$ is an ordinary rank r free group, then the free rank r object in the variety of $\mathbb{Z}[t]$-groups, once a set of free generators for this algebra is fixed, contains a canonical copy of F and so may be viewed as an F-group. $F^{\mathbb{Z}[t]}$ will denote the free rank r $\mathbb{Z}[t]$-group.

Theorem 0.12 ([5]) *Let F be a finitely generated nonabelian free group. Let H be a finitely generated F-group. Then $H \equiv_\forall F$ as F-groups if and only if H F-embeds in $F^{\mathbb{Z}[t]}$.*

Alexei Myasnikov asked the third author if there is a similar result for free nilpotent groups G with the Hall completion G^B with respect to a suitable binomial ring B playing the role of $F^{\mathbb{Z}[t]}$. The main result of this paper is the following.

Theorem 0.13 *Let G be a finitely generated nonabelian free nilpotent group. Then there is a countable locally ω-residually $\mathbb{Z}$ ring A, depending only on G, with the property that if H is any finitely generated G-group, then $H \equiv_\forall G$ as G-groups if and only if H G-embeds in G^B where B is the binomial closure of A.*

Remark 0.14 The theorem could have been stated somewhat more simply since the binomial closure of a countable locally ω-residually $\mathbb{Z}$ ring is also a countable locally ω-residually $\mathbb{Z}$ ring. The authors prefer to state the theorem as is since it was proven by first constructing A and then taking the binomial closure.

1   Proof of the Main Theorem

Throughout this section G will be a fixed finitely generated nonabelian free nilpotent group with (ordered) free generating system $a_1, \ldots, a_r$ relative to N_c and $a_1, \ldots, a_r, c_{r+1}, \ldots, c_N$ will be a fixed system of Hall basic commutators for G beginning with the ordered basis $a_1 < \cdots < a_r$.

Lemma 1.1 *Let B be a locally ω-residually $\mathbb{Z}$ binomial ring and let H be a finitely generated G-group. If H G-embeds in the Hall completion G^B, then $H \equiv_\forall G$ as G-groups.*

Proof Suppose H is generated by $a_1, \ldots, a_r, b_{r+1}, \ldots, b_k$.
  Note that

$$a_i \to a_i, \qquad\qquad 1 \le i \le r$$
$$b_j \to a_1^{\beta_{j,1}} \ldots a_r^{\beta_{j,r}} c_{r+1}^{\beta_{j,r+1}} \ldots c_N^{\beta_{j,N}}, \qquad r+1 \le j \le k$$

determines a G-embedding of H into G^B. Now the subring R generated by the $\beta_{j,l}$, $r+1 \le j \le k$, $1 \le l \le N$ is ω-residually $\mathbb{Z}$; hence, so is its binomial closure S. We may view H as G-embedded in the Hall completion G^S. S is also ω-residually $\mathbb{Z}$. Identifying H with its G-image in G^S, let

$$x_1 = a_1^{\xi_{1,1}} \ldots a_r^{\xi_{1,r}} c_{r+1}^{\xi_{1,r+1}} \ldots c_N^{\xi_{1,N}},$$
$$\vdots$$
$$x_n = a_1^{\xi_{n,1}} \ldots a_r^{\xi_{n,r}} c_{r+1}^{\xi_{n,r1}} \ldots c_N^{\xi_{n,N}}$$

be finitely many nontrivial elements of H. Then there is a retraction $\rho : S \to \mathbb{Z}$ which does not annihilate any of the $\xi_{i,j}$ which are not already 0. This induces a group retraction $\hat{\rho} : G^S \to G$ which does not annihilate any of $x_1, \ldots, x_n$. The restriction ψ of $\hat{\rho}$ to H is a retraction $\psi : H \to G$ which does not annihilate any of $x_1, \ldots, x_n$. Thus, there is a discriminating family of retractions $H \to G$. It follows by Remark 0.8 that $H \equiv_\forall G$ as G-groups.   □

Lemma 1.2 *Let H be a finitely generated G-group such that $H \equiv_\forall G$ as G-groups. Then there is a finitely generated ω-residually $\mathbb{Z}$ ring R (depending on H) such that H G-embeds in G^S where S is the binomial closure of R.*

Proof By Lemma 0.6 there is a G-embedding of H into an ultrapower $^*G = G^I/D$ of G. Form the corresponding ultrapower $^*\mathbb{Z} = \mathbb{Z}^I/D$ of the ring $\mathbb{Z}$ of integers. If $g = [\gamma]_D \in {}^*G$ and $\nu = [n]_D \in {}^*\mathbb{Z}$, we may set $g^\nu = [\xi]_D$ where $\xi(i) = \gamma(i)^{n(i)}$ for all $i \in I$. One checks that this is well-defined and converts $^*G$ into a $^*\mathbb{Z}$-group. Furthermore, every element of $^*G$ is uniquely of the form

$$a_1^{\zeta_1} \ldots a_r^{\zeta_r} c_{r+1}^{\zeta_{r+1}} \ldots c_N^{\zeta_N}$$

where the ζ_j lie in $^*\mathbb{Z}$. It follows that $^*G$ is the Hall completion of G with respect to the Peano ring $^*\mathbb{Z}$. Suppose that H is generated by $a_1, \ldots, a_r, b_{r+1}, \ldots, b_k$. Identifying H with its image in $^*G$, we have $b_j = a_1^{\beta_{j,1}} \ldots a_r^{\beta_{j,r}} c_{r+1}^{\beta_{j,r+1}} \ldots c_N^{\beta_{j,N}}$ where $\beta_{j,l} \in {}^*\mathbb{Z}$, $r+1 \leq j \leq k; 1 \leq l \leq N$. Let R be the subring of $^*\mathbb{Z}$ generated by the $\beta_{j,l}$, $r+1 \leq j \leq k; 1 \leq l \leq N$. Then R is a finitely generated ω-residually $\mathbb{Z}$ ring and H G-embeds in G^S where S is the binomial closure of R.   □

Every finitely generated nilpotent group is finitely presented. More generally, polycyclic groups are finitely presented. See [9]. There are, up to isomorphism, only countably many finitely presented groups. Now, if $H \equiv_\forall G$, then H satisfies $\forall x_1, \ldots, x_{c+1}([x_1, \ldots, x_{c+1}] = 1)$. Hence, such H are nilpotent. It follows that there are only countably many G-isomorphism classes of finitely generated G-groups H such that $H \equiv_\forall G$ as G-groups.

Remark 1.3 That there are infinitely many such H can be seen from the fact that, for every positive integer n, $G \times \langle a\, ;\ \rangle^n \equiv_\forall G$ as G-groups. This is so since there is a discriminating family of retractions $G \times \langle a\, ;\ \rangle^n \to G$ mapping $\langle a\, ;\ \rangle^n$ into the center of G. Moreover, the $G \times \langle a\, ;\ \rangle^n$ are nonisomorphic in pairs.

Proof of Theorem 0.13:

Let $(H_n)_{n<\omega}$ be an ω-sequence of representatives of the distinct G-isomorphism classes of finitely generated G-groups H such that $H \equiv_\forall G$ as G-groups. Let $H_0 = G$. (Here ω, the first limit ordinal is identified with the set of nonnegative integers.) Now each H_{n+1} G-embeds into an ultrapower G^{I_n}/D_n of G. For each H_{n+1} fix a finite set of generators containing $a_1, \ldots, a_r$ and construct the finitely generated ω-residually $\mathbb{Z}$ ring $R_n \subseteq \mathbb{Z}^{I_n}/D_n$ as in the proof of Lemma 1.2. Now define an ω-chain $(A_n)_{n<\omega}$ of Peano rings as follows. $A_0 = \mathbb{Z}$ and, if A_n has already been defined, then $A_{n+1} = A_n^{I_n}/D_n$. (Note that $R_n \subseteq \mathbb{Z}^{I_n}/D_n \subseteq A_n^{I_n}/D_n = A_{n+1}$ for all $n < \omega$.) Form the ultralimit $A_\omega = \lim(A_n)_{n<\omega}$. Now each R_n is contained in the Peano ring A_ω. Let A be the subring of A_ω generated by the images of the R_n in A_ω. Since each R_n is finitely generated and there are only countably many R_n, A is a countable locally ω-residually $\mathbb{Z}$ ring. Moreover, each H_n G-embeds in G^B where B is the binomial closure of A. $\square$

It is tempting to conjecture that A can be taken to be the integral polynomial ring $\mathbb{Z}[t]$. Thus

Question If G is a finitely generated free nilpotent group and H is a finitely generated G-group such that $H \equiv_\forall G$ as G-groups must H G-embed in G^B where B is the binomial closure of $\mathbb{Z}[t]$?

2   Relative Tensor Completion

Suppose A is a nontrivial commutative ring. Myasnikov and Remeslennikov restricted the class of A-groups H to those satisfying the additional axioms $(xy)^\alpha = x^\alpha y^\alpha$ whenever $xy = yx$ $\forall (x,y) \in H^2$; $\alpha \in A$. One can view these as quasi-identities: one for each $\alpha \in A$. This allowed them to view A-groups as non-commutative A-modules and apply various notions of commutative algebra to the nonabelian case. In particular one can define the tensor A-completion of a group G [7]. (More generally, if A_0 is a nontrivial commutative ring and $\mu : A_0 \to A$ is a ring homomorphism they defined the tensor A-completion of an A_0-group G relative to μ. We shall not need this more general notion. Note that $n \to n \cdot 1$ is the only possible ring homomorphism $\mathbb{Z} \to A$. In particular, if A has characteristic 0, then $\mathbb{Z} \to A$ may be viewed as the inclusion map.) Here we relativize their definition to an arbitrary variety V of groups. For simplicity we shall restrict ourselves to the case where A has characteristic 0. Let A be a commutative ring of characteristic 0 and V be a variety of groups. Let V^A be the corresponding variety of A-groups (according to Lyndon) and let V_{MR}^A be the subquasivariety of V^A satisfying the quasi-identities (one for each $\alpha \in A$)

$$\forall x, y((xy = yx) \to ((xy)^\alpha = x^\alpha y^\alpha)).$$

Definition 2.1 Let G be a group lying in the variety V of groups and let A be a commutative ring of characteristic 0. A group $T_V(G, A) \in V_{MR}^A$ is a **tensor A-completion of G relative to V** provided

1. *There exists a group homomorphism $i : G \to T_V(G, A)$ such that $i(G)$ generates $T_V(G, A)$ as an A-group.*

2. *For any $H \in V_{MR}^A$ and any group homomorphism $\varphi : G \to H$ there exists an A-homomorphism $\psi : T_V(G, A) \to H$ such that $\psi i = \varphi$. (Here we write our maps to the left of their arguments and compose accordingly.)*

One can show that (since $\mathbb{Z}$ is a subring of A) $T_V(G, A)$ exists and is unique up to an A-isomorphism.

Example 2.2 ([7]) If F is a free group, then the Lyndon completion $F^{\mathbb{Z}[t]}$ of F with respect to the polynomial ring $\mathbb{Z}[t]$ is the tensor $\mathbb{Z}[t]$-completion of F relative to the variety of all groups.

Example 2.3 ([6]) If G is free in the variety N_c of nil-c groups and B is a binomial ring, then the Hall completion G^B of G with respect to B is the tensor B-completion of G relative to N_c.

References

[1] J. L. Bell and A. B. Slomson, *Models and Ultraproducts: An Introduction*, Second revised printing, North-Holland, Amsterdam, 1971.

[2] C. C. Chang and H. J. Keisler, *Model Theory*, Second edition, North-Holland, Amsterdam, 1977.

[3] B. Fine, A. M. Gaglione, A. G. Myasnikov and D. Spellman, A classification of fully residually free groups of rank three or less, *J. Algebra* **200** (1998), no. 2, 571–605.

[4] P. Hall, *The Edmonton Notes on Nilpotent Groups*, Queen Mary College Mathematics Notes, London, 1969.

[5] O. Kharlampovich and A. G. Myasnikov, Irreducible affine varieties over a free group II: systems in triangular quasi-quadratic form and description of residually free groups, *J. Algebra* **200** (1998), no. 2, 517–570.

[6] A. G. Myasnikov, Private communication.

[7] A. G. Myasnikov and V. N. Remeslennikov, Exponential groups I: foundations of the theory and tensor completion, *Siberian Math. J.* **5** (1994).

[8] V. N. Remeslennikov, ∃-free groups, *Siberian Math. J.* **30** (1989), no. 6, 153–157.

[9] D. J. S. Robinson, *A Course in the Theory of Groups*, Second edition, Springer, New York, 1995.

IRREDUCIBLE WORD PROBLEMS IN GROUPS

ANA R. FONSECA, DUNCAN W. PARKES and RICHARD M. THOMAS

Department of Computer Science, University of Leicester, Leicester LE1 7RH, England
Email: `rmt@mcs.le.ac.uk`

Abstract

In this paper we consider irreducible word problems in groups. In particular, we look at results concerning groups whose irreducible word problem lies in some given class of languages (such as the class of finite languages or the class of context-free languages).

1 Introduction

In this paper we look at irreducible word problems in groups; see Section 4 below for the definition. We are particularly interested in connections with formal language theory; to be more specific, we consider which types of group can have their irreducible word problem lying in some given class of languages (such as the class of finite languages or the class of context-free languages).

We summarize what we need from formal language theory in Section 2. The general question of the connection between irreducible word problems and classes of languages follows on from the analogous question concerning the links between word problems and classes of languages, and we look at some relevant information in Section 3. We come to reduced and irreducible word problems in Section 4, and we talk there about groups with a finite irreducible word problem. We mention some general results about irreducible word problems and languages in Section 5, and then, in Section 6, concentrate on groups whose irreducible word problem is context-free. We finish with some further comments in Section 7.

2 Formal languages

We shall use this section to introduce the concepts we require from formal language theory and to fix our notation. The reader is referred to [5, 11, 16, 17] for further information.

If Σ is an *alphabet* (i.e., a finite set of symbols), then Σ^* denotes the set of all finite words (or strings) over Σ, including the empty word ϵ. The subsets of Σ^* are called *languages* over Σ.

If L_1 and L_2 are languages, we write $L_1 L_2$ for the *concatenation* of L_1 and L_2, i.e., the set $\{w_1 w_2 : w_1 \in L_1, w_2 \in L_2\}$; we abbreviate LL to L^2, LLL to L^3, and so on. If L is a language, then the *Kleene closure* L^* of L is the submonoid of Σ^* generated by L, i.e., the set $\{\epsilon\} \cup L \cup L^2 \cup L^3 \cup \ldots$, and L^+ is the subsemigroup of Σ^* generated by L, i.e., the set $L \cup L^2 \cup L^3 \cup \ldots$.

We will use the expression $v \equiv w$ to mean that v and w are identical as strings of symbols. As we will see later, words will also represent elements of a group G; in that context, we will write $u = v$ or $u =_G v$ if u and v represent the same element of G. A word u is a *prefix* of w if there exists v (possibly ϵ) such that $uv \equiv w$ and u is a *subword* of w if there exist v_1 and v_2 (again, allowing ϵ in each case) such that $v_1 u v_2 \equiv w$. Note that some authors refer to this as a *factor*, and use the term "subword" for a sequence of symbols occurring in w which are in the same order as their occurrence in w but which are not necessary consecutive in w. The length of a word w will be denoted by $|w|$, and $|w|_x$ will denote the number of occurrences of the symbol x in w.

A (*non-deterministic*) *finite automaton* M is a quintuple $(Q, \Sigma, \delta, s, A)$, where Q is a finite set of *states*, Σ is a finite set of *input symbols*, the *transition relation* δ is a subset of $Q \times (\Sigma \cup \{\epsilon\}) \times Q$, the *start state* s is a special element of Q, and the set A of *accept states* is a subset of Q. We will abbreviate the expression "non-deterministic finite automaton" to NFA. An element of δ of the form (q, ϵ, q') is known as an *empty transition*.

The transition relation δ may be extended from a subset of $Q \times (\Sigma \cup \{\epsilon\}) \times Q$ to a subset δ^* of $Q \times \Sigma^* \times Q$ in the following obvious inductive way:

> let (q, ϵ, q) be in δ^* for each $q \in Q$;
> if $(q_1, x, q_2) \in \delta$ then let (q_1, x, q_2) be in δ^*;
> if $(q_1, w, q_2) \in \delta^*$ and $(q_2, x, q_3) \in \delta$ then let (q_1, wx, q_3) be in δ^*.

We say that M *accepts* a word $w \in \Sigma^*$ if $(s, w, f) \in \delta^*$ for some $f \in A$. The set of words from Σ^* which are accepted by M is denoted by $L(M)$, and this is known as the *language accepted by* M. A language is said to be *regular* if it is accepted by an NFA; we let $\mathcal{R}$ denote the class of regular languages.

An NFA is said to be *deterministic* if there are no empty transitions and if δ is a partial function from $Q \times \Sigma$ to Q; we write DFA for "deterministic finite automaton". An important result is that any regular language can be accepted by a DFA.

We can think of an NFA M as a device with a collection of states and an input tape which contains a word w. If we have a transition (q, a, r) in δ, then M may move from state q to state r whilst reading the symbol a on the input tape; if $a = \epsilon$, then we can move from q to r without reading a symbol. We start with M in the start state reading the symbol in the leftmost cell of the input tape, and the word w is accepted if we can be in an accept state once all the input has been read.

An equivalent way of specifying the regular languages is as the languages denoted by regular expressions. The *regular expressions* R over an alphabet Σ, and the languages $L(R)$ they denote, are defined inductively as follows:

- $\emptyset$ is a regular expression with $L(\emptyset) = \emptyset$;
- ϵ is a regular expression with $L(\epsilon) = \{\epsilon\}$;
- for each $a \in \Sigma$, a is a regular expression denoting $\{a\}$;
- if R is a regular expression then R^* is a regular expression denoting $L(R)^*$;

- if R_1 and R_2 are regular expressions then $(R_1 \cup R_2)$ is a regular expression denoting $L(R_1) \cup L(R_2)$ and $(R_1 R_2)$ is a regular expression denoting $L(R_1)L(R_2)$.

In addition, we use R^+ to stand for RR^*, so that $L(R^+) = L(R)^+$.

A *context-free grammar* $G = (N, \Sigma, P, S)$ consists of an alphabet N of *non-terminal* symbols, an alphabet Σ of *terminal* symbols, a finite set P of *productions* and a special symbol $S \in N$ called the *start symbol*; the sets N and Σ are required to be disjoint. The elements of P are of the form $A \to w$ where $A \in N$ and $w \in (N \cup \Sigma)^*$. Let $u_1, u_2 \in (N \cup \Sigma)^*$; we write $u_1 A u_2 \Rightarrow u_1 w u_2$ if $A \to w$ is in P, and then extend $\Rightarrow$ by reflexive transitive closure to $\Rightarrow^*$. The *language* $L(G)$ generated by G is $\{w \in \Sigma^* : S \Rightarrow^* w\}$. A language is said to be *context-free* if it can be generated by a context-free grammar; we let $\mathcal{CF}$ denote the class of context-free languages. If we restrict P to be a subset of $N \times (\Sigma N \cup \{\epsilon\})$, then we say that G is a *regular grammar*; regular grammars generate precisely the regular languages.

A *pushdown automaton* (PDA) M is a septuple $(Q, \Sigma, \Gamma, \delta, s, Z_0, A)$, where Q is a finite set of *states*, Σ is an alphabet of *input symbols*, Γ is an alphabet of *stack symbols*, the *transition relation* δ is finite subset of $Q \times (\Sigma \cup \{\epsilon\}) \cup (\Gamma \cup \{\epsilon\}) \times Q \times \Gamma^*$, $s \in Q$ is the *start state*, $Z_0 \in \Gamma$ is the *bottom marker* for the stack, and $A \subseteq Q$ is the set of *accept states*. We want the symbol Z_0 to appear at the bottom of the stack and nowhere else; therefore we also insist that, if $(q, x, Z_0, r, \gamma) \in \delta$, then $\gamma \in (\Gamma \setminus \{Z_0\})^* Z_0$, and, if $Z \neq Z_0$ and $(q, x, Z, r, \gamma) \in \delta$, then $\gamma \in (\Gamma \setminus \{Z_0\})^*$.

We call the set $Q \times \Sigma^* \times (\Gamma \setminus \{Z_0\})^* Z_0$ the set of *configurations* of M, and write $(q_1, xw, Z_1 v) \rightsquigarrow (q_2, w, uv)$ if $(q_1, x, Z_1, q_2, u) \in \delta$. We let $\overset{*}{\rightsquigarrow}$ denote the reflexive transitive closure of $\rightsquigarrow$. If $(s, w, Z_0) \overset{*}{\rightsquigarrow} (f, \epsilon, w)$ for some $f \in A$ and some $w \in \Gamma^*$, then we say that M *accepts* w. The set of words accepted by M is denoted by $L(M)$. An important result is that a language is context-free if and only if it is accepted by a PDA.

If M is a PDA such that, for any configuration, there is at most one possible move that M can make, then M is said to be a *deterministic pushdown automaton* (DPDA) and $L(M)$ is then said to be a *deterministic context-free* language. We let $\mathcal{DCF}$ denote the class of deterministic context-free languages. Unlike the situation with finite automata, determinism does make a difference here, in that $\mathcal{DCF}$ is a proper subclass of $\mathcal{CF}$; in addition, $\mathcal{DCF}$ properly contains $\mathcal{R}$.

A context-free grammar in said to be in *Greibach normal form* if every production is either of the form $A \to aw$, where A is a non-terminal, a is a terminal, and w is a (possibly empty) string of non-terminals, or is the production $S \to \epsilon$. Every context-free language can be generated by a grammar in Greibach normal form. A grammar in Greibach normal form is said to be *simple* if, whenever we have productions $A \to au$ and $A \to av$, then we must have that $u \equiv v$, and, if $S \to \epsilon$ is a production, then it is the only production; a language which is generated by a simple grammar is said to be *simple*. The simple languages form a proper subclass of $\mathcal{DCF}$; moreover, every regular language not containing ϵ is simple. An alternative characterization of the simple languages is those languages that can be accepted by a DPDA with just one state that accepts by empty stack.

Given a language L in Σ^*, we shall call the set $pf(L)$ of all non-empty words in L which have no non-empty proper prefix in L the *prefix-free* part of L and the set $sf(L)$ of all words from $pf(L)$ which have no non-empty proper subword in L the *subword-free* part of L. We have that $pf(L) = L \setminus (L\Sigma^+ \cup \{\epsilon\})$ and $sf(L) = L \setminus (\Sigma^* L\Sigma^+ \cup \Sigma^+ L\Sigma^* \cup \{\epsilon\})$. Some texts use $Min(L)$ instead of $pf(L)$. We say that a language L is *prefix-free* if $L = pf(L)$ and *subword-free* if $L = sf(L)$.

We also need to consider some closure properties of classes of languages. Let $\mathcal{F}$ be a class of languages; we say that $\mathcal{F}$ is *closed under homomorphism* if

$$L \in \mathcal{F}, L \subseteq \Sigma^*, \phi : \Sigma^* \to \Omega^* \text{ a monoid homomorphism } \Rightarrow L\phi \in \mathcal{F},$$

and that $\mathcal{F}$ is *closed under inverse homomorphism* if

$$L \in \mathcal{F}, L \subseteq \Omega^*, \phi : \Sigma^* \to \Omega^* \text{ a monoid homomorphism } \Rightarrow L\phi^{-1} \in \mathcal{F}.$$

We also say that $\mathcal{F}$ is *closed under intersection with regular languages* if

$$L_1 \in \mathcal{F}, L_1 \subseteq \Sigma^*, L_2 \in \mathcal{R}, L_2 \subseteq \Sigma^* \Rightarrow L_1 \cap L_2 \in \mathcal{F}.$$

There are other similar closure properties we could consider, such as closure under union, intersection, concatenation, Kleene star, or taking the prefix-free or subword-free part of the language; see the table below for the closure properties of the classes of languages we have introduced so far. We shall assume these facts in the remainder of this paper.

Property	$\mathcal{R}$	$\mathcal{DCF}$	$\mathcal{CF}$
union	√	×	√
intersection	√	×	×
intersection with regular sets	√	√	√
concatenation	√	×	√
complementation	√	√	×
Kleene star	√	×	√
homomorphism	√	×	√
inverse homomorphism	√	√	√
prefix-free part	√	√	×
subword-free part	√	×	×

A *cone* is a family $\mathcal{F}$ of languages closed under homomorphism, inverse homomorphism and intersection with regular languages. As can be seen from the table above, $\mathcal{R}$ and $\mathcal{CF}$ are both cones whereas $\mathcal{DCF}$ is not (as it is not closed under homomorphism). We should comment that the term "cone" is not universally used; other terms for this concept include *rational cone* and *full trio*.

3 Word problems of groups

For the purpose of what follows, we have to be a little careful with the definition of a generating set for a group. A set X, where each $x \in X$ represents an element of a group G, is said to be a *monoid generating set* for G if every element of G can be written as a word in X^*; if every element of G can be written as a word in X and the inverses of elements of X, then X is said to be a *group generating set* for G. As we explained in Section 2, we will write $u = v$ or $u =_G v$ if u and v represent the same element of G, and $u \equiv v$ if u and v are equal as strings of symbols.

If X is a group generating set for a group G, let Y be the alphabet consisting of X together with a disjoint set in a one-to-one correspondence with X whose elements represent the inverses of the elements in X. The *word problem* $W_X^g(G)$ of G with respect to X is then the set of all words w in Y^* such that w represents the identity element G. The word problem $W_X^m(G)$ of G with with respect a monoid generating set X is the set of words in X^* which are equal to the identity in G.

These definitions give us a natural link between group theory and the theory of formal languages; in particular, they present us with the question of what relationship exists between the complexity of the word problem as a formal language and the algebraic structure of the group.

With any family of languages $\mathcal{F}$, we can associate the class of groups which have some finite generating set with respect to which the word problem lies in $\mathcal{F}$; an interesting task is to try to find algebraic descriptions of such classes of groups. While, in general, the fact that the word problem lies in $\mathcal{F}$ may depend on our choice of finite generating set, this will not be the case for the families $\mathcal{F}$ we consider here due to the following well known result (see [13] for example):

Proposition 3.1 *If $\mathcal{F}$ is a family of languages that is closed under inverse homomorphism, G is a group with a finite monoid generating set X and $W_X^m(G) \in \mathcal{F}$, then $W_Y^m(G) \in \mathcal{F}$ for every finite monoid generating set Y for G.*

Given Proposition 3.1, one can simply refer to a group "having word problem in $\mathcal{F}$" when $\mathcal{F}$ is closed under inverse homomorphism.

One also has the analogue of Proposition 3.1 for group generating sets by noting that X is a group generating set for G if and only if $X \cup X^{-1}$ is a monoid generating set for G. In general, where results are true for monoid generating sets and the special case for group generating sets follows as a result, we will only state the result in the more general setting. The reason for our making the distinction is that there are situations where we have a result for group generating sets that does not generalize to monoid generating sets.

There have been several such characterizations of the groups having word problem in a particular class of languages. For example, Anisimov proved [1]:

Theorem 3.2 *A finitely generated group G has a regular word problem if and only if G is finite.*

In [22] the class of accessible groups with context-free word problem was shown to coincide with the virtually free groups. Given that context-free groups are finitely presented by [2], the accessibility hypothesis can be removed thanks to the result of [7] that all finitely presented groups are accessible, and we have:

Theorem 3.3 *A finitely generated group G has a context-free word problem if and only if G is virtually free.*

In fact, if the word problem of a group is context-free, then it is deterministic context-free [23]. The reader is also referred to [4] for a further simplification (the word problem of a virtually free group is a so-called *NTS language*) and to [20, 26] for some similar results on context-free groups to those of Muller and Schupp in [22].

One further family of languages that will be of interest to us is that of the *one-counter languages*. A pushdown automaton is said to be *one-counter* if its stack alphabet contains only one symbol other than the bottom marker Z_0. As the one-counter languages form a cone (and so, in particular, are closed under inverse homomorphism), the property of a group having a one-counter word problem is independent of the choice of finite generating set. We let $\mathcal{OC}$ denote the class of one-counter languages.

As with the context-free languages, determinism does make a difference here, as the class of languages $\mathcal{DOC}$ accepted by deterministic one-counter automata is a proper subclass of $\mathcal{OC}$. Clearly $\mathcal{DOC}$ is contained in $\mathcal{DCF} \cap \mathcal{OC}$, but the converse does not hold; for example, it is noted in [3] that the language

$$\{w_1 x w_2 : w_1, w_2 \in \{a, b\}^*, w_1 \not\equiv w_2\}$$

lies in both $\mathcal{DCF}$ and $\mathcal{OC}$ but not in $\mathcal{DOC}$.

As far as the relationship with word problems is concerned, Herbst proved the following result [12]:

Theorem 3.4 *A finitely generated group G has a one-counter word problem if and only if G is virtually cyclic.*

Similar to the case of the context-free languages, if the word problem of a group is a one-counter language, then it is necessarily a deterministic one-counter language [12]. Herbst also showed that the class of groups with word problem in $\mathcal{C}$, where $\mathcal{C}$ is a cone contained in $\mathcal{CF}$, must be either the finite, one-counter or context-free groups. While there are, of course, interesting families of languages which are not cones, these results do suggest that the one-counter languages are of particular interest when considering word problems of groups. The reader is also referred to [8] for some related results.

4   Reduced and irreducible word problems

In [10] Haring-Smith defined the *reduced word problem* of a group G generated by a finite set X to be the subset of the word problem consisting of those non-empty words which have no non-empty proper prefix equal to the identity and the *irreducible word problem* to be the set of all non-empty words w in the word problem such that no non-empty proper subword of w represents the identity. We write $R_X^g(G)$ or $R_X^m(G)$ for the reduced word problem (according as X is a group or monoid generating set respectively) and, similarly, $I_X^g(G)$ or $I_X^m(G)$ for the irreducible word problem. One can think of the reduced word problem as consisting of those words that label a closed path in the Cayley graph of the group which only return to their starting point for the first time at the end of the path, and the irreducible word problem as those words that label closed paths with no repetitions of vertex (other than the initial and final vertices of the path).

In the case of group generating sets, Haring-Smith gave the following elegant characterization [10]:

Theorem 4.1 *If G is a group and X is a finite group generating set for G, then the following are equivalent:*

1. $R^g_X(G)$ *is a simple language;*
2. $I^g_X(G)$ *is finite;*
3. G *is a free product of a finitely generated free group and finitely many finite groups.*

Haring-Smith named this class of groups the *plain* groups (as he reasonably pointed out, "simple groups" would be somewhat confusing!). It should be noted that a plain group need not have a simple reduced word problem (or finite irreducible word problem) with respect to every finite group generating set.

In the same paper Haring-Smith conjectured that the class of groups with a strict deterministic reduced word problem (i.e., a reduced word problem that is a prefix-free deterministic context-free language) is the same as the class of finite extensions of plain groups. This was shown to be the case in [25] where it was proved that both classes coincide with the context-free groups. An important observation in this respect (essentially just reworking the definition) is the following:

Proposition 4.2 *If G is a group with a finite monoid generating set X then $W^m_X(G) = R^m_X(G)^*$ and $R^m_X(G) = pf(W^m_X(G))$.*

Given Proposition 4.2, and the closure of $\mathcal{R}$ under Kleene star and the operation pf, we see that $R^m_X(G)$ is regular if and only if $W^m_X(G)$ is regular; so the groups with a regular reduced word problem are precisely the finite groups by Theorem 3.2.

Now Theorem 4.1 tells us that the groups which have finite irreducible word problem with respect to some group generating set are exactly the plain groups. It is important to note here that we are looking at the irreducible word problem from the point of view of group generating sets, since there are groups which have finite irreducible word problem with respect to some monoid generating set which are not plain group; see [24] where (for example) the following result is proved:

Theorem 4.3 *A finitely generated group G which has an infinite cyclic central subgroup has a finite irreducible word problem (with respect to some finite generating set) if and only if G is of the form $A \times B$ where A is a finite cyclic group and B is an infinite cyclic group.*

We note that the groups $\mathbb{Z}_m \times \mathbb{Z}$ (with $m > 1$) are not plain. So we have the following natural question:

Question 4.4 Which groups have a finite irreducible word problem with respect to some finite monoid generating set?

This is equivalent to a question of Madlener and Otto from [21] about string rewriting systems; see [21, 24] for details.

5   Irreducible word problems and languages

In Section 4 we considered the situation where the irreducible word problem was finite; one could look at the question of determining the groups where the irreducible word problem lies in some other specified class of languages. When studying this, the following notion from [18] is helpful:

Definition 5.1 If $L \subseteq \Sigma^*$ is a language, then L is said to be *insertion closed* if, whenever u, v and w are words in Σ^* such that $v \in L$ and $uw \in L$, then $uvw \in L$.

We will need the following result from [18]:

Proposition 5.2 *Let Σ be a finite alphabet and $L \subseteq \Sigma^*$. Then:*

1. *there is a unique smallest insertion closed language K in Σ^* containing L;*
2. *if L is context-free, then K is context-free.*

We call the language K in Proposition 5.2 the *insertion closure* of L in Σ^*. The relevance of this notion here comes from the following result:

Proposition 5.3 *If G is a group and X is a finite monoid generating set for G, then $W_X^m(G)$ is the insertion closure of $I_X^m(G) \cup \{\epsilon\}$ in X^*.*

Proof Let $W = W_X^m(G)$, $I = I_X^m(G)$ and K be the insertion closure of $I \cup \{\epsilon\}$.

If u, v and w are words such that $v \in W$ and $uw \in W$, then $v =_G 1 =_G w$, and so $uvw =_G 1$, giving that $uvw \in W$. So W is an insertion closed language containing $I \cup \{\epsilon\}$, and therefore $K \subseteq W$ by definition of K.

Suppose that W is not contained in K, and let w be a word of minimal length in $W \setminus K$. If $w \in I$ or $w \equiv \epsilon$, then $w \in K$ by definition; so we may assume that $w \equiv usv$ for some words u and v with $|uv| > 0$ and some non-empty word $s \in I$. Since $uv \in W$ and $|uv| < |w|$, we have that $uv \in K$. Since $s \in I \subseteq K$ and K is insertion closed, we have that $w \equiv usv \in K$, a contradiction. □

To express $I_X^m(G)$ in terms of $W_X^m(G)$, we note that $I_X^m(G) = sf(W_X^m(G))$. If L is an insertion closed language, then $sf(L)$ is referred to as the *insertion base* of L in [18]; so $I_X^m(G)$ is the insertion base of $W_X^m(G)$.

The next class of languages which we would naturally consider is that of the regular languages $\mathcal{R}$. However, it is not hard to see that we must look further than $\mathcal{R}$ if we wish to find a wider class of groups:

Proposition 5.4 *Let G be a group with finite monoid generating set X. If the irreducible word problem of G with respect to X is regular then it is finite.*

Proof Suppose that $I = I_X^m(G)$ is regular. By the pumping lemma for regular languages (see Theorem 4.1 of [16] for example), there is a constant N such that, if $z \in I$ with $|z| \geqslant N$, then there exist words u, v and w with $z \equiv uvw$, $|v| < N$ and $uv^i w \in I$ for all $i \in \mathbb{N}$.

We have that $uw =_G uvw =_G 1$ and hence that $v =_G 1$; but v is a proper subword of uvw, a contradiction. Thus any word in I has length less than N, and there can be only finitely many of them. $\qquad\square$

6 Context-free irreducible word problems

Next we consider the groups which have context-free irreducible word problem with respect to some group generating set; this class clearly contains those groups with a finite irreducible word problem, i.e., the plain groups. However, in contrast to the situation described in Proposition 5.4, we do have a new class of groups here, and we start with an example of a group which is in this class but which is not a plain group.

Proposition 6.1 *The irreducible word problem of the group G defined by the presentation $\langle f : \ \rangle \times \langle a : a^2 \rangle$ with respect to the group generating set $\{a, f\}$ is context-free but not finite.*

Proof We write A for a^{-1} and F for f^{-1}; note that a and A represent the same element of G.

A word w in the symbols $\{a, A, f, F\}$ is equal to the identity in G if and only if it contains an even number of instances of elements of $\{a, A\}$ and, in addition, we have that $|w|_f = |w|_F$. Let L be the language consisting of the words of the following forms, all of which are equal to the identity in G and are irreducible:

- fF, Ff, aA and Aa;
- $f^i b_1 F^j b_2 f^k$, where $b_1, b_2 \in \{a, A\}$ and $i + k = j$, $j > 0$;
- $F^i b_1 f^j b_2 F^k$, where $b_1, b_2 \in \{a, A\}$ and $i + k = j$, $j > 0$.

L is a union of context-free languages and is therefore context-free.

We shall show that any word from the word problem which contains more than two instances of elements of $\{a, A\}$ cannot be irreducible, and hence that L is the irreducible word problem I of G with respect to the group generating set $\{a, f\}$.

Assume, for a contradiction, that there is a word w in I of the form

$$g_1^{i_1} b_1 g_2^{i_2} b_2 g_3^{i_3} \cdots g_{n-1}^{i_{n-1}} b_{n-1} g_n^{i_n},$$

where $n > 3$, each g_j is either f or F, each b_i is a or A, and where $i_j > 0$ for each j such that $2 \leqslant j \leqslant n - 1$. At least one of i_1 and i_n must be non-zero, otherwise, by removing b_1 and b_{n-1} from w, we would produce a proper subword of w equal to the identity.

We assume that $i_1 > 0$ and that g_1 is f (the other cases are similar). If $i_n > 0$, then g_n must also be f (otherwise we could remove the first and last symbol of w to leave a proper subword equal to the identity). We must have $g_j = g_{j+1}$ for $2 \leqslant j \leqslant n - 1$, otherwise, $b_{j-1} g_j^{i_j} b_j g_{j+1}^{i_{j+1}} b_{j+1}$ would contain a proper subword equal to the identity. Thus $g_2 = g_3 = \ldots = g_{n-1} = F$. So we now have that w must be of the form

$$f^{i_1} b_1 F^{i_2} b_2 F^{i_3} \ldots F^{i_{n-1}} b_{n-1} f^{i_n},$$

where $i_j > 0$ for each j such that $1 \leqslant j \leqslant n-1$. We must have that $i_1 < i_2$ in order to prevent $f^{i_1} b_1 F^{i_2} b_2$ from having a subword equal to the identity, and, similarly, we must have that $i_{n-1} > i_n$. From this we deduce that $i_1 + i_n < i_2 + \ldots + i_{n-1}$, a contradiction, since we must have $i_1 + i_n = i_2 + \ldots + i_{n-1}$ for w to be in the word problem.   □

Having shown that the plain groups are a proper subclass of the class of groups with context-free irreducible word problem for some group generating set, our next task must be to put some bound on how complex these groups can be. It turns out that, even if we allow monoid generating sets, such groups must be context-free:

Proposition 6.2 *If G is a group with a finite monoid generating set X, and if $I_X^m(G)$ is context-free, then $W_X^m(G)$ is context-free.*

Proof If $I_X^m(G)$ is context-free, then $I_X^m(G) \cup \{\epsilon\}$ is context-free, and the result follows from Propositions 5.2 and 5.3.   □

The converse of Proposition 6.2 is false; that is, there exists a group G with finite monoid generating set X, such that the word problem of G with respect to X is context-free but the irreducible word problem of G with respect to X is not. In order to demonstrate this we shall make use of the concept of a GSM-mapping.

A *generalized sequential machine* (GSM) is a sextuple $M = (Q, \Sigma, \Delta, \delta, s, A)$ where Q is a finite set of *states*, Σ and, Δ are finite sets (the *input* and *output alphabets*), δ is a function from $Q \times \Sigma$ to the set of finite subsets of $Q \times \Delta^*$, s is the *start state*, and A is the set of *accept states*. The domain of δ is extended to $Q \times \Sigma^*$ by defining $\delta(q, \epsilon)$ to be $\{(q, \epsilon)\}$ and, for $u \in \Sigma^*$ and $a \in \Sigma$, $\delta(q, ua)$ to be the set of all (p, vw) such that, for some $p' \in Q$, $(p', v) \in \delta(q, u)$ and $(p, w) \in \delta(p', a)$. For $u \in \Sigma^*$, let $M(u)$ be the set words $w \in \Delta^*$ such that $(f, w) \in \delta(s, u)$ for some $f \in A$, and let $M(L) = \bigcup \{M(u) : u \in L\}$. We now have a mapping from the set of languages over Σ to the set of languages over Δ; this is the *GSM-mapping* defined by M.

We are now in a position to give the following example, which will be used subsequently to prove that all infinite context-free groups have a non-context-free irreducible word problem for some finite group generating set:

Proposition 6.3 *Let $G = \langle f : \rangle$ be the free group on one generator and let $a = f$, $b = f^2$ and $c = f^6$. Then G has non-context-free irreducible word problem with respect to the group generating set $X = \{a, b, c\}$.*

Proof We let $I = I_X^g(G)$ and $W = W_X^g(G)$, and let A, B and C denote a^{-1}, b^{-1} and c^{-1} respectively. Let Y be the language denoted by the regular expression $abc^* BC^* Bc^* bAC^*$ and let $L = I \cap Y$. If L can be shown not to be context-free then it follows that I is not context-free.

Consider a typical element $w \equiv abc^i BC^j Bc^k bAC^l$ of Y. We need to find conditions which tell us exactly when such an element of Y is also in I. Firstly, we must have that $i + k = j + l$ in order that w is in W. We also need to make sure that no subword of w is in W.

Since any instance of b, B, c or C in w is equivalent to an even number of instances of f, the instance of a can effectively only cancel out with the instance of A; so any subword of w which is equal to the identity in G and contains the instance of a must also contain the instance of A and vice-versa, and hence must be the entire word w. As a result, we need only consider subwords of $bc^i BC^j Bc^k b$. There are not enough instances of b and B in this subword to cancel out an instance of c or C; so, in any subword which is equal to the identity, the instances of b must effectively cancel with those of B, and the instances of c must cancel with those of C.

Hence any proper subword of $bc^i BC^j Bc^k b$ which is equal to the identity must be either a subword of $bc^i BC^j$ or a subword of $C^j Bc^k b$; so, to make sure there are no such subwords, we must have that $i > j$, $j < k$ and that $i + k \neq j$. If $i > j$ and $j < k$ then $i + k > j$, so this last condition may be dropped, and the language L we need to consider is

$$\{abc^i BC^j Bc^k bAC^l : i + k = j + l, i > j \text{ and } j < k\}.$$

Let M be the GSM $(Q, \Sigma, \Delta, \delta, q_0, \{q_5\})$, where

$$Q = \{q_0, q_1, q_2, q_3, q_4, q_5\}, \ \Sigma = \{a, b, c, A, B, C\}, \ \Delta = \{x, y, z\}, \text{ and}$$

$$\begin{aligned}
&\delta(q_0, a) = \{(q_1, \epsilon)\}, \quad \delta(q_1, b) = \{(q_2, \epsilon)\}, \quad \delta(q_2, c) = \{(q_2, x)\}, \\
&\delta(q_2, B) = \{(q_3, \epsilon)\}, \quad \delta(q_3, C) = \{(q_3, y)\}, \quad \delta(q_3, B) = \{(q_4, \epsilon)\}, \\
&\delta(q_4, c) = \{(q_4, z)\}, \quad \delta(q_4, b) = \{(q_5, \epsilon)\}, \\
&\delta(q, d) = \{(q_5, \epsilon)\} \quad \text{for all other } (q, d).
\end{aligned}$$

The image of L under M is the language $L' = \{x^i y^j z^k : i > j \text{ and } j < k\}$. It is known that L' is not context-free, and it follows that L is not context-free, as the class of context-free languages is closed under GSM-mappings. $\square$

Given this result, we have:

Corollary 6.4 *Every infinite context-free group has non-context-free irreducible word problem for some finite group generating set.*

Proof Every infinite context-free group G contains the free group $\langle f : \rangle$ of rank 1 as a subgroup by Theorem 3.3. Let a, b, c, A, B and C be as in Proposition 6.3, and then choose a finite group generating set Z for G which includes a, b and c. We now consider the intersection of $I_Z^g(G)$ with the regular language $\{a, b, c, A, B, C\}^*$. This set is not context-free as in Proposition 6.3; so $I_Z^g(G)$ is not context-free. $\square$

In particular, we see that the insertion base of a context-free language need not be context-free. The situation with the regular languages is different [18]: the insertion base of a regular language is necessarily regular, whereas the insertion closure of a regular language need not be regular.

It is interesting to note that we now have three generating sets for the group $\mathbb{Z}_2 \times \mathbb{Z}$:

- a monoid generating set with respect to which the irreducible word problem is finite (as in Theorem 4.3);
- a group generating set with respect to which the irreducible word problem is context-free, but not finite (as in Proposition 6.1); and
- a group generating set with respect to which the irreducible word problem is not context-free (as in Proposition 6.3).

Although every infinite context-free group has a finite group generating set with respect to which its irreducible word problem is not context-free, there is something we can say about the irreducible word problem for any monoid generating set: it must have a context-free complement.

Proposition 6.5 *Let G be a group with finite monoid generating set X. If $W_X^m(G)$ is context-free then the complement in X^* of $I_X^m(G)$ is context-free.*

Proof Let $I = I_X^m(G)$ and $W = W_X^m(G)$. The complement $X^* \setminus I$ of I is the union of the complement of W and the set Y of all words which contain a proper subword which is in W. Now $Y = X^*WX^+ \cup X^+WX^*$ is context-free since the class of context-free languages is closed under concatenation and union. The complement of W is context-free since W is deterministic context-free. Thus the complement of I must be context-free as it is the union of two context-free languages.   □

Propositions 6.2 and 6.5 together give us the following:

Corollary 6.6 *If G is a group with context-free irreducible word problem for some finite monoid generating set then the complement of the irreducible word problem of G is context-free with respect to any finite monoid generating set.*

It is interesting to contrast Proposition 6.5 with the results in [15] concerning groups whose word problem is the complement of a context-free language.

Several of the results here carry across to the one-counter groups. For instance, we have:

Corollary 6.7 *Let G be a group with finite monoid generating set X. If $W_X^m(G)$ is one-counter then the complement in X^* of $I_X^m(G)$ is one-counter.*

On the other hand, the analogue of Proposition 6.2 in the one-counter case fails. The free group on two generators provides a simple example of a group which has a set of generators for which the irreducible word problem is finite, and therefore certainly one-counter, but the word problem is not one-counter. This reflects the fact that, unlike the case of the context-free languages, the insertion closure of a one-counter language need not be one-counter.

In fact, the irreducible word problem given in Proposition 6.1 is not only context-free but is, in fact, one-counter. However, there do exist groups where the irreducible word problem is context-free but not one-counter.

7 Some further comments

In this paper we have surveyed some connections between formal languages and irreducible word problems; as we have said above, this follows on from a consideration of the connections between formal languages and word problems. In some sense the latter are best understood when our classes of languages are reasonably restrictive; we have precise characterizations of the groups whose word problem is regular, one-counter or context-free, but, in general, we do not have such results for families above the context-free languages. For some families $\mathcal{F}$ of languages well above context-free, however, we do have some results of the form "a group has a word problem in $\mathcal{F}$ if and only if it can be embedded into a particular type of group"; see [6, 14] for example.

As a result, when considering irreducible word problems here, we have concentrated on classes of languages within the context-free languages. There has been some work done of irreducible word problems lying in families above context-free; see [19] for example.

Acknowledgements The third author would like to thank Hilary Craig for all her help and encouragement.

References

[1] A. V. Anisimov, Group languages, *Kibernetika* **4** (1971), 18–24.
[2] A. V. Anisimov, Some algorithmic problems for groups and context-free languages, *Kibernetika* **8** (1972), 4–11.
[3] J. M. Autebert, J. Berstel and L. Boasson, Context-free languages and pushdown automata, in G. Rozenberg & A. Salomaa (eds.), *Handbook of Formal Languages, Volume 1* (Springer-Verlag, 1997), 111–174.
[4] J. M. Autebert, L. Boasson and G. Sénizergues, Groups and NTS languages, *J. Comput. System Sci.* **35** (1987), 243–267.
[5] J. Berstel, *Transductions and Context-free Languages* (Teubner, 1979).
[6] W. W. Boone and G. Higman, An algebraic characterization of groups with a solvable word problem, *J. Austral. Math. Soc.* **18** (1974), 41–53.
[7] M. J. Dunwoody, The accessibility of finitely presented groups, *Invent. Math.* **81** (1985), 449–457.
[8] G. Z. Elston and G. Ostheimer, On groups whose word problem is solved by a counterautomaton, *Theoret. Comput. Sci.* **320** (2004), 175–185.
[9] R. H. Gilman, Formal languages and infinite groups, in G. Baumslag, D. B. A. Epstein, R. H. Gilman, H. Short & C. C. Sims (eds.), *Geometric and Computational Perspectives on Infinite Groups* (DIMACS Series in Discrete Mathematics and Theoretical Computer Science **25**, American Mathematical Society, 1996), 27–51.
[10] R. H. Haring-Smith, Groups and simple languages, *Trans. Amer. Math. Soc.* **279** (1983), 337–356.
[11] M. A. Harrison, *Introduction to Formal Language Theory* (Addison-Wesley, 1978).
[12] T. Herbst, On a subclass of context-free groups, *Theor. Inform. Appl.* **25** (1991), 255–272.
[13] T. Herbst and R. M. Thomas, Group presentations, formal languages and characterizations of one-counter groups, *Theoret. Comput. Sci.* **112** (1993), 187–213.

[14] G. Higman, Subgroups of finitely presented groups, *Proc. Roy. Soc. Ser. A* **262** (1961), 455–475.

[15] D. F. Holt, S. E. Rees, C. E. Röver and R. M. Thomas, Groups with a context-free co-word problem, *J. London Math. Soc.* **71** (2005), 643–657.

[16] J. E. Hopcroft, R. Motwani and J. D. Ullman, *Introduction to Automata Theory, Languages and Computation* (Addison-Wesley, 2001).

[17] J. E. Hopcroft and J. D. Ullman, *Introduction to Automata Theory, Languages, and Computation* (Addison-Wesley, 1979).

[18] M. Ito, L. Kari and G. Thierrin, Insertion and deletion closure of languages, *Theoret. Comput. Sci.* **183** (1997), 3–19.

[19] S. R. Lakin and R. M. Thomas, Context-sensitive decision problems in groups, in C. S. Calude, E. Calude & M. J. Dinneen (eds.), *Developments in Language Theory: 8th International Conference, DLT 2004, Auckland, New Zealand* (Lecture Notes in Computer Science **3340**, Springer-Verlag, 2004), 296–307.

[20] A. A. Letičevskiǐ and L. B. Smikun, On a class of groups with solvable problem of automata equivalence, *Dokl. Acad. Nauk SSSR* **227** (1976); translated in *Soviet Math. Dokl.* **17** (1976), 341–344.

[21] K. Madlener and F. Otto, About the descriptive power of certain classes of finite string-rewriting systems, *Theoret. Comput. Sci.* **67** (1989), 143–172.

[22] D. E. Muller and P. E. Schupp, Groups, the theory of ends, and context-free languages, *J. Comput. System Sci.* **26** (1983), 295–310.

[23] D. E. Muller and P. E. Schupp, The theory of ends, pushdown automata, and second-order logic, *Theoret. Comput. Sci.* **37** (1985), 51–75.

[24] D. W. Parkes, V. Yu. Shavrukov and R. M. Thomas, Monoid presentations of groups by finite special string-rewriting systems, *Theor. Inform. Appl.* **38** (2004), 245–256.

[25] D. W. Parkes and R. M. Thomas, Groups with context-free reduced word problem, *Comm. Algebra* **30** (2002), 3143–3156.

[26] L. B. Smihun, Relations between context-free groups and groups with a decidable problem of automata equivalence, *Kibernetica* **5** (1976), 33–37.

[27] I. A. Stewart and R. M. Thomas, Formal languages and the word problem for groups, in C. M. Campbell, E. F. Robertson, N. Ruškuc & G. C. Smith (eds), *Groups St Andrews 1997 in Bath, Volume 2* (LMS Lecture Note Series **261**, Cambridge University Press, 1999), 689–700.

RECENT GROWTH RESULTS

ERIC M. FREDEN and TERESA KNUDSON

Southern Utah University, Cedar City UT, U.S.A.
Email: `freden@suu.edu`, `teresa@math.utah.edu`

Abstract

In this survey, we summarize the calculation and estimation of growth functions for Baumslag–Solitar groups and related subgroups.

0   Introduction

The Baumslag–Solitar groups, first introduced in [1], form a fascinating family of objects in the area of geometric and combinatorial group theory. They all share two-generator, one relator presentations,

$$BS(p,q) = \langle\, a,b \mid ab^p a^{-1} = b^q \,\rangle$$

yet such simplicity belies the incredibly complicated structure of some of these groups. Thurston showed that when $p \neq q$ these groups have no automatic structure, but all are asynchronously automatic [7]. About the same time, Edjvet and Johnson computed the rational growth functions for the automatic family [6]. The solvable family ($p = 1$) was also shown to have rational growth by Brazil [2], with the explicit growth functions determined in [3]. It is well-known that all Baumslag–Solitar groups (except for the trivial cases of the fundamental group of the Klein bottle, $1 = p = -q$, and the integer lattice, $1 = p = q$) have uniformly exponential growth [5]. Also [5] lists the computation of the growth series for $BS(2,3)$ as an open problem. Evidently this computation is difficult because the growth functions for the remaining (non-automatic, non-solvable) Baumslag–Solitar groups are still to be determined.

We announce a method to estimate the asymptotics of growth series for all Baumslag–Solitar groups, with details forthcoming [9]. We can compute the exact growth series for the horocyclic subgroup in the groups $BS(2,4)$, $BS(2,6)$, $BS(3,6)$, and we note that the latter two are non-hopfian groups. Again, further details will become available in the future [10]. We express our thanks to Alisha McCann for help with figures and Jennifer Schofield for additional geometric insight.

1   Growth Series

Let $B(z) = \sum_{r=0}^{\infty} \beta_r z^r$ be a power series. Recall that the *radius of convergence* is defined as the reciprocal of

$$\omega = \limsup_{r \to \infty} (\beta_r)^{\frac{1}{r}}.$$

Let $G = \langle \mathcal{X} | \mathcal{R} \rangle$ be a presentation for a finitely generated group and let β_r (respectively σ_r) be the number of elements g in G whose word length is at most (respectively equal to) r. Then recall that the above sum $B(z)$ (or $S(z)$, respectively) is called the *cumulative* (respectively *spherical*) growth series for the presentation. For either growth series, the above limit superior is actually an ordinary limit [5], and is called the *exponent of growth*. This ω is a number in $[1, +\infty)$ and is the reciprocal of the modulus of the smallest singularity for $B(z)$. Recall the formal convolution product

$$\sum_{r=0}^{\infty} c_r z^r = \Big(\sum_{r=0}^{\infty} a_r z^r \Big) \Big(\sum_{r=0}^{\infty} b_r z^r \Big)$$

has coefficients defined by $c_r = \sum_{j=0}^{r} a_j b_{r-j}$. By definition, $\beta_r = \sum_{j=0}^{r} \sigma_j \cdot 1$. We see that the cumulative growth series $B(z)$ is the convolution product of the spherical growth series $S(z)$ with the geometric series $\sum_{r=0}^{\infty} z^r = 1/(1-z)$. Therefore, $B(z) = S(z)/(1-z)$. The smallest singularity of any meromorphic $B(z)$ is completely determined by $S(z)$ since $1/(1-z)$ contributes a singularity of 1. This is the largest radius of convergence possible since its reciprocal must lie in $[1, +\infty)$. Thus, both series have the same radius of convergence, the same exponent of growth ω, and each is a rational function if and only if the other is a rational function.

It is well-known that $B(z) = \sum_{r=0}^{\infty} \beta_r z^r$ is a rational function if and only if the coefficients satisfy a linear, recurrence relation with constant coefficients for all sufficiently large r. In this case, the denominator of $B(z)$ is completely determined by the recurrence, i.e.,

$$\beta_r = k_1 \beta_{r-1} + k_2 \beta_{r-2} + \cdots + k_j \beta_{r-j}$$

$$\Longleftrightarrow$$

$$B(z) = \frac{\text{polynomial}(z)}{1 - k_1 z - k_2 z^2 - \cdots - k_j z^j}.$$

The coefficients of the numerator polynomial are determined by the values of the initial terms β_0, $\beta_1, \ldots$ prior to the recursion.

2 Baumslag–Solitar groups

For the general group $BS(p, q)$ in this genre, the defining relator $ab^p a^{-1} b^{-q}$ is referred to as a "horobrick" by [8]. The Cayley graph consists of "sheets" each of which is endowed with a coarse euclidean geometry (when $p = q$) or coarse hyperbolic geometry (when $p < q$) glued along $\langle b \rangle$-cosets referred to as "horocycles". This geometry is quasi-isometric with the upper half space model of the hyperbolic plane, and induces a natural vertical orientation. Paths with labels a^n, $n > 0$ go "up", labels $b^{\pm n}$ are "horizontal", while paths labelled a^{-n} go "down" (see Figure 1).

Recall that the Cayley 2-complex of a presentaion is obtained from the Cayley graph by filling in each basic relator and its conjugates with a topological disk. For $BS(p, q)$ the Cayley 2-complex is homeomorphic to the product of the real line with a simplicial tree (see Figure 2).

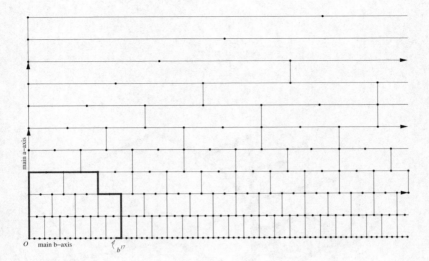

Figure 1. Partial main sheet for $BS(2,3)$, along with a geodesic path representing the word b^{17}. Missing vertical edges correspond to (deleted) branching half-sheets that are shown in Figure 2.

In contrast to [4], we use a restricted definition of half-sheet: we require each upper half-sheet to contain some half coset $\{wa^n : n > 0\}$ where w is a geodesic word with suffix b or b^{-1}. We think of w as a shortest connecting segment from the origin (identity vertex) to the half-sheet. Similarly each lower half sheet must contain a half coset of the form $\{wa^{-n} : n > 0\}$ where w is a geodesic word with suffix b or b^{-1}. There are only countably many half-sheets using these definitions. The geodesic word w above is termed the *stem* associated with the half-sheet. The notation $|w|$ refers to the geodesic length of w.

Since the defining relator must be maintained in each half-sheet, each $\langle b \rangle$-coset of $BS(p,q)$ gives rise to q upper half-sheets and p lower half-sheets. If Figure 2 is rotated so that the horocycles extend along the reader's line of sight, the Cayley graph projects to the simplicial tree mentioned earlier. For example Figure 3 illustrates part of the tree for $BS(1,3)$ with each dot representing a horocycle.

The family $BS(q,q)$ has euclidean sheets joined along geodesic $\langle b \rangle$-cosets. The upper and lower half-sheets merge together into copies of the integer lattice (with some vertical edges missing). Each of these groups is automatic and has rational growth explicitly computed by [6]. The next simplest family are the solvable groups $BS(1,q)$. These also have rational growth as shown by [2] and [3]. The methodology used to prove these results is standard combinatorial group theory: find a geodesic normal form for each group element and count these forms. This idea has been unsuccessful so far when applied to $BS(p,q)$ where $1 < p < q$.

We propose the use of geometry to compute the growth series for the horocyclic subgroups $\langle b \rangle$ (as suggested by L. Bartholdi), as well as the entire growth series for certain groups. Even when a growth series cannot be exactly computed, our

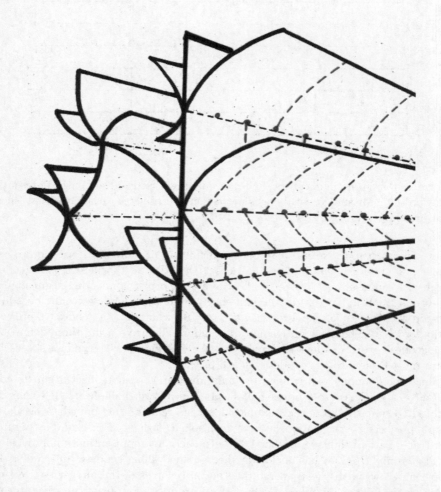

Figure 2. Partial Cayley graph for $BS(2,3)$. *Rendered by Alisha McCann.*

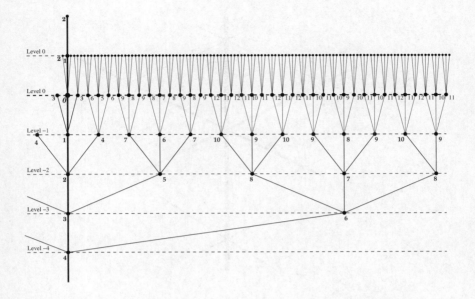

Figure 3. Side view of the half-sheet branching model for $BS(1,3)$. Each dot represents a horocycle. Levels and horocycle distances from the identity are labelled.

methods allow us to determine or estimate the exponents of growth (as suggested by A. Talambutsa). The latter strategy is based on the following key idea: estimate the growth of each half sheet (or horocycle) and count the various sheets (or horocycles). Counting the branching OF half-sheets (or horocycless) is equivalent to counting the different geodesic stems w used to define the half-sheets (or horocycles). The growth ON the half sheets (or horocycles) we consider depends directly on the growth of the corresponding horocyclic subgroup $\langle b \rangle$. As a reality check, both aspects of our program can be applied to the automatic and solvable groups and compared with the already known growth functions.

3   The automatic case

Each sheet of $BS(q,q)$ has some form of quadratic growth bounded above by the growth of $\mathbf{Z} \times \mathbf{Z}$, which is $\beta_r = 2r^2 + 2r + 1$. Let C_r denote the number of sheets at distance r from the origin vertex. Here r is the length of the stem w associated to the sheet. Counting these stems is elementary and illustrated in Figure 4. This defines a sequence with exponential growth. Since each sheet has growth of order $\Theta(r^2)$ where r is the radius, we get the cumulative growth estimate for $BS(q,q)$

$$\beta_r \approx \sum_{k=0}^{r-1} C_k \Theta\big((r-k)^2\big) = \sum_{j=1}^{r} C_{r-j}\Theta(j^2).$$

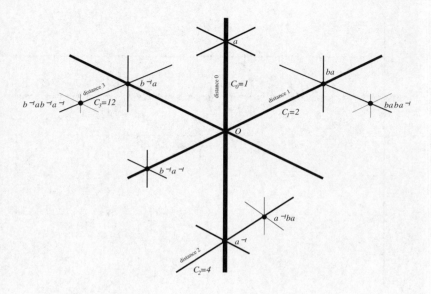

Figure 4. Counting the branching process for $BS(3,3)$.

The convolution product arises because k is the length of the stem associated to a sheet and consequent penetration into the sheet is closer to $r - k$ rather than r.

The exponent of growth is the limit $\omega = \lim_{r \to \infty} (\beta_r)^{\frac{1}{r}}$ which can be estimated via the inequalities

$$C_{r-1} < \beta_r < \sum_{j=1}^{r} C_{r-1} \Theta(r^2) = r C_{r-1} \Theta(r^2).$$

Take r^{th} roots and let $r \to \infty$ to see that ω is

$$\lim_{r \to \infty} (C_{r-1})^{\frac{1}{r}} = \lim_{r \to \infty} (C_r)^{\frac{1}{r}}.$$

Consequently the growth on a sheet is irrelevant, ω is the exponent of growth for the sequence of sheet branchings $C_0, C_1, C_2, \ldots$. This latter sequence satisfies a linear, constant coefficient recurrence, and sums to a rational function $\mathcal{C}(z)$. The exponent of growth, ω, for $BS(q,q)$ is the reciprocal of the modulus of the smallest singularity for this rational function.

Examples   For $BS(2,2)$, $C_r = 2 \sum_{j=0}^{r-2} C_j$ is the recurrence and

$$\mathcal{C}(z) = \frac{z-1}{2z-1}, \quad \omega = 2.$$

For $BS(3,3)$, $C_r = 4 \sum_{j=0}^{r-2} C_j$ is the recurrence (see Figure 4) and

$$\mathcal{C}(z) = \frac{(z-1)(2z+1)}{4z^2+z-1}, \quad \omega = \frac{1+\sqrt{17}}{2}.$$

For $BS(4,4)$, $C_r = 4C_{r-2} + 6\sum_{j=0}^{r-3} C_j$ is the recurrence and

$$\mathcal{C}(z) = \frac{(z-1)(z+1)}{2z^2 + 2z - 1}, \quad \omega = 1 + \sqrt{3}.$$

The denominator of the rational function $\mathcal{C}(z)$, hence ω, is determined strictly by the recurrence relations above. The growth exponent problem reduces to finding the smallest root of a polynomial.

In general, we have for odd q and $r > (q-1)/2$ the recurrence relation

$$C_r = 4C_{r-2} + 8C_{r-3} + 12C_{r-4} + \cdots + 4\left(\frac{q-1}{2} - 1\right) C_{r-\left(\frac{q-1}{2}\right)}$$
$$+ 2(q-1)\left(C_{r-\left(\frac{q+1}{2}\right)} + \cdots + C_1 + C_0\right)$$

For even q and $r > \frac{q}{2}$ the relation is

$$C_r = 4C_{r-2} + 8C_{r-3} + 12C_{r-4} + \cdots + 4\left(\frac{q}{2}\right) C_{r-\left(\frac{q}{2}\right)}$$
$$+ 2(q+1)\left(C_{r-\left(\frac{q+2}{2}\right)} + \cdots + C_1 + C_0\right)$$

In each case above, replace r with $r-1$ and calculate

$$C_r - C_{r-1} = 4C_{r-2} + 4C_{r-3} + \cdots + 4C_{r-\frac{q-1}{2}}$$
$$C_r - C_{r-1} = 4C_{r-2} + 4C_{r-3} + \cdots + 2C_{r-\frac{q}{2}}$$

for q odd and even, respectively. These recurrences have rational generating functions with denominators

$$1 - z - 4z^2 - 4z^3 - \cdots - 4z^{\frac{q-1}{2}} \quad \text{and}$$
$$1 - z - 4z^2 - 4z^3 - \cdots - 4z^{\frac{q-2}{2}} - 2z^{\frac{q}{2}}$$

for odd and even q, respectively.

The smallest root of each must be larger than $\frac{1}{3}$ in modulus, since the growth exponent of a non-free, two-generator group is strictly less than 3 (see [5]). On the other hand, each denominator above is a partial sum for the geometric series

$$1 - z - 4z^2\left(\sum_{j=0}^{\infty} z^j\right) = \frac{(1+z)(1-3z)}{(1-z)}.$$

Consequently, the radius of convergence approaches $\frac{1}{3}$ geometrically as $q \to \infty$. Thus the growth exponent $\omega \uparrow 3$ FAST. Our growth recurrence relations above are reflected as factors in the denominators of the rational spherical growth functions found by [6]. The actual growth functions can also be explicitly re-computed using the methods of the next section. The details are left to the reader.

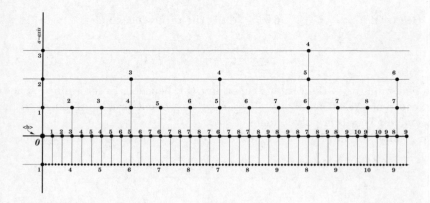

Figure 5. Partial main sheet for $BS(1,3)$. Quadrant I is everything above the $\langle b \rangle$ axis and right of the a-axis.

4   The solvable case

The sheets in these groups, $BS(1,q)$, have a discrete hyperbolic geometry. The subgroup $\langle b \rangle$ is a horocycle that divides the main sheet into one lower and several upper half-sheets. The growth function (either spherical or cumulative) on any of these grows exponentially with respect to radius. It is not difficult to see that the spherical growth of "Quadrant I" on the upper main half-sheet (see Figure 5) is exactly the cumulative growth of half the horocycle $\{b^n : n > 0\}$ (in fact, this holds for all the Baumslag–Solitar groups). This is because the cosets $a\langle b \rangle$, $a^2 \langle b \rangle$, $a^3 \langle b \rangle, \ldots$ comprising the main upper half-sheet have spherical growth counts that are exactly offset by $1, 2, 3, \ldots$ from that of $\langle b \rangle$. The lower half-sheet consists of horocycles that eventually follow the growth pattern of $\langle b \rangle$, but different behavior is exhibited near the origin (we discuss this in terms of level in the next paragraph). The growth on the lower half-sheet always outpaces that of any upper half-sheet (consider a circle in the hyperbolic plane divided through its center by a horocycle—the convex piece of this disk has smaller area). Since growth on each half-sheet is no longer sub-exponential, the naive estimates used for the automatic case need to be sharpened.

Jennifer Schofield has suggested dispensing with half-sheets entirely in favor of $\langle b \rangle$-cosets. We follow this suggestion in the solvable case. The growth function of such a horocycle depends recursively on that of an adjacent horocycle. Here the notion of "coset level" comes into play. Recall that the generator a defines a vertical direction in the Cayley graph. We define the horocyclic subgroup to have level zero. Any horocycle adjacent to and immediately higher than a horocycle of level zero, also has level zero. Any horocycle adjacent to and immediately lower than a horocycle of level $-n \leq 0$, is defined to have level $-n-1$. Any horocycle adjacent to and immediately higher than a horocycle of level $-n < 0$, is defined to have level $-n+1$. Thus the horocycles $a^4 ba \langle b \rangle$, $ba^{-3}ba \langle b \rangle$ have levels 0 and -2, respectively.

The first step in computing the growth function is to determine a recursion for the horocyclic subgroup, then find the growth recursion along any $\langle b \rangle$-coset as a function of level. We then count how many horocycles there are at a given level, form a convolution product with the recursion along such a coset, and finally sum over all levels.

As an example, consider the case of $BS(1,3)$. Define $b(0,r)$ to be the number of b^k such that $|b^k| = r$. This defines the spherical growth series for the horocyclic subgroup, which satisfies

$$b(0,0) = 1, \ b(0,1) = b(0,2) = b(0,3) = 2, \ b(0,4) = 4 \ \text{ and }$$
$$b(0,r) = b(0,r-2) + 2b(0,r-3)$$

for all $r \geq 5$. Multiplying both sides of the recursion by z^r and summing over all $r \geq 5$ gives

$$\sum_{r=5}^{\infty} b(0,r)z^r = z^2 \sum_{r=5}^{\infty} b(0,r-2)z^{r-2} + 2z^3 \sum_{r=5}^{\infty} b(0,r-3)z^{r-3}.$$

Now use the definition $B_0(z) = \sum_{r=0}^{\infty} b(0,r)z^r$ and initial conditions to rewrite the equation above:

$$B_0(z) - 1 - 2z - 2z^2 - 2z^3 - 4z^4 = z^2\left(B_0(z) - 1 - 2z - 2z^2\right) + 2z^3\left(B_0(z) - 1 - 2z\right)$$

and solve to get the generating function for the horocyclic subgroup

$$B_0(z) = \frac{1 + 2z + z^2 - 2z^3 - 2z^4}{1 - z^2 - 2z^3} = \frac{(1-z)(1+z)(1+2z+2z^2)}{1 - z^2 - 2z^3}. \qquad (4.1)$$

Given any other horocycle K of level zero, there is a (unique) element w of K with minimal geodesic length such that $K = w\langle b \rangle$. Treat w as a relative origin for K. Then it is true that $b(0,i)$ also counts the number of elements g of K satisfying $|g| - |w| = i$. It turns out that the growth coefficients for any such level zero coset K are the same as those of the horocyclic subgroup except for a shift of the indices by $|w|$.

A relation between spherical counts on horocycles is $b(-n,i) = b(-n+1,i) + 2b(-n+1, i-1)$ where the arguments $-n$ and $-n+1$ refer to level and the other argument refers to the difference of distances as explained in the previous paragraph. This recursion is valid for all $n, \ i \geq 1$. Again, multiply each side by z^i and sum over all $i \geq 1$, and use the initial condition $b(-n,0) = 1$ to derive

$$B_{-n}(z) = (1 + 2z)B_{-n+1}(z).$$

This equation is valid for all $n \geq 1$ and immediately implies

$$B_{-n}(z) = (1+2z)^n B_0(z) = \frac{(1-z)(1+z)(1+2z+2z^2)(1+2z)^n}{1 - z^2 - 2z^3} \qquad (4.2)$$

which is valid for all $n \geq 0$.

Now define $\chi(-n, r)$ as the number of cosets at level $-n$ whose closest point projection to the identity element is r. In other words, $\chi(-n, r)$ counts all of the cosets $w\langle b\rangle$ with $|w| = r$. A glance at Figure 3 will confirm that $\chi(0, 0) = 1$, $\chi(0, 1) = 1$, $\chi(0, 2) = 3$ and more generally, $\chi(-n, r) = 0$ for all $0 \leq r < n$, and that $\chi(-n, n) = 1$, $\chi(-n, n + 1) = 0 = \chi(-n, n + 2)$, $\chi(-n, n + 3) = 2$. Furthermore Figure 3 illustrates the branching recursion

$$\chi(-n, r) = \chi(-n + 1, r - 1) \quad \text{for all } n > 1 \text{ and } r \geq n + 3. \tag{4.3}$$

Define the coset generating functions $X_{-n}(z) = \sum_{r=0}^{\infty} \chi(-n, r) z^r$. Multiply each side of the recursion by z^r, sum over all $r \geq n + 3$, and use the initial conditions to derive

$$X_{-n}(z) = zX_{-n+1}(z) = z^{n-1} X_{-1}(z) \quad \text{valid for all } n \geq 1. \tag{4.4}$$

In order to determine $X_{-1}(z)$, Figure 3 again shows how horocycles propogate from a horocycle directly underneath:

$$\chi(-n, r) = \chi(-n - 1, r - 1) + 2\chi(-n - 1, r - 2) \quad \text{for all } n > 0 \text{ and } r \geq n + 3. \tag{4.5}$$

Put $n = 1$ and use relation (4.3) to obtain

$$\chi(-1, r) = \chi(-1, r - 2) + 2\chi(-1, r - 3) \quad \text{for all } r \geq 4 \tag{4.6}$$

which is the same recursion as that of the horocyclic subgroup, albeit with different initial conditions. The (by now) familiar method yields the generating function

$$X_{-1}(z) = \frac{z(1 - z)(1 + z)}{1 - z^2 - 2z^3}$$

and hence

$$X_{-n}(z) = \frac{z^n(1 - z)(1 + z)}{1 - z^2 - 2z^3} \quad \text{for all } n > 0. \tag{4.7}$$

In order to find $X_0(z)$, we examine how new zero-level horocycles arise from already counted horocycles of level zero and from level minus one. In fact,

$$\chi(0, r) = \chi(0, r - 1) + 2\chi(0, r - 2) + \chi(-1, r - 1) + 2\chi(-1, r - 2) \quad \text{for all } r \geq 3.$$

Substitute the branching recursion $\chi(-1, r - 1) + 2\chi(-1, r - 2) = \chi(-1, r + 1)$ stated earlier and solve for

$$X_0(z) = \frac{1}{z(1 + z)(1 - 2z)} X_{-1}(z) = \frac{1 - z}{(1 - 2z)(1 - z^2 - 2z^3)}. \tag{4.8}$$

Define $\sigma(-n, r)$ as the number of vertices in the Cayley graph at level $-n$ whose radius from the identity is r. For fixed $n \geq 0$ we count $\sigma(-n, r)$ via the convolution

$$\sigma(-n, r) = \sum_{i=0}^{r} \chi(-n, i) b(-n, r - i).$$

This is easily visualized: if g is an element at level $-n$ and has geodesic length r, then g lies on some horocycle $w\langle b\rangle$ where w is a geodesic stem of length i. There is always a geodesic path for g having prefix w. The number $b(-n, r - i)$ counts the number of elements on the horocycle $w\langle b\rangle$ at distance $r - i$ from w. There are exactly $\chi(-n, i)$ such cosets.

Let $S_{-n}(z)$ denote the generating function for this sequence. Combining (4.1) and (4.8), we see

$$S_0(z) = X_0(z)B_0(z) = \frac{(1 - z)^2(1 + z)(1 + 2z + 2z^2)}{(1 - 2z)(1 - z^2 - 2z^3)^2}$$

and for $n > 0$, combining (4.2) with (4.7) yields

$$S_{-n}(z) = X_{-n}(z)B_{-n}(z) = \frac{(1 - z)^2(1 + z)^2(1 + 2z + 2z^2)z^n(1 + 2z)^n}{(1 - z^2 - 2z^3)^2}.$$

The entire spherical count for the group is $\sigma_r = \sum_{n=0}^r \sigma(-n, r)$, which implies that the overall generating function is

$$
\begin{aligned}
S(z) &= \sum_{n=0}^{\infty} S_{-n}(z) = S_0(z) + \sum_{n=1}^{\infty} S_{-n}(z) \\
&= S_0(z) + \frac{(1 - z)^2(1 + z)^2(1 + 2z + 2z^2)}{(1 - z^2 - 2z^3)^2} \sum_{n=1}^{\infty} z^n(1 + 2z)^n \\
&= S_0(z) + \frac{(1 - z)^2(1 + z)^2(1 + 2z + 2z^2)}{(1 - z^2 - 2z^3)^2} \left(\frac{1}{1 - z(1 + 2z)} - 1 \right) \\
&= \frac{(1 + z)(1 - z)^2(1 + z + 2z^2)(1 + 2z + 2z^2)}{(1 - 2z)(1 - z^2 - 2z^3)^2}.
\end{aligned}
$$

This is in agreement with [3]. Note that the exponent of growth is 2, coming from the $(1 - 2z)$ factor in the denominator. This factor arises from the branching process of horocycles at level zero. Just as in the automatic case, we see that the exponent of growth depends only on the branching of the coset tree. Our method extends with only minor modification to $BS(1, q)$, for any odd q. For even q the branching of cosets is somewhat more involved but still far faster than the traditional means used in [3]. Another benefit of the geometric method is that the exponent of growth can be explicitly computed even when the horocyclic subgroup growth function cannot be explicitly computed. In fact, for $BS(1, q)$ it is easy to show that $\omega \to 1 + \sqrt{2}$ as $q \to \infty$.

5   Horocyclic subgroups

In both the automatic and solvable cases, the growth series for the corresponding horocyclic subgroup is known. We have successfully found the growth series of $\langle b\rangle$ for $BS(p, q)$ in the case of (p, q) pairs $(2, 4)$, $(2, 6)$, $(3, 6)$, $(4, 8)$, and $(5, 10)$. All of these are rational. Investigations for the cases where p does not divide q are ongoing. There is an algorithm written by Alisha McCann and revised by Jennifer Schofield that can find the geodesic length of any word b^k. In terms of k, the program runs in polynomial time. This McCann–Schofield reduction algorithm has been proved to give correct output and is extremely useful for generating raw data. Indeed, our procedure for establishing the horocyclic growth series has been to produce data using the McCann–Schofield reduction algorithm, use *Maple* to check for possible linear recursions, then find a proof for the validity of a candidate recursion. Such proofs seem to be *ad hoc* at this point—no general method has yet been found. We sketch the ideas for $BS(2,4)$ with a series of lemmas. The proofs are not at all difficult but are omitted for brevity (full details will be forthcoming in [10]). Although we use some geometric intuition, the idea is the usual method of counting normal forms.

It should be noted (and can be proved) that all reductions can take place in the main upper half-sheet of $BS(2,4)$. The first lemma indicates that a standard geodesic for b^d goes up in the main sheet, across, then repeatedly down and across.

Lemma 5.1 *Given any word of form b^d with $d \geq 6$, there is a standard geodesic representative having form $a^k b^l a^{-1} ws$ where $k \geq 1$, $l \in \{4, 6\}$, w is a word in the alphabet $\{a^{-1}, b^2, b^{-2}\}$ and s is a word in the alphabet $\{a^{-1}, b, b^{-1}, b^2, b^{-2}\}$.*

The next two lemmas show how geodesics propogate.

Odd Lemma 5.2 *If $n \geq 11$ is odd and w is a standard geodesic of length $n - 3$ representing b^d then*

(1) *$awa^{-1}b$ and $awa^{-1}b^{-1}$ are geodesics (of length n)*

(2) *every standard geodesic of length n for any b^c has the form given in (1).*

Even Lemma 5.3 *If $n \geq 10$ is even and γ is a standard geodesic for b^d where $|\gamma| = n - 2$, then $a\gamma a^{-1}$ is geodesic (of length n). Further, if γ is geodesic of length $n - 3$ having form $\gamma = a^k b^l \cdots a^{-m} b^{\pm 1}$, then $a^{k+1} b^l \cdots a^{-m-1} b^{\pm 2}$ is geodesic (of length n). Finally, every length n standard geodesic for b^c has one of the forms above.*

The Odd Lemma implies that the number of odd-length standard geodesics v for which $|v| = n \geq 11$ is twice the number of standard geodesics having length $n - 3$. The Even Lemma implies that the number of even-length standard geodesics γ for which $|\gamma| = n \geq 10$ equals the number of standard geodesics having length $n - 2$ plus twice the number of standard geodesics having length $n - 3$. In terms of recursions this says

$$b_r = 2b_{r-3} \text{ if } r \geq 11 \text{ is odd} \quad \text{and} \quad b_r = b_{r-2} + 2b_{r-3} \text{ if } r \geq 10 \text{ is even}$$

where b_r is the rth spherical growth coefficient for the horocyclic subgroup $\langle b \rangle$. Taken together these recurrences imply

Theorem 5.4 *The growth series for the horocyclic subgroup of* $BS(2,4)$ *satisfies the recursion* $b_r = b_{r-2} + 2b_{r-6}$, *for all* $r \geq 11$, *and has generating function*

$$B(z) = \frac{1 + z + z^5 - z^6 - z^7}{1 - z^2 - 2z^6}.$$

As remarked earlier, there are no obvious patterns in the growth series for the horocyclic subgroup of $BS(p,q)$ when $p \nmid q$. To end this section we propose

Conjecture 5.5 *The growth series of the horocyclic subgroup for* $BS(2,3)$ *is not rational.*

6   The general case

Our method of estimating exponents of growth can be generalized to $BS(p,q)$ for $2 < p < q$, but there are further complications. As conjectured earlier, the growth series for the horocyclic subgroup is almost surely not rational when $p \nmid q$. Also, there are many lower half-sheets to contend with. Lower half-sheets (respectively, horocycles at negative level) can have a great deal of initial "spread" where the closest point projections from the origin to the half-sheet (or horocycle) are several and far apart, thus inflating the initial growth estimate on such a half-sheet. These factors can be estimated by judicious use of the following observations:

- the non-solvable groups branch more vigorously than the solvable groups, with exponent of branching growth quickly approaching 3 as in the automatic case.
- there is negligible spread in upper half-sheets.
- the exponent of growth for $\langle b \rangle$ and every half-sheet is bounded above by $1 + \sqrt{2}$.
- those lower half-sheets (resp., horocycles) with large initial spread are relatively sparse.

Full details for these estimates will appear in [9], for now we will sketch the case for $BS(3,6)$. Let $\Phi(r), \Psi(r)$ denote the actual growth functions on the main upper and lower half-sheets, respectively. The actual branching formula L_r for the number of lower half-sheets at radius r from the origin is bounded below by a formula satisfying the recursion

$$L'_r = L'_{r-1} + 4L'_{r-2} + 2L'_{r-3} + L'_{r-4}, \quad r \geq 6.$$

The branching of upper half-sheets is also bounded below by the same recursion (that the upper and lower half-sheet branchings have the same asymptotic growth rate is true more generally). The exponent of growth for this recursion is $\omega_{L'} \approx 2.576$. The spherical growth for the horocyclic subgroup satisfies $b_r = b_{r-2} + 2b_{r-7} + c_r$, $r \geq 8$, where

$$c_r = \begin{cases} 0 & \text{if } r \equiv 0,3,7 \pmod{7}; \\ +1 & \text{if } r \equiv 1,4 \pmod{7}; \\ -1 & \text{if } r \equiv 2,5 \pmod{7}. \end{cases}$$

This implies the homogeneous, constant-coefficient, linear recursion

$$b_r = b_{r-14} + 2(b_{r-7} + b_{r-9} + b_{r-11} + \cdots + b_{r-19}),$$

which has exponent of growth $\omega_\Phi \approx 1.268$ (details to appear [10]). This is the exponent of growth for the main upper half-sheet as well as the horocyclic subgroup.

An initial estimate for the total spherical growth coefficient at radius r involves the convolutions

$$\sigma_r \approx b_r + \sum_{j=0}^{r-1} U_j \Phi_{r-j} + \sum_{j=0}^{r-1} L_j \Psi_{r-j},$$

where the first term reflects the horocyclic subgroup, the middle summation is the growth on all relevant upper half-sheets, and the final summation is the growth on all relevant lower half-sheets. This latter summation neglects the spreading of stems and is seriously deficient (unlike the former terms which are asymptotically accurate). Ignoring this undercount for the moment, the generating function decomposes as

$$S(z) = \sum_{r=0}^{\infty} \sigma_r z^r \approx S_{\langle b \rangle}(z) + S_U(z)S_\Phi(z) + S_L(z)S_\Psi(z),$$

where the constituent functions are the respective growth functions for the subgroup, branching of half-sheets, and growth on half-sheets. The decomposition implies that the radius of convergence for $S(z)$ is the modulus of the smallest pole of any of the constituent functions, and consequently the total exponent of growth is the maximum of $\omega_{\langle b \rangle} = \omega_\Phi \approx 1.268$, $\omega_\Psi \leq 2.414$, $\omega_U = \omega_L \geq 2.576$.

In actuality, things are more complicated. The idea of level as defined in the solvable case plays a role in the estimate and can be adapted to the general case. Those half-sheets based at a horocycle with large negative level need to be adjusted for. The frequency of half-sheets as a function of level forms a probability distribution with exponentially decreasing tail. The relative paucity of half-sheets at large negative level allows the asymptotic estimate of the previous paragraph to be valid. The exponent of growth for $BS(3,6)$ is in fact the branching exponent ω_L which satisfies $2.576 \leq \omega_L < 3$. Unfortunately, the exact branching sequence is almost surely not rational. Tighter estimates of the branching exponent will appear in [9].

References

[1] G. Baumslag and D. Solitar, Some two-generator one-relator non-Hopfian groups, *Bull. Amer. Math. Soc.* **68** (1962), 199–201.

[2] M. Brazil, Growth functions for some nonautomatic Baumslag–Solitar groups, *Trans. Amer. Math. Soc.* **342** (1994), 137–154.

[3] D. J. Collins, M. Edjvet and C. P. Gill, Growth series for the group $\langle x, y \mid x^{-1}yx = y^l \rangle$, *Arch. Math. (Basel)* **62** (1994), no. 1, 1–11.

[4] B. Cook, E. M. Freden and A. McCann, A simple proof of a theorem of Whyte, *Geom. Dedicata* **108** (2004), 153–162.

[5] P. de la Harpe: *Topics in Geometric Group Theory*, The University of Chicago Press, 2000.

[6] M. Edjvet and D. L. Johnson, The growth of certain amalgamated free products and HNN extensions, *J. Austral. Math. Soc. Ser. A* **52** (1992), no. 3, 285–298.

[7] D. B. A. Epstein, J. W. Cannon, D. F. Holt, S. V. Levy, M. S. Paterson and W. P. Thurston, *Word Processing in Groups*, Jones–Bartlett, Boston, 1992.

[8] B. Farb and L. Mosher, A rigidity theorem for the solvable Baumslag–Solitar groups, *Invent. Math.* **131** (1998), 419–451.

[9] E. M. Freden, T. Knudson and J. Schofield, Growth in Baumslag–Solitar groups I: Asymptotics, in preparation.

[10] E. M. Freden, T. Knudson and J. Schofield: Growth in Baumslag–Solitar groups II: Subgroups and rationality, in preparation.